Science
and Philosophy
in the
Soviet Union

Science
and Philosophy
in the
Soviet Union

Loren R. Graham

 Alfred A. Knopf / New York / 1972

Acknowedgment is gratefully extended to the following to reprint from their works:

Academic Press, Inc.: From THE ORIGINS OF PREBIOLOGICAL SYSTEMS AND OF THEIR MOLECULAR MATRICES, edited by Sidney W. Fox. Copyright © 1965 by Academic Press, Inc. From LIFE: ITS NATURE, ORIGIN AND DEVELOPMENT, by A. I. Oparin. Copyright © 1961 by Academic Press, Inc.

Atherton Press: From SCIENCE AND IDEOLOGY IN THE SOVIET UNION, edited by George Fischer (1967).

Commonwealth Agricultural Bureaux, England: From THE NEW GENETICS IN THE SOVIET UNION, by P. S. Hudson and R. H. Richens.

Humanities Press Inc. and D. Reidel Publishing Company: From THE PHILOSOPHY OF QUANTUM MECHANICS, by D. I. Blokhintsev.

International Publishers Co., Inc.: From READER IN MARXIST PHILOSOPHY, edited by Howard Selsam, *et al.* Copyright © 1963 by International Publishers Co., Inc. From DIALECTICS OF NATURE, by Friedrich Engels. Copyright 1940 by International Publishers Co., Inc.

The Macmillan Company: From THE ORIGIN OF LIFE, by A. I. Oparin (1938). From THE MYSTERY OF THE EXPANDING UNIVERSE, by William Bonnor (1964).

Pergamon Press Limited: From THE THEORY OF SPACE, TIME AND GRAVITATION, by V. A. Fock (1959). From PROCEEDINGS OF THE FIRST INTERNATIONAL SYMPOSIUM ON THE ORIGIN OF LIFE ON THE EARTH, edited by A. I. Oparin, *et al.* (1959). From PRESENT DAY RUSSIAN PSYCHOLOGY, edited by Neil O'Connor (1966).

Universe Books, Inc. and George Weidenfeld & Nicolson Ltd.: From THE ORIGIN OF LIFE, by J. B. Bernal (1967).

The article entitled "Cybernetics" appeared in different form in the July 1964 issue of *Survey* (London).

For
P.P.A.

Contents

Preface

The writing of this book began in the fall of 1959, although at the time I did not know that a full-length study would be the result of my research. I was a chemical engineer who had re-entered university studies in order to become a historian. As a member of a seminar at Columbia University under Professor Henry Roberts, I found my attention attracted to a specific topic in science that had caused some discussion in the Soviet Union: the theory of resonance. In the course of investigating that issue, I followed one different thread after another until an entire web of scientific, philosophical, and political issues became visible. This web constituted a single historical problem, yet it encompassed far more than any one historian could hope to master. The problem proved to be one of the most interesting and unexplored questions I could ever anticipate encountering, but it was thoroughly intimidating in its dimensions. In the years that followed, I worked on another book, a history of the Soviet Academy of Sciences, but I continued to collect information on the relation of Soviet Marxism to specific problems of scientific interpretation. Until 1964 I did not believe it possible or proper for a single person to try to bring together in one frame issues requiring competence in such diverse technical backgrounds. Nonetheless I was unconsciously building toward this larger goal. Through study of the Soviet Academy I had gained a better understanding of the political and institutional framework of science in the Soviet Union. During a year as an exchange graduate student at Moscow University in 1960–61 I frequently discussed dialectical materialism with Soviet students of science and philosophy, and Soviet scientists themselves. After returning from the Soviet Union, I began teaching at Indiana University, where I soon found myself investigating Soviet interpretations of cybernetics and quantum mechanics. My study of cybernetics led to my returning to the Soviet Union in the spring of 1963 as a member of a two-man delegation from the United States Office of Education. This visit gave me additional opportunities for discussions with Soviet engineers, philosophers, and educational specialists. At Indiana University my association with colleagues and students in both the departments of history and of the history and philosophy of science enriched my knowledge of Russian history and the history of science. Meanwhile my interest in dialectical materialism continued to grow.

Not until late 1964, however, did I decide to commit myself to a full-length study of the relation of Soviet Marxism to natural science. The event crystallizing that decision was the offer of a senior fellowship at Columbia University's Russian Institute, which gave me the opportunity of eighteen months of uninterrupted study. Without that generous award I doubt that this book would ever have been attempted. Alexander Dallin, then director of the Russian Institute, saw that possibilities became actualities. I am deeply grateful to him and to the Russian Institute for providing what can only be described as a labor-loving environment. I was also encouraged in my efforts at a very early stage by Angus Cameron of Alfred A. Knopf. In more recent years, as a member of the faculty of the history department at Columbia University and of the Russian Institute, my appreciation of my university colleagues, my students, and the library facilities of the New York area has grown. New York has helped me to greater awareness of both intellectual and social issues. Particularly important in my development has been the experience of teaching each year general courses in the history of science at Columbia College. My membership since 1968 in Columbia's Institute for the Study of Science in Human Affairs has also been helpful.

The completion of this book was made possible by an appointment in 1969–70 as a member of the Institute for Advanced Study in Princeton, New Jersey, and a Guggenheim Fellowship for the same year. The facilities of the Institute, the association there with scholars from the natural, social, and mathematical sciences, and the material support were all vital contributions to this study.

In the summer of 1970, while in Moscow, I gave a report to the Institute of Philosophy concerning the methodological problems contained in this study and showed parts of the draft to a number of Soviet philosophers of science, who made many helpful suggestions. No single Soviet scholar read the entire book at that time, however, and of course none is responsible for my interpretations. My three-month visit was sponsored by the exchange program between the American Council of Learned Societies and the Soviet Academy of Sciences, a program administered on the U.S. side by the International Research and Exchanges Board.

My earlier misgivings concerning the scope of this project have never disappeared. No one knows better than I that it is quite impossible to discuss such diverse topics as cybernetics, quantum mechanics, genetics, and psychology while achieving equal levels of scholarly quality in all. What overcame my reluctance to embark on the project was the opinion, which others will have to judge, that the gain from exploring the strands of all these issues, finding at least some of their common sources, and attempting analysis where possible was great enough to justify the risk that my competence, unequal in the various fields, would result in serious distortion.

I have not included in this volume a discussion of the controversy in the Soviet Union over formal logic, on the ground that the study is restricted to the topic of the relation of Soviet philosophy to the natural sciences.

It is quite possible that in subsequent years I shall publish additional material on the subject of dialectical materialism and natural science. With that eventuality in mind I would be particularly grateful for any suggestions for corrections or additions to the present work, especially from Soviet scientists and philosophers who have been close to the events described.

Of the many people who have given me assistance in writing this study, two entered into the effort so deeply that I often felt we were working on a common project. They are Joseph G. Brennan, Department of Philosophy, Barnard College, Columbia University, and Bert Hansen, Bensalem College, Fordham University. Professor Brennan's contribution was long-term and constant; in several different cities and countries we read, discussed, and argued together, occasionally vehemently. Clarification of an important issue in philosophy was the invariable result. Professor Hansen's enthusiasm for the history of science often kept me up later at night in Princeton than I wished, but with him, one learned to do important things first.

In addition to the people and institutions already mentioned, I would like to thank the Fulbright-Hays Fellowships and the U.S. Office of Education for a grant in 1966 for work in the libraries of the Soviet Union and western Europe. I am grateful to Christopher Wright and the Institute for the Study of Science in Human Affairs, Columbia University, for a travel grant to the Twelfth International Congress of the History of Science in Paris, August 1968, where I gave a paper based on this research. I would also like to thank the following persons for reading parts of the manuscript or for giving other valuable assistance: M. D. Akhundov, Bette Alexander, L. G. Antipenko, Kendall Bailes, Donald Bakker, Valentine Bargmann, L. B. Bazhenov, Daniel Bell, Severyn Bialer, Elof Carlson, E. M. Chudinov, Stephen Cohen, David Comey, Ruth Cowan, Richard De George, Theodosius Dobzhansky, Freeman Dyson, Paul Feyerabend, George Fischer, V. A. Fock, Kenneth Freeman, Felice Gaer, Lars Gårding, Patricia Albjerg Graham, David Joravsky, V. V. Kaziutinskii, George Kennan, Karl Koecher, P. V. Kopnin, Edwin Levy, Jr., Louise Luke, Woodford McClellan, Rex Martin, Alan Mayer, Mrs. Herman J. Muller, Ernest Nagel, M. E. Omel'ianovskii, Linus Pauling, L. A. Petrushenko, L. L. Potkov, Helen Powers, Wesley Salmon, Marshall Shulman, James Swanson, Robert Tucker, and Gustav Wetter.

Parts of chapters of the present book have appeared in earlier and different versions in the following publications: *Isis, Survey, The State of Soviet Science, Transactions of the XII Congress of the History of Science, Slavic Review, Science and Ideology in the Soviet Union* (George Fischer, ed.; Atherton Press). I am grateful to these publications and their editors for permission to reproduce this material.

L.R.G.

Institute for Advanced Study
Princeton, New Jersey
July 1970

Science
and Philosophy
in the
Soviet Union

I   Introduction: Background of the Discussions

The origin of the philosophic schools of materialism and idealism is to be found in the basic questions, What is the world made of? and How does man learn about the world? These questions are among the most important ones that philosophers and scientists ask. They have been posed by man for at least twenty-five hundred years, from the time of such pre-Socratic philosophers as Thales and Anaximenes.

Materialism and idealism were two of the schools of thought that developed as attempts to answer these questions. Materialists emphasized the existence of an external reality, defined as "matter," as the ultimate substance of being and the source of man's knowledge; idealists emphasized man's mind as the organizing source of knowledge, and often found ultimate meaning in religious value. Both schools of thought have usually been connected with political currents and often been supported by political establishments or bureaucracies. This political element has not, however, always destroyed the intellectual content of the writings of scholars addressing themselves to important philosophical questions. For example, the support of the Catholic Church for the scholastic system of the Middle Ages, despite the well-known restrictions of that system, was one of the causes for the innovations in Aristotelian thought in Oxford and Paris in the fourteenth century. This new scholastic thought had an impact on subsequent scientific development, leading to a new concept of impetus, or inertia. It is the thesis of this book that despite the bureaucratic support of the Soviet state for dialectical materialism, a number of able Soviet scientists have created intellectual schema within the framework of dialectical materialism that are sincerely held by their authors and that, furthermore, are intrinsically interesting as the most advanced developments of philosophical materialism. These natural scientists are best seen, just as in the case of the fourteenth-century scholastic thinkers, not as rebels against the prevailing philosophy, but as intellectuals who wish to refine the system, to make it more adequate as a system of explanation.

The history of materialism is to a large degree a story of exaggerations built on assumptions that in themselves have been quite valuable to science. Those assumptions have been that explanations of nature and natural events should avoid reference to spiritual elements or divine intervention,[1] should be based on belief in the sole existence of something called matter or (since relativity) matter-energy, and should to a maximum degree be verifiable by means of man's perception of that matter through his sense organs. The exaggerations based on these assumptions have usually been attempts to explain the unknown in terms of materialistic knowns that were quite inadequate for the task at hand. Forced to rely upon that portion of man's constantly developing knowledge accepted by science at any one point in time, materialists have frequently posed hypotheses that were later properly judged to be simplistic. Examples of such simplifications—materialists' description of man as a machine in the eighteenth century, or their defense of spontaneous generation in the mid-nineteenth—are often taken by readers of a later age as no more than amusing naïvetés. But the oversimplicity of these explanations, now evident, should not cause us to forget that the accepted science of today, from which we look back upon these episodes, does not contradict the initial materialistic assumption upon which these exaggerations were constructed. It is this continuity of initial assumption that continues to sustain the materialist view.

Materialism, like its denial, is a philosophical position based on assumptions that can neither be proved nor disproved in any rigorous sense. The best that can be done for or against materialism is to make a plausible argument on the grounds of adequacy. Important scientists of modern history have included both supporters and detractors of materialism, as well as many who consider the issue irrelevant. The sophistication of a scientist's attitude toward materialism is probably more important than the actual position—for, against, or undecided—he chooses to take. Yet it is also probably safe to say that since the seventeenth century, supporters of materialism have forced its detractors to revise their arguments in a more fundamental way than the reverse.

Within the Soviet Union in the period since World War II there have been a number of important discussions concerning the relation of dialectical materialism to natural science. Many outside the U.S.S.R. are familiar with the genetics controversy and the part played in it by Lysenko, but few are aware of the details of other debates over relativity physics, quantum mechanics, resonance chemistry, cybernetics, cosmology, the origin of life, and psychophysiology. The present volume, which treats each of these topics in its own full chapter, is a first attempt, an initial sketching-out, of what is the largest, most intriguing nexus of scientific-philosophical-political issues in the twentieth century. The thousands of Soviet books, articles, and pamphlets on dialectical materialism and science contain all sorts of questions deserving discussion. Historians and philosophers of science will long argue over the issues raised in these publications: Were they real issues, or were they only the artificial creations of politics? Did Marxism actually influence the thinking of scientists in the Soviet Union, or were their statements to this effect mere window-dressing? Did the controversies have effects that historians and philosophers of science outside the field of Russian studies must take into account? I have posed tentative answers to these questions, based on information I have been able to obtain in the Soviet Union and elsewhere: Much of this voluminous Soviet discussion was the immediate result of political causes, but the debates have now gone far beyond the political realm into the truly intellectual sphere. The political influence is neither surprising nor unique in the history of science; it is, rather, part of that history. Marxism is taken quite seriously by some Soviet scientists, less seriously by others, disregarded by still others. There is even a category of Soviet philosophers and scientists who take their dialectical materialism so seriously that they refuse to accept the official statements of the Communist Party on the subject; they strive to develop their own dialectical materialist interpretations of nature, using highly technical articles as screens against the censors. Yet these authors consider themselves dialectical materialists in every sense of the term. They are criticized in the Soviet Union not only by those scientists who resist any intimation that philosophy affects their research (a category of scientist that exists everywhere), but also by the official guardians of dialectical materialism, who believe that philosophy has such effects but would leave their definition to the Party intellectual spokesmen. I am convinced that dialectical materialism has influenced the work of some Soviet scientists, that in certain cases these influences helped them to arrive at views that won them international recognition among their foreign colleagues. All of this is important to the history of science in general, and not simply to Russian studies.

One of the more specific conclusions issuing from this research is that the controversy known best outside the Soviet Union—the debate over genetics—is the least relevant to dialectical materialism in a philosophical sense. Nothing in the philosophical system of dialectical materialism lends obvious support to any of Lysenko's views. On the other hand, the controversy known least well outside the Soviet Union—that over quantum mechanics—touches dialectical materialism very closely as a philosophy of science. Not surprisingly, the terms of this particular controversy most closely approach those of discussions of quantum mechanics that have taken place in other countries.

In the genetics debate, Lysenko advanced the position that affirmed the inheritance of acquired characteristics, together with a vague theory of the "phasic development of plants." Nowhere in systematic dialectical materialism can support for these views be found.[2] The claims advanced by Lysenko were staked outside the small circle of Marxist biologists in the Soviet Union as well as outside the established groups of Soviet philosophers. Contrary to the views of a number of non-Soviet authors, there did not exist a peculiarly "Marxist" form of biology from Marx and Engels onward.[3] The concept of the inheritance of acquired characteristics was part of nineteenth-century biology, not an aspect of Marxism.[4] True, an assumption of the inherent plasticity of man was consonant with the desire of Soviet leaders to create a "new Soviet man," and the inheritance of characteristics acquired during one's lifetime might seem a promising function of such plasticity. Surprisingly, however, the application of Lysenkoism to human genetics was not supported in the Soviet Union; this was a common interpretation of Lysenkoism outside the Soviet Union rather than the justification for it given within that country. Close reading of Soviet sources lends no support to the view that Lysenkoism prospered because of its implications for eugenics. During the entire period of Lysenko's influence the shaping of the heredity of man was a subject frowned upon in the Soviet Union. The rise of Lysenkoism was the result of a long series of social, political, and economic events rather than connections with Marxist philosophy. These events, together with their results, have been well described in the work of David Joravsky.[5]

In the Soviet discussions over quantum mechanics, however, an approach was made to the heart of dialectical materialism as a philosophy of science. Because of different political factors, however, the result was quite unlike that of the genetics affair. The core of dialectical materialism consists of two parts: an assumption of the independent and sole existence of matter-energy, and an assumption of a continuing process in nature in accordance with dialectical laws. Quantum mechanics, in the opinion of some scholars, undermined both parts: Its emphasis on the important role played by the observer seemed to favor philosophical idealism, while the impossibility of predicting the path of an individual particle called into question the concept of causality implicit in the assumption of a continuing process in nature. In the years after 1947 a great debate on the subject was mounted in the Soviet Union. Unfortunately, philosophers, scientists, and historians of science outside the Soviet Union tended to take the most simple and dogmatic Soviet pronouncements on the argument as representative of all Soviet positions on quantum mechanics, rather than examining the more careful and rigorous statements of major participants in the debate, such as D. I. Blokhintsev, M. E. Omel'ianovskii, and V. A. Fock. During the course of the discussions several interpretations of quantum mechanics were evolved in the Soviet Union that were considered consonant with dialectical materialism. They also have interest from a scientific point of view. The Soviet theoretical physicist Fock, a frequent writer on science and dialectical materialism, debated the issue with Niels Bohr and, according to Fock, helped to shift Bohr's opinions away from emphasis on measurement to a more "realist" view.[6]

One of the most notable characteristics of the Soviet controversy over quantum mechanics was its similarity to the worldwide discussion on this topic. If Omel'ianovskii objected to the idea that the macrophysical system surrounding the microparticle somehow causes the particle to display the particular properties with which we describe it, so did many non-Soviet authors, such as the American philosopher Paul Feyerabend. If Blokhintsev rejected von Neumann's claim to have refuted the possibility of hidden parameters, so did some scientists elsewhere, including David Bohm. If Fock refused to accept the idea that quantum theory implied a denial of causality, so did the French scientist de Broglie and (for different reasons) the American philosopher Ernest Nagel.[7] What seems most striking in the quantum controversy is the similarity between views advanced by Soviet scientists and dialectical materialists on the one hand, and those opinions advanced by non-Soviet scholars with rather different philosophies of science on the other. From this one might be tempted to conclude that dialectical materialism is meaningless. But one may also conclude that the concerns of dialectical materialists in the Soviet Union and those of philosophers of science in other parts of the world are in many ways similar, and one of the reasons for this is the essential character of the problem of materialism. One should not forget the fact that the debate between materialism and idealism did not arise with the Soviet Union but is, instead, more than two thousand years old. Soviet and non-Soviet interpreters of nature frequently ask the same questions, and occasionally they give very similar answers.

Great harm was done to science in the Soviet Union, particularly to genetics, by the wedding of centralized political control to a system of philosophy with claims to universality. Observers outside the Soviet Union have frequently placed the blame for this damage on the philosophy concerned rather than on the system of political monopoly that endeavored to control it. As a philosophy of science, dialectical materialism has been significant in the Soviet Union, not in promoting or hindering fields of science as a whole, but rather in subtle areas of interpretation. Occasionally a certain formulation of Marxist philosophy of science has been converted to an official ideological statement by endorsement from Party organs. Then harmful effects have indeed occurred. The genetics controversy was the most tragic of these events.

Yet it is clear that man, whether in the Soviet Union or elsewhere, will never be content without asking the kinds of ultimate questions that universal systems of philosophy attempt to answer. Dialectical materialism is one of these philosophical systems. If we admit the legitimacy of asking fundamental questions about the nature of things of which man is a part, the approach represented by dialectical materialism—science-oriented, rational, materialistic—has some claims of superiority to available rival universal systems of thought, claims it is appropriate to receive with respect. If dialectical materialism were allowed to develop freely in the U.S.S.R., it would no doubt evolve in a direction consistent with the common assumptions of a broad nonmechanistic, nonreductionist materialism.[8] Such results would be fruitful and interesting. We can hope, therefore, that the day will come when this further development of dialectical materialism can take place under conditions of free debate; such conditions would contrast both with the official protectionism found in the Soviet Union, which makes it difficult to revise dialectical materialism substantially, and the informal hostility to it existing in the United States, a hostility that makes it difficult to speak of its strengths.

Historical and Political Background

The revolutions of 1917 occurred in a nation that was in an extremely critical position: On a gross scale the Soviet Union was a backward and underdeveloped country in which a quick solution of the major problems of poverty and suffering was inconceivable. The U.S.S.R. inherited a tradition of autocratic government that would exercise influence long after the overthrow of the old regime. The new nation was subject to overwhelming pressures of military and economic rivalries. On the European scene it was viewed jealously before the successful Bolshevik Revolution and with quite extraordinary hostility after that event. The new Soviet Union possessed an able group of intellectuals, heir to a distinguished scientific and cultural tradition, whose members were, however, forcefully opposed to the new government. The political leaders of that new government were products of a conspiratorial tradition, hardened to the use of terror by having been previously the objects of terror; they were men who possessed a world view persuasive as an explanation of their role in history and convenient as a method of discipline.

Within this troubled context it should not have been surprising that the degree of intellectual freedom that developed in Soviet Russia after the Revolution was substantially less than in those countries in western Europe and North America with which the new nation would be most frequently compared. The possibility of unusual controls over intellectual life was heightened soon after the Revolution by the elimination of all political parties other than that of the Bolsheviks, later renamed the Communist Party of the Soviet Union. The Party soon developed a structure paralleling the government's on every level and controlling the population in almost every field of activity. This population did not object to the controls nearly so much as non-Soviet observers have proclaimed; the government enjoyed the support or toleration of a large portion of the lower classes, who made up the vast majority of the Soviet citizenry. The existence of this support strengthened the freedom of action of the Party leaders in intellectual fields, although the intellectuals themselves, a relatively small group, were frequently opposed to Party policies. The possibility of intervention in intellectual fields was further strengthened by the Party leaders' past expressions of strong opinions and preferences on certain issues in the arts and sciences.

Nonetheless, in the years immediately after the Revolution almost no one thought seriously that the Communist Party's supervision of intellectuals would extend from the realm of political activity to that of scientific theory itself. Party leaders neither planned nor predicted that the Party would approve or support certain viewpoints internal to science; indeed, such endorsement was fundamentally opposed by all the important leaders of the Party. Even when Party judgment on scientific theory became an evident possibility, there still remained at the time other attitudes or courses of action that the Party could well have taken. Throughout the 1920's there were in the Soviet Union no obvious grounds for predicting the enforcement of ideological strictures in the natural sciences. A specific Soviet Marxist philosophy of nature does not necessarily entail official pronouncements on scientific issues; indeed, a condition free of such entailment actually obtained in the late fifties and sixties for all the sciences except genetics, and for genetics as well since 1965. Besides, among Soviet scientists and philosophers there never was a *single* interpretation of Marxist philosophy of science.

The Revolution had less immediate effects on the tone and style of work in the natural sciences than in any other field of Soviet intellectual activity. After overcoming their initial feelings of pessimism about prospects for research—a feeling that spread wide after the Revolution—most Russian scientists continued the lines of work on which they had embarked earlier in their careers. Some were even invigorated by the prospect of an unusual union of science and government for the benefit of the working class. The Revolution may also have had liberating effects on scientific theory in a few areas where materialistic viewpoints had earlier been more controversial, such as the study of the origin of life.[9] But on the whole, during the early years science retained the same intellectual tone it had possessed before the Revolution, though on the physical level it suffered badly from the effects of civil strife in the years 1918–21 and from the scarcity of funds and materials for research. The emigration of a number of leading scientists was also a serious loss.

During the early period of Soviet history known as that of the New Economic Policy (N.E.P.), which lasted from 1921 to 1926, the intellectual scene was relatively relaxed. Usually, so long as scholars and artists refrained from political activity offensive to the Communist Party, they did not need to fear persecution by the police or interference from ideologists. Those persons whose backgrounds or previous political activities were considered particularly incriminating might be exceptions to this generalization. But even people who previously had been members of non-Bolshevik political parties, as well as those with past connections to the tsarist bureaucracy, were able to maintain positions in cultural and educational institutions. The universities, the Academy of Sciences, health organizations, archives, and libraries all served as relatively secure refuges for "former people," most of whom sought no more than living out their lives uneventfully under the drastically new conditions.

Control of scientific institutions remained in the hands of pre-Revolutionary academicians throughout most of the twenties. As late as January 1929, not a single member of the prestigious Academy of Sciences (founded in 1724) was also a member of the Communist Party. A political rival to the Academy of Sciences was the Communist Academy (founded in 1918 as the Socialist Academy; renamed in 1923), which made some effort in the twenties to bring the natural sciences under Bolshevik tutelage. But even here there was no attempt to force particular interpretations of scientific theory upon scientists. The journal of the Communist Academy in these years contained articles that were interesting, diverse in interpretation, and occasionally of genuine originality. The work of the Communist Academy in the twenties was called by a non-Soviet historian of the Communist Party, Leonard Schapiro, "the golden age of Marxist thought in the U.S.S.R." [10] Most of the academy's work was confined to the social sciences, but it occasionally sponsored publications on natural science.

In the second half of the 1920's, there emerged two developments of crucial significance for the future of the Soviet Union: the struggle between the leaders of the Party culminating in the ascendancy of Stalin, and a decision to embark on ambitious industrialization and collectivization programs. The story of the rise of Stalin to supreme power has been told innumerable times (although there are many aspects of it that are still unclear), and no attempt will be made to retell that story here. But Stalin's personal influence on subsequent developments in the intellectual world of the Soviet Union proved to be of tremendous importance. His intellectual predilections caused impact on a number of fields, including certain areas of science. Most foreign historians of the Soviet Union have doubted that ideology played an important role in determining Stalin's actions, preferring to believe that power considerations dominated his choices. These historians have noticed how Stalin retreated from ideological positions when such shifts seemed desirable from a practical standpoint, and they cite as an example the turn in the Soviet government's attitude toward the Church. More recent study of Stalin has indicated, however, that a simple interpretation of the man in terms of power is insufficient to explain him. Leader of the Soviet Union for a quarter of a century, Stalin was governed by a complex mixture of motivations. These drives may well have been power-oriented in many respects, but they also contained ideological elements. Important leaders often combine ideological and power factors in their decisions; the history of the popes of the Catholic Church, of many crowned rulers of Europe, and of leaders of modern capitalist countries illustrates this interplay of power and idea. In Stalin the ideological and power-oriented factors combined; moreover, the actual political power he possessed was truly extraordinary, and he used it with increasing arbitrariness.

The traumatic break that occurred in the years 1927–29, the abrupt shock of an industrial, agricultural, and cultural revolution, will always be causally linked with Stalin. True, it was not only Stalin but almost all of the Soviet leaders who had declared the need to industrialize rapidly and to reform cultural institutions. The international realities of the late 1920's exerted strong pressures upon Soviet Russia to modernize itself in order to maintain its international and economic sovereignty. But it was Stalin who in large part determined the specific forms and tempos of these campaigns, and these in the end became as important as the campaigns themselves. Of the varieties of rapid industrialization programs proposed in the second half of the twenties, Stalin supported the most strenuous course; his choice required forcible methods for enactment. Similarly, Stalin's collectivization program in agriculture was breathtaking in its tempo and staggering in its violence. Ten years after Stalin's death, Soviet historians permitted themselves to observe on occasion that Stalin's agricultural collectivization program had been premature and coercive, however much one would agree with its goal of creating large farms tilled by collective labor.[11]

Accompanying the industrial and agricultural campaigns was the cultural revolution. Personnel of educational and scientific institutions were submitted to political examinations and purges. Here "purge" must be taken to mean not only imprisonment or execution, but the almost equally tragic dismissal of personnel from academic positions. Functionally, the purge had begun in Soviet academic institutions as a means of personnel replacement. In the late 1920's, this renovative technique was used to oust bourgeois academicians of certain institutions in order to replace them with supporters of the Communist Party. These replacements were frequently persons of inferior scholarship whose enthusiasm for social reconstruction commended them to preferment. Later, under Stalin's complete control, the purge became quite arbitrary and violent. Dismissal and exile to labor camps were more common among social scientists than among natural scientists, but even in the institutions of the natural sciences a structure of control was created. In the period 1929–32, the Academy of Sciences was thoroughly renovated and brought under the control of the Communist Party.[12] Even at this time, however, no attempt was made to impose ideological interpretations upon the work of scientists; nonetheless, the precedent of forcing submission to specific political, social, and economic campaign pressures would later prove to be significant, especially after World War II, when the ideological issues in the sciences became most aggravated. The passing of the twenties into the thirties in the Soviet Union was marked by a growing tendency to classify science itself as "bourgeois" or "idealistic"—clearly something beyond the distinguishing of certain philosophical interpretations of science. While this tendency is now the subject of sharp criticism on the part of several leading Soviet philosophers of science, it had a long and harmful influence on Soviet science.[13] The attribution of political character to the body of science itself eased the way for Lysenko's concept of "two biologies," as well as for ideological attacks on the substance of other branches of natural science. As early as 1926, V. P. Egorshin, writing in *Under the*

Banner of Marxism, an influential philosophy journal of the time, declared that "modern natural science is just as much a class phenomenon as philosophy and art. . . . It is bourgeois in its theoretical foundations." [14] And an editorial in the journal *Natural Science and Marxism* in 1930 asserted that "philosophy and the natural and mathematical sciences are just as politically partisan as the economic and historical sciences." [15]

Not all Soviet philosophers and very few Soviet scientists accepted the assumption that the natural sciences contained political elements in themselves and the corollary that Western science was implicitly distinct from Soviet science. Many scientists and philosophers of the strongest Marxist persuasion were still capable of drawing distinctions between science and the uses made of it, whether moral or philosophical. Even those who thought, with justification, that the theoretical body of science cannot be completely separated from philosophical issues usually realized that any attempt to determine those issues by political means would be quite harmful. The prominent Marxist scientist O. Iu. Schmidt, who will appear as an important participant in the cosmology debate, declared in 1929 that:

Western science is not monolithic. It would be a great mistake indiscriminately to label it "bourgeois" or "idealistic." Lenin distinguished "unconscious materialists," who included most experimenters of his time, from idealists. . . . An unconscious attraction to the dialectic is growing. . . . There are no conscious dialectical materialists in the West, but elements of the dialectic appear among very many scientific thinkers, often in idealistic or eclectic garb. Our task is to find these kernels and to refine and use them.[16]

The debates over the nature of science in the late twenties and in the thirties did not touch most practicing Soviet scientists of the period. The majority of researchers tried to remain as far from considerations of philosophy and politics as scientists elsewhere. The importance of these discussions was not their immediate impact but the precedent they provided for the much sharper ideological debates of the postwar period, when Stalin accepted Lysenko's definition of "two biologies" and intervened personally in choosing between them. Without Stalin's arbitrary action the actual suppression of genetics in the Soviet Union would not have occurred, but the discussions of the thirties helped to prepare the way for the suppression by strengthening the suspicion in which Western science was held by some Soviet critics.

Another characteristic of Soviet discussions of the thirties that re-emerged after the Second World War was the emphasis on utility. In a nation rapidly modernizing in the face of external threats, the priority of practical concerns was not only understandable but necessary. As is often the case with underdeveloped nations that nonetheless possess a small highly educated stratum, Russia's past scientific tradition had been excessively theoretical. The emphasis on industrial and agricultural concerns in the thirties was doubtless a needed correction of this tradition. At its root, the high priority given to practice had a positive moral content, since the ultimate results of a growing economy were a higher standard of living, greater educational opportunities, and better social welfare. So long as the value of theoretical science was also recognized, a relative shift toward applied science was a helpful temporary stage. The new priority was carried to an extreme, however, and had results that were philistine and anti-intellectual. In art and literature the stress on industrial expansion buttressed "socialist realism," the art style supplanting the earlier experimental forms that sprouted immediately after the Revolution. Socialist realism commended itself to the bureaucrats who were gradually replacing the more sophisticated and cosmopolitan older revolutionaries. The situation in the arts in these years was only indirectly related to that of the sciences, but it was nonetheless a significant aspect of the general environment of the Soviet intellectual. Analogous to the desired artistic concentration on themes calculated to inspire the workers aesthetically and emotionally was the role assigned to scientists as discoverers of new means to speed industrialization. Many scientists who had been trained in highly theoretical areas in the thirties found themselves closely involved with the industrialization effort. In addition to their research duties, they began to serve as industrial consultants.

Thus a result of the industrialization and collectivization efforts in the Soviet Union was an increase in pressure upon scientists and intellectuals to mold their interests so that their work would benefit the construction of "socialism in one country." One of the effects of this pressure was the growth of nationalism in science, as in other fields. The very possibility of constructing socialism in one country had, of course, been the subject of one of the great debates among Stalin and his fellow leaders. The original revolutionaries had believed that the Revolution in Russia would fail unless similar revolutions occurred in other more advanced countries. The Bolshevik leaders assumed that the capitalist world was implacably hostile toward the Soviet state, and indeed concrete evidence of this animosity could be seen in the aid extended by leading capitalist countries to anti-Bolshevik forces during the civil war and in the diplomatic isolation of Soviet Russia in the early twenties. Stalin announced that socialism *could* be constructed in one country and called for reliance upon native resources, scientific and otherwise. This shift in emphasis represented a weakening of the internationalist strain in the Communist movement that historians have linked with the name of Trotsky and that resulted, among other things, in a greater isolation of Soviet scientists. Stalin called for a maximum effort by all Soviet workers, including scientists, to achieve the nearly impossible—to make the Soviet Union a great industrial and military power in ten or fifteen years. An intrinsic part of this effort, Soviet nationalism, gradually gained strength in the thirties as the possibility of a military confrontation with Nazi Germany grew.

During World War II, as a result of stress upon patriotism and heroism, the nationalist element in Soviet attitudes emerged all the more clearly. In science, this emphasis on national achievement had many effects. Into controversies over scientific interpretation it introduced an element, national pride, that was totally absent from the dialectical materialism derived from Marx, Engels, and Lenin. It resulted in claims for national priority in many fields of science and technology. Many of these claims have now been abandoned in the Soviet Union, where they are regarded as consequences of the "cult of the personality." Others have been retained. Of these, some are justified or at least arguable in light of the long years in which appreciation of Russian science and technology by non-Russians was obstructed by linguistic barriers, ethnic prejudices, and simple ignorance.

Perhaps the most important characteristic of Soviet society contributing to the peculiar situation that developed in the sciences after the war was the very high degree of centralization of control over public information, personnel assignment and promotion, academic research and instruction, and scientific publishing. This system of control had been completed long before Stalin decided to intervene directly in the biology dispute after the war. Indeed, any effort actively to oppose this awesome

accumulation of power became unthinkable during the great purges of
the thirties, when it became clear that not even the highest and most
honored officials of the Party were immune to Stalin's punitive power.
The atmosphere created by these events permeated all institutions of
Soviet society. People on lower levels of power looked to those above
for signals indicating current policy; as soon as these signals were discernible, the subordinates hurried to follow them. By the late thirties,
for example, no local newspaper would have thought of contradicting
or questioning a policy announced in *Pravda,* the official publication of
the Central Committee of the Communist Party. Censorship was not
left, however, to voluntary execution; it was officially institutionalized
and extended even to scientific journals, although the limits of toleration there were usually greater and varied from time to time somewhat
more than elsewhere. Appointment of officials influential in science and
education—ministers of education and agriculture, presidents of the All-
Union Academy of Sciences and of other specialized academies, rectors
of the universities, editorial boards of journals—all were under the control of Party organs. Approval of science textbooks for use in the school
system and even the awarding of scientific degrees to individual scholars
were also under close political supervision. All these features of the Soviet power structure help explain the way in which Stalin was able, after
the war, to give Lysenko's interpretation of biology official status despite the opposition of established geneticists, men of science who fully
recognized the intellectual poverty of Lysenkoism.

The description above of the centralization of power in Soviet society
is familiar to all students of Soviet history. What is much less well
known, and indeed frequently entirely overlooked, is that beneath the
overlay of centralized political power there existed among the Soviet
population rather widespread support for the fundamental principles of
the Soviet economy, and among intellectuals, increasing support for a
materialist interpretation of the social and natural sciences. Studies of
refugees from the Soviet Union during the Second World War have
shown that despite a large degree of disaffection toward the political
actualities of the Soviet Union, these people remained convinced, by
and large, of the superiority of a socialist economic order.[17] Similarly
there is much evidence that Soviet intellectuals of genuine ability and
achievement found historical and dialectical materialist explanations of
nature to be persuasive on conceptual grounds. O. Iu. Schmidt, I. I. Agol,

S. Iu. Semkovskii, A. S. Serebrovskii, A. I. Oparin, L. S. Vygotsky and S. L. Rubinshtein are examples of distinguished Soviet scholars who made clear their belief, in diverse ways, that Marxism was relevant to their work *before* statements of the relevance of Marxism were required of them. The views of Schmidt, Oparin, Vygotsky, and Rubinshtein will be discussed in some detail later in the present study, since their views continued to be influential after 1945. In the concerns of these men, science came first, politics second. But one should not assume that the presence of strong political motivation necessarily undermines the intellectual value of a man's views. Nikolai Bukharin, a Party leader, was a Soviet politician to whom a materialistic, naturalistic approach to reality was far more than rhetoric; portions of his writings are remarkable for the degree to which they draw upon a materialist interpretation of natural science and for the intellectual clarity with which this view is presented.[18]

Several of the persons named above, and many more of their type, disappeared in the purges and had their writings banned in the Soviet Union. But unless one remembers that there existed before the forties a category of Soviet scholars who took dialectical materialism seriously, it will be difficult to understand why, after the passing of the worst features of Stalinism, scientists re-emerged in the Soviet Union who combined a dialectical materialist interpretation of nature with normal standards of scientific integrity.

Immediately after the Second World War many intellectuals in the Soviet Union hoped for a relaxation of the system of controls that had been developed during the strenuous industrialization and military mobilizations. Instead, there followed the darkest period of state interference in artistic and scientific realms. This postwar tightening of ideological controls spread rather quickly from the fields of literature and art to philosophy, then finally to science itself. Causal factors already mentioned include the prewar suspicion of bourgeois science, the extremely centralized Soviet political system, and the personal role of Stalin. But there was another condition that exacerbated the ideological tension: the Cold War between the Soviet Union and certain Western nations, particularly the United States. This struggle, which many contemporary American historians now see as much less the sole responsibility of the Soviet Union than previously, was rising to a peak in the years immediately following the war.[19] In the very period when Soviet politicians were finding "bourgeois idealism" lurking in the minds of Soviet scientists, many American politicians were convinced that the State Department was infested with Communists. These were years in which ideological sensitivity ran feverishly high in both the United States and the Soviet Union; the two great countries reinforced each other's fears and prejudices. The Cold War involved passions of a sort reminiscent of past quarrels over religion. The Soviet suppression of genetics in 1948 has often been compared to the Catholic condemnation of Copernicus in 1616. The Catholic sensitivity to the astronomy issue at the time was in part a reaction to pressures upon the Church brought about by the Protestant Reformation.[20] Similarly (although allowing for enormous differences) in the late 1940's the Soviet Union considered itself in the midst of a global ideological struggle, and the Cold War produced emotions not unlike those current during the Counter Reformation.

"Zhdanovshchina" is the name by which the postwar ideological campaign came to be known; it was named for Andrei A. Zhdanov, Stalin's assistant in the Central Committee of the Party. Most Western historians of the Soviet Union believe that Zhdanov was in some personal way responsible for the ideological restrictions in all areas of culture, including science. There is, however, reason to doubt that Zhdanov was responsible for ideological interference in the sciences. Evidence exists that Zhdanov actually opposed the Party's intervention in Lysenko's favor, and even attempted to stop it.[21] In any event, we know that Zhdanov carried out a campaign of intimidation and proscription in literature and the arts. A series of decrees laid down ideological guides for fiction writers, theater critics, economists, philosophers, playwrights, film directors, and even musicians. Until the month of Zhdanov's death, however, natural scientists escaped the rule by decree that obtained in other cultural fields.

When Lysenko's views of biology were officially approved in August 1948—an event to be reviewed in some detail in the present volume's analysis of the genetics controversy—a shock wave ran through the entire Soviet scientific community. No longer could it be hoped that Party organs would distinguish between science and philosophical interpretations of science. Evidently Stalin had no intention of making such distinctions, and he was in control of the Party. It soon became clear that other scientific fields, such as physics and physiology, were also objects of ideological attack, and Soviet scientists were genuinely fearful that each field would produce its own particular Lysenko.

Soviet scientists now found themselves in a difficult dilemma. By this time the Party's control over scholarly institutions was almost absolute. Open resistance to the Party's supervision was possible only if the resisters were prepared to sacrifice themselves entirely; opposition to Party control usually meant professional ruin and imprisonment in labor camps. A few scientists resisted openly and met the fate of the geneticist N. I. Vavilov, who was destroyed even before the war. Another approach was taken by a relatively small but quite influential group of scientists who decided to meet the ideological onslaught by defending their respective sciences from *within* the framework of dialectical materialism. Their subsequent accomplishment was genuine, significant, and intellectually interesting; a good part of this book concerns their feat. What many non-Soviet observers have failed to see is that this defense of science from the position of dialectical materialism was not merely a tactic, not an intellectual deceit; the leaders of this movement—whose names will be mentioned many times in this book—were sincere in their defense of materialism. As Soviet observers frequently say, "Their dialectical materialism was internal." They included Soviet scientists with eminent international reputations. A few associates of this group may have been initially hypocritical in their approach, willing to use any terminology or any philosophical system that would save their science from the fate of Lysenkoism. But the majority, certainly including those who had even before the war been interested in dialectical materialism, as well as a new group now making their previous materialistic views explicit, saw no contradiction between science and a sophisticated form of materialism. In speaking of dialectical materialism and science as congenial intellectual frameworks, they did not think they were compromising their professional integrity. Indeed, they strove to increase the sophistication of both Soviet natural science and Soviet philosophy, and in both goals they eventually had genuine success. They were assisted by those professional philosophers who saw the validity of this defense of scholarship and who greeted the work of these scientists as a contribution to a philosophical understanding of science.[22]

The scientists of the immediate postwar period began reading Marx and Engels on philosophical materialism in order better to answer their ideological critics. They developed arguments more incisive than those of their Stalinist opponents; they constructed defenses that exposed the fallacies of their official critics yet were in accord with philosophical materialism and—most important of all—preserved the cores of their sciences. They were even willing to examine the methodological principles and terminological frameworks of their sciences, revising them if necessary. As scientists they now had a stake of self-interest in the philosophy of science. They took heart in the defeat, even while Stalin was alive, of G. V. Chelintsev, a mediocre chemist who tried to win the position of a Lysenko in chemistry at an all-Union conference that bore some resemblance to the 1948 biology conference.[23] They turned back the ideological campaigns in relativity physics and quantum mechanics by developing materialist interpretations of these unsettling developments in physical theory and stoutly resisting attempts to displace them. Some eventually became personally committed to these interpretations, continuing to defend them in the sixties and beyond, long after Stalin's death. During these later years, younger scholars, both scientists and philosophers, joined the discussions. For them, motives of self-defense were no longer overriding in importance. The intellectual issues themselves emerged more fully. A comprehensive and cogent philosophy of science was being created.

Since the scientists were frequently men of genuine intellectual distinction and deep knowledge of their fields, and since science does contain serious and legitimate problems of philosophical interpretation, it was only natural that the entrance of the scientists into the debates would result in discussions important in their own right. Outside the field of genetics—where the issues remained on a very primitive level until the final overthrow of Lysenko—many of the discussions in the Soviet Union contained authentic issues of philosophical interpretation. These issues included, in the physical sciences, the problem of causality, the role of the observer in measurement, the concept of complementarity, the nature of space and time, the origin and structure of the universe, and the role of models in scientific explanation. In the biological sciences, relevant problems included those of the origin of life, the nature of evolution, and the problem of reductionism. In physiology and psychology, discussions arose concerning the nature of consciousness, the question of determinism and free will, the mind-body problem, and the validity of materialism as an approach to psychology. In cybernetics, problems concerned the nature of information, the universality of the cybernetics approach, and the potentiality of computers. This is by no means a complete list of the problems that emerged in Soviet discussions of the philosophy of science, although in the following sections of the present study attention is concentrated on them.

Occasionally the Soviet philosophers made genuine contributions to the discussions, even though Soviet scientists often directed well-deserved criticism against them. It is worth noticing that the worst threats to Soviet science in the late forties and early fifties did not come, as is often thought, from professional philosophers, but from third-rate scientists who tried to win Stalin's favor. These people included T. D. Lysenko in genetics, G. V. Chelintsev in chemistry, A. A. Maksimov and R. Ia. Shteinman in physics, and O. B. Lepeshinskaia in cytology.[24] These persons were criticized by *both* scientists and philosophers whenever political conditions permitted. What was going on in the worst period of ideological invasion of science was not primarily a struggle between philosophers and scientists. It was a struggle, crossing these academic lines on both sides, between genuine scholars on the one hand and ignorant careerists and ideological zealots on the other.

As the ideological campaign of 1948–53 receded into the past, it became less and less a determining factor in Soviet discussions of the relationship of science and philosophy. To be sure, censorship is still a universal fact in the Soviet Union. Genetics did not regain full status until 1965, and even now that science suffers the effects of its years of suppression. But from the mid-fifties onward, areas of science other than genetics regained that large degree of autonomy they enjoyed before the war. The political interference of the fifties and sixties that reached to the heart of literature and the arts—so offensive in the cases of Pasternak and Solzhenitsyn—was not duplicated in the natural sciences. A fairly normal intellectual life prevailed in the natural sciences, and this normality extended, on many technical issues, to the philosophy of science as well.

A scholar outside the Soviet Union might assume that a normal intellectual life among Soviet scientists would mean their dropping all interest in dialectical materialism. A number of Soviet scientists who were earlier involved in ideological discussions have, indeed, returned entirely to research work or scientific administration. But the most striking characteristic of the recent period has been the degree to which discussions in the philosophy of science have continued and even expanded. The professional philosophers have played a larger and wiser role than previously, but natural scientists have also continued to be involved in the discussions. Moreover, the circumstances surrounding the scientists' contributions to the discussions lead one to the belief that the performances cannot possibly all be command ones. In recent years Soviet scientists have on occasion volunteered contributions to foreign journals on the subject of dialectical materialism and science when they easily could have remained silent. Since a scientist can almost never improve his international reputation by writing on philosophy—indeed, the usual result is to harm it—one must assume that the Soviet scientists who continue to defend dialectical materialism as an approach to science have strong reasons for doing so, and that these motives are by no means entirely political.

II Dialectical Materialism: The Soviet Marxist Philosophy of Science

Dialectical Materialism: Soviet or Marxist?

Contemporary Soviet dialectical materialism as a philosophy of science is an effort to explain the world by combining these principles: All that exists is real; this real world consists of matter-energy; and this matter-energy develops in accordance with universal regularities or laws. A professional philosopher would say, therefore, that dialectical materialism combines a realist epistemology, an ontology based on matter-energy, and a process philosophy stated in terms of dialectical laws.

Dialectical materialism incorporates features of both absoluteness and relativity, of both an Aristotelian commitment to the immutable and independent and a Heraclitean belief in flux. To its defenders, this combination of opposite tendencies is a source of flexibility, strength, and truth; to its detractors, it is evidence of ambiguity, vagueness, and falseness.

Dialectical materialism has usually been discussed as if it were a uniquely Soviet creation, far from the traditions of Western philosophy. It is true that the term "dialectical materialism" comes from a Russian and not from Marx, Engels, or their west European followers. It is also true, of course, that Soviet dialectical materialism has acquired characteristics that are only explicable in terms of, first, its revolutionary, and later, its institutional setting. But the roots of dialectical materialism extend back to the beginning of the history of thought, at least to the Milesian philosophers and continue forward as subdued, changing, but reappearing strands in the history of philosophy. It is impossible to present here a discussion of the origins of dialectical materialism, which would constitute a large book in itself and would belong to the province of the professional philosopher; nonetheless, many similarities between dialectical materialism and previous currents of European philosophy will appear in the following pages.

The term "dialectical materialism" was first used, to the best of our knowledge, in 1891 by G. V. Plekhanov, a man frequently called the father of Russian Marxism.[1] Marx and Engels utilized terms such as "modern materialism" or "the new materialism" to distinguish their philosophical orientation from that of classical materialists such as Democritus or thinkers of the French Enlightenment such as La Mettrie or Holbach. Engels did speak, however, of the dialectical nature of modern materialism.[2] Lenin adopted the phrase used by Plekhanov, "dialectical materialism."

The basic writings of Marx, Engels, and Lenin on the philosophic and social aspects of science are Engels's *Anti-Dühring,* printed first as a series of articles in 1877; his *Dialectics of Nature,* written in 1873–83 but not printed until 1925; his *Ludwig Feuerbach and the End of Classical German Philosophy,* published as a series of articles in 1886 and as a pamphlet in 1888; Marx's doctoral dissertation, written in 1839–41 and first published in 1902; pieces of the correspondence of Marx and Engels; a few sections of Marx's *Capital;* Lenin's *Materialism and Empirio-Criticism,* published in 1908; his *Philosophical Notebooks,* published in 1925–29 and later, in a complete form, in 1933; and fragments from his correspondence and speeches.[3] Marx also left a number of unpublished manuscripts concerning science, technology, and mathematics, most of which are now in the Institute of Marxism-Leninism in Moscow. Some of these appeared in print only in the late 1960's.[4] Together all these writings establish the basis of dialectical materialism as it is usually discussed in the Soviet Union, with the older writings obviously playing a more formative role than the newer ones. In this rather large body of material written over a period of many decades by different authors for different purposes one can find a considerable diversity of viewpoints and even contradictions on fairly important questions. The dates of publication of the various works and the context in which each was composed are quite important in gaining an understanding of the evolution, modification, and structure of Soviet Marxist thought on the nature of science.

Although the primary interests of Marx and Engels were always in economics, politics, and history, they both devoted a surprisingly large segment of their time to the scrutiny of scientific theory, and cooperated in publishing their views on science. Engels described their background in the sciences:

Marx and I were pretty well the only people to rescue conscious dialectics from German idealist philosophy and apply it in the materialist conception of nature and history. But a knowledge of mathematics and natural science is essential to a conception of nature which is dialectical and at the same time materialist. Marx was well versed in mathematics, but we could only partially, intermittently and sporadically keep up with the natural sciences. For this reason, when I retired from business and transferred my home to London, thus enabling myself to give the necessary time to it, I went through as complete as possible a "moulting" as Liebig calls it, in mathematics and natural sciences, and spent the best part of eight years on it.[5]

Engels was much more important in elaborating the Marxist philosophy of nature than was Marx. This commitment to the study of the natural sciences as well as the social sciences was, in Engels's mind, a necessary consequence of the fact that man is, in the final analysis, a part of nature; the most general principles of nature must, therefore, be applicable to man. The search for these most general principles, based on knowledge of science itself, was a *philosophic* enterprise. Engels believed that by means of a knowledge of a philosophy that was materialistic, dialectical, and grounded in the sciences, both natural scientists and social scientists would be aided in their work. Those natural scientists who maintained that they worked without relying on philosophical principles were deluded; better to form consciously a philosophy of science, Engels thought, than to pretend to avoid one:

Natural scientists may adopt whatever attitude they please, they will still be under the domination of philosophy. It is only a question whether they want to be dominated by a bad, fashionable philosophy or by a form of theoretical thought which rests on acquaintance with the history of thought and its achievements.[6]

Engels's interest in the philosophy of science was so much more evident than Marx's that many scholars have maintained that it was Engels, not Marx, who was responsible for the concept of dialectical materialism; and that in bringing the natural sciences into the Marxist system, Engels violated original Marxism. Among the scholars holding this view are those who emphasize the young Marx as a theorist interested, not in universal systems, but specifically in man and his sufferings, a person whose first achievement was to present an explanation of the role of the proletariat in the modern world through the concept of alienated labor. Examples of exponents of this view are George Lichtheim, who wrote that dialectical materialism is a "concept not present in the original Marxian version, and indeed essentially foreign to it, since for the early Marx the only nature relevant to the understanding of history is human nature," [7] and Z. A. Jordan, who maintained that dialectical materialism was a "conception essentially alien to the philosophy of Marx." [8]

Scholars such as Lichtheim and Jordan have correctly emphasized the humanitarian ethic of the young Marx and the anthropological nature of his analysis, but they have erred in saying or implying that the idealistic young Marx was interested only in human nature, not physical nature. Marx's doctoral dissertation, written in 1839–41, several years before the now noted *Economic and Philosophical Manuscripts,* was suffused with the realization that an understanding of man must begin with an understanding of nature.[9] Entitled "The Difference between the Nature Philosophy of Democritus and the Nature Philosophy of Epicurus," the dissertation was a long discussion of the physics of the ancients, of the deviations from straight line descent in the atomic theory of Epicurus, of the nature of elementary substances and of elementary concepts such as time. Marx's attention to physical nature for an understanding of philosophy as a whole was entirely within the context of much of European thought; it was, further, an advantage rather than a disadvantage of his approach. Those recent writers who have tried to divest Marxism of all remnants of inquiry into physical nature have not only misrepresented Marx but have also deprived Marxism of one of its intellectual strengths. It is not necessary to restrict Marx's interests to ethics and economics to free him from vulgar materialism of the type of Vogt or Moleschott. Indeed, one of the points of Marx's dissertation was to show that Epicurus, although like Democritus a believer in atoms and the void, was not a strict determinist. The twenty-three-year-old Marx saw the atom as an abstract concept containing a Hegelian contradiction between essence and existence.[10] Marx would later discard the philosophic idealism underlying this formulation, but there is no evidence that he ever abandoned his interest in physical nature itself.

As a young student of philosophy, Marx was affected by the metaphysical aspirations of almost all great philosophical systems prior to his time and accepted the necessity of making certain epistemological and ontological assumptions. In later years, he attempted to move away from metaphysics, a tendency of some significance since materialism (like all other philosophical systems, including pragmatism) is in the final analysis founded on metaphysical assumptions. At no known point, however, did he resist Engels's effort to bring nature explicitly into their intellectual system. Engels read the entire manuscript of his *Anti-Dühring* to Marx, who presented no objections and even contributed a chapter himself (not on natural philosophy, however) for inclusion in the book. At least as early as 1873, ten years before Marx's death, Engels began work on what became many years later *Dialectics of Nature;* their correspondence illustrated that the mature Marx shared Engels's interest in "modern materialism" in nature notwithstanding the fact that he usually yielded to Engels on issues concerning science. Another spot where Marx indicated his agreement with Engels's efforts occurred in *Capital;* Marx observed that the dialectical law of the transition of quantity into quality, applicable to economics, also applied to the molecular theory of modern chemistry.[11] The thought behind Marx's somewhat casual observation on chemistry was his belief that in nature the whole is greater than the sum of its parts. It is a defensible and valuable principle defended by a large variety of thinkers, including contemporary dialectical materialists.

The point here is not that Marx's and Engels's views were identical, which has been maintained in the past both by Soviet scholars who wished to preserve the unity of dialectical materialism and by anti-Soviet scholars who wished to condemn Marx with the albatross of Engels; rather, the main point is that to emphasize primarily the differences of two men whose views have a great many affinities and who did both consider themselves modern materialists is as much, if not more, of an inaccuracy as crudely to lump them together. It is one thing to say that Marx never committed himself to finding dialectical laws in nature to the extent to which Engels did; it is quite another to say that such an effort contradicted Marx's thought, particularly when Marx is known to have supported the effort on several occasions. Jordan called Marx a "naturalist" rather than a "materialist," meaning that Marx wished to avoid a metaphysical commitment to matter as the sole source of knowledge, but Jordan acknowledged elsewhere:

The materialist presuppositions which were shared by Marx might have included the principle of the sole reality of matter ("matter" being the term used to denote the totality of material objects, and not the substratum of all changes which occur in the world), the denial of the independent existence of mind without matter, the rule of the laws of nature, the independent existence of the external world, and other similar assumptions traditionally associated with materialism.[12]

Jordan pointed out that Marx did not regard knowledge as the mere passive reflection of external matter in the human brain; rather, Marx saw knowledge as a result of a complex interaction between man and the external world. This epistemology does not deny materialism if one assumes that man is a part of the material world, but neither does it absolutely require it. There is a certain leeway in Marx's thought, permitting the supposition of naturalism instead of materialism, just as there is room in the Lenin of the *Philosophical Notebooks* (but not in the Lenin of *Materialism and Empirio-Criticism*) for several epistemologies. But despite these elements of latitude in Marx's thought, he never disclaimed "modern materialism," frequently accepted or used the term, supported Engels's elaboration of it, and in consequence, is, I believe, more accurately described as a materialist than as a naturalist.

The recent effort by many non-Soviet scholars to eliminate from Marxism an interest in physical nature can be explained, on the one hand, by their distaste for the ideological restrictions on science that were imposed in the Soviet Union, and on the other, by the general trend of philosophical thought in western Europe and North America. The interference of ideology with science in the Soviet Union, culminating in most people's minds in the Lysenko episode, led to a discrediting of the claims of Marxist philosophy in the natural sciences. Meanwhile, in the countries of western Europe and North America, metaphysical and ontological studies were out of fashion; dialectical materialism as an approach to nature was often seen as a vestige of archaic *Naturphilosophie,* an attempt to invade a realm that now belonged exclusively to the specific sciences. Many writers in Europe in the postwar generation were concentrating upon the individual man contained in existentialism rather than the soulless determinism that most people assumed was contained in dialectical materialism. Even those deeply interested in Marxism, such as Sartre, usually paid little attention to detailed studies of physical nature.

Scholars still committed to Marxism often attempted to save it from *Naturphilosophie* by trying to separate the writings of Engels on science from those of Marx, an operation that is technically possible but that, as we have already seen, usually resulted in conclusions incorrectly restricting the breadth of Marx's interests. On the other hand, those scholars who were opposed to Marxism used the ideological incursions on science in the Soviet Union as important supports in their efforts to prove that Marxism was essentially a perversion of science, antirational and even anti-Western, ignoring the deeply Western origins of Marxism and the fact that the Lysenko affair had little to do with Marxism as a philosophy of science.

Frankly recognizing Marx's interest in science would not only be more accurate in a historical sense but would also eventually have a salutary effect upon Marxism as a political doctrine. The same man who wrote the *Economic and Philosophical Manuscripts* also inquired into atomism, the history of the concept of mathematical differentiation, criticized Newton's method of quadratures, and discussed the mathematics of Lagrange, Maclaurin, and Taylor. Science was not only an area of knowledge that Marx attempted to bring within his purview, but also a model to him as a methodology of investigation. As Engels said at Marx's graveside, Marx tried to do for human history what Darwin did for the organic world; indeed, Marx offered to dedicate *Capital* to Darwin as a sign of his esteem. Marx believed that his interpretation of history was scientific, not in a rigidly deterministic sense but in the morphological Darwinian sense. On the basis of his theory of evolution Darwin could not predict when a new species would arise, and on the basis of his theory of history Marx could not predict when a new form of society would arise. If Marxist materialism, either as an explanation of nature or of social history, is to be discussed in terms of science, the comparison must be with nondeterministic theories of low predictive power, such as the theory of evolution or probability theories. The analogy between Darwin's and Marx's theories could be easily exaggerated into a misleading form of "biological sociology," but there is reason to believe that Marx himself saw a similarity in terms of the predictive reliability of the two approaches.

One of the conclusions issuing from the observation that Marx considered his scheme scientific (in roughly the way Darwin's was) is that modification of the doctrine must therefore be a desired end, not a feared one. Nothing is more un-Marxist than a pathological fear of revision.[13] Science, which must stand prepared to revise its conclusions in the light of new data, is in essence revisionist.[14] A Marxism that took its goal of being scientific seriously instead of dogmatically would be a constantly evolving Marxism, an openly revised Marxism.

The current turn to humanism within Marxism is a development that all people aware of the importance of ethics and aesthetics—topics not given their due in the Marxism of the past—can only greet with enthusiasm. Marxism has been one of the great influences of the past century and will no doubt continue to be significant in the future; its contribution will be greater if its humanization is thorough. However, just as the humanism of the Renaissance is seen by historians of science as a development with contradictory effects on the development of science,[15] so also might humanism in Marxism eventually have detrimental effects if it pushed science outside that realm of activities to which Marxism gives attention. Indeed, science is an intensely human activity; a sophisticated humanistic Marxism would be eager to bring it within its purview, not for the purpose of imposing a point of view, but for the derivation of additional knowledge of reality and for the appreciation of the crucial importance of science to the ethical and aesthetic system that it is attempting to create. Any conceptual scheme that hopes to be relevant to the modern world cannot afford to be uninterested in science. Science not only has philosophical content, but also requires careful evaluation —not blind criticism—from an ethical standpoint. Marxism has in the past possessed the advantage of being, as Sartre said, a totalizing "philosophy of our time," the single contemporary philosophical system that takes man in "the materiality of his condition." [16] Whether Marxism as a philosophic system can lose its dogmatism without simultaneously losing its comprehensiveness is an unanswered question.

In the Soviet Union philosophers have not attempted to divest Marx of his interest in all of reality, including physical as well as human nature; they have not followed the trend elsewhere in abandoning the effort to construct comprehensive explanations of reality based on studies of nature itself. They have recognized that one of the most intellectually attractive aspects of Marxism is its explanation of the organic unity of reality; according to Marxism, man and nature are not two, but one. Any attempt to explain either will inevitably have implications for the other. But Soviet philosophers have frequently squandered this intellectual advantage by supporting a dogmatic philosophy, by raising it to a status of a political ideology used for the rationalization of the existing governmental bureaucracy. They have failed to recognize the essential intellectual revisionism contained within Marxism's claim to a scientific approach. As a result they have not been adept in connecting dialectical materialism with new interests arising in non-Soviet philosophy with which it is potentially compatible, such as process philosophy.

Engels and Lenin on Science

Although both Marx and Engels were interested in science from early ages, it is nonetheless true that Engels turned most seriously to science only after the Marxist philosophy of history had been fully developed. By 1848 their political and economic views were well formed, but Engels did not begin systematic study of the sciences, nor did Marx initiate his most detailed studies of mathematics, until some time later. Engels remarked that he took up the study of science "to convince myself in detail—of what in general I was not in doubt—that amid the welter of innumerable changes taking place in Nature, the same dialectical laws are in motion as those which in history govern the apparent fortuitousness of events." [17]

Just what the term "law" (*Gesetz*) meant to Engels is not altogether clear. He did not attempt a philosophical analysis of the many different meanings that have been given to such terms as "law of nature," "natural law," or "causal law," and he did not clearly indicate what he meant by "dialectical law." Engels's dialectical laws were considerably different from those laws of physics that, within the limits of measurement, permit empirical verification. Engels saw, for example, the dialectical law of the transition of quantity into quality in the observed phenomenon that water, after absorbing quantities of heat, experiences a qualitative change when it comes to a boil at 100 degrees centigrade. Such a change can, indeed, be empirically verified by heating many samples of water to 100 degrees. But Engels believed (and Marx agreed in *Capital*) that the same law describes the fact that "not every sum of money, or of value, is at pleasure transformable into capital. To effect this transformation, in fact, a certain minimum of money or of exchange-value must be presupposed in the hands of the individual possessor of money or commodities." [18] The latter case of the dialectical law of the transition of quantity into quality is rather different from the former, even though both are described as instances of the same law. The boiling of water seems to be simply a phenomenon adequately described by a normal physical law; the quantitative-qualitative dialectical relationship is best used in describing the much more abstract phenomenon that a large aggregate of an entity, taken as a whole, displays characteristics not displayed by its parts. In the case of economics, there is no way in which the law can be verified in every instance; if a certain accumulation of money occurred without its conversion into capital, one could merely say that the correct poinnt had not yet been reached. In the case of the water, one not only possesses the description of *what* change is to occur, but information about *when* it is to occur.

Engels believed that nothing existed but matter and that all matter obeys the dialectical laws. But since there is no way of deciding, at any point in time, that this statement is true, the laws that he presupposed are not the same as usual scientific laws. It should be admitted that even in the case of "usual" laws in natural science the stated relationship, as a universal statement, is not subject to absolute proof. One cannot say, for example, that there will *never* be a case in which a standard sample of water heated to 100 degrees centigrade fails to boil. But when the violation of such laws does occur, it is, within the limits of measurement, apparent that something remarkable has happened.

The definition of "law" is a very controversial and difficult topic within the philosophy of science,[19] and I shall not pursue it beyond noting that Engels's concept of dialectical laws was quite broad, embracing very different kinds of explanations. Indeed, he referred to the dialectical relationships not only as "laws," but also as "tendencies," "forms of motion," "regularities," and "principles."

Engels is known for two major works on the philosophy of science, *Anti-Dühring* and *Dialectics of Nature*. Since only the first of these was a finished book and appeared almost fifty years earlier than the second, it exercised the greatest influence on the formation of the Marxist view of nature. In *Anti-Dühring* Engels criticized the philosophic system advanced by Eugen Karl Dühring in his *Course in Philosophy*.[20] Dühring was a radical lecturer on philosophy and political science at the University of Berlin, a critic of capitalism who was gaining influence among German social democrats. Engels disagreed with Dühring's claim to "a final and ultimate truth" based on what Dühring called "the principles of all knowledge and volition." The object of Engels's criticism was not Dühring's goal of a universal philosophic system, but the method by which he derived it and his claims for its perfection. Dühring's "principles" were to Engels a product of idealistic philosophy: "What he is dealing with are *principles,* formal tenets derived from *thought* and not from the external world, which are to be applied to nature and the realm of man and to which therefore nature and man have to conform. . . ."[21] Engels believed, contrary to Dühring, that a truly materialistic philosophy is based on principles derived from matter itself, not thought. The principles of materialism, said Engels, are

not the starting-point of the investigation, but its final result; they are not applied to nature and human history, but abstracted from them; it is not nature and the realm of humanity which conform to these principles, but the principles are valid in so far as they are in conformity with nature and history. That is the only materialistic conception of matter, and Herr Dühring's contrary conception is idealistic, makes things stand completely on their heads, and fashions the real world out of ideas, out of schemata, schemes or categories existing somewhere before the world, from eternity— just like *a Hegel*.[22]

A number of writers have commented that this desire to counteract Dühring's idealistic philosophy pushed Engels's first philosophical work toward the positivistic position of maintaining that all knowledge must be composed of verifiable data derived from nature.[23] They have frequently cited *Dialectics of Nature,* the later work, as containing an opposite metaphysical tendency, and have observed that the tension between these two strains in Marxist thought—positivistic materialism and metaphysical dialectics—has been present throughout its history. A tension between materialism and the dialectic has indeed existed within Marxism, and will be commented upon later, but the extent to which *Anti-Dühring* is positivistic and *Dialectics of Nature* metaphysical has been overdrawn. True, Engels in *Anti-Dühring* directed his chief criticism against a philosopher (Dühring) for not being materialist, while in *Dialectics of Nature* he, more in passing, admonished scientists (such as Karl Vogt, Ludwig Büchner, and Jacob Moleschott) for not being dialectical. But in both works Engels attempted to locate a balance point between reliance on the empirical findings of science, on the one hand, and the dialectical structure inherited from Hegel on the other. In *Anti-Dühring*, the reputedly positivistic work, Engels also presented some of his best-known discussions of the dialectic in nature, while in *Dialectics of Nature,* the work supposedly heavily Hegelian in inspiration, he stoutly defended the concept of the materiality of the universe.[24]

If one turns from Engels's works on the philosophy of science to a consideration of his knowledge of science itself, one is likely to conclude that although essentially a dilettante, he was a dilettante in the best sense. For a person of his background he possessed a remarkable knowledge of the natural sciences. Engels's formal education never went beyond the *gymnasium,* but he immersed himself in the study of science at certain periods of his life; he was able, for example, to write a long chapter on the electrolysis of chemical solutions, including computations of energy transformations.[25] He was familiar with the research of Darwin, Haeckel, Liebig, Lyell, Helmholtz, and many other prominent nineteenth-century scientists. In retrospect his errors do not draw so much attention as his unlimited energy and audacity in approaching any subject and the high degree of understanding that he usually achieved. Even if one is not willing to accept J. B. S. Haldane's observation that Engels was "probably the most widely educated man of his day," he was, indeed, a man of impressive knowledge.[26] Elements of naïveté and literalness are easily found, but they are less significant than his conviction that an approach to all of knowledge, and not just one portion, was necessary for a new understanding of man.

Indeed, a re-evaluation of Engels by historians of science is overdue. Engels's "errors" in science, as seen from the vantage of today—his quaint descriptions of electricity, his discussions of cosmogony, his descriptions of the structure of the earth, and his assertion that mental habits can be inherited—were usually the "errors" of the science of Engels's time. Engels was a materialist, and suffered from the tendency toward simplification that has plagued many materialists, but he was far more sophisticated than the popularizers of materialism of his day, who were usually scientists, such as Büchner and Moleschott. Those recent writers who have dismissed Engels's writings on science have usually forgotten the context of nineteenth-century materialism in which they were written. Against the background of this materialism Engels appears as a thinker with a genuine appreciation of complexity and an awareness of the dangers of enthusiastic reductionism. He was, for example, convinced that life arose from inorganic matter, but he ridiculed the simple approach of the supporters of spontaneous generation who had in the 1860's suffered a defeat at the hands of Pasteur. Engels's attitude toward the origin of life has been praised by biologists even recently.[27]

Lenin's writings on science are similar to Engels's, not only in terms of philosophical commitment, but also in several other secondary respects: He came to science after formulating his political and economic views; he first entered the field of philosophy of science for polemical reasons; he was responsible for two major works with somewhat different emphases; and his later, more sophisticated period is much less well known than his earlier, relatively untutored phase.

The particular viewpoints of Lenin on the philosophy of science, as expressed in *Materialism and Empirio-Criticism* and in the *Philosophical Notebooks* will be discussed in the following sections on epistemology and dialectics, but at this point it is necessary to mention the fact that most non-Soviet discussions of Lenin's philosophic views are based on *Materialism and Empirio-Criticism*. The *Philosophical Notebooks*, which consist of abstracts, fragments, and marginal notes, were not published until the end of the twenties, and did not appear in English until 1961. Consequently, they have been neglected by Anglo-American students of Leninism. Yet to the extent that Lenin achieved sophistication in philosophy, that stage is revealed in the *Philosophical Notebooks*, where we have his comments on Hegel, Aristotle, Feuerbach, and other writers. As two recent editors of Marxist philosophy commented:

We are seeing a non-professional philosopher, but a professional social revolutionary, studying some of the most technically difficult and theoretically advanced philosophical works of all time. He seeks better to be able to handle complex concepts in the analysis of everchanging and moving social forces. In these notes we are able to see Lenin's mind at work, trying out every abstract idea he comes across in the effort to test it in the crucible of his own rich experience. His main concern was to reconstruct the Hegelian dialectics on a thoroughly materialist foundation. . . . While Lenin was always the enemy of idealism, he opposed the offhand dismissal of this type of philosophy. As against vulgar materialism, he insisted that philosophical idealism has its sources in the very process of cognition itself. His conclusion was that "intelligent idealism is closer to intelligent materialism than stupid materialism." Thus, these *Philosophical Notebooks* are an indispensable supplement to Lenin's previous philosophical works and observations. Indeed, they constitute a plea for a richer and fuller development of dialectical materialism.[28]

The interpretation of the *Philosophical Notebooks* and their integration into Lenin's thought present particular problems for the historian. Lenin composed these fragments for himself alone, jotting down what first came to mind, and did not rewrite or rethink them. Obviously such materials must be treated more carefully than his published *Materialism and Empirio-Criticism*. Yet to rely upon the published work entirely would mean underestimating the full development of Lenin's thought. Lenin was quite aware of his shortcomings in philosophy in the earlier years; his efforts to overcome these deficiencies and his subsequent viewpoints emerge impressively in the *Philosophical Notebooks*.

The *Philosophical Notebooks* have exercised increasing influence in Soviet dialectical materialism since their publication, although they are still considered secondary to *Materialism and Empirio-Criticism*. As we shall see, this influence was usually in the direction of a greater appreciation of the subtleties of epistemology and of the dangers of reductionism. When the *Philosophical Notebooks* first appeared in the Soviet Union, they became elements in the debates between the dialecticians and the mechanists. In later years, they were frequently considered to be particularly suited for advanced students of dialectical materialism, partly because of their fragmentary and unsystematized nature, but even more, no doubt, because of the greater awareness that Lenin displayed there of the alternatives of epistemology.

Materialism and Epistemology[29]

In the Marxist philosophy of science as presented by Engels, there is nothing in the objective world other than matter and its emergent qualities. This matter has extension and exists in time; as Engels remarked,

"The basic forms of all being are space and time, and being out of time is just as gross an absurdity as being out of space." [30] (This view was somewhat modified in the Soviet Union after the advent of relativity, as will be shown.) The material world is always in the process of change, and all parts of it are inextricably connected. All matter is in motion. Furthermore, Engels agreed with Descartes's assertion that the quantity of motion in the world is constant. Both motion and matter are uncreatable, indestructible, and mutually dependent: "Matter without motion is just as inconceivable as motion without matter." [31]

It is important to note that Engels did not think of matter as a substratum, a *materia prima*. Matter is not something that can be identified or defined as a unique and most primitive substance that enters into an infinite number of combinations resulting in the diversity of nature. Rather, matter is an abstraction, a product of a material mind referring to the "totality of things." Engels commented:

Matter as such is a pure creation of thought and an abstraction. We leave out of account the qualitative difference of things in comprehending them as corporeally existing things under the concept matter. Hence matter as such, as distinct from definite existing pieces of matter, is not anything sensuously existing. If natural science directs its efforts to seeking out uniform matter as such, to reducing qualitative differences to merely quantitative differences in combining identical smallest particles, it would be doing the same thing as demanding to see fruit as such instead of cherries, pears, apples, or the mammal as such instead of cats, dogs, sheep, etc., gas as such, metal, stone, chemical compound as such, motion as such.[32]

According to Engels, abstractions such as matter are parts of thought and consciousness, the emergent products of a material brain. In discussing the materiality of the brain, Engels carefully dissociated himself from simple materialists such as Büchner, Vogt, and Moleschott. He agreed with them that thought and consciousness are products of a material brain, but he disagreed with simple analogies such as "the brain produces thought as the liver produces bile." On the contrary, on the basis of Hegelian quantitative-qualitative relationships Engels believed that each level of being has its own qualitative distinctiveness; to compare in a reductionist manner the thought produced by the brain to bile produced by the liver or motion produced by a steam engine conceals more than it reveals. Yet for all the distinctiveness of motion on each level of being, the carrier of that motion is matter: "One day we shall certainly 'reduce' thought experimentally to molecular and chemical motions in the brain; but does that exhaust the essence of thought?" [33]

According to Engels, man's knowledge flows from nature, the objective, material world. He saw two different epistemological schools in the history of philosophy: the materialists, who believe knowledge to derive from objective nature, and the idealists, who attribute primacy in cognition to the mind itself. As Engels said, "The great basic question of all philosophy, especially of more recent philosophy, is that of the relation of thinking and being." [34] At this point in *Ludwig Feuerbach* Engels proceeded to link the epistemological problem of knowledge to the ontological one of the existence of God:

The question of the relation of thinking to being, the relation of spirit to nature—the paramount question of the whole of philosophy—has, no less than all religion, its roots in the narrow-minded and ignorant notions of savagery. . . . The question: which is primary, spirit or nature—that question, in relation to the Church was sharpened into this: Did god create the world or has the world been in existence eternally? [35]

A number of Engels's critics have pointed to what they call a "fatal flaw" in his reasoning at this point: his confusion of epistemology and ontology.[36] There is no reason, they have said, for identifying idealism with a belief in God, or realism with atheism. A person could believe in objective reality and withhold judgment on the question of God or even consider God to be "objectively real." Within the framework of the problem of epistemology alone the critics are correct; there are more than two camps on the issue of cognition. In describing how man comes to know, one can emphasize the role of objective reality (realism); the role of matter (materialism); the role of the mind (idealism); or one can maintain that it is impossible to know how man comes to know (agnosticism). Furthermore, one's religious views are not determined by one's epistemology. But for Engels the ontological principle that all that exists is matter came before all others. Therefore, for him a God who could be objectively real to a person in terms of his epistemology but nonmaterial in terms of ontology was nonsense.

The key to the Marxist philosophy of science is not its position on cognition, which contains considerable flexibility, as evidenced not only by Lenin's writings in the *Philosophical Notebooks* but even more so by subsequent developments (particularly in countries such as contemporary Yugoslavia), but its position on matter itself. What justification do we have for assuming that an ill-defined "matter" (later "matter" was equated with "energy") alone exists? The more thoughtful Russian Marxists such as Plekhanov (and perhaps Lenin at moments) have veered toward the position that the principle of the sole existence of matter is a simplifying assumption necessary for subsequent scientific analysis. Other Marxists, such as Engels, the Lenin of *Materialism and Empirio-Criticism,* and most Soviet philosophers, have maintained that the principle of materialism is a fact presented by scientific investigation. But as a result of the sensitivity of the subject, the issue of the justification for the belief in materialism has not been thoroughly investigated by philosophers in the Soviet Union.

To return to Engels's treatment of the opposition of idealism and materialism, we can see his merging of the problems of existence and cognition in the following quotation:

Contrary to idealism, which asserts that only our mind really exists, and that the material world, being, Nature, exists only in our mind, in our sensations, ideas and perceptions, the Marxist materialist philosophy holds that matter, Nature, being, is an objective reality existing outside and independent of our mind; that matter is primary, since it is the source of sensations, ideas, mind, and that mind is secondary, derivative, since it is a reflection of matter, a reflection of being. . . .[37]

Engels's last phrase, "mind . . . is a reflection of matter," strikes to the heart of the mind-matter relationship. In Russian Marxist philosophy the description of this relationship has been a major issue. Engels's term "reflection" was followed by Plekhanov's "hieroglyphs," Bogdanov's "socially-organized experience," and Lenin's "copy-theory." The copy-theory of Lenin, to be subsequently discussed, became the most influential model for Soviet philosophy. It will also be important in the discussions of physiology and psychology in this volume.

Connected with Engels's view of the nature of the material world was his opinion on the attainability of truth about that world. Parallel to the existence of matter apart from mind was the existence, potentially, of truth about that matter. Scientists strive toward complete explanations of matter even though these explanations are never reached. The relationship between man's knowledge and truth, according to Engels, is asymptotic (knowledge approaches truth ever more closely, but will never reach its goal).[38] It is not correct to say that Engels believed in the attainability of absolute truth. Only at the unattainable point of infinity in the relationship between man's knowledge and truth does an intersection obtain. Nonetheless, Engels believed in a cumulative, almost linear, relationship of knowledge to truth. Lenin, on the contrary, saw many more temporary aberrations in the upward march, and used the image of a "spiral movement" to describe the process.

Reinterpretations of Russian Marxist Views on Materialism and Epistemology

Among Russian Marxists the problems of epistemology and the philosophy of nature attracted considerably more attention than among west European Marxists. G. V. Plekhanov, Lenin's tutor in Marxism and later an opponent of the Bolsheviks, developed his "hieroglyphic" theory of knowledge in 1892 in his notes to his translation of Engels's *Ludwig Feuerbach*. Plekhanov wrote:

Our sensations are sorts of hieroglyphs informing us what is happening in reality. These hieroglyphs are not similar to those events conveyed by them. But they can completely truthfully convey both the events themselves and—and this is important—also those relationships existing between them.[39]

The analysis presented by Plekhanov was an attempt to go beyond the common-sense realism implied by Engels's writings to a recognition of the difference between objects-in-themselves and our sensations of them. In Plekhanov's view there was a distinct difference—so much so that he felt that these sensations "are not similar to those events conveyed by them." Nonetheless, he said, there is a correspondence between these events and our sensations. Thus Plekhanov went from a "presentational" theory of perception to a "representational" one.[40] His epistemology was dualistic in a Lockean sense, but without the introduction of subjective qualities. It was still materialistic in that it assumed the existence of material objects outside the mind that reveal themselves in an indirect but trustworthy fashion by means of man's sensations.

It was important to Plekhanov that to each of man's sensations in the process of perceiving an object there be a materialistic correlate, and to each of the changes in a material object there be a sensational correlate. He said that one should imagine a situation in which a cube is casting a shadow on the surface of a cylinder:

This shadow is not at all similar to the cube: The straight lines of the cube are broken; its flat surfaces are bulged. Nevertheless for each change of the cube there will be a corresponding change of the shadow. We may assume that something similar occurs in the process of the formation of ideas.[41]

Plekhanov was aware that his epistemology was not scientifically provable, as the above words "we may assume" indicate. He discussed respectfully Hume's view that there was no way of proving that physical objects are anything more than mental images.[42] Plekhanov's writings implied that by assuming the primacy of matter in cognition, he considered himself to be making a plausible and useful philosophic choice rather than coming to a scientific conclusion.

In the early twentieth century a controversy arose among Russian Marxists that ultimately led to the entry of Lenin into the field of epistemology. In the resulting *Materialism and Empirio-Criticism* Lenin criticized not only his immediate disputants, the "Russian Machists," but also Plekhanov. In order to introduce the controversy, some mention must be made of Ernst Mach (1830–1916).

The late nineteenth century's most formidable criticism of the philosophic belief in a material world independent of man's mind was contained in Mach's sensationalism. Mach was an Austrian physicist and philosopher who provided much of the impetus to the development of logical positivism and who prepared the way for the acceptance of relativity and quantum theory. His antimetaphysical views were equalled by those of his contemporary, the German philosopher Richard Avenarius, the proponent of the theory of knowledge known as empirio-criticism. Mach and Avenarius occupy a special place in Soviet Marxist philosophy, since they are the objects of copious criticism in Lenin's *Materialism and Empirio-Criticism*.

In *Analysis of Sensations* Mach defended the view, already ancient among philosophers but now made particularly relevant to modern science, that the "world consists only of our sensations." [43] According to Mach, space and time were as much sensations as color or sounds.[44] A physical object was merely a constant sensation (or "perception," taken as a group of sensations). Mach followed Berkeley, then, in denying the dualism of sense perceptions and physical objects. But while Berkeley was a realist in the sense of assuming the reality of mental images and of an external God, Mach endeavored to introduce no elements into his system that were not scientifically verifiable. Therefore, he made no pronouncements about ultimate reality. According to his "principle of economy," scientists should select the simplest means of arriving at results and should exclude all elements except empirical data.[45] Mach's approach employed on the practical scientific level, where he intended it to be utilized, would mean that a scientist would cease worrying about the "real" or "actual" nature of matter and would merely accept his sense-perceptions, working as carefully and thoroughly as he could. A theory that found a pattern in the data would be judged entirely on the basis of its usefulness rather than its plausibility in terms of other existing considerations. There might even be more than one "correct" way of describing matter (a concept to have influence later in discussions of quantum mechanics). Two explanations, working from opposite directions, could both be useful and could supplement each other, even if there seemed to be a contradiction between the two approaches.[46]

Mach had shifted the emphasis from matter *reflecting* in the mind to the mind *organizing* the perceptions of matter. A group of Marxist philosophers soon followed Mach's lead. This school of Russian empiriocritics included A. Bogdanov (pseudonym of A. A. Malinovskii); A. V. Lunacharskii, the future commissar of education; V. Bazarov (V. A. Rudnev); and N. Valentinov (N. V. Vol'skii). Bogdanov, a medical doctor, was swayed by the lucidity and scientific nature of Mach's arguments, but dissatisfied with what he saw as their inconsistencies. If, as Mach maintained, sensations and objects are the same, why do two different realms of experience—the subjective and the objective—continue to exist? [47] Why are there two different sets of principles or "regularities" (*zakonomernosti*) in these different realms? Thus, in the objective world, there are such sensations as sight, sound, and smell. In the subjective realm are emotions and impulses: anger, desire, and so forth. Bogdanov defined objective sensations as those that are universally perceived,[48] and subjective sensations as those that may be apparent to only one person or a small group of persons. Bogdanov then attempted to find the roots of this dualistic system and thereby unite them in a philosophical system called empiriomonism. The key to this development is the concept of the "organization of experience." To Bogdanov, the physical world equals "socially organized experience," while the mental world is "individually organized experience." Therefore, "if in the single stream of human experience we find two principally different conformities of law (*zakonomernosti*), then nevertheless, both of them arise in equal measure from our own organization: They convey two biological-organizational tendencies. . . ." [49] Parenthetically, it is worthwhile to note that Bogdanov's emphasis upon organizational structure and the means of transmitting information would cause a new surge of interest in his work in the Soviet Union many years later when cybernetics and information theory were applied to psychology and epistemology.

Lenin's original entry into the field of philosophy was the result of his being disturbed by the views of Russian Marxist writers such as Bogdanov and Plekhanov. His first motivation was a tactical one; he wished to protect the Bolsheviks' claim to a materialistic view of nature and history. Only many years later did he become genuinely interested intellectually in problems of philosophy.

At first Lenin hesitated to commit himself in the philosophical fray; in order to avoid commenting on these difficult problems, he attempted to set a policy of philosophical neutralism within the Bolshevik faction of the Russian Social Democratic Party. As late as February 1908 the Bolsheviks continued to give equal treatment in their legal press to "Machism" and orthodox dialectical materialism. [50] This position became intolerable for Lenin, however, as Bogdanov, Bazarov, Lunacharskii, and their friends published works increasingly sympathetic to the views of Mach; Lenin found the Bolsheviks being labeled "revisionists" while the rival Marxist faction, the Mensheviks, were winning the designation "orthodox." [51]

In 1908 Lenin set himself to the task of writing a major work on philosophy in order, as he put it, "to find out what was the stumbling block to these people who under the guise of Marxism are offering something incredibly muddled, confused, and reactionary." [52] The stumbling block, he found, was the influence of the latest developments of science upon philosophers, including Marxist philosophers such as Bogdanov.

By the early twentieth century many people believed that the foundations of materialism were being undermined by scientists themselves. [53] The relative confidence of scientists of Marx's and Engels's time in their knowledge of nature had been replaced by perplexity. The investigation of the radiations of radium and uranium, resulting in the identification of alpha rays (helium nuclei) and beta rays (high speed electrons), had discredited the concept of nondivisible atoms. Such scientists as L. Houllevigue remarked, "The atom dematerialises, matter disappears." [54] Henri Poincaré observed that physics was faced with "a debacle of principles." [55]

The rise of philosophical schools such as empirio-criticism on the continent and phenomenalism in England was largely a response to these and other developments in science. In Lenin's opinion, the philosophers following these trends were subordinating the search for truth about matter to attempts to provide convenient explanations of isolated perceptions. Idealism was again a threat, and Bishop Berkeley's theories were reborn, in the name of science rather than God.

In countering these new movements, Lenin stressed two tenets of his interpretation of dialectical materialism: the copy-theory of the mind-matter relationship and the principle that nature is infinite. It seems clear that Lenin regarded these principles as minimum requirements in order for dialectical materialism to have philosophical consistency or significance. He was not attempting to impose philosophy upon science, but to locate the bedrock of the materialist philosophy of science; he believed it impossible for science to contradict these principles.

By the "copy-theory" of matter Lenin meant that materialism is based on recognition of "objects-in-themselves" or "without the mind." According to him, "ideas and sensations are copies or images of these objects." Just *how* similar these ideas are to the objects themselves was left unsaid. There is some reason to believe that at the time of the writing of *Materialism and Empirio-Criticism* Lenin considered man's mental images to be quite similar to the corresponding objects. His epistemology was at this time close to that of common-sense realism; he criticized the "vagueness" of Plekhanov's hieroglyphic epistemology. Yet even in some of his remarks of *Materialism and Empirio-Criticism* Lenin indicated that the essential aspect of dialectical materialism was the principle of objectively existing matter, not the degree of correspondence between man's images and the objects of the material world. Indeed, he approached reducing the fundamentals of materialist epistemology to one principle: "Only one thing is from Engels's viewpoint immutable—the reflection by the human mind (when the human mind exists) of a world *existing and developing independently of the mind.*" [56] Lenin added that this independent objective world can be known by man; "To be a materialist is to acknowledge objective truth, which is revealed to us by our sense organs." [57]

It should be noticed that if a person accepted literally the last sentence in the above paragraph as Lenin's definition of materialism, he would be fully justified in saying that Lenin confused realism ("all that exists is real") with materialism ("all that exists is material") and that Lenin was, in fact, not a materialist, but a realist. For one could take Lenin's statement "To be a materialist is to acknowledge objective truth, which is revealed to us by our sense organs" and with complete justification change it to read "To be a realist is to acknowledge objective truth, which is revealed to us by our sense organs." Was Lenin, then, a realist rather than a materialist? An accurate answer to this question, one which took into consideration all of Lenin's writings, would have to be negative. Lenin always spoke of materialism, not realism, and he saw the difference, particularly in his later works; he supplemented his realist epistemology with an *assumption* of ontological materialism resulting from his belief in the conceptual value of such an assumption. The fact that Lenin's materialism was founded on an assumption has not been openly discussed in the Soviet Union, where dialectical materialism is usually portrayed as a provable doctrine, even an inevitable conclusion of modern science. Yet the best argument for materialism starts out with a recognition of its assumptive or judgmental character and a defense of such a minimal assumption or judgment as being consonant with all available evidence and persuasive to many scientists. This argument must, of course, leave room for the person who wishes to make a different initial assumption. Individual scientists are likely to have preferences for one or the other. The noted American biologist Hermann J. Muller recognized and approved the assumptive origin of Lenin's materialism in an article that he wrote in 1934 and that is included as an appendix in this book. To Muller, Lenin's assumption could be defended, further, on the basis of inductive judgment:

To those scientists who would protest that we should not make such prejudgements regarding scientific possibilities, on the basis of a prior "philosophical" assumption of materialism, but should rather follow in any direction in which the empirical facts of the case seem to be leading, we may retort, with Lenin, that all the facts of daily life, as well as those of science, together form an overwhelming body of evidence for the materialistic point of view . . . and that therefore we are justified, in our further scientific work, in taking this principle as our foundation for our higher constructions. It too is ultimately empirical, in the better sense of the word, and it has the overwhelming advantage of being founded upon the evidence of the whole, rather than upon just some restricted portion of the latter.[58]

But Lenin in 1908 was not yet able to recognize the judgmental or preferential bases of materialism, although he would approach them later in his *Philosophical Notebooks*. In 1908 he was, instead, concentrating on a criticism of the Russian followers of Mach, and he naturally was more interested in revealing the vulnerable points in their analysis than in his own. He asserted that Bogdanov's idealistic philosophy actually concealed a belief in God, in spite of his repudiation of all religion. If, said Lenin, the physical world equals merely "socially organized experience," then the door is opened to God as an aspect of experience. "Idealism says that physical nature is a product of this experience of living beings," added Lenin, and therefore "idealism is equating (if not subordinating) nature to God. For God is undoubtedly a product of the socially organized experience of living beings." [59]

In Lenin's opinion, the epistemological problem was not separated from the question of the nature of matter itself. The mental realm of experience is not distinct from the material realm, but is a result of it on a higher level. Matter itself is not at all threatened by Rutherford's dismantling of the atom because "the electron is as *inexhaustible* as the atom, nature is infinite, but it infinitely exists." [60] Lenin believed that the expression "matter disappears" was an indication of philosophical immaturity by scientists and philosophers who did not understand that science will constantly discover new forms of matter and new principles of motion. The forms and principles of matter are taken as immutable only by naïve materialists who have fallen prey to metaphysics, Lenin asserted:

The recognition of immutable elements, "of the immutable substance of things," and so forth, is not materialism, but *metaphysical,* i.e., anti-dialectical, materialism. That is why the "subject-matter of science is endless," that not only the infinite, but the "smallest atom" is immeasurable, unknowable to the end, *inexhaustible.* . . . In order to present the dialectical materialist standpoint, we must ask: Do electrons, ether *and so on* exist as objective realities outside the human mind or not? The scientists will also have to answer this question unhesitatingly; and they do invariably answer it in the *affirmative,* just as they unhesitatingly recognize that nature existed prior to man and prior to organic matter. Thus the question is decided in favor of materialism. . . .[61]

Lenin believed that philosophies opposing science were based either on idealism or simple materialism, not dialectical materialism. He attempted to make dialectical materialism less vulnerable to criticism and less likely to retard science by drawing a line between it and simple materialism. Yet, if one judges by *Materialism and Empirio-Criticism,* one must conclude that this line was not drawn with any degree of clarity. Lenin did not even discuss in this work the laws of the dialectic, the principles that distinguish dialectical materialism from simple materialism. He merely maintained that dialectical materialism, a philosophical viewpoint, cannot be affected by the vacillations of scientific theory. Lenin labored to reforge the bond between the theory of dialectical materialism and the practice of science. While insisting on the materialist copy-theory, he also affirmed that nature is infinite. The division of particles into smaller particles could go on forever, he believed, but matter would never disappear.

In the notations that Lenin made six or seven years later during his further study of philosophy, he revealed a greater appreciation of the alternatives of epistemology. Although he did not repudiate his earlier copy-theory, he now saw the link between the material objects of the world and man's images as much more indirect. Indeed, he seemed to believe that in the highest forms of their development, materialist and idealist theories of cognition were linked in a unity of contradiction; they tended to pass into one another. Thus Hegel, whom he believed to be the greatest idealist in philosophy, arrived unwittingly at the threshold of dialectical materialism. Lenin's evolution was from another direction, from the side of materialism, but he approached the same spot as Hegel, the moment of unity between two philosophies. As Lenin observed, "the difference of the ideal from the material is not unconditional." [62] The area where the distinction between idealism and materialism became nearly imperceptible was that of mental abstraction; in order to understand nature it is necessary for man not only to perceive matter but to construct a series of concepts that "embrace conditionally" eternally moving and developing nature. And these abstractions may include elements of fantasy:

The approach of the (human) mind to a particular thing, the taking of a copy (= a concept) of it *is not* a simple, immediate act, a dead mirroring, but one which is complex, split in two, zigzag like, which *includes in it* the possibility of the flight of fantasy from life; more than that: The possibility of the *transformation* (moreover, an unnoticeable transformation, of which man is unaware) of the abstract concept, idea, into a *fantasy* (in the final analysis = God). For even in the simplest generalization, in the most elementary general idea ("table" in general), *there is* a certain bit of *fantasy*. (Vice versa, it would be stupid to deny the role of fantasy, even in the strictest science; cf. Pisarev on useful dreaming, as an impulse *to* work, and on empty daydreaming.) [63]

The Lenin who is revealed in the above passage is not the one who is known to most students of Leninism; this Lenin recognizes the painful, halting, indirect path of knowledge, a path that includes clear reversals. He grants the useful role of fantasy "even in the strictest science." He sees this fantasy as an *inherent possibility* in scientific thought and is aware that in the final analysis it can lead to a belief in God. In his statement that the possibility of fantasy is included in the approach of the human mind to nature, he, like Plekhanov before, seemed to recognize that the rejection of idealism is not a matter of scientific proof but philosophic choice. And Lenin continued, of course, to choose materialism. The possibility of fantasy in science was, he believed, a result of the indirectness of man's knowledge. The fantasy in a given scientific theory was progressively reduced in scope, and the useful role it might play was temporary. But Lenin recognized that with the advent of new theories the possibilities of new fantasies, including useful ones, would arise.

This more flexible view of materialism, which sees it as a result of choice and not of proof, opened up room for its potential accommodation with some other philosophic currents, a development that, however, still has not occurred. If one considers, for example, some of the recent writings of W. V. O. Quine, it becomes apparent that there are similarities of argument. Quine wrote in his "Two Dogmas of Empiricism":

As an empiricist I continue to think of the conceptual scheme of science as a tool, ultimately, for predicting future experience in the light of past experience. Physical objects are conceptually imported into the situation as convenient intermediaries—not by definition in terms of experience, but simply as irreducible posits comparable, epistemologically, to the gods of

Homer. For my part I do, qua lay physicist, believe in physical objects and not in Homer's gods; and I consider it a scientific error to believe otherwise. But in point of epistemological footing the physical objects and the gods differ only in degree and not in kind. Both sorts of entities enter our conception only as cultural posits. The myth of physical objects is epistemologically superior to most in that it has proved more efficacious than other myths as a device for working a manageable structure into the flux of experience.[64]

The possibility for the convergence of the epistemological view represented by the quotation from Quine above and the epistemology of dialectical materialism is considerable. Holders of both views prefer a concept of "physical objects," and both find justification for this concept in pragmatic success. To be sure, what Quine calls "superior myth," the dialectical materialist has called "truth." But does the word "myth" here mean "that which is false" or "that which can not be proved"? And does the dialectical materialist really believe that the "truth" of his position can be illustrated, or is his position one that he assumes to be true for reasons similar to those for which Quine assumes his myth to be "superior"? And if the dialectical materialist may have some difficulty in defining "matter-energy," falling back eventually to Engels's reference to the "totality of things," Quine says with equal indefiniteness that "physical objects are postulated entities which round out and simplify our account of the flux of experience." [65] In the final analysis there are differences between the positions, to be sure, but hardly of the type that would place dialectical materialism outside the realm of philosophy while leaving Quine's epistemology within.

The Laws of the Dialectic

The discussion of dialectical materialism has so far centered on the last half of the term: materialism. The other half, the dialectic, concerns the characteristics of the development and movement of matter.

There are two rather different views of the dialectic that Soviet thinkers have, at different moments, taken; one is the belief not only that matter-energy obeys laws of a very general type, but that these laws have been identified in the three laws of the dialectic to be discussed below. This has been the official, ideological view expressed in the Soviet textbooks on dialectical materialism. The other view is that matter-energy does indeed obey general laws, but that the three laws of the dialectic are provisional statements to be modified or replaced, if necessary, as science provides more evidence. This has been an unofficial, philosophical view, appearing in the Soviet Union from time to time, particularly among younger scientists and philosophers.[66]

The dialectic as applied by Engels to the natural sciences was based on his interpretation of Hegelian philosophy. This interpretation involved not only the well-known conversion of Hegelian philosophy from idealism to materialism, but also the reduction of Hegel's thought to a simple scheme of dialectical laws and triads.

In his *Science of Logic* Hegel spoke of "dialectic" as "one of those ancient sciences which have become most misjudged in modern metaphysics and in popular philosophy of ancients and moderns alike." [67] Hegel believed that the way in which dialectic had previously been used had involved only two terms (dualisms, antinomies, opposites); he referred to Kant's discussion of "Transcendental Dialectic" in his *Critique of Pure Reason,* in which Kant advanced the view that human reason is essentially dialectical in that every metaphysical argument can be opposed by an equally persuasive counterargument. Hegel saw a means of transcending this contradiction in a third position, "the second negative," which is "the innermost and most objective moment of Life and Spirit, by virtue of which a subject is personal and free." [68]

Contrary to much opinion, Hegel never used the terms "thesis-antithesis-synthesis" in the neat fashion so often attributed to him; he recognized, however, the importance of the thesis-antithesis contradiction in the writings of Kant, Fichte, and Jacobi, and he did sparingly use the term "synthesis" to indicate the moment of transcendence of such a polarity.[69] But Hegel opposed reducing his analysis to a triadic formula, and warned that such a scheme was "a mere pedagogical device," a "formula for memory and reason." "The distinctions of the Concept of philosophical cognition can be realized," he said, "only by going through the whole process of self-knowledge." [70]

Hegel did not provide a straightforward method of analysis that merely had to be "turned on its head" in order to become dialectical materialism. Engels's use of Hegel involved not only inversion, but also codification, a dubious reduction of rather obscure complexity. Nonetheless, many of the elements of Engels's dialectical materialism were indeed present in Hegel's thought. The very fact that Engels sought to simplify Hegel is not surprising—many men, including Goethe, have condemned the great Prussian philosopher for being unnecessarily complex—but Engels's centering of attention upon the laws of the dialectic had the unfortunate effect of tying Marxism to three codified laws of nature rather than simply to the principle that nature does conform to laws more general than those of any one science, laws that may, with varying degrees of success, be identified.

To Engels the material world was an interconnected whole governed by certain general principles. The great march of science in the last several centuries had, as a regrettable by-product, so compartmentalized knowledge that the important general principles were being overlooked. As he observed, the scientific "method of work has also left us as legacy the habit of observing natural objects and processes in isolation, apart from their connection with the vast whole; of observing them in repose, not in motion; as constants, not as essentially variables; in their death, not in their life." [71]

By "dialectics" Engels said that he meant the laws of all motion, in nature, history, and thought. He named three such laws: the Law of the Transformation of Quantity into Quality, the Law of the Mutual Interpenetration of Opposites, and the Law of the Negation of the Negation. These dialectical principles or laws were supposed to represent the most general patterns of matter in motion. Like Heraclitus, dialectical materialists believe that nothing in nature is totally static; the dialectical laws are efforts to describe the most general uniformities in the processes of change that occur in nature. The concept of the evolution or development of nature is, therefore, basic to dialectical materialism. The dialectical laws are the principles by which complex substances and concepts evolve from simple ones.

According to Engels, the laws were equally valid in science and human history. This universal applicability of the laws has served both as a source of strength and of weakness for dialectical materialism. On the one hand, the possession of the dialectic has given Marxists a conceptual tool of considerable power; many thinkers have been attracted by the Hegelian framework of dialectical materialism. The urge to possess a key to knowledge has been perhaps the strongest motivation in the history of philosophy. To oppose dialectical materialism to "Western thought," as many commentators have done in recent years, is to ignore the metaphysical foundation of most of European philosophy during the past several thousand years.

On the other hand, the universality of dialectical materialism has been frequently a disadvantage for its adherents. Many non-Soviet philosophers have turned away from it in the belief that it contains precisely those elements of Western philosophy that should have been abandoned before they were; dialectical materialism is a vestige, they say, of scholasticism. Rather than describing *how* matter moves, in the post-Newtonian sense, it attempts, in the Aristotelian sense, to explain *why* matter moves. Furthermore, the generality of the dialectic is achieved at the price of such diffuseness that to many critics its usefulness seems negligible. As one critic of the dialectic, H. B. Acton, remarked, the Law of the Negation of the Negation is "already general almost to the point of evanescence" when it is applied to such very different things as mathematics and the growing of barley; when the law is then extended to the transition from capitalist to communist society, "the only point of likeness appears to be the words employed." [72] To this criticism dialectical materialists would reply that if one accepts the existence of one real world of which all aspects of man's knowledge are derivative parts, one should expect there to be at least a few general principles that are applicable to all those parts. Some of the more sophisticated dialectical materialists of the post-Stalin period would add that they are in principle prepared to reject the three laws of the dialectic enunciated by Engels if superior substitutes can be found, and there have been a few attempts to achieve this through the application of information theory.

The principle of the Transformation of Quantity into Quality derived from Hegel's view that "quality is implicitly quantity, and conversely quantity is implicitly quality. In the process of measure, therefore, these two pass into each other: each of them becomes what it already was implicitly. . . ." [73]

Engels pointed to what he considered numerous examples of the operation of this law in nature. These were the cases when quantitative succession in a natural phenomenon is suddenly interrupted by a marked qualitative change. One example given by Engels was the homologous series of carbon compounds. The formulas for these compounds (CH_4, C_2H_6, C_3H_8, and so on) follow the progression C_nH_{2n+2}. The only difference among members of the progression,

Engels observed, is the *quantity* of carbon and hydrogen. Nevertheless, the compounds have greatly differing properties, some being gases and others solids in their natural states. In these diverse properties Engels saw the Law of the Transformation of Quantity into Quality at work.[74]

Perhaps the most unusual case that Engels cited as an example of the transformation of quantity into quality concerned Napoleon's cavalry during the Egyptian campaign. During the conflicts between French and Mameluke horsemen a curious relationship became apparent. Whenever a small group of Mamelukes would come upon a small group of Frenchmen in the desert, the Mamelukes would always win, even if somewhat outnumbered. On the other hand, whenever a large group of Mamelukes would come upon a large group of Frenchmen, the Frenchmen would always win, even if somewhat outnumbered. Engels's description can be represented in the following table:

Number of Mamelukes	Number of Frenchmen	Victors
2	3	Mamelukes
100	100	Even match
1500	1000	Frenchmen

The reason for the results was that the Frenchmen were highly disciplined and trained for large-scale maneuvers, but were not veteran horsemen. The Mamelukes had been on horses from the earliest age, but knew very little about discipline and tactics. Hence, a qualitative-quantitative relationship existed that yielded differing results at different quantitative levels.[75]

To Marx and Engels, Darwin's theory of evolution was an important illustration of the principle of the transition of quantity into quality. This tenet as a part of the Hegelian dialectic preceded Darwin, of course, but Marx and Engels considered Darwinism a vindication of the dialectical process. In the course of natural selection, different species developed from common ancestors; this transition could be considered an example of accumulated quantitative changes resulting in a qualitative change, the latter change being marked by the moment when the diverging groups could no longer interbreed.[76]

In the interpretation of science, the principle of the transition of quantity into quality has been important in the Soviet Union as a warning against reductionism. Reductionism here means the belief that all complex phenomena can be explained in terms of combinations of simple or elemental ones. A reductionist would maintain that if a scientist wishes to understand a complex process (growth of crystals, stellar evolution, life, thought), he must build up from the most elemental level. Reductionism tends to emphasize physics at the expense of all other sciences. It is a view that was often supported by nineteenth-century materialists and even today continues to have considerable strength. Reductionism is highly criticized by Soviet dialectical materialists, who carefully distinguish themselves from earlier materialists. In the biological sciences in particular, the quantity-quality relationship has been interpreted in the Soviet Union as foreclosing the possibility of explaining life processes—most of all, thought—in elementary physico-chemical terms. The development of matter during the history of the earth from the simplest nonliving forms up through life and eventually to man and his social organizations is regarded as a series of quantitative transitions involving correlative qualitative changes. Thus, there are "dialectical levels" of natural laws.[77] Social laws cannot be reduced to biological laws, and biological laws cannot be reduced to physico-chemical laws. In dialectical materialism the whole is greater than the sum of its parts. This principle has been a valuable guard against simplistic explanations in materialistic terms, but it has also, on occasion, skirted the opposite danger of organicism or even vitalism.

The principle of the transition of quantity into quality, therefore, distinguishes dialectical materialism from mechanistic materialism. A materialist, similar to Democritus, would say, for example, that the human brain is essentially the same as an animal's brain but is organized in a more efficient manner. According to this line of thinking, then, the difference is merely quantitative. The Marxist materialist, however, would say that the human brain is distinctly different from that of an animal and that this difference is a qualitative change resulting from accumulated quantitative changes during the course of the evolution of man from lower primates. Therefore, human mental processes cannot be reduced to those of other animals. Indeed, life processes in general cannot be totally reduced to physical and chemical processes if the latter are defined in contemporary terms. This emphasis on the qualitative distinctiveness of complex entities from simple ones has led dialectical materialists in recent years to become interested—although cautiously—in such concepts as "integrative levels" and "organismic biology," approaches widely discussed in Europe and America in the thirties and forties and displaying new vigor with the advent of cybernetics. The views of Soviet scholars toward these concepts will be discussed more fully in a subsequent chapter.[78]

The attitudes of Soviet philosophers toward explanation of organic processes illustrate the complex and perhaps even contradictory nature of dialectical materialism. The dialectical materialist says, in effect, "There is nothing but matter, but all matter is not the same." Some critics have seen this expression as a paradox existing at the very root of dialectical materialism. Berdyaev wrote, for example, "Dialectic, which stands for complexity, and materialism, which results in a narrow one-sidedness of view, are as mutually repellent as water and oil." [79] Of course, one can easily note that almost every philosophical or ethical system contains tensions: The strain existing between the ideals of individual freedom and social good has existed in much of Western thought without destroying the value of that body of thought. Similarly, the tension between complexity and simplicity in dialectical materialism is by itself of small consequence in judging the adequacy of the system for the problems to which it is put. For the practicing scientist this tension has the merit of providing him with the confidence that nature can be fruitfully approached, while warning him against assuming that success in one area or at one level will answer ultimate questions.

Thus, the tension between complexity and simplicity that inheres in the principle of the transition of quantity into quality is simply a permanent feature of dialectical materialism that, operating at different times in different ways, both strengthens and weakens it. This tension was an important source of the disputes that arose in Soviet philosophy in the 1920's.

A partial rationalization of this dichotomy is offered by the second principle of the dialectic, the Law of the Mutual Interpenetration of Opposites, sometimes called the Law of the Unity and Struggle of Opposites. Hegel gave his views on this principle in terms of "positive" and "negative":

Positive and negative are supposed to express an absolute difference. The two however are at bottom the same: the name of either might be transferred to the other. Thus, for example, debts and assets are not two particular, self-subsisting species of property. What is negative to the debtor, is positive to the creditor. A way to the east is also a way to the west. Positive and negative are therefore intrinsically conditioned by one another, and are only in relation to each other. The north pole of the magnet cannot be without the south pole, and *vice versa*. If we cut a magnet in two, we have not a north pole in one piece and a south pole in the other. Similarly, in electricity the positive and the negative are not two diverse and independent fluids. In opposition, the different is not confronted by any other, but by its other.[80]

By the principle of the unity of opposites Engels meant that harmony and order are found in the resulting synthesis of two opposing forces.[81] Engels saw the operation of this law in the rotation of the earth around the sun, which resulted from the opposing influences of gravitational and centrifugal forces. The same law governed the formation of a salt by the chemical interaction of an acid and a base. Other examples of the unity of opposites cited by Engels were the atom as a unity of positive and negative particles, life as a process of birth and death, and magnetic attraction and repulsion.[82]

The Law of the Unity and Struggle of Opposites is an explanation within dialectical materialism for the energy inherent in nature. To the question, How did the matter in the world acquire its motion? dialectical materialism answers that matter possesses the property of self-movement as a result of the contradictions or opposites present in it. Thus, it is not necessary for dialectical materialists to postulate the existence of a First Mover who set the planets, molecules, and all other material objects in motion. This concept of self-movement derived from internal contradictions was also present in the thought of Hegel, who commented in his *Science of Logic,* "contradiction is the root of all movement and life, and it is only in so far as it contains a contradiction that anything moves and has impulse and activity." [83]

The Law of the Negation of the Negation is closely connected with the second law, since negation is supposedly the process by which synthesis occurs. Negation, according to Hegel, is a positive concept, an affirmation. The constant struggle between the old and the new leads to superior syntheses. In its most general sense, the principle of the negation of the negation is merely a formal statement of the belief that nothing in nature remains constant. Everything changes; each entity is eventually negated by another. Engels commented, "Negation in dialectics does not mean simply saying no, or declaring that something does not exist, or destroying it any way one likes, and long ago, Spinoza said:

Omnis determinatio est negatio—every limitation or determination is at the same time a negation." [84] Engels considered the principle of the negation of the negation extremely important to dialectical and historical materialism; he wrote that it is a law of development of "Nature, history and thought," and a law that "holds good in the animal and plant kingdoms, in geology, in mathematics, in history and in philosophy." [85] He gave a number of examples of the law: the negation of capitalism (which was a negation of feudalism) by socialism; the negation of plants such as orchids by artificially altering them through cultivation, yielding better seeds and more beautiful blooms; the negation of butterfly eggs by the birth of the butterflies, which then lay more eggs; the negation of barley seed by the growth of the plant, which then yields more barley grains; differentiation and integration in calculus; and the mathematical process of squaring negative numbers.[86]

It is clear that "negation" meant a number of different things to Engels: replacement, succession, modification, and so forth. The last example above deserves more comment. Engels proposed taking the algebraic symbol a and negating this quantity by making it $-a$, then "negating the negation" by multiplying by $-a$, obtaining a^2. He said that a^2 represented a "synthesis of a higher level" since it was a positive number of the second power.[87] One may ask, Why did Engels multiply in order to negate the negation, instead of adding, subtracting, or dividing? Why did he multiply by $-a$ instead of some other figure? The obvious reply is that Engels picked a particular example that suited his purpose from the myriad of examples available. One of his discussions of the $\sqrt{-1}$ (evidently to prove the Law of the Mutual Interpenetration of Opposites) caused one anguished mathematician to write to Marx complaining about Engels's "wanton attack on the honor of -1." [88]

The dialectical laws in Marxist philosophy remained virtually as Engels elaborated them for many years. In the period immediately after the Russian Revolution, Soviet philosophers neglected the dialectical laws. At that time neither Engels's *Dialectics of Nature* nor Lenin's *Philosophical Notebooks* were known to them. The latter work, published separately in 1933, introduced one change in the Soviet treatment of the dialectical laws: Lenin considered the Law of the Unity of Opposites the most important of the three. Engels had listed the laws in the following order: (1) the Law of the Transformation of Quantity into Quality, (2) the Law of Mutual Interpenetration of Opposites, and (3) the Law of the Negation of the Negation. Lenin chose the order (2), (3), (1). Lenin even hinted that the transformation of quantity into quality was really only another description of the unity of opposites; if the two were truly synonymous, only two of the three primary laws would remain.[89]

Before this discussion leaves the subject of the dialectic, a few words must be said about the "categories." In dialectical materialism the term "categories" is employed to refer to those basic concepts that are necessary in order to express the forms of interconnection of nature. In other words, while the laws of the dialectic just discussed are attempts to identify the most general uniformities of nature, the categories are those concepts that must be employed in expressing these uniformities. Examples of categories given in the past in Soviet discussions of the dialectic have been concepts such as matter, motion, space, time, quantity, and quality.

Nowhere does dialectical materialism reveal its affinities with traditional philosophy more clearly than in its emphasis on categories, although dialectical materialists have given the classical categories a new formulation and meaning. The word "category" was first used as a part of a philosophical system by Aristotle. In his treatise *Categories* Aristotle divided all entities into the following ten classes: substance, quantity, quality, relation, place, time, posture, state, action, and passion. Objects or phenomena that belonged to different categories were considered to have nothing in common and, therefore, could not be compared. In his writings Aristotle frequently listed only some of the above ten categories, with no indication that others had been omitted. Aristotle apparently considered the exact number of categories and the terminology best suited for describing them open questions. Following Aristotle, many thinkers relied on a system of a priori categories of varying number and nature as a base for their philosophical systems. Medieval philosophers usually considered the original ten categories of Aristotle complete, ignoring Aristotle's latitude on the issue.

The two greatest modifiers of the Aristotelian concept of categories were Kant and Hegel. Kant based his categories not on particular subjects or entities, but on different types of judgments or propositions. To him, categories applied to logical forms, and not to things in themselves. "Quality" to Kant meant not "bitter" or "red," as it did to Aristotle, but logical relationships such as "negative" or "affirmative." "Quantity" meant to him not "five inches long" but "universal," "particular," and "singular." The Kantian radical reform of the Aristotelian categories is a complex subject of continuing discussion to philosophers.

Soviet philosophers have borrowed heavily from the approaches of Aristotle and Kant, adding to them Hegel's belief that the categories are not absolute. They consider Aristotle's writings amenable to the interpretation that the categories are reflections of the general properties of objectively existing objects and phenomena, although, as one Soviet text comments, "he did not always hold to this view, and, moreover, he did not succeed in revealing the inner dialectical connection of the categories." [90] Kant's major contribution to an understanding of the categories, in the opinion of Soviet philosophers, was his research into the logical functions of the categories and into the role of thought in refining sense perceptions. But Kant, they continue, made the great error of eliminating all connection between the categories and the objective world, looking upon them as the product of reason. Hegel's achievement in understanding the categories, according to Soviet dialectical materialists, was his realization that the categories are not static, but are in the process of development and are connected with each other. Thus, for example, the category "quantity" can grow into "quality."

The major distinction of the dialectical materialist approach to the categories is its heavy emphasis on natural science. Since, according to Marxism, being determines consciousness and not consciousness being, the material world, as reflected in man's consciousness, determines the very concepts by which he thinks—that is, the categories. Thus, "In order for the materialist dialectic to be a method of scientific cognition, to direct human thought in search of new results, its categories must always be located at a level with modern science, with its sum total of achievements and needs." [91]

Since man's knowledge of the material world changes with time, so then will his definition of the categories. The *Short Philosophical Dictionary*, published in Moscow in 1966, defined the categories as "the most general concepts, reflecting the basic properties and regularities of the phenomena of objective reality and defining the character of the scientific-theoretical thought of the epoch." [92] The same source listed as examples of the categories matter, motion, consciousness, quality and quantity, cause and effect, and so on.[93]

The inclusion of the "and so on" at the end of the list of categories is an important indication of the flexibility of the categories within dialectical materialism. As with Aristotle, the list is purposely kept open and subject to revision. Lenin remarked that "if *everything* develops, then doesn't this refer to the most general *concepts* and *categories* of thought as well? If it doesn't, then that means thinking is not connected with being. If it does, then it means that there is a dialectic of concepts and a dialectic of knowledge having an objective significance." [94] The same approach is reflected in the following description in the *Short Philosophical Dictionary:* "The categories are regarded as flexible and changing because the very properties of objective phenomena are also mobile and changing. The categories do not appear at once in a completed form. They are formed during the long historical process of the development of knowledge." [95] Thus, the categories develop with science itself.

The avowed elasticity of the categories gives, indirectly, room for interpretation of the dialectical laws themselves, since the categories are the terms in which the laws are expressed. In this study the possibility of revising the categories will be particularly relevant in the discussion of cosmology, where certain authors reinterpreted the term "infinity" after 1956 by examining the categories. "Time" and "space" had been listed as categories in the texts of the fifties, and these concepts were re-examined.[96] Another area where the categories were scrutinized was quantum mechanics: Here the concept of causality, or the category of "cause and effect," was actually modified.

The Unity of Theory and Practice

Another aspect of dialectical materialism that has relevance for science is not so much an integral part of the intellectual structure of the system as it is a methodological principle; this aspect is the unity of theory and practice. During a considerable portion of Soviet history the unity of theory and practice meant for scientists that they should give their research a clear social purpose by tying it to the needs of Soviet society. The strength of this recommendation has varied greatly in different fields and at different times. The unity of theory and practice can be traced back to Marx's opposition to speculative philosophy; he hoped to transcend philosophy by "actualizing" it. One of his best-known sentences referred to this effort to build a conceptual theory that would result in concrete achievements: "The philosophers have only *interpreted* the world in various ways; the point however is to *change* it." [97]

Engels believed that the unity of theory and practice was connected with the problem of cognition. The most telling evidence, he believed, against idealistic epistemologies was that man's knowledge of nature resulted in practical benefits; man's theories of matter "worked" in the sense that they yielded products for his use. As Engels commented, "If we are able to prove the correctness of our conception of a natural process by making it ourselves, bringing it into being out of its conditions and using it for our own purposes into the bargain, then there is an end of the Kantian incomprehensible 'thing-in-itself.' " [98] Thus, practice becomes the criterion of truth. Of course, Engels admitted that many theories or explanations "worked" while being incomplete or based on false assumptions. The Babylonians were able to predict certain celestial phenomena through the use of tables, with almost no knowledge of the location and movement of the bodies themselves. Every scientific theory at any point in time contains false assumptions and lacks important evidence; many useful theories, such as Ptolemaic astronomy, are "overthrown." But Engels maintained that the successful application of a theory about nature indicated that it contained somewhere within it a kernel of truth.[99]

During the Soviet period the principle of the unity of theory and practice acquired an influential moral significance. When the Soviet Union was attempting to industrialize rapidly, scientists were under considerable pressure to pursue research that would help in the effort. For geologists to study the strata of the earth merely to speculate about its age was considered immoral at a moment when they could have been identifying valuable mineral deposits. For mathematicians to develop theories that seemed to have no conceivable application was believed to be a squandering of the investment that society made in their educations.

This emphasis on practical return, highly typical in a different form among businessmen in other countries, would have significance in the controversy over genetics. Yet its operational meaning is unclear. Engels's belief that the approach to truth is evidenced by the benefit from man's theories did not provide a method for choosing among new hypotheses.[100] It is clear that a time element is involved in the unity of theory and practice. Almost no scientific theory gives immediate practical benefit. Time is required not only for the development of industrial processes that would utilize the theory, but for the exploration of the theory's implications that have practical consequences. When a theory is first developed, it may have no apparent applications at all but may turn out much later to be an invaluable approach to a practical problem. In the years after Stalin's death an appreciation of the indirectness of the link between theory and practice grew in the Soviet Union.

It should be recognized that every country, and particularly those such as the Soviet Union that feel the existence of external threats, will sacrifice certain luxuries in order to meet the most urgent practical needs. The emphasis that the Soviet Union placed in science upon the unity of theory and practice in the thirties and forties was more a result of these considerations than the opposition to speculative philosophy contained in Marxism.

Looking back over the system of Soviet dialectical materialism, we see, on the most general level, that it represents a natural philosophy based on the following quite reasonable principles and opinions:

The world is material, and is made up of what current science would describe as matter-energy.

The material world forms an interconnected whole.

Man's knowledge is derived from objectively existing matter.

The world is constantly changing, and, indeed, there are no truly static entities in the world.

The changes in matter occur in accordance with certain over-all regularities or laws.

The laws of the development of matter exist on different levels corresponding to the different subject matters of the sciences, and therefore one should not expect in every case to be able to explain such complex entities as biological organisms in terms of the most elementary physicochemical laws.

Matter is infinite in its properties, and therefore man's knowledge will never be complete.

The motion present in the world is explained by internal factors, and therefore no external mover is needed.

Man's knowledge grows with time, as is illustrated by his increasing success in applying it to practice, but this growth occurs through the accumulation of relative—not absolute—truths.

The history of thought clearly shows that no one of the above principles or opinions is original to dialectical materialism, although the total is. Many of the above opinions date from the classical period and have been held by various thinkers over a period of more than two thousand years. Today many working scientists operate, implicitly or explicitly, on the basis of assumptions similar to the above principles (hence the Soviet view that outstanding non-Marxist scientists are often at least implicit dialectical materialists). Yet the common currency of many of the most general principles of dialectical materialism does not devaluate it. First, these principles have been more fully developed and more closely linked to science in the writings of dialectical materialists than in any other corpus of literature. Furthermore, unexceptional as some of their opinions may at first glance seem, dialectical materialists have their opponents even on these broadest principles. Their commitment to the primacy of matter is rejected by many, probably most, philosophers. Materialism has never been a philosophy of the majority of philosophers at any point in the history of Western philosophy; its most ardent advocates have not usually been professional philosophers. In addition, dialectical materialists disagree not only with their most obvious opponents—theists and idealists—but also with materialists of the old type, the thoroughgoing reductionists who believe that all science will eventually be absorbed by physics. In sum, dialectical materialism —despite what some observers regard as the unobjectionable character of its most general principles—is still a controversial world view, one that enjoys the explicit support of only a small minority of philosophers and scientists in the world. When one adds to these intellectual obstacles the political liability deriving from dialectical materialism's support by the bureaucracy of a nondemocratic state, it is not surprising that dialectical materialism has won relatively few supporters outside the Soviet Union. Yet it should be noticed that in intellectual terms dialectical materialism is a legitimate and valuable point of view, far more interesting than non-Soviet scientists and philosophers have usually assumed.

Soviet dialectical materialism as a philosophy of science draws upon both Russian sources and traditional European philosophy. The Soviet contribution has been primarily one of emphasis on the natural sciences as determining elements of philosophy. In the opinion of Soviet philosophers, dialectical materialism both helps scientists in their research and, in turn, is ultimately affected by the results of that research. Their critics have occasionally maintained that such a description of the relation of science and philosophy is no description at all. Exactly what meaning is carried in the statement "Philosophy influences science and is, in turn, influenced by it"?

An answer that accurately weighs the mutual influence of philosophy and science is difficult, but it is clearly true that such a mutual influence exists. Furthermore, that interaction is an important element in the genesis and elaboration of scientific schemes. Questions may revolve around the degree of influence of philosophy upon science, or the mechanisms by which such influence is transmitted, but the existence of the interaction can not be questioned. Throughout the history of science, philosophy has significantly affected the development of scientific explanations of nature, and, in turn, science has influenced philosophy. Scientists inevitably go beyond empirical data and proceed, implicitly or explicitly, on the basis of one or another philosophy. Philosophers, on the other hand, have been forced by the evolution of science to revise basic concepts underlying their philosophic systems, such as the concepts of matter, space, time, and causality.

Moments when philosophy has importantly influenced science can be found from the earliest points in the history of science, and similar influences continue today in all countries. The early teachings of the Ionian natural philosophers, based on a naturalistic or nonreligious approach to nature, were heavily modified by the post-Socratic Greeks in the name of philosophic viewpoints that assumed the necessity of a divine being for an understanding of the cosmos. Benjamin Farrington commented that astronomy was "Pythagoreanized and Platonized within a few generations of the Ionian dawn." He also observed that "astronomy did not really make its way with the Greek public until it had been rescued from atheism." [101] Here was an example in which a philosophic world view influenced a scheme of scientific explanation.

Many others have followed. The historian of science Alexandre Koyré maintained that Galileo was a Platonist in his understanding of nature, and that this view had important influences on his scientific development. Koyré's view has been criticized, but not from the standpoint of a denial of philosophic influence on Galileo.[102] Newton's explanation of nature was presented within a religious framework that made it more acceptable to the society of his time and yet also revealed something important about Newton's internal convictions and presuppositions. Descartes actually postponed publication of his *Principia Philosophiae* in an effort to fit an orthodox and religious interpretation to his view of nature; this orthodoxy accorded with his views, but the fitting process was not obvious. The impact of German nature philosophy upon European scientists in the early decades of the nineteenth century is well known, and such historians of science as L. Pearce Williams would even see nature philosophy as an important ingredient in the origin of field theory, maintaining that convertibility of forces "was an idea that was derived from nature philosophy and one to which the Newtonian system of physics was, if not hostile, at least indifferent." [103] In each of these cases interaction between science and philosophy is one of the prime topics of study for the historian of science.

The impact of philosophy upon science has continued through the present day; it should not be regarded as a vestige of the past that will hopefully soon be overcome if it has not already been vanquished. Einstein even wrote, "In our time physicists are forced to concern themselves with philosophical questions to a greater degree than physicists of previous generations." [104] Einstein frequently acknowledged his personal debt to philosophical criticisms of science; out of such criticisms arose a revolution in twentieth-century science.

We stand so close to the development of contemporary science that we may not at first discern the interaction of philosophy and science, but it is certainly there. As an example, the new concepts in quantum mechanics and relativity physics of this century not only had a philosophic background, but in turn exerted a considerable influence on subsequent philosophy in Western countries in the first half of this century. These were countries in which many different philosophic viewpoints were expressed but the most popular ones favored religion over atheism and idealism over materialism. Consequently, it is not surprising that a number of influential scientists and philosophers in these countries seized upon the new physics and attempted to build philosophical systems justifying their religious and epistemological points of view. The uncertainty principle was to some of them an opportunity for the defense of freedom of will, while the rise of relativity physics signaled to many of them the end of materialism. In the Soviet Union the defenders of materialism answered back in full measure—indeed, in more than full measure—criticizing idealistic and religious points of view. Each side went far beyond conclusions that were intellectually justified, treating the opposition as if its position were groundless. The outcome of this debate was to illustrate that neither side possessed the clear superiority of argument that it claimed. This gradually became apparent to many writers, and the quality of their arguments began to improve. The Soviet authors, the main concern of this study, were able to develop a dialectical materialist interpretation of the universe based on the very principles of contemporary science that their opponents attempted to use against them.

The fact that emerges from these considerations is that science and philosophy have interacted at all times and places, not merely in the ancient past or in the contemporary Soviet Union. Soviet science is a part of world science, and the type of interaction of philosophy and science that can be found in Soviet scholarly writings (those of intellectuals, not of Party activists) is not essentially different from the interaction of science and philosophy elsewhere. But since the philosophical tradition in the Soviet Union is different from the tradition in western Europe and the United States, the results of that interaction have not been identical with the results of similar interactions in other geographical areas.

Thus, the significance of dialectical materialism is not so much its insistence on this mutual interaction of philosophy and science—many of its critics would readily grant such a relation—but the way in which this interaction has actually occurred in the Soviet setting. The dialectical materialism in the Soviet Union today is not the same as the dialectical materialism there thirty years ago, and the impact of developments in science is one of the reasons for its change. But then neither is science itself in the Soviet Union the same as it was thirty years ago, and one of the many influences upon it has been dialectical materialism. Although the Communist Party has attempted at times to control this interaction —much more so twenty years ago than now—it quite obviously was not able to do so. An independent intellectual process of significance and interest to historians and philosophers of science was also at work. The following chapters contain discussions in detail of the mutual influence of science and dialectical materialism in the Soviet Union. The most helpful sources for these discussions were the writings of individual Soviet scientists.

III   Quantum Mechanics

Of all the philosophic issues posed by modern scientific theory, those involving quantum mechanics have been the most pressing and obstinate. Several problems in the philosophy of science of the past generation—such as the interpretation of special relativity—held the attention of scholars for a decade or more but have now lost much of their allure; other issues—such as the interpretation of cybernetics and information theory—have gained prominence only recently. But more than forty years after the publication of the essential mathematical apparatus of quantum mechanics the controversy swirls unabated, even intensified.[1] It is a debate in which the scholars of many nations have participated, including those of the U.S.S.R.

The structure of quantum mechanics may be divided into a mathematical formalism and a physical interpretation of that formalism. The mathematical formalism, which is the core of quantum mechanics, is a differential wave equation, the solution of which is usually termed the *psi* (ψ) function; the wave equation was first developed by Erwin Schrödinger, who pursued Louis de Broglie's extension of the concept of wave-corpuscle duality from light to elementary particles of matter. The advantage of this formalism is that it yields, on a probabilistic basis, numerical values permitting a more complete mathematical description of microphysical states, including prediction of future states, than has any other formalism so far. The disadvantage of the mathematical apparatus of quantum mechanics is that the only widely accepted (some would say the only possible) physical interpretation for it contradicts several of man's most basic intuitions concerning matter. Specifically, quantum mechanical computations, in contrast to the classical laws of the macrophysical realm, do not yield arbitrarily exact values for position and momentum coordinates of microparticles. According to the well-known uncertainty relation, the more exactly the position of a microparticle is known, the less exactly the momentum, and vice versa.[2]

In view of the success of the mathematics of quantum mechanics for the derivation of useful physical values, the obvious question arises, What is the physical significance of the wave function? Can it be that matter is, indeed, undulatory? It is over this question of the physical interpretation of the mathematics of quantum mechanics that scores of philosophers and scientists have splintered their pens.[3]

The evolution of quantum mechanical theories is a trail littered with unsatisfactory explanations. De Broglie originally proposed that matter *is* wavelike and that the waves described by quantum mechanics do not "represent" the system; they *are* the system.[4] This explanation encounters enormous difficulties, far too complex to enter into here, but the nature of some of them may be indicated by noting that a literal acceptance of the physical reality of the wave function would involve such concepts as that of physical space with an almost unlimited number of dimensions. And most graphically, such an interpretation cannot explain satisfactorily the fact that a single micro-object upon impact on a sensitive emulsion leaves a spot, not the imprint of a wave front.[5] Max Born originally suggested the alternative: Matter is corpuscular, and the wave function describes not the particles but our knowledge about them. This ingenious theory unfortunately runs into equally disastrous physical facts, which are best illustrated by reference to the now classic two-slit interference experiment. When particles are allowed to pass through two narrow slits in a barrier before striking a sensitive emulsion, the impacts form an interference pattern that can be explained only on the basis of the wavelike characteristics of microbodies.

The Copenhagen Interpretation, developed by Niels Bohr and Werner Heisenberg, resolves the contradictions of previous interpretations by postulating that no observable has a value before a measurement of that observable has been made. As Heisenberg declared, "The path comes into existence only when we know it." [6] Thus, it becomes meaningless to speak of the characteristics of matter at any particular moment without empirical data in hand relating to that moment. It is senseless to speak of the position of a particle ("position" is a property of the corpuscular theory) without a measurement of position; it is equally unjustified to speak of the momentum (a wave property) without a measurement of momentum. This reconciling of classically incompatible properties by granting them existence only at the moment of measurement is usually called complementarity and is the heart of the most critical discussions of quantum mechanics.

Physicists and philosophers of science do not agree on a single definition of complementarity, but a satisfactory definition is the one just given—that is, contradictory properties of a microbody are reconciled by granting these individual properties existence only at separate moments of measurement. Another formulation, one that evades the question of "existence" of properties but that nonetheless is commonly given, is to say that the quantum description of phenomena divides into two mutually exclusive classes that complement each other in the sense that one must combine them in order to have a complete description in classical terms. This latter view was the one accepted by Oppenheimer when he stated that the notion of complementarity "recognizes [that] various ways of talking about physical experience may each have validity, and may each be necessary for the adequate description of the physical world, and may yet stand in a mutually exclusive relationship to each other, so that to a situation to which one applies, there may be no consistent possibility of applying the other." [7] It must also be added that even such early leaders in quantum mechanics as Bohr and Wolfgang Pauli did not entirely agree in their definitions of complementarity.[8] The essential problem was the perennial one in the history of science: giving a verbal interpretation to a mathematical relationship.

Before World War II the views of Soviet physicists on quantum mechanics were quite similar to those of advanced scientists elsewhere. Russian physics was in many ways an extension of central and west European physics. The work of such men as Bohr and Heisenberg influenced scientists in the Soviet Union as it did everywhere. Indeed, Soviet physicists spoke of the "Russian branch" (*filial*) of the Copenhagen School, composed of a group of highly talented theoretical physicists, including M. P. Bronshtein, L. D. Landau, I. E. Tamm, and V. A. Fock. And yet behind this exterior of agreement with scientists everywhere on quantum mechanics (or, more accurately, disagreements similar to those everywhere), as early as the 1920's certain Soviet physicists were aware that dialectical materialism might some day be interpreted in a way that could influence their research.[9] Lenin had, after all, devoted an entire book, *Materialism and Empirio-Criticism,* to the crisis in interpretations of physics and had particularly criticized the neopositivism of Ernst Mach, out of which much of the philosophy of modern physics grew. Lenin's assertion that a dialectical materialist must recognize the existence of matter separate and independent from the mind, while not inherently contradictory to quantum mechanics, could be regarded as at least uncongenial to the Copenhagen School's disinclination to comment upon matter in the absence of sensible measurement. And the extension of the concept of complementarity beyond physics to other realms, including ethical and cultural problems, by certain members of the Copenhagen School almost guaranteed some conflict with representatives of Marxism.[10] As early as 1929 the leading Soviet philosopher at that time, A. M. Deborin, gave a lecture on "Lenin and the Crisis of Contemporary Physics" to the Academy of Sciences.[11] But the first serious Soviet critique of the customary interpretation of quantum mechanics in a physics journal, rather than a philosophy journal, occurred in 1936, written by K. V. Nikol'skii.[12] In the dispute that developed between Nikol'skii and V. A. Fock, a leading interpreter of quantum mechanics in the Soviet Union to the present day and originally an adherent of the Copenhagen School, Nikol'skii called the Copenhagen Interpretation "idealistic" and "Machist," [13] two appellations that were to be frequently utilized after World War II by Soviet Marxist critics. Nikol'skii's own view of quantum mechanics deserves examination for still another reason: It was a purely statistical approach, with only a few differences from D. I. Blokhintsev's postwar "ensemble" interpretation, which will be discussed in greater detail below.

With mention of Nikol'skii's "purely statistical" approach it is appropriate at this point to insert a few remarks on the concept of probability, which is crucial to any interpretation of quantum mechanics. Probability in quantum mechanics has been interpreted by different scholars in both epistemological and statistical senses. The statistical, or frequency, approach, used by Nikol'skii, was an attempted objective interpretation in which probability was seen as inherent in nature. On the other hand, a number of scholars have seen quantum mechanics, particularly through Born's original interpretation, as containing probability because of its epistemological assumptions, and have even discussed such peculiar things as "waves of knowledge." The distinction between these two approaches, often blurred in discussions of quantum mechanics, is absolutely necessary in attempting to decide whether a theory that is irreducibly probabilistic is also necessarily idealistic.

Fock's interpretation in 1936 of the physical significance of the wave function was essentially the same as that of the Copenhagen School, which combined Born's emphasis on the mathematical description of man's knowledge of the microworld with its own emphasis on the role of measurement; Fock stated in an introduction to a Russian translation of the 1935 debate of Einstein, Podolsky, and Rosen versus Bohr:

In quantum mechanics the conception of state is merged with the conception of "information about the state obtained as a result of a specific maximally accurate operation." In quantum mechanics the wave function describes not the state in the usual sense, but rather this "information about the state." [14]

The importance of this prewar position of Fock's lies in its subtle difference from his stated views after the war, when he was placed under heavy pressure to desert the Copenhagen School.[15] Nevertheless, Fock's change in position was small compared to the swerves that occurred in the views of several other prominent Soviet philosophers and scientists.

The debate of the 1930's did not, however, leave a permanent imprint on Soviet attitudes toward quantum mechanics. Even many philosophers accepted much of the Copenhagen view. Early in 1947 M. E. Omel'ianovskii, a Ukrainian philosopher who with Fock and Blokhintsev completes the triumvirate whose views will be examined in detail here, argued a position on quantum mechanics close enough to the Copenhagen orientation to cause him intense embarrassment only a few months later. His 1947 book has become more interesting in recent years, since it represents a view that Omel'ianovskii later returned to and developed further.[16] In this work, *V. I. Lenin and Twentieth-Century Physics,* Omel'ianovskii accepted much of the common interpretation of quantum mechanics. He recognized and used such terms as "the uncertainty principle" and "Bohr's principle of complementarity." (A year later Omel'ianovskii's terminology became "the so-called 'principle of complementarity.'") He guarded against using these concepts in a way that might deny physical reality, as he said certain people (including Bohr on occasion) had done, but his major thesis in this book was a defense of the surprising but necessary concepts of modern physics against adherents of the determinism of Laplace, by then clearly outdated.[17] Buried within Omel'ianovskii's arguments, however, one may observe, at least in retrospect, the core of his own interpretation of quantum mechanics and of his later criticisms of the Copenhagen School. Although he acquiesced in the vocabulary of Copenhagen, he emphasized that the correct interpretation of quantum mechanics began with a recognition of the peculiar qualities of microparticles, not with problems of cognition: "And so we have come to the conclusion that Heisenberg's uncertainty principle, like Bohr's principle of complementarity, is a generalized expression of the facts of the dual (corpuscular and undulatory) nature of microscopic objects." [18] Thus, the uncertainty principle was not actually an epistemological limitation or a limitation of knowledge, but a direct result of the combined wavelike and corpusclelike nature of micro-objects, which was the *material* reason why classical concepts could not be applied to the microworld. In view of this material source of the phenomenon of canonically conjugate parameters, one could not expect ever to possess simultaneous exact values of position and momentum of elementary particles. For his recognition of the basic position of contemporary views on quantum mechanics, Omel'ianovskii was soon criticized severely, and eventually produced a second edition of his book, in which, most notably, he repudiated the principle of complementarity.[19]

The most important event of the postwar years for Soviet scholarship was A. A. Zhdanov's speech on June 24, 1947, at the discussion of G. F. Aleksandrov's *History of Western European Philosophy,* an event well known to historians of the Soviet Union. Only near the end of that speech did Zhdanov mention specific issues in science, and less than a sentence referred directly to quantum mechanics: "The Kantian vagaries of modern bourgeois atomic physicists lead them to inferences about the electron's possessing 'free will,' to attempts to describe matter as only a certain conjunction of waves, and to other devilish tricks." [20]

Although Zhdanov's speech is now known as the beginning of the most intense ideological campaign in the history of Soviet scholarship, the Zhdanovshchina, the first few issues of the new journal *Problems of Philosophy* that appeared after the speech were surprisingly unorthodox.[21] Evidently taking seriously the slogan of the journal—"to develop and carry further" Marxist-Leninist theory—the editors promoted vital discussions of several philosophic questions. In no field was this vitality more apparent than in the philosophy of physics; the second issue of *Problems of Philosophy* contained an article by the outstanding theoretical physicist M. A. Markov, a specialist in the relativity theory of elementary particles, which may well still be the most outspoken presentation of the Copenhagen point of view to appear in a Soviet philosophy journal since World War II.[22] Just why Markov chose this moment, after Zhdanov's condemnation of Aleksandrov and during the tightening of ideological controls, to expose himself to criticism so extensively may never be known, but there are several hints available. Markov was a research scientist in the Physics Institute of the U.S.S.R. Academy of Sciences, the organization that in the past had most stoutly defended international viewpoints in science and that would do so in the future, incurring sharp criticism from political activists.[23] It is probable that the theoretical physicists in the Academy, aware since the 1930's that, given the will, dialectical materialism could be used against prevalent interpretations of quantum mechanics, decided that the nascent ideological campaign meant that an official position on quantum mechanics was very likely to be imposed and felt that an early attempt to make that official position compatible with contemporary quantum theory was necessary. Markov probably knew well just how controversial his article would be, but hoped, first, that it would be vindicated, and second, that even if his point of view was rejected, the final compromise would be more palatable to the physicists as a result of his strong stand. Furthermore, Markov was able to capitalize on a feud among the professional philosophers. As the course of the debate illustrated, the chief editor of the new philosophy journal *Problems of Philosophy* was disliked by the old guard, which had published *Under the Banner of Marxism,* the major Soviet journal of philosophy from 1922 to 1944. The debate over Markov consequently contained many dimensions: It was an effort by the physicists to protect quantum mechanics, it was a volley in a feud among philosophers, and it was a decisive struggle over whether physicists or philosophers would have the ultimate influence on the philosophy of science in the postwar period.

Markov accepted modern quantum theory completely and agreed with Bohr's position in Bohr's debate with Einstein, Podolsky, and Rosen. Thus, Markov considered quantum mechanics to be complete, in the technical sense that no experiment that did not contradict it could yield results not predicted by it; and he consequently rejected all attempts to explain the behavior of microparticles on the basis of "hidden parameter" theories that would later permit restitution of the concepts of classical physics: "It is impossible to regard quantum mechanics as a classical mechanics that has been corrupted by our 'lack of knowledge.' " [24] Such complementary functions as "momentum" and "position" simply did not have simultaneous values; to suggest that they did would mean contradicting quantum theory.[25]

Not only was Markov's view on conjugate parameters typical of the Copenhagen School, but his approach to science bore few traits of dialectical materialism despite his initial quotations from Marxist classics. He asked that no statements be made that could not be empirically verified; he accepted relativity theory, including relativity of spatial and temporal intervals; he used the term "complementarity" without hesitation. To be sure, he affirmed that his view of science was "materialist" and criticized James Jeans and other non-Soviet commentators on science, but nowhere in his article did he make any effort to illustrate the relevance of dialectical materialism to science.

Markov maintained that "truth" is obtained from many sources; when we speak of knowledge of the microworld, which we gain with instruments, we are speaking about knowledge that has come from three sources: nature, the instrument, and man. The language we use to describe our knowledge is perforce always "macroscopic" language, since this is the only language we possess. The measuring instrument performs the role of "translating" the microphenomenon into a macrolanguage accessible to man. "We consider physical reality to be that form of reality in which reality appears in the macroinstrument." [26] Thus, according to Markov, our concept of physical reality is subjective to the extent that it is formed in macroscopic language and is "prepared" by the act of measurement, but it is objective in the sense that physical reality in quantum mechanics is a macroscopic form of the reality of the microworld.

The role of the measuring instrument is one of the thorniest issues in quantum mechanics. Markov's view was essentially in agreement with that of the Copenhagen Interpretation, according to which the wave describing a physical state spreads out over larger and larger values until a measurement is made, when a reduction of this spread (wave packet) to a sharp value occurs. Such an interpretation does indeed imply that complementary microphysical quantities have no inherent sharp values but that such values instead result from, or are "prepared by," the measurement. The most imaginative attempt, no doubt, to illustrate the striking results of some interpretations of this view of microphysical measurements was made by Schrödinger: A live cat is imprisoned in a cage in such a fashion that it will be poisoned if a light-sensitive trigger is activated by a photon. In order to strike the trigger, however, the photon, whose position is mathematically described by a wave function spread through space, has to pass through a half-silvered mirror. If the emitter sends one photon toward the half-mirror at a certain time, say, twelve o'clock, and then no one looks at the cat until one o'clock, are we still to assume, Schrödinger asked, that the cat is neither dead nor alive until someone actually looks in the cage and thus "prepares" the cat in a state of life or death? [27]

The cat paradox raises a basic question about the difference in quantum mechanics between relational aspects and subjective aspects. It is quite possible to interpret the cat paradox from a relational standpoint in which subjective considerations play no role. The moment of interaction between the microentity and the macroentity is not the moment when the observer looks in the cage, but the moment when the photon either was reflected by the half-silvered mirror or passed through it, poisoning the cat. As Hilary Putnam has pointed out, contemporary physicists explain the cat paradox by saying that all macro-observables always have sharp values, and therefore the poisoning of the cat (if it occurred) was in itself a "measurement." [28] Similarly, the moment of measurement when a photographic plate is used to record a quantum phenomenon is the moment of exposure, not the moment a human being looks at the plate. Thus, although measurement theory is still an uncertain area in the philosophy of science, measurement can be made an essential part of quantum mechanics without necessarily including subjective considerations.

Whatever Markov's views on the cat paradox, his opinion that the instrument "prepares" the state of microphysical reality, together with his acceptance of the Copenhagen Interpretation in general, exposed him to criticisms from a number of quarters, ranging from dogmatic ideologues to ordinary physicists with hopes for the eventual replacement of the views of Bohr and his colleagues by an interpretation more agreeable to common-sense intuition. The Markov article very quickly became the occasion for a full-blown controversy, involving several dozen participants, on the nature of physical reality and the dialectical materialist interpretation of quantum mechanics.

The polemic began with the appearance of an article by A. A. Maksimov in the *Literary Gazette,* an unusual place for a commentary on the philosophy of science.[29] The article, entitled "Concerning a Philosophic Centaur," contained very serious allegations against Markov. As the title indicates, Maksimov considered Markov a strange species, a creature combining Western idealistic views on the philosophy of science with professions of loyalty to dialectical materialism. The background of the article reveals that it was more than a mere alternative viewpoint on quantum mechanics. Originally a physicist, Maksimov was a veteran of the struggles in philosophy of the 1920's and 1930's, and had been a member of the editorial board of the journal *Under the Banner of Marxism* when it ceased publication in 1944. After the war he became a defender of the most dogmatic positions in the philosophy of science; he accused such scientists as Einstein and Bohr not only of committing grievous mistakes in the interpretation of science, but even of factual errors. Maksimov originally submitted his article to *Problems of Philosophy,* whose editor, B. M. Kedrov, disagreed strongly with it but decided to print it "for discussion purposes" along with a rebuttal. After Maksimov had already returned galley proofs to the journal, another article of his with an identical title and very similar content appeared in the *Literary Gazette,* prefaced with a condescending note from the editors about the questionable quality of the new Soviet theoretical journal *Problems of Philosophy.* Kedrov immediately canceled Maksimov's article in *Problems of Philosophy* and in the following issue accused both Maksimov and the editors of the *Literary Gazette* of bad faith.[30] Why were the editors of the *Literary Gazette* carrying on a vendetta against another publication? Here again the facts may never be known, but any attempt to explain this very rare disagreement between editorial boards in Stalinist Russia must begin with the fact that the member of the board of the *Literary Gazette* responsible for philosophy was none other than M. B. Mitin, who had been one of the editors of the defunct *Under the Banner of Marxism* and was obviously discontented with the new philosophy journal that had appeared in the wake of Zhdanov's speech.

The central point of Maksimov's article was that Markov was a supporter of the Copenhagen Interpretation of quantum mechanics and was trying to whitewash this view with a few statements about its agreement with dialectical materialism. According to Maksimov,

M. A. Markov, following directly behind Bohr, asserts that in physical experiments there is a mutual influence of the microworld and the instrument, which in essence can never be overcome. However, this argument in no way touches upon the basic epistemological question, Does microreality with such properties exist before the application of an instrument by man? M. A. Markov answers that it does not exist, but is "prepared" by the instrument.[31]

After the appearance of Maksimov's article in the *Literary Gazette*, the editors of *Problems of Philosophy* published a discussion of quantum mechanics. A number of authors (D. S. Danin, M. V. Vol'kenshtein, and M. G. Veselov) gave Markov strong support, revealing the numerous errors of Maksimov, and the noted D. I. Blokhintsev also took a fairly positive view of Markov's interpretation of quantum mechanics. Other critics, however, pointed out Markov's "anthropomorphism" in science, a result of his emphasis on the "observer" (L. I. Storchak) and his disregard of Party loyalty, or *partiinost'* (I. K. Krushev, V. A. Mikhailov).[32] But the factor that determined the eventual disapproval of Markov's article was a decision, beyond any doubt promoted by the Party, to replace Kedrov as editor of *Problems of Philosophy* by D. I. Chesnokov. It is clear that Maksimov's attack on Markov played an important role in Kedrov's downfall. Maksimov's *Literary Gazette* article was a clear criticism of Kedrov, and in a statement that appeared after Kedrov's dismissal, Maksimov commented:

Only a decisive rejection of the idealistic inventions of N. Bohr and M. A. Markov, only a decisive repudiation of the position taken on this question by the editorial board of the journal *Problems of Philosophy* [in No. 1, 1948] can lead our philosophical organ out of this blind alley into which it attempted to lure several sections of our intelligentsia, those inclined to waver on the basic questions of Marxist-Leninist ideology.[33]

In a note in the third issue of 1948 the reformed editorial board of *Problems of Philosophy* admitted that the journal had not taken the correct position on quantum mechanics and particularly on Markov's article, which had "weakened the position of materialism." The article had contained "serious mistakes of a philosophic character" and was in essence a departure from dialectical materialism "in the direction of idealism and agnosticism." [34]

In terms of personnel, the immediate casualty of the Markov affair was Kedrov, but in terms of the philosophy of science, the casualty was the principle of complementarity. The period from 1948 to roughly 1960 may be called, with respect to discussions of quantum mechanics in the Soviet Union, the age of the banishment of complementarity.[35] Only a few scientists in this period, most notably V. A. Fock, attempted to include complementarity as an integral part of quantum theory.

This critical attitude toward complementarity after 1948 was made clear in an article by Ia. P. Terletskii that immediately preceded the final statement on the Markov controversy by the editors of *Problems of Philosophy*. Terletskii observed that Markov's article was actually an attempt to justify the acceptance of complementarity by maintaining that as a result of the role of measuring instruments as "translators" of reality, statements about microphysics often contradict each other. Such a view, thought Terletskii, was merely a restatement of Mach's opinion that scientists must describe nature in terms of sensations. A true dialectical materialist approach, however, showed, Terletskii continued, that the principle of complementarity was in no way a basic physical principle and that quantum mechanics could very well "get along without it." [36]

The result of the Markov affair, then, was a victory for dogmatic ideologists. Maksimov, an ideologist, had triumphed over Markov, an active theoretical physicist in the Academy of Sciences. But it also became fairly clear that Maksimov was not capable of advancing an interpretation of quantum mechanics that held any chance of official acceptance.[37] His articles on quantum mechanics revealed all too clearly his ignorance of the subject. And it was the same Maksimov who was simultaneously opposing not only Einsteinian relativity but even Galilean relativity, maintaining that every object has an absolute trajectory and that a meteorite inscribes this trajectory on the earth upon collision with it.[38] Maksimov clearly represented pseudoscience, and his role in both quantum mechanics and relativity theory was a purely destructive one, isolating the "Machists" and "idealists" among Soviet scientists and winning a certain support for that service, but presenting no tenable alternatives to current interpretations of physical theory. As in the case of relativity theory, Maksimov soon lost his influence among

Soviet interpreters of quantum theory. The period after 1948 was dominated instead by physicists and a small number of philosophers with some knowledge of physics, all of whom, however, were influenced by the atmosphere created by the Markov affair. Until approximately 1958 the major interpreter of quantum mechanics was the philosopher of science Omel'ianovskii, who drew upon the theories of the physicist Blokhintsev, advocate of the "ensemble" interpretation. Also important was Fock, who termed his interpretation a recognition of the "reality of quantum states." And a good many others, including A. D. Aleksandrov, Ia. P. Terletskii, B. G. Kuznetsov, as well as the foreign scholars Louis de Broglie, J.-P. Vigier, and David Bohm, influenced the discussions of dialectical materialism and quantum mechanics.

Soviet philosophers and scientists in the early fifties were restricted in their writings on quantum mechanics by certain reigning dogmatic interpretations of Marxist texts. For example, Marx in one passage of *Capital* had observed that "the properties of a given thing do not arise out of its relations to other things, but are only identified in such relations. . . ." This obscure passage, taken out of its context of economic analysis and employed in the analysis of physics, was interpreted by some Soviet philosophers as revealing the incorrectness of the Copenhagen School's emphasis on measurement (the relation of the microparticle to an instrument); these ultraorthodox writers believed that scientists should emphasize the inherent properties of the microparticle, not its relations with other objects.[39] However, with the passing of Stalinism in the late fifties, other Soviet philosophers began clearing away these obstacles standing in the way of a genuine Marxist philosophy of science. An important book in overcoming the dogmatic interpretation of the above passage of Marx was A. I. Uemov's *Things, Properties and Relations,* published in Moscow in 1963.[40] By the late sixties Soviet scientists and philosophers were able to defend materialism, as many of them without any question wished to do, without being hobbled by quotations torn out of the context of the writings of Marx and Engels.[41]

D. I. Blokhintsev

D. I. Blokhintsev, one of the best known Soviet specialists in quantum mechanics and after 1956 director of the Joint Institute of Nuclear

Research at Dubna, as well as winner of Lenin and Stalin prizes, published in 1944 a university textbook on quantum mechanics.[42] He produced a revised second edition in 1949—that is, after the Markov affair and after quantum mechanics had become a subject of heightened ideological interest.[43] The two editions, straddling the beginning of the debate, differ on several interesting points. As Blokhintsev himself commented:

The chapter that concerns the concept of state in quantum mechanics has been changed, and the clarity of the discussion of the uncertainty relation has been improved. In the new edition of the book ideological questions connected with quantum mechanics are also considered, and the idealistic conceptions of quantum mechanics that are now widespread abroad are subjected to criticism.[44]

In the 1944 publication, Blokhintsev's interpretation of the physical significance of the wave function was a version of the original Born interpretation, according to which the wave does not represent the state of the system but the scientist's *knowledge* of the state. As Blokhintsev observed, "The de Broglie waves give . . . only statistical information [*svedenie*] concerning the movement of particles." [45] According to Blokhintsev, quantum mechanics describes the actions of microparticles in a purely statistical fashion. The wave function is an instrument for ascertaining probabilities. The intensity of the de Broglie waves in a certain location of space is proportional to the probability "of finding" the particle at that particular spot. The important phrase here is "of finding," as it obviously places emphasis on the process of measurement, since only by measuring does one "find." Blokhintsev's attitude toward measurement was crucial if he was to stake out an independent view on quantum mechanics. Yet in his 1944 publication he made no effort to characterize the solution of the wave equation (the ψ function) as something inherently objective, as he did later, but instead discussed it in terms of the results of individual measurements. "Knowing the ψ function," he remarked, "permits one to predict the *results of different measurements* carried out on particles." [46]

An exhaustive statement of Blokhintsev's position in 1944, which can be characterized as a probability interpretation of the ψ function, would have led him to a normal particle interpretation of the electron, similar to Born's original position.[47] Blokhintsev attempted to avoid the causal anomalies that arise from this view (see page 70 above) by maintaining that microparticles were inherently different from macroparticles. His statistical description did not imply, he pointed out, that each microparticle had a normal trajectory (possessed simultaneous values of momentum and position):

A priori one might think that such trajectories also exist in the realm of the microworld but that quantum mechanics pays attention only to a certain statistical average of those movements along trajectories, similar to what occurs in statistical mechanics. Simple considerations illustrate that such is not the case. In the area of the microworld mechanical values are in different relationships than in the realm of the macroworld.[48]

Blokhintsev was, in effect, refusing to accept microparticles as either corpuscular or undulatory, although his approach favored the corpuscular view. By refusing to go beyond the mathematical formalism any further than absolutely necessary, he avoided answering such questions explicitly, and he did not refer to "complementarity."

In his statistical interpretation of quantum mechanics Blokhintsev put great emphasis on "ensembles." He noted that the probability yielded by the wave function was derived from a series of repeated measuring operations. Therefore, when one talked about the wave function of one particle, or one system, what was actually being talked about was a large number of such particles or systems. A collection of such particles that were independent of one another and that could serve as material for repeated independent experiments was called an ensemble. The Heisenberg uncertainty relation, which was often discussed in terms of one particle, was actually a result, according to Blokhintsev, of measuring operations carried out on particles belonging to an ensemble. If all the particles in an ensemble could be described by one wave function, it was a "pure ensemble." If, however, an ensemble consisted of subensembles, each of which was described by a wave function, then the total was a "mixed ensemble." The relevance of this breakdown of the ensembles for the question of the nature of the wave function was the following: If a measurement was carried out on a pure ensemble, according to Blokhintsev, that very operation caused the ensemble to become mixed, since the act of measurement placed those few (perhaps one) microparticles affected by the measurement in a different state, described by a different wave function.[49]

In the 1949 edition of his textbook, Blokhintsev altered the wording of his interpretation of the wave function, eliminated the section that described the wave function as a prediction of measuring operations, and added a section on "methodological questions" in which for the first time he explicitly criticized Bohr's complementarity and the Copenhagen School.[50] In the section entitled "The Statistical Interpretation of the de Broglie Waves," Blokhintsev changed only a few words, but in exactly the most crucial area: In 1944, he had said that the de Broglie waves gave only "statistical *information* concerning the movement of particles"; in 1949, the sentence was changed to say that the waves gave a "statistical *description* of the movement of microparticles." [51] The latter statement emphasized the objective nature of the microparticles. The words "description of" were more easily accommodated to Lenin's copy-theory of epistemology than "information about."

The main difficulty with the Copenhagen School, Blokhintsev thought, was its attachment to complementarity, or the belief that the quantum description of phenomena divided into two classes complementary to each other in the sense that one must combine them to have a complete description in classical terms. From complementarity the adherents of the Copenhagen School drew conclusions favoring the denial of causality, the liquidation of materialism, and the subjective nature of the wave function.[52] According to Blokhintsev, the Copenhagen followers considered the wave function not an objective characteristic of the quantum ensemble but an expression of the information possessed by the observer. The irony of Blokhintsev's position is revealed in his criticism of the Copenhagen "idealists' " and "Machists' " use of a description of the wave function that he himself had used in the earlier edition of the same book.[53] Now, however, he no longer considered the wave function a representation of "information," but instead a "completely objective characteristic of the quantum ensemble, independent of the observer." [54]

The most complete statement of Blokhintsev's criticism of the Copenhagen School and the philosophic significance of his alternative ensemble interpretation was a long article that appeared in a leading Soviet physics journal in 1951.[55] Blokhintsev set himself the task of proving that quantum statistics had objective reality and in no way depended on the observer, in contrast to Bohr's early belief that the statistics could be considered a result of the uncontrollable influence of the instrument upon the object. He noted that radioactive atoms decayed according to statistical laws that were independent of observers and instruments. Blokhintsev considered radioactivity a phenomenon of a "certain statistical ensemble of radioactive atoms, existing independently in nature." [56] Cosmic rays were similarly dependent on objective statistical laws. And, he observed, the microlevel of matter was an area where such statistical laws were inherently "objective" (did not derive from underlying causal factors) and therefore commonplace. In contrast,

the Copenhagen School relegates to secondary importance the fact that quantum mechanics is applicable only to statistical ensembles and concentrates on analysis of the mutual relations of a single phenomenon and the instrument. This is an essential methodological error: In such an interpretation all quantum mechanics takes on an "instrumental" character, and the objective aspect of things is extinguished.[57]

Blokhintsev's definition of the quantum ensemble underwent a slight but interesting change between 1949 and 1951. Whereas in 1949 the ensemble was defined as a large number of microparticles in a certain state, or combination of states—and therefore was definable in terms of phenomena on the microlevel—by 1951, the ensemble included part of the macrolevel also, namely, the connection of the microsystem to the macroscopic environment, of which the measuring instrument was a special case.[58] This change was evidently prompted by two motivations, one of which, interestingly enough, derived from an interpretation of dialectical materialism, while the other derived from physics. One of the elements of dialectical materialism is the belief that all of reality forms an interconnected whole; this principle received a particularly blunt enunciation in the Stalinist *Short History of the Communist Party,* which stated that "not a single phenomenon in nature can be understood if it is considered in isolation, disconnected from the surrounding phenomena." [59] In this light Blokhintsev's earlier statements about the "sharp border" between the microlevel and the macrolevel seemed suspect. By defining the ensembles in terms of both the microsystems themselves *and* their connections with the macroworld, Blokhintsev mended this rift. At the same time, he was able to present the ψ function as a statement of values for potential measuring operations, as did quantum theorists all over the world, including those of the Copenhagen School.

The important question after this shift in the definition of the ensembles was, How would Blokhintsev preserve a uniquely dialectical materialist position toward quantum mechanics? Blokhintsev attempted to answer this question by maintaining that quantum mechanics was inapplicable to individual micro-objects, since no individual micro-object could be studied in isolation from its environment. In this way, by studying large numbers of microparticles, knowledge of objective reality could "in principle" be attained: "Quantum mechanics studies the properties of a single microphenomenon by means of the study of the statistical laws of the collective of such phenomena." [60] Blokhintsev readily granted that a measuring operation would change the state of a particular particle, placing the particle in a different ensemble, but asserted that all the other particles in the old ensemble would still be in their previous states. Therefore the scientist could conceive of objective reality through the concept of the totality, or ensemble.

Blokhintsev also indicated that a "hidden parameter" theory of quantum mechanics might at some future date permit a numerical description of the individual microparticle, although at the present time he considered such a description to be impossible. He dismissed John von Neumann's and Hans Reichenbach's well-known attempts to disprove hidden parameter theories by pointing out that both rested their cases on the existing mathematical apparatus of quantum mechanics, which would surely be changed if a new theory were devised.[61] He also dismissed the position of Einstein, Podolsky, and Rosen, noting that these authors based their views on the application of the wave function to individual particles, whereas he believed it should be applied only to groups or ensembles.[62]

The central weakness of Blokhintsev's interpretation was his definition of ensembles. He failed in his goal of separating the quantum description of matter from the process of measurement, as can be seen by analyzing his definition of ensembles: Blokhintsev defined an ensemble as a combination of the microsystem (a collection of particles) and its macroenvironment. But what did the "macroenvironment" include? According to Blokhintsev, the macroenvironment included measuring instruments as "special cases." He then defined the wave function as the "association" of a particle with an ensemble.[63] But his chain of reasoning had led him full circle, since he had started with the desire to separate quantum mechanics from measurement and ended by including measurement in his definition of the ensemble. Thus, the ψ function became, as before, a probabilistic statement of the results of measurement.

In the controversy between Blokhintsev and Fock that soon followed, the concept of ensembles became a basic issue. Fock very quickly located the weakness at the bottom of Blokhintsev's discussions of the ensemble. He extracted the fundamentals of quantum mechanics that Blokhintsev had defined in 1949: (a) an ensemble is a collection of particles that independently of one another are in a state such that the ensemble can be characterized by the wave function ψ; (b) it follows that the state of a particle should be understood as the association of that particle with a definite ensemble, so that (c) the wave function does not concern an individual particle. Fock then demonstrated that these propositions contradict each other:

In assertion (a) the state of an individual particle is defined by means of its wave function, but in assertion (c) it is denied that the wave function concerns the individual particle. This is a contradiction. Furthermore, in assertion (a) the ensemble is defined by means of the wave function, but in assertion (b) the wave function is defined through the ensemble. This is a vicious circle.[64]

Furthermore, continued Fock, Blokhintsev could not treat the ensembles as statistical collectives, as he intended to do, unless they met the standard criteria of such collectives in accordance with established theory of statistics. By this theory, a statistical collective is a collection of elements that may be sorted out in accordance with a certain indicator (*priznak*). Such an indicator would be the value of a certain physical magnitude, or a group of physical magnitudes simultaneously measured. But according to quantum mechanics, microparticles do not possess definite values that would permit the sorting out of a definite collective. Therefore, Blokhintsev, said Fock, had no way of even denoting the members of his much touted ensembles, which were really only "speculative constructions." Instead, he should frankly state that his quantum ensembles concealed a reference to a statistical statement of the results of measurements on a micro-object, conducted with the aid of a classical instrument designed for measuring a given magnitude. Fock concluded that Blokhintsev's incorrect position was connected with that of Bohr:

We see the basic cause of all difficulties in the fact that a purely statistical point of view is incorrect in a philosophic sense. In contrast to what dia-

lectical materialism teaches us, the statistical point of view issues not from
the objects of nature but from observations, not from the micro-object and
its state but from the statistical collective of the results of observations. This
draws it toward the positivist point of view of Bohr, which also denies that
the wave function relates to the micro-object, and attributes to the wave
function only a purely symbolic significance.[65]

A reply to this criticism was no easy task for Blokhintsev, who must
have felt somewhat uneasy about the definition of his ensembles, to
judge from the waverings in his writings on the subject. Much of his
answer to Fock was a criticism of the latter's own belief that the wave
function is an objective description of individual microbodies. This
aspect of their debate will be considered in the following section, which
concerns Fock's own interpretation of quantum mechanics. On the
question of the definition of the ensembles, Blokhintsev merely affirmed
his previous views, defending himself from Fock's criticism by saying
that as long as it is possible to conceive of a pure ensemble, it is possible
to separate conceptually the quantum description of matter from measurement and therefore from subjectivism or idealism. This hypothetical
ensemble would be one on which no measuring operation had been carried out and which, therefore, could in principle be described by one
wave function. But since no measurements had in fact been made, almost nothing could be said about such an ensemble other than "it
exists" according to Blokhintsev.[66]

In 1966 Blokhintsev published an interesting book entitled *Questions
of Principle in Quantum Mechanics,*[67] which was later translated into
English and published jointly in Europe and the United States under the
title *The Philosophy of Quantum Mechanics.*[68] The book was Blokhintsev's most complete treatment of philosophical issues in quantum mechanics; as the work of a distinguished Soviet professional physicist, it
deserves careful examination by people seeking to learn more about
Soviet philosophy and science. Blokhintsev's approach was highly technical, and he warned his readers that "the present monograph is concerned more with theoretical physics than with philosophy." [69] Yet it is
clear that Blokhintsev recognized fully the interaction of physics and
philosophy, and addressed himself to several major aspects of this interaction. Blokhintsev's study was both enlightened and tolerant; if philosophers have found certain unclear points of definition, it should be
remembered that scientists and philosophers everywhere agree that the

interpretation of quantum mechanics is an exceedingly difficult problem. There is no agreement anywhere on these matters.

Blokhintsev's 1966 book was essentially an attempt to clarify and support the ensemble interpretation of quantum mechanics that he had earlier developed. True, there were certain small changes of emphasis and aspiration, particularly in his opinion on the possibility of finding latent parameters in quantum mechanics. His description of the ψ function also changed a bit, in the direction of his 1944 position although not a complete return. But the differences between him and Fock on the validity of the ensemble approach and the applicability of the ψ function to the individual particle remained. The real significance of his new book, however, was not its discussion of these issues, since his opinions here remained essentially unchanged, but his fuller description of causality and his criticism of determinism. Even though the 1966 book was written and published at a time when there was little pressure from ideologists upon scientists in the Soviet Union, Blokhintsev's views were still essentially a continuation and further development of the earlier debates. This continuity derived not as much from politics as from the attractiveness to Blokhintsev of the underlying philosophical issues of interpretation of nature.

Blokhintsev began his 1966 book with a criticism of what he called "the illusion of determinism." He thought that the advance of science, and particularly the new understanding of nature issuing from quantum mechanics, illustrated the weakness of a belief in strict determinism. The fallacies of this "worship of ideal determinism" were seen, albeit incompletely, Blokhintsev thought, by even a few nineteenth-century critics of Laplacian mechanism, such as Engels, who said in *Dialectics of Nature* that "necessity of this kind does not take us outside the theological view of nature." [70]

For a long time humanity believed in divine predestination, and afterwards in rigid causal connection. Engels appreciated the philosophical resemblance and narrowness of these viewpoints, while failure to appreciate this affinity has over the Centuries been the cause of tragedy to many outstanding men.[71]

Determinism in the classic sense meant, Blokhintsev observed, that "the state of a system at a preceding instant completely determines the state at a subsequent instant." [72] Even before the development of quantum mechanics, however, there was reason to doubt the validity of such a conception of the universe. Any attempt to rigidly predict the future of a system, Blokhintsev noted, is influenced by the inaccuracy of the initial data, the unpredictability of accidental forces, and the impossibility of keeping any system completely isolated. All these three limitations on classical physics were usually ignored, Blokhintsev commented, although the philosophic interpretation of the world that sees it as an interconnected whole should have revealed more fully at least the impossibility of isolating any system. This feature is very important, Blokhintsev thought:

The future of a mechanical system may be predicted only if we can be sure that the system is isolated. The guarantee required here is not implied by the equations of motion but is an additional condition, which produces a great reduction in the reliance on determinism. A vast and depressing "if" arises in the way of the prophet who sets out to predict the future of a real mechanical system.[73]

Thus, he commented, "the input data must from time to time be corrected even in a science as precise as celestial mechanics, in order to eliminate cumulative errors." [74] And he implied that this necessity was not a practical one, but a theoretical one, since there exists in the interconnected universe an infinite number of potential influences.

All these considerations applied to classical mechanics. With the development of quantum mechanics and the emergence of the necessity of probabilistic descriptions of nature for reasons apparently quite intrinsic to microbodies, the erroneousness of the whole classical approach to determinism became quite apparent.

Does the abandonment of rigid determinism mean a surrender of the principle of causality? Blokhintsev answered this question negatively, as did Fock and other Soviet commentators. He agreed that it was necessary to take a new look at definitions of causality, but he felt that such a redefinition was fully justified as a part of man's constant effort to find order in nature. And Blokhintsev defined causality in the following way:

Causality is a definite form of order in events in space and time; this ordering imposes restrictions even on the most chaotic events, and it makes itself felt in two ways in statistical theories. Firstly, the statistical laws themselves are fully ordered, and the quantities that characterize an ensemble are themselves strictly determined. Secondly, the individual elementary events are also so ordered that one may influence another only if their relative location in space and time allows this without violating casuality (i.e. the rule ordering the events).[75]

Within the context of this understanding of causality Blokhintsev considered quantum mechanics causal; to him, Schrödinger's equation expresses causality in quantum theory, since it describes the motion of the quantum ensemble "in a causal fashion, i.e. so that the state earlier in time determines the subsequent state of the ensemble." [76] Thus one could save not only a concept of causality, but even a concept of determinism, although on a much different level than previously.

On the question of the validity of the term "complementarity," Blokhintsev in his 1966 book wrote that Bohr had formulated this concept in a way that reflected his philosophical concepts, "which were far from those of materialism." [77] Blokhintsev regretted this aspect of Bohr's philosophy, which "has been the origin of the far-reaching conclusion that the current mechanics of the atom cannot be compatible with materialism." [78] In order to oppose this conclusion, Blokhintsev would have preferred another term for "complementarity," but he felt that it was now too well established to eliminate:

It would seem generally better to speak of a principle of exclusiveness rather than complementarity: dynamic variables should be divided into mutually exclusive groups, which do not coexist in real ensembles. However, out of respect for Bohr and his tradition we shall retain the usual terminology.[79]

One of the points on which Blokhintsev's 1966 book differed slightly from his earlier work was his description of the ψ function. From the previous discussion in this chapter it should be clear that the question of the physical significance of the mathematics of quantum mechanics was one of the main features of the past discussions. In 1966, Blokhintsev entitled one of his chapters "The Wave Function as the Observer's Notebook," and at first glance it appears that he had moved back to his position of 1944, when he described the wave function in terms of the observer's knowledge rather than the physical state of the ensemble. But a careful examination of Blokhintsev's book shows that his views had genuinely changed as a result of the intervening discussions; although one could speak of the wave function as the "observer's notebook" for the sake of convenience, philosophically Blokhintsev considered such a description misleading. Because of Blokhintsev's simultaneous desires to defend the physicists' "common jargon" and yet to criticize the philosophic subjectivism that it often contains, he displayed a certain inconsistency in his description of the wave function:

The wave function is simply a routine record of his [the observer's] information on the state of the ensemble of microsystems. There is nothing wrong in this very common view of the wave function and of the contraction process, and it is convenient as a formulation against which it is difficult to raise objection. In using the words "observer," "measurement," "information," and so on we merely pay respect to the physicists' common jargon, which is in no way a lucid tongue for the discussion of major philosophical and methodological topics in physics. The entire theory of measurement thereby acquires a dubious taint of subjectivism, which becomes quite insupportable if we have to answer the question whether quantum mechanics is applicable to the description of physical phenomena occurring in the absence of an observer.[80]

Thus, while on the one hand Blokhintsev said he thought "there was nothing wrong" in considering the wave function as the observer's notebook, on the other hand such a view was "insupportable" from a philosophic standpoint. Indeed, Blokhintsev thought that "the observer is not an entirely obligatory being in the world" [81] and that "the quantum laws would scarcely be altered one iota if this restless observer were removed from the scene." [82] In the final analysis, then, Blokhintsev thought the wave function was something more than a record of information in the observer's notebook, despite the title of his chapter. As in 1949, so also in 1966, Blokhintsev thought that from a philosophic standpoint the wave function must be seen as "an objective characteristic of a quantum ensemble." [83]

Blokhintsev took up the topic of the future of quantum mechanics in his chapter "Is the Wave Function Avoidable?" And on this issue he noted a change in his own position. He said, "The present author himself once hoped that the striking similarities" between the equations for the density matrix and the equations of classical physics "would allow the formulation of quantum mechanics as the statistical mechanics of quantities not simultaneously measurable." [84] But he had now almost completely abandoned this hope. In fact, he said that it was impossible to point to a single experimental fact to indicate that quantum theory is incomplete within the range of atomic phenomena. Nonetheless, he admitted that it was still at least theoretically possible that quantum mechanics would be substantially revised. He wrote that one

possibility to examine is the introduction of latent parameters such as to give meaning to a proportion of the form

$$\frac{x}{\text{quantum mechanics}} = \frac{\text{kinetic theory of matter}}{\text{thermodynamics}} \tag{1}$$

in which x is some unknown (more complete) theory.

It cannot be denied that the symbolic equation of (1), or some similar one, might be soluble, at least in this extremely general and purely methodological formulation of the problem.[85]

Yet Blokhintsev was skeptical of all such attempts, comparing the person who made them with "the seeker for unwettable gunpowder." He thought that the present structure of quantum mechanics was admirably adequate for physicists and need not be disquieting to philosophers—indeed, it had enriched man's understanding of causality and objective reality.

V. A. Fock

Academician V. A. Fock has already been mentioned in the discussion of the views of Blokhintsev. A separate consideration of Fock's own interpretation of quantum mechanics will be the subject of the following section.

Fock, a theoretical physicist at Leningrad University, was elected to the Academy of Sciences of the U.S.S.R. in 1939, won the Stalin Prize in 1946 and the Lenin Prize in 1960. His research has been on problems of mathematical physics, and particularly relativity theory and quantum mechanics. He is also deeply interested in the philosophical implications of modern physics. Both his scientific and his philosophical writings have attracted attention abroad in recent years.

Throughout a number of controversies Fock has been noted for his intense sense of independence, defending himself on numerous occasions against both Soviet and non-Soviet critics. In quantum mechanics, Fock may be correctly defined as a follower of Bohr's Copenhagen

Interpretation if one defines the Copenhagen Interpretation in terms of its minimum rather than its maximum claims. (This "core meaning" of the Copenhagen Interpretation was once described by N. R. Hanson as "a much smaller and more elusive target to shoot at than the *ex cathedra* utterances of the melancholy Dane.")[86]

Fock described his entry into the discussions of the philosophic significance of quantum mechanics in the following fashion:

Having begun, like many physicists, with a formal application of the mathematics of quantum mechanics, I later, especially after 1935, began to think about questions of principle; I finally came to the conclusion that Bohr's formulation could be completely separated from the positivistic coating that at first glance seemed to be intrinsic to it. At the same time I never doubted that Bohr was essentially correct.[87]

The most accurate evaluation of Fock's position might be to say that with a few temporary waverings, his thinking has undergone transitions quite similar to the shifts in Bohr's thinking. In several cases these shifts, all toward de-emphasis of the role of measurement and stress on a realist point of view, occurred first in Fock's interpretation, then in Bohr's, and it is possible that Fock may have been one of the influences on Bohr. The two scientists were aware of each other's work, and in February and March 1957 they held a series of conversations on the philosophic significance of quantum mechanics. The discussions took place in Copenhagen, both in Bohr's home and at his Institute of Theoretical Physics. Fock later reported on the conversations in the following way:

From the very beginning Bohr said that he was not a positivist and that he attempted simply to consider nature exactly as it is. I pointed out that several of his expressions gave ground for an interpretation of his views in a positivistic sense that he, apparently, did not wish to support. I emphasized the necessity to "legalize" all quantum-mechanical concepts as reasonable abstractions based on one interpretation of experiments. He said that he did not deny this legal character in any fashion. Our views constantly came closer together; in particular it became clear that Bohr completely recognized the objectivity of atoms and their properties, recognized that it was necessary to give up determinism only in the Laplacian sense, but not causality in general; he further said that the term "uncontrollable mutual influence" was unsuccessful and that actually all physical processes are controllable.[88]

It was after this exchange that Fock commented, "After Bohr's correction of his formulations, I believe that I am in agreement with him on all basic items." [89] This observation followed a period in which Fock had been rather critical of what he considered Bohr's carelessness on philosophic issues.

In the 1930's, however, when Bohr had been even less cautious in his statements, Fock was one of the leaders of the "Copenhagen branch" in the U.S.S.R. and repeatedly defended its viewpoint in the journals. His agreement with Bohr in the latter's debate with Einstein over the completeness of quantum theory is quite clear. During and shortly after the war Fock retreated a bit in the terminology of his defense of Copenhagen, but never abandoned its position. Indeed, one of the remarkable aspects of Fock's career, and of the history of Soviet philosophy of science, is that he was able to defend the concept of complementarity during a long period when it was officially condemned in the philosophy journals. During this time Fock occupied an anomalous position: His view on quantum mechanics was disapproved, but his interpretation of relativity theory, which did not include the concept of general relativity, became more and more influential. Nothing illustrates better the subtlety of Soviet controversies in the philosophy of science—a subtlety greater than most non-Soviet observers are willing to grant—than Fock's views being simultaneously under ban and approval. After 1958 Fock's interpretation of quantum mechanics gained greater acceptance and was finally adopted by the philosopher Omel'ianovskii, who had previously supported Blokhintsev. Ironically, in this period Fock's interpretation of relativity, although still very influential, was coming under more and more criticism from such people as M. F. Shirokov.[90] If the shifts seem confusing, some consistency may be perceived in the fact that these latter changes (away from Fock in relativity, toward him in quantum mechanics) both put Soviet science in a closer position to dominant non-Soviet interpretations, which had themselves undergone certain changes.

Most of Fock's effort in interpreting quantum mechanics has been directed toward establishing the fact that the Copenhagen Interpretation, including the principle of complementarity, did not violate dialectical materialism. As early as 1938 he maintained that "the thesis that a contradiction exists between quantum mechanics and materialism is an idealistic theory." Bohr's principle of complementarity was, to Fock, "an integral part of quantum mechanics" and a "firmly established objectively existing law of nature." [91] Throughout the years he has defended the essential Copenhagen position, although he carefully dissociated himself from certain of Bohr's views, such as the latter's early attribution of primary importance to the process of measurement. Nevertheless, his interpretation of the physical significance of the ψ function was the same as that of Bohr. Before the war Fock did not consider the wave function to be a description of the state of matter. This was, he noted, the position of Einstein, who then became involved with paradoxes. Fock, along with Bohr, considered the ψ function to be a description of "information about the state" (*svedeniia o sostoianii*).[92] It is not surprising, then, that Fock engaged in two particularly bitter exchanges with Maksimov, which were separated by a period of fifteen years. Maksimov advertised Fock as a conscious partisan of the idealistic, bourgeois Copenhagen School, while Fock observed that Maksimov was a wonderful example of how *not* to defend materialism.[93]

The most difficult period for Fock was immediately after the Markov affair. The new position, advanced by Terletskii and quickly supported by Omel'ianovskii, was that Heisenberg's uncertainty relation was, indeed, an integral part of quantum mechanics and must be retained, but that complementarity in no way followed from uncertainty.

According to Fock, on the contrary, there was no essential difference between the Heisenberg uncertainty relation and complementarity.[94] Both were the result of crossing the dividing line between the macrolevel and the microlevel. It was quite conceivable, Fock indicated in a preface to the works of N. S. Krylov, that if it were possible to give a description of the microlevel of matter in terms appropriate to that level (microlanguage), then there might exist a new kind of "complementarity" that would arise when one attempted to describe the macrolevel in microlanguage. This new complementarity would be analogous to, but different from, the complementarity of existing quantum mechanics, which was based on description in macrolanguage.[95] In this view the kernel of objective reality that dialectical materialism demands as a minimum in every physical description becomes very elusive indeed.[96]

Fock's identification of uncertainty and complementarity brought him under very heavy criticism. In the famous 1952 "Green Book" on philosophic problems of science, edited by a group headed by the ultra-conservative Maksimov, Omel'ianovskii observed: "Unfortunately several of our scientists . . . have not yet drawn the necessary conclusions from the criticism to which Soviet science subjected the Copenhagen School. For example, V. A. Fock in his earlier works did not essentially distinguish the uncertainty relationship from Bohr's principle of complementarity." [97]

It was this kind of criticism that caused Fock to alter his terminology and temporarily to hesitate in his advocacy of complementarity. While previously he had considered the ψ function to be a description of "information about the state," he now called the ψ function a characterization of the "real state" of the micro-object.[98] In 1951 Fock indicated that as a result of the blurring of the original meaning of complementarity, he might abandon it altogether:

At first the term complementarity signified the situation that arose directly from the uncertainty relation: Complementarity concerned the uncertainty in coordinate measurement and in the amount of motion, . . . and the term "principle of complementarity" was understood as a synonym for the Heisenberg relation. Very soon, however, Bohr began to see in his principle of complementarity a certain universal principle . . . applicable not only in physics but even in biology, psychology, sociology, and in all sciences. . . . To the extent that the term "principle of complementarity" has lost its original meaning . . . it would be better to abandon it.[99]

One of the most complete statements of Fock's interpretation of quantum mechanics appeared in a collection of articles on philosophic problems of science published in Moscow in 1959.[100] Written at a time of relative freedom from ideological restriction, it is a statement of both scientific rigor and philosophic conviction. Fock began his discussion by considering and then dismissing attempts to interpret the wave function according to classical concepts. De Broglie's and Schrödinger's attempts originally to explain the wave function as a field spread in space, similar to electromagnetic and other previously unknown fields, were examples of classical interpretations, as was also de Broglie's later view that a field acts as a carrier of the particle and controls its movement (pilot-wave theory).[101] Bohm's "quantum potential" was essentially the same type of explanation, since it attempted to preserve the concept of trajectory.[102] Similarly, Vigier's concept of a particle as a point or focus in a field was an attempt to preserve classical ideas in physics.[103] All these interpretations, according to Fock, were extremely artificial and had no heuristic value; not only did they not permit the solution of problems that were previously unsolvable, but their authors did not even attempt such solutions.

Fock believed that the true significance of the wave function began to emerge in the statistical interpretation of Max Born, especially after Bohr combined this approach with his own view of the importance of the means of observation. This emphasis on measuring instruments was essential for quantum mechanics, Fock agreed, but it was exactly on this point that Bohr also slipped:

In principle it seems that it is possible to reduce a description to the indications of instruments. However, an excessive emphasis on the role of instruments is reason for reproaching Bohr for underrating the necessity for abstraction and for forgetting that the object of study is the properties of the micro-object, and not the indications of the instruments.[104]

Bohr then compounded the confusion, said Fock, by utilizing inexact terminology—terminology he was forced to invent in order to cover up the discrepancy that arose when he attempted to use classical concepts outside their area of application. One of the most important of these uses of inexact terminology was his opposition of the principle of complementarity to the principle of causality. According to Fock, if one defines terms with the necessary precision, no such opposition exists. The complementarity that *does* exist in quantum mechanics is a complementarity between *classical descriptions* and causality. But this does not deny causality in general because classical descriptions of macroparticles are necessarily inappropriate for microparticles. Using classical descriptions (macrolanguage) is merely a necessary method since we do not have a microlanguage. Realizing that a microdescription of microparticles would be different from a classical description of the same particles, we can say that on both levels (micro- and macro-) the principle of causality holds. Since we always use a macrodescription, however, we should redefine causality in such a way that it fits both levels. Our new approach, said Fock, should be to understand causality as an affirmation of the existence of laws of nature, particularly those connected with the general properties of space and time (finite velocity of action, the impossibility of influencing the past). Causal laws can, therefore, be either statistical or deterministic. The true absence of causality in nature would mean to Fock that not even probabilistic descriptions could be given; all outcomes would be equally probable. Fock concluded his remarks on causality by commenting that in his recent conversations he had found Bohr in agreement with these observations. Thus, a few redefinitions of complementarity and causality would go far toward strengthening the Copenhagen Interpretation.

Fock's opinion of the role of measurement in quantum mechanics was based on a recognition of objective reality. He accepted Heisenberg's uncertainty relation as a factual statement of the exactness of measurements on the microlevel. But this relativity with respect to the means of measurement in no way interfered with objectivity: "In quantum physics the relativity that arises from the means of observation only increases the preciseness of physical concepts. . . . The objects of the microworld are just as real and their properties just as objective as the properties of objects studied by classical physics." [105] The instrument in quantum mechanics plays an important role, Fock observed, but there is no reason to exaggerate that role since the instrument is merely another part of objective reality, obeying physical laws. The importance of the instrument is that it necessarily gives its description in classical terms.

The root of quantum mechanics, according to Fock, is, however, something radically new in science: the potential possibility for a microobject to appear, in dependence on its external conditions, either as a wave, a particle, or in an intermediate form.[106] This new concept, coupled with the statistical characteristics of the state of an object, leads us to a different understanding of causality and of matter. Bohr tried to find his way to this new understanding by way of emphasizing the role of the instrument and by stressing the concept of complementarity. Fock preferred a slightly different way: "I try to bring in new concepts, for example, the concept of potential possibilities inherent in the atomic object, and it seems to me that the mathematical apparatus of quantum mechanics may be correctly understood only on the basis of these new concepts." [107] Fock, then, considered his essential contribution to the interpretation of quantum mechanics to be the idea of "potential possibilities" and the consequent distinction between the potentially possible and the actually realized results in physics. As will be illustrated below,

Fock's approach differed sharply from hidden parameter interpretations, since he did not believe it was possible, in principle, to arrive at an exact description of microparticles.

In experiments designed to study the properties of atomic objects, Fock distinguished three different stages: the preparation of the object, the behavior (*povedenie*) of the object in fixed external conditions, and the measurement itself. These stages might be called the "preparatory part" of the experiment, the "working part," and the "registering part." In diffraction experiments through a crystal, the preparatory part would be the source of the monochromatic stream of electrons, as well as the diaphragm in front of the crystal; the working part would be the crystal itself; and the registering part would be a photographic plate. Fock emphasized that in such an experiment it is possible to change the last stage (the measurement) without changing the first two, and he would build his interpretation of quantum mechanics on this recognition. Therefore, by varying the final stage of the experiment, it is possible to make measurements of different values (energy, velocity, position) all of which are derived from the same initial state of the object:

To each value there corresponds its own series of measurements, the results of which are expressed as a distribution of probabilities for that value. All the indicated probabilities may be expressed parametrically through one and the same wave function, which does not depend on the final stage of the experiment and consequently is an objective characteristic of the state of the object immediately before the final stage.[108]

In the last sentence, then, is the meaning of Fock's often-quoted statement that the wave function is an objective description of quantum states, a position that he adopted after World War II. The wave function is objective, said Fock, in the sense that it requires an objective (independent of the observer) description of the *potential possibilities* of mutual influences of the object and the instrument. Therefore, the scientist is correct, Fock believed (contrary to Blokhintsev), in saying that the wave function relates to a given single object. But this objective state is not yet actual, he continued, since none of the potential possibilities has yet been realized. The transition from the potentially possible to the existing occurs in the final stage of the experiment. Thus, Fock completed his interpretation of quantum mechanics with an affirmation of a realist (he would say dialectical materialist) position on the philosophy of science. Nevertheless, the extension of a concept of realism to statements concerning potential situations rather than actual situations was open to a number of logical objections.

M. E. Omel'ianovskii

M. E. Omel'ianovskii, a member of the Ukrainian Academy of Sciences, has been one of the most influential Soviet philosophers of science in recent years. In the 1940's Omel'ianovskii helped to create a school in the philosophy of science in Kiev that has remained strong to this day; after his shift to Moscow in the mid-fifties he was the most important figure in the largely successful effort to repair the damages of Stalinism in the philosophy of science and to create a tighter union between scientists and philosophers.[109] Although Omel'ianovskii yielded to the political pressures of the 1948–56 period, he understood modern physical theory and fully appreciated its significance for the philosophy of science, as his pre-1948 publications indicated.[110] Soon after the denunciation of Stalinism at the Twentieth Party Congress in 1956 Omel'ianovskii published an important article calling for a new approach to dialectical materialism.[111] He later described the article as one of the most important turning points in his professional development. As a leader of the sector on the philosophical problems of science at the Institute of Philosophy of the U.S.S.R. Academy of Sciences, Omel'ianovskii has been instrumental since 1956 in arranging frequent conferences and publishing collections of articles in which both prominent philosophers and well-known natural scientists participated. As an example, in 1970 Omel'ianovskii edited an interesting volume of original articles on the philosophy of science entitled *Lenin and Modern Science,* in which a number of eminent Soviet scientists and several prominent foreign scientists published articles.[112] Omel'ianovskii also succeeded in attracting to the Institute of Philosophy a number of outstanding young specialists with science backgrounds who approached the problems of the philosophy of science with a much more open mind than many of the older philosophers.

Omel'ianovskii published in 1956 his most significant independent contribution to a Soviet Marxist interpretation of quantum mechanics, his *Philosophic Problems of Quantum Mechanics.*[113] Although this book was later superseded by Omel'ianovskii's modified views, as had also been the case with his 1947 volume, it established him for the remainder of the 1950's as the major Soviet interpreter of quantum mechanics. The work was an extremely ambitious one; Omel'ianovskii, a philosopher, not a physicist, was attempting to outline a clearly independent position on quantum mechanics. He agreed completely with no major physicist, Soviet or non-Soviet, although his interpretation was closest to that of Blokhintsev. Among physicists, he set himself apart most markedly, of course, from the Copenhagen School (to which he implied Fock primarily belonged), much less strongly but still significantly from "materialist" non-Soviet physicists such as Bohm and Vigier, and least of all but still perceptibly from Blokhintsev.

Omel'ianovskii viewed the controversy in quantum mechanics as one of the latest developments in the ancient struggle between materialism and idealism, a contest directly connected to class interests. He maintained that the "conception of complementarity grew out of the reactionary philosophy of Machism-positivism. This conception is foreign to the scientific content of quantum mechanics. It is not accidental that P. Frank, H. Reichenbach, and other modern reactionary bourgeois philosophers joined with Jordan, who, invoking Bohr and Heisenberg, 'liquidated materialism.' " [114] Having delivered this simple analysis of the relationship of philosophy and the economic order, however, Omel'ianovskii proceeded to the theoretical problems of a physical interpretation of quantum mechanics according to dialectical materialism.

Omel'ianovskii believed that such an interpretation must proceed from the following basic points, considered by him to be intrinsic to any dialectical materialist view of the microworld: (1) Microphenomena and their regularities (*zakonomernosti*) exist objectively; (2) macroscopic and microscopic objects are qualitatively different; (3) although they are qualitatively different, there is no impassable gulf between the microworld and the macroworld, and all properties of micro-objects appear in one form or another on the macrolevel; (4) there are no limits to man's knowledge of microphenomena. Omel'ianovskii attempted to utilize points one and four as his main criticisms of the "physical idealists" of the Copenhagen School, and point two against misguided but good-hearted critics of Copenhagen who hoped for a return to the laws of classical physics.

According to Omel'ianovskii, the physical significance of the wave function is its "representation" (*otobrazhenie*) of the peculiar statistical laws of microphenomena, laws that are not the same as the statistical laws of macrophenomena (statistical mechanics). The peculiarity of these new statistical laws on the microlevel consists in the fact that micro-objects simultaneously possess both corpuscular and wave properties. To look upon micro-objects first as particles and then as waves would be to fall prey to complementarity, a concept that Omel'ianovskii rejected. Instead, one must always consider micro-objects as *simultaneously* possessing wavelike and particlelike properties. Micro-objects thus represent a dialectical unity of contradictory properties. Consequently, the wave function cannot be applied to individual micro-objects (here the Copenhagen School and Fock were wrong, thought Omel'ianovskii) but only to the quantum ensemble, developed by Blokhintsev and also favored on several occasions by Einstein, who, however, "failed" to understand the qualitative differences between the statistical laws of classical mechanics and those of quantum mechanics. The difference between these two classes of statistical laws, said Omel'ianovskii, can be illustrated by the fact that in classical ensembles the distributions of momenta and coordinates are not connected with each other, whereas in quantum ensembles they are. In his definition of quantum ensembles Omel'ianovskii differed with his colleague Blokhintsev, who in the second edition of his textbook said that quantum ensembles must be defined in relation to macroinstruments, which "fix" or "settle" the ensemble. Omel'ianovskii, on the contrary, believed that the question of the measuring instrument was not relevant to the definition of the quantum ensemble. But by so stating his position, he ran into the very serious problem of isolating the ensemble, which had been one of the reasons that led Blokhintsev to bring the measuring instrument into the definition in the first place. Omel'ianovskii's only way out was through the substitute of defining the ensemble in terms of what it is not and in terms of what it "represents," not in terms by which it could be rigorously identified. To quote from Omel'ianovskii: "The quantum ensemble is not a 'collective of experiments,' not a 'collective of results of measurements.' It is not a speculative formulation; it is a concept which reflects the association of a sufficiently large number of equal, in this or that measure, micro-objects which under definite conditions belong to one and the same species (vid.)." He capped this very loose definition of ensembles with the observation that "the problem of the corpuscular-wavelike nature of micro-objects is still insufficiently worked out. This circumstance is of importance also in the exposition of the conception of quantum ensembles." [115]

Omel'ianovskii believed that the concept of complementarity arose from Bohr's and Heisenberg's exaggeration of the meaning of the uncertainty relation. The first step in this exaggeration was the raising of the uncertainty relation to a higher rank, the "uncertainty principle." Omel'ianovskii accepted the uncertainty relation as a fact of science, but this physical fact in itself said nothing, he maintained, about the "uncontrollable influence" of the instrument, upon which Heisenberg in particular based the "uncertainty principle." [116] Omel'ianovskii believed this view of the role of the instrument to be directly responsible for complementarity. While he used the term "uncertainty relation," he refused to use the phrase "uncertainty principle," substituting the term "Heisenberg relation." Omel'ianovskii's opinion of the "Heisenberg relation" is revealed clearly by his remark that "the relation established by Bohr and Heisenberg by means of the analysis of several thought experiments—we call it the Heisenberg relation—has no physical significance and is a 'principle' changing the content of quantum mechanics in the spirit of the subjective concept of complementarity." [117] The error of complementarity, in turn, was that it does not emphasize the characteristics of atomic objects, which are the proper subject of study of quantum mechanics, as much as it does the role of the measuring instrument. Omel'ianovskii's position, which ignored the tendency of many members of the Copenhagen School, including Bohr, to attribute the uncertainty relation not to the measuring instrument but to the simple nonexistence of physical values of conjugate parameters, was thus primarily a criticism of alleged subjectivism in measurement.

Omel'ianovskii devoted the last section of his book to a discussion of determinism and statistical laws. In his opinion, determinism, a basic principle of nature, was in no way threatened by quantum mechanics. On this issue he agreed with P. Langevin that "what is understood at the present time as the crisis of determinism is really the crisis of mechanism." [118] Determinism is perfectly compatible, according to Omel'ianovskii, with statistical laws. Furthermore, Omel'ianovskii considered the statistical laws of quantum mechanics to be not the result of the uncontrollable influence of measurement (Heisenberg), not the result of indeterminism governing the individual micro-object (Reichenbach), not the result of hidden parameters (Bohm), not the result of the relationship of the micro-ensemble and its macro-environment (Blokhintsev), but instead the result of what he called the "peculiar wave-corpuscular properties of micro-objects." Such a position, according to Omel'ianovskii, does not preclude the existence of hidden parameters (contrary to Neumann), although it does not promise them, and does not suppose that their discovery would result in a classical description of micro-objects, as Omel'ianovskii believed Bohm, Vigier, and the latter-day de Broglie hoped. Thus, Omel'ianovskii completed the edifice of his interpretation of quantum mechanics, a structure consisting almost entirely of statements telling what quantum mechanics *is not* but very rarely hinting what it *is*. In answer to the question, What is quantum mechanics? Omel'ianovskii could cite only the first of his original four points, that it is the study of objectively existing micro-objects and their regularities, a point on which all Soviet interpreters of quantum mechanics agreed.

In recent years a number of changes have occurred in Soviet views on quantum mechanics, although no new theoretical positions have been developed. The most heartening change has been the improvement in tone of most Soviet writings on the subject; at the present time, almost all articles and books published by scholarly presses are truly philosophical in approach, and not ideological. Blokhintsev's 1966 book was a prominent example.

Another change was the shift of Omel'ianovskii from relying primarily on Blokhintsev to relying on Fock. His shift can be traced in two steps: first, his acceptance of the view that quantum mechanics can be applied to the individual micro-object, and second, his rehabilitation of the term "complementarity," although with continuing reservations.

The beginning of these changes could be seen at the October 1958 all-Union conference in Moscow on the philosophic problems of modern science. This conference was convened, in the words of E. N. Chesnokov, as a result of "some instances of insufficiently profound appreciation by certain philosophers of the achievements of modern science." [119] The reports concerned relativity theory, cybernetics, cosmogony, biology, and physiology as well as quantum mechanics. In the discussion that followed the reports, the scientists, including A. D. Aleksandrov, V. A. Fock, S. L. Sobolev, V. A. Ambartsumian, and A. I. Oparin, clearly dominated the philosophers. In his report entitled "V. I. Lenin and the Philosophic Problems of Modern Physics" Omel'ianovskii changed his position on the significance of the wave function. Whereas earlier he had believed that it could be applied only to Blokhintsev's ensembles, he said at the conference that "the wave function characterizes the probability of action of an individual atomic object." This description was very similar to Fock's statements on the significance of the wave function, and in expanding on his interpretation, Omel'ianovskii revealed that he had also accepted Fock's distinction between the "potentially possible" and the "actually existing."

In 1958 Omel'ianovskii had not accepted the term "complementarity," still considering it to be synonymous with the Copenhagen Interpretation. At the Thirteenth World Congress of Philosophy held in 1963 in Mexico City, however, he agreed even further with Fock by accepting complementarity and even maintaining that it is based on a dialectical way of thinking through its assertion that "we have the right to make two opposite mutually exclusive statements concerning a single atomic object." [120] Thus, Omel'ianovskii believed that the link between dialectics and the notion of complementarity "lies at the center of the Copenhagen Interpretation in quantum mechanics." [121] A vestige of his old views could be seen in his comments about the remaining "deficiencies" in the concept of complementarity, such as its insistence on applying classical notions in the new realm of atomic objects.

In a 1968 article on philosophical aspects of measurement in quantum mechanics Omel'ianovskii emphasized in an interesting and helpful way that contrary to much common belief, it is not really proper to speak of the "uncontrollable influence of the measuring instrument on the micro-object." [122] If we think of a crystalline lattice as the measuring instrument for an electron, before passing through the lattice, the electron is located in a state with a definite momentum and an indefinite position; after passing through the lattice, the electron is in a state with a definite position and an indefinite momentum. Measurement therefore changes the state of the micro-object, but this change is not a result of a force acting on the object, such as gravitational or electromagnetic force. The lattice itself did not exert any force on the electron that passed through it. Rather, the influence of measurement arises from the very corpuscle-wave nature of the micro-object. Omel'ianovskii explained his position most graphically through an analogy: "The change of quantum state under the influence of measurement is similar to the change of mechanical state of a body in classical theory when one makes the transition from one system of reference to another moving relative to the first." [123] This clarification by Omel'ianovskii, which is in agreement with Bohr's views shortly before his death,[124] goes a long way toward resolving many debates over the "uncontrollability" of measuring instruments in quantum mechanics.

At the same time that Omel'ianovskii redefined his interpretation of quantum mechanics, a number of other Soviet scholars became interested in the philosophic problems of quantum mechanics. Some of them displayed interest in de Broglie's "theory of double solution," a hidden-parameter approach replacing his earlier "pilot-wave theory." [125] Others were seeking a unified theory that would combine the realms of quantum theory and relativity theory. Such attempts have been made in other countries as well, where similarly they have not been successful although they continue to be interesting. Soviet authors discussing new approaches have become relatively accustomed to handling ideas that in the late forties or early fifties would automatically have been considered suspect, such as the theory of a finite universe or the hypothesis that in the "interior" of microparticles future events might influence past events. In a 1965 article in *Problems of Philosophy* the veteran philosopher E. Kol'man pleaded that Soviet scientists be granted permanent freedom to consider such theories; naturally, he observed, these viewpoints

give idealists cause for seeking arguments in favor of their point of view. But this does not mean we should reject these "illogical" conceptions out of hand, as several conservative-minded philosophers and scientists did with the theory of relativity, cybernetics, and so forth. These conceptions are not in themselves guilty of idealistic interpretations. The task of philosophers and scientists defending dialectical materialism is to give these conceptions a dialectical materialist interpretation.[126]

Yet it should not be thought that as Soviet discussions of quantum mechanics became more and more free from political influence, all Soviet interpretations moved closer to the reigning Copenhagen Interpretation. Some Soviet writers renewed their criticism of the Copenhagen School, although on a much higher intellectual level than in the early fifties. One of these was the Soviet physicist A. A. Tiapkin, who in 1970 published an interesting chapter in a book based on reports given several years before at a conference at the well-known United Institute of Nuclear Research in Dubna.[127] This conference included physicists from Dubna, philosophers from the Institute of Philosophy of the U.S.S.R., and scholars from various Soviet universities. Like Blokhintsev in his most reflective moments, Tiapkin believed that it was possible to create an unknown, more complete theory of quantum mechanics. The advantage of this new theory, however, would be, according to Tiapkin, largely philosophical; it would not predict a single new effect or result of measurement that existing quantum theory does not already produce.[128] Tiapkin's ambitions were at once both great and modest; on the one hand, he wanted to do the seemingly impossible—to give a statistical description of phenomena that he agreed were in principle "unobservable"; on the other hand, he admitted that if he achieved his goal, it would not directly affect present quantum mechanical computations in any way. Its main advantage, he said, was that it would help to eliminate from physics the positivist slogan "If you can't measure it, it doesn't exist." [129] Tiapkin maintained that Marxist philosophers and physicists should seek to explain the unmeasurable interphenomena of quantum physics in objective terms even though Bohr had been quite correct in demonstrating to Einstein that present quantum mechanics is complete in the sense of predicting all data from measurement.[130] But it was still incomplete, said Tiapkin, in another, broader sense: It made no attempt to describe the movement of micro-objects between moments of measurement. Tiapkin remained convinced, like Einstein, that some kind of movement occurred in those intervals and that the task of a physicist would not be complete until he had given a description of that movement.

Tiapkin believed that a broader theory was not only needed but possible. The one criterion that it must meet, he said, was that it must have a single-valued compatibility with the whole structure of predictions of measurement generated by present theory.[131] He suggested then a "reverse course" of seeking the function of the unobservable distribution of probabilities by taking the existing apparatus of quantum mechanics and working backward.[132] Such attempts had been made several times in the past by scientists such as Wigner, Blokhintsev, and Dirac, but because of mathematical difficulties they had not succeeded. Tiapkin thought such a solution was still possible and might ultimately be given a physical interpretation. One possibility was dividing the micro-object into a discrete particle, on the one hand, and a continuous wave process in a vacuum that has a statistical influence on the microparticle, on the other.[133] Such an interpretation should not be confused, said Tiapkin, with de Broglie's pilot-wave hypothesis, since de Broglie's goal was a dynamic, causal, nonstatistical description of the results of measurement.[134] Tiapkin remained convinced that von Neumann was correct in considering such attempts impossible. To Tiapkin, the description of both the measurable and the unmeasurable movement of micro-objects was inherently statistical.

The interpretation of quantum mechanics is still a very open question, not only in the Soviet Union, but in all countries where there is an active concern with current problems of the philosophy of science. As I have earlier noted, the Soviet discussions of causality, the influence of the observer, and the possibility of hidden parameters were quite similar to the worldwide controversy.[135] In the Soviet Union the main participants in the debate—Fock, Blokhintsev, and Omel'ianovskii—all had disagreements with each other, and outside the Soviet Union the interpreters of quantum mechanics also have had intense disputes.

All scientists in the course of their investigations must proceed beyond physical facts and mathematical methods; such theorization is one of the bases of scientific explanation. Choices among alternative courses that are equally justifiable on the basis of the mathematical formalism and the physical facts must be made. The choice will often be based on philosophic considerations and will often have philosophic implications. Thus, Fock in his interpretation of quantum mechanics defined "complementarity" as a "complementarity between classical descriptions of microparticles and causality." [136] In his subsequent choice between retaining either a classical description or causality, he chose causality, and thereby lost the possibility of a classical description. He could have gone the other way. This decision inevitably involved philosophy.

The Soviet scientists and philosophers drew attention to a significant and fruitful concept when they observed that as long as even probabilistic descriptions of nature are possible, the principle of causality can be retained. To them, the nonexistence of causality in quantum mechanics would mean that all possible values of position and momentum for a micro-object would have equal probability. In such a world, a science of quantum mechanics would be impossible.

No one knows if quantum mechanics will retain its present mathematical formalism, or gain a new formalism permitting a more deterministic interpretation of quantum mechanics; the present evidence is not very reassuring for those people who want to find a new realm of strict determinism below the one with which we now work.[137] If the present opinion of most scientists is confirmed and it becomes increasingly clear that causality must be interpreted probabilistically if it is to be retained at all, the resulting discussions could lead to refreshing developments in the age-old debates over determinism and free will, particularly in the Marxist framework in which freedom is seen as knowledge of natural laws; Marxists could allow room for a given situation to generate a range of possible outcomes without resorting to any factors outside the natural world. This concept was advanced by several Soviet physiologists, and will appear in the discussion of physiology and psychology.[138] But the full significance of quantum mechanics in its present form has not yet been adequately absorbed by specialists in other fields, Marxists or non-Marxists.

Whether the future of quantum mechanics will reassure the probabilists or the determinists will depend on science. In the meantime, Soviet philosophers and scientists have found an interpretation—or rather, several interpretations—that makes the world seem more intelligible to them and that could handle either eventuality.[139]

IV   Relativity Theory

The special theory of relativity (STR), as elaborated by Einstein, flows from two postulates: (1) the principle of relativity, which asserts that physical processes occurring in a closed system are unaffected by non-accelerated motion of the system as a whole, and (2) the principle that the velocity of light is independent of the velocity of its source. The first postulate was accepted in classical mechanics long before Einstein, and is perhaps best illustrated by comparing physical phenomena, such as falling objects, in two different inertial systems (systems within which bodies unaffected by outside forces move at constant speed in straight lines). If a given inertial system is moving at a constant velocity in a straight line relative to another given system, then the laws of mechanics must have the same form in both systems. The common illustration of this relationship is the fact that to an observer in a train moving at a constant velocity, a falling object describes a path identical to the one he would see if he and the object were on the ground. To an observer alongside the moving train, however, the falling object in the train describes a parabola. In this case, a transformation from one reference system to another has been made, and in accordance with classical mechanics, the Galilean transformation equations would provide the means of plotting the equation of the parabola from data obtained from inside the railroad car.[1]

Einstein in his development of STR extended the principle of relativity to cover electromagnetic phenomena as well as mechanical ones. This extension necessitated the derivation of new transformation equations, since the Galilean equations could not account for the constancy of the velocity of light in all inertial frames, a constancy that had been illustrated prior to Einstein's work by the noted Michelson and Morley experiment. In order to preserve the principle of the constancy of the velocity of light in different reference frames and to maintain the existence of equivalent reference frames, Einstein modified the rules of transformation from one system to another. The new equations, known as the Lorentz transformations, accomplish this accommodation by providing that clocks in different inertial systems run at different speeds, and that spatial distance between points varies in different reference systems.[2]

Until the end of the Second World War, professional physicists in the Soviet Union were largely unconcerned with dialectical materialism, despite the attention that Lenin devoted to physics in his *Materialism and Empirio-Criticism*. To be sure, there had been a debate of moderate proportions over relativity physics among Soviet philosophers in the 1920's and 1930's.[3] In addition to the philosophers, several physicists had been involved, but the debate did not attain a very high technical level.[4] Certainly the Soviet reception of relativity did not seem uniquely hostile when compared with that in other countries, even though some opposition to relativity was expressed. Relativity physics was, after all, a topic of discussion and occasional polemic among the literate public all over the world. S. Iu. Semkovskii, the first Soviet Marxist writer to give a careful analysis of relativity physics, declared in 1926 that Einstein's new physics not only did not contradict dialectical materialism, but brilliantly confirmed it.[5] Semkovskii emphasized that space and time according to relativity theory were not products of "pure reason" but "forms of the existence of matter."[6] David Joravsky, an American historian of Russia, even commented that "as for active opposition to the new physics, one might even argue that there was less in the Soviet community of physics than elsewhere."[7]

Russian physicists before the war were fully aware of the controversies over the relation between science and philosophy that had occurred as a result of the widespread acceptance of the views of such scientists as Ernst Mach and Henri Poincaré, and they knew that these new conceptual approaches had been important in Einstein's development of relativity theory. Those Soviet physicists who knew that Mach was the object of lengthy criticism by Lenin may have felt reticent about discussing the philosophical background of relativity, but as scientists they could find reassurance in Lenin's careful distinction between science and philosophical interpretations of science. In university lectures, monographs, and textbooks of the prewar years one finds much evidence that Russian physicists and mathematicians were responding to the same scientific and even philosophical currents as scientists in all countries.

Examples of the typically international attitudes of Soviet physicists can be found in the university lectures of the well-known physicist L. I. Mandel'shtam (1879–1944), who from 1932 until 1944 taught theoretical physics at Moscow University, and who deeply influenced a generation of Soviet physicists. Among his students were G. S. Landsberg and I. E. Tamm. Mandel'shtam, educated in Novorossiisk University and the University of Strasbourg, was greatly interested in and attracted to Western philosophical thought, from Mach onward through the whole trend of the Viennese circle and logical positivism. Mandel'shtam taught his students that there was an essential difference between the logical structure of a scientific theory and the empirical data to which it related, and he believed that links between the two were created on the basis of definitions, which were neither true nor false in themselves, but merely convenient or inconvenient. This approach, one of the cornerstones of the logical empiricists in the philosophy of science, was apparent in Mandel'shtam's discussions of the metric of length and time. He commented that "the physicist must have a recipe [*retsept*] in order to find out what length is. He must indicate that he does not discover that recipe, but defines it." [8] Similarly, thought Mandel'shtam, time is defined in relation to some kind of periodic physical phenomenon, such as the rotation of the earth or the movement of the hands of a chronometer; this stipulation is also merely a definition without absolute content: "Let us take for the sake of simplicity the definition of time by means of a chronometer. In this fashion, time (that is, what I insert in Newtonian formulae in the place of *t*) is that which is indicated by the hands of my watch." Without such definitions, thought Mandel'shtam, such equations as those of Newton and Maxwell express only mathematical relationships and are not directly relevant to physical experience.

Mandel'shtam's viewpoints, familiar to physicists and philosophers of science everywhere, and yet not without controversial aspects, were not published during his lifetime even though they were well known among his students and fellow physicists. The appearance of the fifth volume of his works in 1950, in which these statements appeared, caused quite a sensation among philosophers of science in the Soviet Union. (See p. 118.) The case of L. I. Mandel'shtam will serve as evidence, which could be easily supplemented, that physicists in the Soviet Union were familiar, although perhaps somewhat incompletely, with the dominant trends before World War II in the interpretation of the philosophical foundations of relativity theory. Indeed, it would have been quite impossible for them not to have been aware of the discarding of Kantian concepts of space and time that was necessary for the development of relativity theory.

In a 1948 physics textbook approved by the Ministry of Higher Education for use in the universities, the following statement left no doubt about the authors' belief in the conventionality of spatial and temporal congruency. Here one found stated clearly what many Soviet philosophers of science and some distinguished scientists (for example, A. D. Aleksandrov) criticized in later years:

Einstein showed that simultaneity of spatially distant events is a question of *definition*: It is necessary simply *to agree* what distant events *by definition* will be considered simultaneous, just as we agree to understand *length* as a number indicating how many times a definite rigid rod (standard of length) can be laid down between two given points. It is possible to give *other* definitions of length and of an interval of time, based on *other standards and possible uses of these standards*.[9]

Soon after World War II the increasingly restrictive intellectual environment of the Soviet Union permitted the militant ideologists to attempt a direct influence on the physicists. In his speech of June 24, 1947, A. A. Zhdanov did not mention the issue in science that was already becoming the most heated—biology—but he did criticize certain interpretations of physical theories:

Not understanding the dialectical path of cognition, the mutual relation of absolute and relative truth, many followers of Einstein, transferring the results of research on the laws of movement of a finite, bounded part of the universe to the whole infinite universe, have begun speaking about a finite world, about its temporal and spatial boundaries; the astronomer Milne even "calculated" that the world was created two billion years ago.[10]

Zhdanov's remarks, although directed more against cosmological interpretations of general relativity than against the basic positions of either special or general relativity theory, prefaced a new debate on the philosophic foundations of relativity theory that lasted until 1955, and that in altered and much more sophisticated forms has continued to the present time. The cosmological aspect of the debate will be considered separately in the following chapter.

Most of the Soviet articles on the philosophic aspects of relativity theory that appeared in the next few years were thoroughly hostile to non-Soviet interpretations, and not a few were opposed to the theory itself, referring to it by such terms as "reactionary Einsteinism." [11] Not until 1951 did the major philosophical journal of the Soviet Union carry an article that presented the theory of relativity in a generally positive fashion, and this article was roundly criticized, not only by individual philosophers, but by the editorial board of the journal itself.[12] As late as 1953, an article appeared in *Problems of Philosophy* that termed the theory of relativity "obviously antiscientific." [13] Because of the protracted life of such objections a historical view of Soviet attitudes toward relativity theory must include a description of their content. However, to equate the positions of the early Soviet opponents of relativity theory with the views of such later prominent critics and interpreters of relativity in the Soviet Union as V. A. Fock, A. D. Aleksandrov, and M. F. Shirokov would be a serious error, since the later writers were genuine intellectuals firmly grounded in the field.

Ironically, one of the first articles on the philosophical implications of relativity theory to appear following Zhdanov's speech was by the same G. I. Naan who later came to the defense of relativity and thereby incurred a great deal of criticism. This article appeared in an issue of *Problems of Philosophy* dedicated to the recently deceased Zhdanov. The article was directed against the "physical idealists" of the United States and England, the physicists and philosophers of science who, according to the author, had questioned the materiality of the world and denied the "regularities" (*zakonomernosti*) of nature. Naan included among the physical idealists a heterogeneous group of Western scientists and philosophers, including A. S. Eddington, James Jeans, Pascual Jordan, E. T. Whittaker, E. A. Milne, Bertrand Russell, and Philipp Frank. Frank was particularly criticized for commenting that neopositivism takes its starting point from Mach but so formulates its position that confusion with idealistic or solipsist doctrines is impossible, since the question of the existence of a real world behind our sensations is only a "pseudoquestion." Naan concluded from this that "the basic problem of philosophy has been christened a 'pseudoproblem.' " [14]

The following issue of *Problems of Philosophy* (Number 3, 1948) was an important one for the philosophy of science in the Soviet Union. It contained several articles on modern physics and also one on biology, as well as an editorial calling for ideological militancy in science. The articles on physics, by M. E. Omel'ianovskii, A. A. Maksimov, and R. Ia. Shteinman followed Naan's example by denouncing many of the prominent non-Soviet interpreters of science; in addition to the other names, they added Poincaré, Bohr, Heisenberg, Schrödinger, Reichenbach, and Carnap.[15] Omel'ianovskii was particularly exercised over Rudolf Carnap, an "open enemy" of materialism, for his belief that he had "risen above" the conflict of idealism and materialism. Eddington was criticized for maintaining that many of the constants of physics must be introduced a priori, and Frank for trying to build a bridge between dialectical materialism and logical empiricism.[16]

These Soviet critics of non-Soviet views of physics often utilized as sources the popular and philosophical writings of non-Soviet scientists, which, especially in the cases of people such as Jeans and Eddington, often sacrificed scientific rigor for colorful language and lucidity. But a serious error of the Soviet critics was to proceed from this criticism of informal interpretations to a condemnation of relativity theory itself; it was as if one could hold the theory responsible for all statements, professional and nonprofessional, uttered by its adherents. This was done most flagrantly by A. A. Maksimov, who ended up by denying not only Einsteinian relativity but even Galilean relativity. Maksimov commented:

A. Einstein wrote in his book about the theory of relativity: ". . . trajectories in themselves do not exist, but each trajectory can be related to a definite reference body." This judgment that a body does not have an objective, given trajectory existing independently from the choice of system of coordinates is completely antiscientific.[17]

The dimensions of this malapropism were so great that the editors of the journal could not refrain from adding a footnote to Maksimov's text explaining that although they shared his desire to criticize idealistic views of modern physics, they felt his discussion of trajectory did not "embrace this problem in all its complexity." [18] Not deterred, Maksimov tried to buttress his position with the observation that the objective characteristics of a meteorite's trajectory are revealed when it plows a path into the earth's surface, from which a cast can be made suitable for research. Maksimov admitted that the mathematical relations of the Lorentz transformations are valid, but maintained that such concepts as length, time, and simultaneity have objective meanings. He did not, however, attempt to give serious definitions of these concepts.

A considerable amount of time passed before Maksimov received the physics education that his article made inevitable. Several subsequent authors, such as G. A. Kursanov, tried to find a more defensible middle ground without specifically denying Maksimov's argument; they agreed that motion cannot be related to any absolutely motionless body, system, or ether—as evidently Maksimov would have it—but they pointed out that this relativity did not contradict the movement of bodies independent from the consciousness of man. Such a view of relativity certainly does not permit, said Kursanov, the consideration of concepts such as "space," "time," "volume," and "movement" as "pseudoconcepts," a position that he attributed to Carnap and the Viennese circle. Kursanov realized, nevertheless, that the relativity of times and lengths is not in the process of observation, but is inherent in the characteristics, as defined by modern science, of physical phenomena themselves. To this extent he chastened certain Soviet misinterpreters of relativity theory. But he held to a belief in the existence of absolute simultaneity.[19]

An outright rejection of relativity theory was, of course, highly improbable. At this time physicists utilized certain aspects of special relativity as comfortably and frequently as engineers employed Newtonian mechanics. But now that the topic had been raised to the level of ideological discussion, there *were* certain embarrassing facts about relativity theory. Aside from the basic questions concerning materialism and objectivity was the secondary but quite troublesome historical fact that Einstein had been heavily influenced by Mach, had repeatedly acknowledged his debt to Mach; and yet Mach was the object of criticism of Lenin's *Materialism and Empirio-Criticism*.[20] Could relativity be separated from "Machist idealism"? It was a question that troubled Soviet philosophers of science for some time, although by the end of the fifties it was resolved with an affirmative answer. One possible exit from the situation lay in finding important precursors to Einstein's work other than Mach. Frequent attempts were made by Russian authors to emphasize the importance of Nikolai Lobachevskii, the Russian creator of the first non-Euclidean geometry. Thus, L. I. Storchak commented, "The establishment of the priority of Lobachevskii in formulating the principle of relativity debunks the old myth that the invention of this principle belongs to Mach." [21] But this attempt to employ Lobachevskii as a replacement for Mach was not convincing even in the Soviet Union, although the brilliant Lobachevskii stood in no need of additional honors to assure his place in the history of mathematics.[22]

In early 1951 the Estonian scholar G. I. Naan submitted Maksimov's 1948 article to a thorough criticism, scornfully commenting that for Maksimov to maintain simultaneously that the equations of the STR were correct but that absolute trajectories exist was equivalent to commenting that the multiplication tables are correct but denying that $8 \times 11 = 88$.[23] Since his 1948 article, which decried many non-Soviet interpretations of relativity physics, Naan's views had evidently changed greatly. True, he did not directly contradict his previous statements, but while the earlier article had been a militant critique of non-Soviet philosophers of science, the later one was a sober course in elementary relativity theory for philosophers. His only criticism of the physical idealists was now restricted to those who had stated that the relativity of trajectory, kinetic energy, mass, space, and time intervals depends on the observer. In the manner of Kursanov, Naan pointed out that relativity is not a subjective phenomenon, but is inherent in the physical processes themselves. His insistence on the absolute nature of acceleration, however, revealed that he had not fully accepted the usual interpretation of general relativity. Naan's article could be summarized as a critique of the vulgar materialists such as Maksimov combined with an outline of modern relativity theory. The article was tolerant on philosophic questions to a striking degree in Stalinist Russia, considering its place and time of publication.

Shortly before Naan's article the fifth volume of L. I. Mandel'shtam's works, the one containing his views on relativity theory, was published by the Academy of Sciences. This volume was based on notes taken by his students during his lectures and presented for publication after his death. When combined with Maksimov's articles, the total spectrum of viewpoints on philosophic interpretations of relativity theory now available to Soviet readers was surprisingly broad, considering the intensity of the ideological scene in those years. In Mandel'shtam's works, one could find the interpretation of those scientists and philosophers who greeted the revisions in epistemological thought that originated largely in central Europe at the end of the nineteenth and the early part of the twentieth century. Naan's view, while not of the same scale of importance as that of Mandel'shtam, represented that of Soviet scientists who wished most of all to get on with the work of physics and who were quite impatient with the intrusions of philosophers.

This spectrum, although rather diverse, presented little choice for a Soviet Union that would emerge from Stalinism and yet retain a commitment to a universal Marxist philosophy. Maksimov's position was contrary to much of modern physics, Naan's was nearly neutral to dialectical materialism, and Mandel'shtam's was even implicitly opposed to Soviet dialectical materialism in the sense that it drew all its inspiration from non-Soviet and non-Marxist sources and disagreed with current Soviet Marxist interpretations.

Authors who followed G. I. Naan's 1951 entry into the discussion were almost universally critical of his approach. The year 1952 marked one of the most intense periods of the debate; in the first issue of *Problems of Philosophy* for that year three different writers disagreed with Naan's exposition of relativity theory, each of them defending, incredibly, a different version of the thesis that objects in movement describe absolute paths.[24] G. A. Kursanov, rejoining the argument, tried to have it both ways: Bodies have "objectively real" trajectories, but these trajectories are somehow connected with reference systems. He maintained:

Every material body moves in real, objective space, occupying at each given moment of time a completely defined place at a given point of space. . . . At the same time, the movement of each body takes place in interaction with other bodies. Therefore, the objectively real property of a body— its trajectory—can be looked upon only from the point of view of the movement of other bodies, in connection with definite "reference systems." [25]

V. Shtern, a scholar from East Berlin, also criticized Naan's view that movement has no absolute context. The concepts of absolute space and time are in themselves legitimate, said Shtern, since one can constantly enlarge one's reference frames (from a room, to the surrounding land area, to the globe, to the solar system, to the galaxy, to the universe, and so on) and each time obtain new definitions of the movement of a certain object. But each such enlargement is an improvement, said Shtern, and absolute movement would be movement within the largest possible reference frame.[26] Shtern failed to answer the logical objections to his view, the most obvious of which is that in the "largest" reference frame, an infinite one, different paths for the same object would result, depending on where the origin of co-ordinates is located. If the arbitrary concept of "center" is used, where is the center of a reference frame of infinite dimensions, or of an unbounded but finite reference frame?

Most surprising of all was the attempt of D. I. Blokhintsev, the well-known Soviet specialist on quantum mechanics, to defend a concept of absolute movement. Blokhintsev's article, hardly more than a page in length, was a statement of his belief that while there are no truly inertial systems of reference, at any moment in the development of science there is one frame of reference that seems more "inertial" in character than any other, and that absolute movement may be understood as motion in a reference system with a maximum property of inertialness. Only in this way, thought Blokhintsev, could spatial concepts be correlated to Lenin's understanding of the relationship of relative and absolute truth.[27]

A genuine improvement in the intellectual quality of Soviet discussions of relativity theory began to occur even before Stalin's death in March 1953. One of the reasons for this change seems to have been the decision of several eminent Soviet physicists and mathematicians to enter the philosophical debate in order to protect relativity theory from attacks by ideologically militant philosophers and mediocre physicists. This decision eventually resulted in a strengthening of both the scientific content of Soviet philosophy and the philosophical perceptivity of Soviet scientists. The danger to relativity had been made clear in 1952 articles by the philosopher I. V. Kuznetsov and the physicist R. Ia. Shteinman;[28] the articles appeared in the same "Green Book" (edited by the ultraconservative A. A. Maksimov) mentioned in the previous chapter on quantum mechanics. Shteinman and Kuznetsov proceeded from a criticism of Einstein's philosophy to a call for the overthrow of relativity theory itself. Kuznetsov wrote that a truly materialist understanding of the physical laws of bodies moving at rapid velocities would result in a repudiation of Einstein's theory of relativity (STR) and the development of an essentially different physical theory.[29] The only alternative, however, that Kuznetsov and Shteinman could present was a return to a prerelativity interpretation of the Lorentz contractions within a framework of absolute space and time. In an article published several months before Stalin's death, V. A. Fock called this approach an attempt to deny the most important achievements in physics of the twentieth century.[30] According to Fock, both special relativity and quantum mechanics had been "brilliantly confirmed" by experiment and, in turn, they were confirmations of dialectical materialism.[31]

Fock defended relativity physics from *within* the intellectual system of dialectical materialism. Even in the thirties he had written on physics and philosophy in the major Soviet Marxist journal of philosophy.[32] In the political atmosphere of Stalinist Russia no other choice than a dialectical materialist approach was available to him if he wished to defend relativity physics. But one should not be too quick to assume that the attempts of Fock and like-minded scientists to develop new dialectical materialist understandings of nature were merely pretense or entirely tactically motivated. A number of them continued to write on philosophy and science long after the passing of the Stalinist period. Fock is still—after several decades—doing sophisticated writing on dialectical materialism and relativity. There seems to be reason to believe that once committed to dialectical materialism, some Soviet scientists such as Fock decided that its most essential principles accorded with their own and that it had serious potential for development.

The Fock-Aleksandrov interpretation of relativity theory has occasionally been presented as a unitary scheme not divisible into the parts for which each author is responsible. This unitary approach is not, however, the most revealing one. Aleksandrov and Fock supported each other, and their views were not contradictory on major points, but each followed a rather different path and emphasized different portions of relativity theory. Aleksandrov focused his attention on interpreting STR, while Fock concentrated on general relativity (GTR). Furthermore, Aleksandrov wrestled more thoroughly with the problem of spatial and temporal congruency definitions and with definitions of simultaneity than did Fock, who, in the manner of many physicists, covered this topic—crucial from the standpoint of the philosophy of science—rather hurriedly.[33] As a result of the different approaches, Aleksandrov was more vulnerable to criticism by those philosophers who refuse to accept the view that space-time has an inherent metric prior to the assumption of conventions than was Fock, who did not express himself so clearly on the questions of metric. Because of these differences, I will consider Fock and Aleksandrov separately.

A. D. Aleksandrov

Aleksandr Danilovich Aleksandrov (1912–) is an internationally known and respected Soviet mathematician who was for some years rector of Leningrad University. He has traveled abroad, and in 1959 visited the University of California at Berkeley. Among mathematicians he is best known for his book *Intrinsic Geometry of Convex Surfaces,* which was translated into English by the American Mathematical Society in 1967.[34] He is considered to be the founder of the Soviet school of geometry in the large and has published many articles on this subject. He has also published articles with titles such as "The Dialectics of Lenin and Mathematics"[35] and "On Idealism in Mathematics."[36] In 1966 Aleksandrov commented:

My professional activity involves mainly the proof of new theorems. And for me Marxist-Leninist philosophy is an unquestioned guide in comprehending general questions of my science. Dialectical materialism, needless to say, does not offer methods for solving specific problems in mathematical science, but it indicates true reference points for searches for scientific truth and arms one with methods for elucidating the true import of theories and the content of scientific concepts. I could cite examples showing how philosophy helps one master the mathematical theory of infinite numbers, Einstein's theory of relativity or quantum mechanics, but this would require the introduction of complicated, specialized concepts. I shall say only that as a student studying in a physics department I was able to understand quantum mechanics to a significant degree thanks to the fact that at the same time I was studying philosophy which helped me to comprehend this difficult theory in the spirit of dialectical materialism.

Unfortunately, in our scientific milieu one still encounters a pragmatic or, to put it more simply, a narrow-minded view of science in which people turn away from its general and philosophical questions, reducing it to the solution of individual particularized problems. This view is a decided hindrance to the posing of fundamental theoretical problems and to searches for new lines of research.[37]

Aleksandrov often began his statements of his position on relativity with a recognition of the great genius of Einstein, a man who Aleksandrov believed was more importantly influenced by his inherent materialist understanding of natural laws and of the concept of causality than he was by Mach and the school of neopositivism. Aleksandrov was one of the prominent Soviet scientists who came to the defense of Einstein at a crucial moment in Soviet history. Aleksandrov, Fock, and other Soviet scholars maintained that most of Einstein's views were an illustration of the relevance of materialism, not its irrelevance. The success of the efforts of such scientists as Aleksandrov and Fock can be in part measured by the great esteem in which Einstein is held in the Soviet Union today. The first (and still only) comprehensive edition of the collected scientific works of Einstein to be published in the world appeared in Russian translation in the Soviet Union in the 1960's.[38] And yet both Aleksandrov and Fock disagreed with Einstein on a number of points, primarily ones of philosophical interpretation.

In fact, Aleksandrov thought that the effects of positivism, coming to Einstein from Mach, were sufficiently strong to lead Einstein into a number of errors. If Einstein had been left to follow his own inclinations, he would have emphasized even more, thought Aleksandrov, the "deep essence" of the theory of relativity, namely that a new conception of absolute space-time (as distinguished from space and time) reveals the objectivity of nature and, even more importantly, establishes the material and "causal-consequential" (*prichinno-sledstvenno*) structure of the world: "When the theory of relativity presents itself not as a theory of relativity, but as a theory of absolute space-time, determined by matter itself, it is a theory in which relativity clearly and necessarily becomes secondary and subordinate." [39]

The absolute character of the space-time continuum became the cornerstone of Aleksandrov's system. He noted that Einstein had arrived at the concept of absolute space-time after passing through and then ultimately discarding Newtonian space and time. He had, thus, proceeded from the relative to the absolute. But, Aleksandrov asked, would not a better conceptual approach be based on the reverse transition, from the absolute to the relative, now that, thanks to Einstein, the absolute nature of space-time has been established? In this sense, the relative character of, respectively, time and space could be explained as "only aspects of the absolute space-time manifold." [40] Here Aleksandrov was following a terminology very similar to that of Hermann Minkowski many years before.[41]

Aleksandrov's further development of his view on the necessity of proceeding backward from absolute space-time reveals that his goal was no less than an affirmation of the inherent objectivity of reference systems:

The principle of relativity is formulated not as a physical law, but as a principle of the dependence of the laws of nature on an arbitrary choice of the system of reference. . . . But the system of reference is something objective. It is in essence an objective coordination of phenomena with relation to material bodies and processes, serving as a base for a system of reference, a coordination, determined, in the final analysis, by material interactions.[42]

Aleksandrov's statement that "a system of reference is something objective" can be taken in two different ways. If he is speaking of a system of reference actually utilized in physical space and time, then the "something objective" may carry a meaning in the same sense as that denoted by such non-Soviet philosophers of science as Adolf Grünbaum, who, after a long consideration of whether there is an empirical warrant for ascribing a particular metric geometry to physical space and time, concluded: "Once the physical meaning of congruence has been stipulated by reference to a solid body and to a clock respectively for whose distortions allowance has been made . . . then the geometry and the ascriptions of durations to time intervals is determined uniquely by the totality of relevant empirical facts." [43] In other words, once a definition for metrical simultaneity has been adopted, then the geometry of physical space and the chronometry of science are determined by experiment.

Was this the intended meaning of Aleksandrov? An analysis of his views on the subject reveals that he differed from Grünbaum's approach in the following way: Grünbaum would make an initial arbitrary definition of a congruency standard; on the contrary, Aleksandrov would select as a congruency standard the physical phenomenon that he considered to possess a universal, objective significance—light. He believed that congruency standards may be obtained by empirical means. He granted that no one would maintain that there are "sets of coordinates etched in the universe," [44] but he nevertheless believed that congruency standards may be established without merely "defining" rigid rods and isochronous clocks. After discussing several varieties of opinion on congruency standards, Aleksandrov commented:

Another variant of an erroneous opinion on the coordinates x, y, z, t in an inertial system consists in the belief that they are defined conditionally by accepted measuring operations. Here the coordinates are again deprived of their objective significance, which is determined by the laws of nature. Every definition of a concept has scientific significance only to the degree that it reflects something objective. . . . [45]

How, then, did Aleksandrov establish his congruency standards— that is, how did he know that his rods are truly rigid and his clocks are truly isochronous? He advanced several attempts to establish such standards.

Aleksandrov followed a path familiar to many students of relativity theory, that of constructing a light-geometry. First, he noticed that if one assumes that the geometry of space is Euclidean, as physicists did before Einstein, then the law of the constancy of the speed of light is expressed by the formula

$$(1) \qquad \sqrt{(x - x_0)^2 + (y - y_0)^2 + (z - z_0)^2} = c(t - t_0)$$

where x_0, y_0, z_0 are the coordinates of the source of light, and t_0 is the moment of light emission. Proceeding from this formula, Aleksandrov and V. V. Ovchinnikov showed (as had been done many times before) that the Lorentz transformations could be obtained merely by keeping (1) unchanged during transformations of the spatial and temporal coordinates. [46]

Still, such a formulation requires congruency standards, or if these standards are inherent, a derivation of the standards. Following a system reminiscent of that of E. A. Milne, Aleksandrov maintained that the "background of radiation," or the "exchange of signals" between bodies, defines their mutual coordination in space and time. These signals should not be thought of as the results of hypothetical experiments conducted by fictitious observers, as Einstein often implied, but as objective results of natural processes. The "background of radiation" was thus a constantly existing objective reality:

Radio-location is precisely based on this experimental method of defining distance. . . . It is exactly in this way that the famed definition of the simultaneity of spatially distant events given by Einstein is based on the sending, reflection, and return of electromagnetic signals. All these processes take place constantly in a natural way because the smallest perturbation of a given body gives rise to an electromagnetic radiation—however weak —which is dispersed by bodies it encounters, even if part of it returns to its source. In other words, the processes responsible for radio-location in the comparison of clocks according to Einstein proceed constantly in a natural way. They establish the mutual coordination of bodies and their phenomena in space and time, and this occurs without any kind of observer. Therefore, the coordination of bodies and processes with regard to a given body is an objective fact, and thus, the system of reference connected with this body is, in the full sense of the word, real.[47]

Aleksandrov believed that such a view of relativity theory eliminated the necessity for descriptions of temporal and spatial congruency standards by means of conventions.[48] The background of radiation played something of the role of the old ether in providing a preferred reference frame, but Aleksandrov insisted that there was no genuine similarity: "The ether was only a medium. . . . Waves expanded *in* the ether. Radiation . . . is the waves themselves." [49]

It was through the concept of the background of radiation that Aleksandrov's views were conjoined with Fock's, who placed great emphasis on the equation for the expansion of the front of an electromagnetic wave. Both Aleksandrov and Fock believed that the speed of such a wave front has universal significance since it establishes the existence of a universal bond between spatial distances and time increments. This relationship is established in the homogeneous space of the special theory of relativity, and they thought that therefore the general theory of relativity cannot be an expansion of the special theory, since the general theory denies the homogeneity of space.

The reference to E. A. Milne's system above indicated that Aleksandrov's view was not original with him; many systems of light-geometry have been constructed in the past. One writer who anticipated many of Aleksandrov's views was the Irish physicist Alfred A. Robb, who as early as 1914 developed an optical geometry of motion in which he attempted to prove that congruency relationships were not assigned, but were inherently contained within the system.[50]

Aleksandrov has himself acknowledged the similarity of his system to that of Robb. He was unaware of Robb's work until 1954, when his attention was directed to it by a member of a seminar in the physics faculty at Leningrad University. After studying Robb's work, Aleksandrov maintained that the reason for the obscurity into which it had fallen was the imposition of positivistic viewpoints upon the theory of relativity.[51]

V. A. Fock

In our discussion of quantum mechanics it was mentioned that V. A. Fock is an internationally known theoretical physicist who has been honored in many countries of the world. In the late 1950's Fock established himself as the most authoritative interpreter of the dialectical materialist position on relativity physics, and continued to hold this position in the sixties despite the existence of other Soviet interpretations. On numerous occasions he expressed his debt to Marxism as an approach to science. In the preface to the 1955 edition of his *The Theory of Space, Time and Gravitation* Fock commented:

The philosophical side of our views on the theory of space, time and gravitation was formed under the influence of the philosophy of dialectical materialism, in particular, under the influence of Lenin's "Materialism and Empirio-criticism." The teachings of dialectical materialism helped us to approach critically Einstein's point of view concerning the theory created by him and to think it out anew. It helped us to understand correctly, and to interpret, the new results we ourselves obtained.[52]

Such statements were not restricted to the fifties nor merely appended to his scientific works. In 1966, in a reply by mail to a request from an American journal for his comments on dialectical materialism and science, Fock wrote:

The essence of dialectical materialism is just the combination of the dialectical approach with the acceptance of the objectivity of the external world. Without a dialectical approach materialism would reduce to mechanical materialism, which was obsolete even at the beginning of the twentieth century and is still more obsolete now. On the contrary, application of the laws of dialectics permits materialistic philosophy to develop with the devel-

opment of science. Even such statements of classical materialism as complete
independence of existence from the possibility of perception can be reconsidered and, if necessary, revised without altering the essence of dialectical
materialism. The ability of this form of philosophy to keep pace with science
is one of its characteristic features. Dialectical materialism is a living and
not a dogmatic philosophy. It helps to give to experience obtained in one
of the domains of science a formulation of such generality it may be applied
to other domains.[53]

Fock developed an interpretation of relativity theory that retained
the mathematical core of Einstein's work but that led him to several
novel concepts. Fock discarded the terms "general relativity," "general
theory of relativity," and "general principle of relativity." Instead, he
called the theory of Galilean space[54] the "theory of relativity" (rather
than "special theory of relativity"), and the theory of Einsteinian space-
time the "theory of gravitation" (rather than the "general theory of relativity").

Yet it would be a great mistake to emphasize only Fock's criticisms of
general relativity. As a matter of fact, he considered general relativity
(he would say the theory of gravitation) to be in need primarily of
interpretative clarifications and methodological amendments. In other
respects he defended Einstein's approach stoutly, and, indeed, it is quite
possible that his initial motivation for writing on relativity and philosophy was a defensive one—that is, to prevent the theory of relativity
from being discredited. But he discussed and defended relativity within
the framework of dialectical materialism; there is considerable evidence
that in the process he became sincerely interested in philosophical problems of the sciences. His emphasis on the necessity for physical content
in scientific explanations—and not just mathematical forms—is clear in
many of his writings. This emphasis may well have been linked to his
materialism.

Fock distinguished carefully between physical theories as they appeared in their completed forms and the methods by which these theories were developed. Fock thought that there might even be a difference
in principle between the initial ideas on the basis of which a theory was
created and the essential ideas contained in the theory after it had been
completed.[55] Such, he thought, was the case with general relativity.
When Einstein created the theory of general relativity, the "principle of
relativity" (mathematically expressed by the covariance of the equations

of physics in all reference frames) and the "principle of equivalence" (mathematically expressed by the identity of inertial and gravitational mass) played important roles in his thought; but Fock believed that these principles were not at the base of relativity in a physical sense. Indeed, according to Fock, the principle of equivalence was only approximately valid, while the principle of relativity (general covariance) was actually contradicted by the characteristics of the existing field of gravitation. The principles of equivalence and of relativity could be derived from the completed structure of general relativity as Einstein presented it, but, said Fock, they were not essential to it as a theory of gravitation. Let us consider his analysis in more detail.

The key to Fock's view of general relativity (always to be distinguished from special relativity, which Fock fully accepted) was his opinion that Einstein failed to see the importance of space-time "as a whole," concentrating instead on local areas within the space-time continuum. This emphasis caused Einstein to fail to see, said Fock, that his GTR is not a generalization of STR at all, but is instead its restriction. Rather than generalizing the concept of relativity, said Fock, Einstein merely generalized certain geometrical concepts and simultaneously violated his original relativization of space and time.

Fock began his discussion of relativity theory by noting that the theory of space and time may be divided into two parts: the theory of homogeneous (Galilean) space and the theory of inhomogeneous (Riemannian, Einsteinian) space. The first half occupied Einstein's attention in his development of STR, and he then attempted (unsuccessfully, said Fock) to generalize his theory into GTR.

The essential characteristic of Galilean space is its homogeneity, which can be illustrated by the equivalency of all points, directions, and inertial systems within it. Both Newtonian physics and special relativity physics were based on an assumption of homogeneous (Galilean) space. Mathematically, the homogeneity of the space of Newtonian physics was expressed in the Galilean transformations; the homogeneity of the space of special relativity was expressed in the Lorentz transformations. It was only in the transition from special relativity to general relativity that the assumption of Galilean space was discarded, and, said Fock, for very good reason.

Einstein correctly demonstrated, Fock continued, that the universal theory of gravitation (GTR) could not be contained within Galilean space. The most essential reason for the inadequacy of Galilean space, said Fock, was the one given by Einstein: Not only the inertial mass of a body but also its gravitational mass depends on its energy. Einstein found a way of describing the new physics by replacing, mathematically, Galilean space with Riemannian space. In so doing he created what is usually known as the general theory of relativity, a new physical theory. But according to Fock, the new theory, though extremely valuable as a theory of gravitation, was not a physical theory of general relativity at all. Fock later summarized his criticism in what he called "two short phrases": (*1*) *La relativité physique n'est pas générale;* (*2*) *la relativité générale n'est pas physique.*[56] Fock's view has been considered seriously by many scientists, both Soviet and non-Soviet. The discussion is continuing even today.

What did Fock consider to be Einstein's conceptual error? The root of it may be found in Einstein's understanding and use of the principle of equivalence, which states that in an infinitely small locality, the gravitational field is equivalent to an acceleration. Einstein illustrated this by a famous thought experiment: If a mass m is suspended by a spring from the top of a compartment (visualize an elevator) in the following fashion,

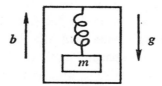

it is then impossible from within the compartment to decide whether an extension of the spring is caused by an upward acceleration of the compartment in the direction b or a downward gravitational field in the direction g.[57] A more ordinary illustration is that an airplane pilot flying in a cloud "by the seat of his pants" may be unable to distinguish a pull toward his seat caused by gravitation and one caused by flying in a loop, a similarity now illustrated algebraically by describing both forces in terms of g.

Einstein thus explained graphically his principle of equivalence, which may be simply stated as the principle of the equivalence of inertial and gravitational mass. Einstein then proceeded to apply this concept to terrestrial gravitation. At first this might seem an impossibility, since any acceleration involving the earth as a whole would have very different effects at different spots on the earth's surface. However, gravitational forces can be "transformed away" if we consider infinitesimal regions only, permitted through the use of differential equations. Thus if we think of a grid of cells surrounding the earth, with each cell representing an infinitesimal region, in the following fashion,

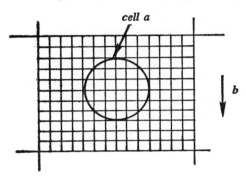

it becomes apparent that the gravitational force at any spot on the earth's surface may be transformed away by imagining an appropriate acceleration of the grid. If we let the above system accelerate at 32 ft/sec² in direction *b*, the gravitational field at cell *a* will disappear in the same way in which the force of gravity disappears in a freely falling elevator.[58]

The above examples of the principle of equivalence help one to understand that according to Einstein's theory of gravitation, in any given point of space the gravitational field can be replaced by an appropriate acceleration. The same relationship is conveyed by the observation that even though Einsteinian space as a whole is not homogeneous, in any infinitesimal region it is homogeneous, and the Lorentz transformations are valid.

It is exactly at this point that Fock objected to the Einsteinian view. He maintained that the local equivalence of acceleration and gravitation was not an adequate justification for concluding a complete equivalence of the fields of acceleration and of gravitation in all space. Indeed, Fock considered the principle of equivalence to be valid only in a restricted, local sense. According to Fock, the principle of equivalence in Einstein's completed theory had an "approximate character and was not a general principle." [59]

Fock noted that the physical basis of the principle of equivalence is the law of falling bodies, by which all unobstructed objects fall with equal acceleration. But this law is a general law, said Fock, not a local one, and if it is to be used for the foundation of another general law (of relativity), some way of considering space as a whole must be found:

In order to construct a theory of gravitation or to apply it to physical problems it is . . . insufficient to study space and time only locally, i.e., in infinitely small regions of space and periods of time. One way or another one must characterize the properties of space as a whole. If one does not do this, it is quite impossible to state any problem uniquely. This is particularly clear in view of the fact that the equations of the gravitational, or any other field, are partial differential equations, the solutions of which are unique only when initial, boundary or other equivalent conditions are given. The field equations and the boundary conditions are inextricably connected and the latter can in no way be considered less important than the former. But in problems relating to the whole of space, the boundary conditions refer to distant regions and their formulation requires knowledge of the properties of space as a whole. One should note that Einstein did not fully appreciate the inadequacy of a local description and the importance of boundary conditions. This is why it is necessary to change substantially Einstein's statement of the basic problems of gravitational theory; this has been done in the author's research. . . .[60]

Fock characterized the boundary conditions in two different ways. In the first case, he assumed space to be homogeneous at infinity in the sense of being characterized by the Lorentz transformations. Masses and their associated gravitational fields were then envisioned as being implanted in homogeneous Galilean space (note, not in finite but unbounded space-time). The second case assumed a space-time that is only partially uniform, with the spatial part of it conforming to Lobachevskiian geometry. Usually termed the space of Friedmann-Lobachevskii, it contains well-defined gravitational fields when the mean density of matter contained within it is not equal to zero.

The important conclusions from these considerations and the ones that reveal most graphically Fock's unorthodox position concern the question of preferred or privileged systems of coordinates. In each of the types of space considered by Fock—that is, Galilean space, space uniform at infinity, and Friedmann-Lobachevskii space—there "probably" is, according to him, a preferred system of coordinates.[61] The word "probably" here indicated Fock's continuing hesitation in the case of Friedmann-Lobachevskii space; in the case of Galilean space and space uniform at infinity he was confident of the existence of preferred systems of coordinates. The existence of such preferred systems of coordinates in each case would be, of course, contrary to Einstein's concept of the complete relativization of motion. Just as STR is associated with the relativization of inertial motion (and therefore the equivalence of inertial reference frames), so GTR is associated with the relativization of accelerated motion (and therefore the equivalence of accelerated reference frames). But now Fock questioned whether GTR actually was a generalization of STR in this sense.

One of the conclusions that some people drew from the extension of relativity by Einstein was the equivalence of the Copernican and Ptolemaic descriptions of the solar system, a result of the fact that circular motion is a case of accelerated motion.[62] But Fock's defense of the existence of preferred systems of coordinates was relevant here; as he commented, "Einstein's gravitational theory retains for the heliocentric Copernican system a preferred position compared to the geocentric Ptolemaic one."[63] He attempted to show the superiority of the Copernican view in terms of harmonic coordinates.[64]

As already mentioned, Fock was certain of the existence of preferred systems of coordinates in the case of Galilean space and in space uniform at infinity. In Galilean space the advantages of the customary Cartesian coordinates, supplemented with time, are illustrated by the fact, said Fock, that the Lorentz transformations when expressed in Cartesian terms are linear. Fock devoted much of his research to the task of proving that in space uniform at infinity there is also a preferred system that is well defined apart from a Lorentz transformation. It is, as already noted, the system formed by harmonic coordinates, which Fock believed reflected "certain intrinsic properties of space-time."[65] Yet it should be noticed that Fock's reliance on harmonic coordinates was one of the most controversial aspects of his approach; a number of physicists who accepted his criticism of the concept of general relativity remained dubious of the preferential status of harmonic coordinates.[66] Fock recognized this criticism in his statement "The above remarks concerning the privileged character of the harmonic system of coordinates should not be understood, in any case as some kind of prohibition of the use of other coordinate systems. Nothing is more alien to our point of view than such an interpretation. . . . The existence of harmonic coordinates, defined apart from a Lorentz transformation, though a fact of primary theoretical and practical importance, does not in any way preclude the use of other, non-harmonic coordinate systems." [67]

Fock believed that many physicists had lost sight of the importance of preferred or privileged systems of coordinates as a result of their exaggeration of the significance of the covariance of equations and, particularly, their belief that this covariance reflects some sort of physical law. For example, using the concepts of tensor analysis, physicists may write equations for space-time intervals without presupposing any coordinate system.[68] Such equations are extremely convenient since they permit an enormous economy in the mathematical description of space-time. However, said Fock, the significance of such covariant expressions of physical facts is *not* that all coordinate systems (in nature) are truly equal. An indication of the essential insignificance (from a physical viewpoint, Fock always emphasized) of covariance is the fact that practically any equation can be stated in covariant form if sufficient auxiliary functions are introduced.[69] In the covariant expression of infinitesimal space-time intervals the auxiliary function that is introduced is the coefficient $g_{\mu\nu}$, a tensor. The important fact is that this introduced function $g_{\mu\nu}$ is the only function used to describe the gravitational field. But one should see, said Fock, that what has happened in the process is that an appropriate theory of gravitation has been introduced into a theory that is then inappropriately dubbed a general theory of relativity, as if the results were a further expression of the relativity of motion. As Fock expressed it,

When Einstein created his theory of gravitation, he put forward the term "general relativity," which confused everything. This term was adopted in the sense of "general covariance," i.e., in the sense of the covariance of equations with respect to arbitrary transformations of coordinates accompanied by transformations of $g_{\mu\nu}$. But we have seen that this kind of covariance has nothing to do with "relativity as such." At the same time the latter received the name "special" relativity, which purports to indicate that it is a special case of "general" relativity. . . . The term "general" relativity or "the general principle of relativity" is also used, beginning with Einstein, in the sense of "theory of gravitation." Einstein's fundamental paper on the theory of gravitation (1916) is already entitled "Foundations of the General Theory of Relativity." This confuses the issue still further. In the theory of gravitation, space is assumed non-uniform whereas relativity relates to uniformity so that it appears that in the general theory of relativity there is no relativity.[70]

No agreement exists among prominent world physicists on Fock's criticisms of "general relativity." Fock's interpretation has been challenged both in the Soviet Union and abroad. Yet it continues to command respect and attention as a defensible and interesting point of view. In 1964 Fock presented a paper in Florence, Italy, in which he summarized the analysis presented above for an audience of distinguished scholars. In the following discussion certain aspects of Fock's scheme attracted considerable praise, while others proved more controversial. Hermann Bondi, professor of applied mathematics at King's College, University of London, commented, "I agree with Fock that the general principle of relativity is empty. We know of course that there is no *physical* equivalence between inertial and accelerate observers. . . . I feel confident that given any laws, mathematicians could find a way of writing these laws in a *mathematically* equivalent way." [71] Professor André Lichnerowicz of the Collège de France also supported Fock's criticism of the principle of equivalence, and Stanley Deser of Brandeis University commented that Fock's analysis of the concept of covariance had been very helpful in understanding general relativity more fully.[72] But a number of the members of the audience, including both Lichnerowicz and Deser, were less enthusiastic about Fock's use of harmonic coordinates. A considerable number of theoretical physicists have not believed harmonic coordinates to be so appropriate for a description of the gravitational field as Fock has indicated.

In the Soviet Union at the present time there are many different shadings of interpretation of general relativity, of which Fock's is only one, although probably still the most prominent one. P. S. Dyshlevyi wrote in 1969 that Soviet philosophers and scientists could be roughly classified into three different groups in terms of their attitudes toward general relativity.[73] The first group was those scholars who considered GTR as expressed by Einstein to be essentially complete. They would introduce modifications here and there, but on the whole they fully accepted the Einsteinian interpretation of relativity, believing that it presented neither scientific nor philosophical problems of a serious nature. They considered Fock's criticism (Fock is not in this group) of general relativity too unorthodox in both its terminology and its conceptions. These scholars accepted the use of the term "general theory of relativity" (in contrast to Fock), and they had little criticism of Einstein's use of the principle of equivalence. They were generally skeptical of efforts to add a "third stage of relativity," such as a "unified field theory." These scientists were willing to accept the present edifice of relativity theory with its two stories of STR and GTR. Among the Soviet scholars whom Dyshlevyi identified as belonging to this group were: in the past, M. Bronshtein, Ia. Frenkel', A. Fridman (Friedmann), V. Frederiks; at the present time, A. F. Bogorodskii, V. L. Ginzburg, Ia. B. Zel'dovich, Kh. P. Keres, A. S. Kompaneets, and M. F. Shirokov.

The second group of interpreters of general relativity in the Soviet Union was the one to which Fock belonged and of which he was the best-known spokesman. The chief characteristic of this second group was its opinion that the foundations of general relativity needed to be given a thorough re-examination in order to make corrections in the conceptual structure of the theory as presented by Einstein.[74] I have already considered the views of this group in detail in the discussion of Fock. Other scientists whom Dyshlevyi placed in this group were A. Z. Petrov and N. V. Mitskevich.

The third group of Soviet interpreters of relativity theory hoped to achieve a new formulation of general relativity by uniting quantum and relativistic physics in a new quantum theory of gravitation. They approached gravitation from the standpoint of the field theory that had been worked out for the fields in physics other than the gravitational field. Dyshlevyi named as members of this group D. D. Ivanenko, O. S. Ivanitskaia, M. M. Mirianashvili, V. S. Kiriia, A. B. Kereselidze, A. E. Levashev, and V. I. Rodichev.

Of these groups, the second is the only one at the present time calling for specific alterations in the interpretations of general relativity. The first group would accept general relativity very nearly in its present form, including those customary philosophic interpretations of it that can be accommodated within the tradition of materialism. The third group proposes a program for the future that, if successful, would no doubt have philosophic implications, but that has so far been discussed only in elementary forms. The second group, however, continues to advance the criticisms initially voiced by Fock, and it is this group and its commentators that have produced the larger part of the philosophical literature on relativity theory.

Indeed, many of the members of groups one and three have avoided philosophical questions of science. With the exceptions of M. F. Shirokov (group one) and D. D. Ivanenko (group three), their names have only rarely appeared in bibliographies of articles and books on dialectical materialism.[75] Of these two men, the one whose ideas most directly bear on the discussion of general relativity is M. F. Shirokov.

Shirokov has supported the validity of the term "general relativity" against the criticisms of the second group, and he has done so—in contrast to some of his colleagues—within an explicit dialectical materialist framework. He maintained that Einstein's interpretation of relativity fully accords with dialectical materialism and is, in fact, a further confirmation of it. In 1964 he wrote of general relativity, "This theory . . . is a great achievement in the materialist understanding of nature, contrary to the numerous idealistic (especially in the spirit of Machism) interpretations of it by several foreign authors." [76] Shirokov thought that Fock and Aleksandrov underrated GTR and greatly simplified its meaning by reducing it to a theory of gravitation. He acknowledged the importance of their work, however, in "confirming" that relativity theory reflects the "objectivity and reality" of nature. Their error was failing to see that when they denied GTR, they also denied the objective reality of fields of inertial forces.[77] Shirokov, like Fock, however, clung to the idea of a preferred reference frame within GTR, relying on his particular view of the concept of "center of inertia." In this sense he agreed with Fock in giving grounds for preferring the Copernican view to the Ptolemaic, but while Fock based his argument on his harmonic coordinates of space uniform at infinity, Shirokov pointed out that the sun represents an approximate center of inertia of the solar system.[78]

One question connected with general relativity on which there is great disagreement among scientists and philosophers in the Soviet Union is, What is gravitation? Answers in many different subtle shades have been given.[79] The members of the first group have frequently equated the gravitational field with curved space-time. Some of their critics, however, have said that this answer comes close to draining gravitation of physical or material content, to identifying nature with geometry. M. F. Shirokov, a member of the first group, has therefore stated his position very carefully. According to him gravitation "reflects the geometric properties of space-time"; the gravitational field does not possess mass or energy; gravitation is not, therefore, matter itself, but is, instead, "a form of existence of matter." D. D. Ivanenko defined gravitation a little differently; it was to him a curvature of space-time caused by matter and the gravitational field itself. Thus, gravitation was to Ivanenko not quite the same as space-time, but instead an independent aspect of the material world. A. Z. Petrov, a member of the second group, described the gravitational field as a "specific form of moving matter." N. V. Mitskevich shared this view and warned against reducing gravitation to geometry. In his opinion, geometry is a manifestation of the gravitational field rather than the reverse. Thus, we find a considerable diversity of views among Soviet scholars. The attempt to define "gravitation" is in the Soviet Union today a subject of discussion in a way very similar to attempts to define "information" and "consciousness" in other fields. The latter terms will be topics of discussion in subsequent chapters.

The present discussions of general relativity in the Soviet Union are in many ways similar to discussions elsewhere, even if terminological distinctions remain. The re-examinations of general relativity theory by such non-Soviet scholars as J. Wheeler, R. H. Dicke, J. L. Anderson, and J. L. Synge have attracted much attention in the Soviet Union. The dimensions of debate in the Soviet Union, including the philosophical dimensions, are fully sufficient for consideration of all such views. Indeed, in the person of such scientists as Fock the Soviet scholars are making their own important contributions to the discussions of the broader significance of relativity theory and the phenomenon of gravitation.

V   Cosmology and Cosmogony

The various answers to the basic questions that cosmology and cosmogony ask about the origin and structure of the universe have always contained implications for philosophic and religious systems. Usually the connections between empirical investigations of the universe on the one hand and metaphysical systems on the other have been much less direct than the defenders or opponents of the systems have supposed, but intense controversies have arisen nonetheless. It is quite difficult to imagine, for example, any scientific evidence that could "prove" or "disprove" the position of a person asserting the existence of God, given at least a moderate degree of sophistication in that person's arguments. Similarly, it would be difficult to imagine a confirmation or refutation of the position of a knowledgeable materialist asserting an entirely naturalistic origin and evolution of the cosmos. Nonetheless, certain kinds of evidence have, with time, significantly affected the plausibility of versions of these differing arguments, and they have, in turn, evolved in response to the challenges thrown up to them. Here I would like to examine the responses of certain Soviet astronomers and philosophers—those who have actively defended the position of dialectical materialism—to astronomical evidence of recent decades. This attempt will require a very brief review of some of the most important findings of astronomers and of several resulting hypotheses.

Although modern cosmological theories are frequently discussed in popular articles as if there were only two competing models—"big bang" and "steady state"—there have been proposed in the last half-century a great multitude of models of which more than a dozen have achieved sufficient currency among cosmologists to have common designations. All of the architects of the models have been forced to take into consideration several fundamental theoretical developments and astronomical findings that are totally new to this century. The most important theoretical innovation was the general theory of relativity as advanced by Einstein in 1916. Contrary to the Newtonian concept of an infinite universe situated in Euclidean space, Einstein's theory proposed the determination of the metric of a space-time continuum by the matter existing in the universe. Rather than yielding a unique space-time, however, Einstein's equations opened the door to several types with curvatures of different signs: positive (Riemannian geometry), zero (Euclidean geometry), or negative (Lobachevskiian geometry). The choice among the three types would be made on the basis of undetermined characteristics of matter within the universe, specifically, its average density. Determining the average density of matter within the whole universe was obviously an impossibility, since at any point in time man can see only so far into the universe. Furthermore, in this century many basic measurements affecting density calculations, such as the distances to stars and nebulae, have been highly questionable; they have, in fact, been drastically revised on several occasions. Therefore, the determination of the average density of matter has been a very difficult task. Nonetheless, it should be noted that the evidence has pointed toward a density in the range that would yield a positive curvature of space-time; consequently, by far the most popular relativistic models have possessed "closed" Riemannian space-time. This conclusion has been based, of course, only on the evidence that astronomers can gather from the parts of the universe accessible to their observations; it is possible that evidence from some portion of the universe so far unobserved would result in a change in the proposed space-time curvature. Astronomers have continued to work in the face of this possibility by simply assuming that the universe is homogeneous throughout; according to this assumption, known as the cosmological principle, the universe everywhere is similar in its gross outlines to that portion on which we have gathered evidence. In answer to the expected question about the arbitrariness of this assumption, cosmologists have replied, by and large, that it would be even more arbitrary to assume differences, or inhomogeneities, for which there is currently no evidence. Thus, the most important theoretical innovation of the century in cosmology—general relativity theory—has resulted in the currency of a number of closed ("finite" in some definitions) models of space-time.

The most important astronomical finding affecting cosmology so far in this century was the shift of the spectral lines of extragalactic nebulae toward the red end of the spectrum. This phenomenon was first observed by V. M. Slipher in 1912, but it was most thoroughly investigated by Edwin Hubble in the 1920's. Hubble and M. Humason in 1928 formulated a relationship of the red-shift to distance that has become known as Hubble's law. This well-known but sometimes misunderstood relationship says that the red-shift of a particular nebula is directly proportional to the distance of the nebula from the observer. When interpreted in terms of the Doppler effect, the red-shift yields a large recessional velocity of the distant nebulae; in some cases this velocity is a significant fraction of the velocity of light. Hubble was cautious in applying to his law the interpretation provided by the Doppler effect, but if such application is made, the law can be understood as saying that the recessional velocity of a nebula is directly proportional to its distance from us. This interpretation has gained increasing acceptance among astronomers and cosmologists throughout the world. It is the basis of the various expanding cosmological models. When an expanding model is accompanied by the hypothesis of an original explosion, a moment when the expansion began to occur, the model becomes a "big-bang" type.

Immediately after World War II the steady-state model was developed by Hermann Bondi, Thomas Gold, and Fred Hoyle. Originally created as a result of conflict between the time scale of the galaxy and that of the universe according to big-bang models, the steady-state theory soon acquired a rationale of its own that continued to be persuasive to some cosmologists after the original conflict was eased. While all relativistic models were based on the cosmological principle (the universe is the same in every direction), the steady-state model was based on what its advocates called the perfect cosmological principle (the universe is the same not only in every direction but at every moment in time). It incorporated the red-shift data by assuming that all galaxies recede from each other in accordance with the Hubble relationship, but that a steady state of the distribution of matter is retained despite this recession as a result of the constant creation of matter in the places of the old galaxies that have moved away. This violation of the law of the conservation of matter had not been detected by scientists, said the steady-state advocates, because it occurs at an extremely low rate, below the level of man's experimental error (as Bondi phrased it, "the steady state theory predicts the creation of only one hydrogen atom in a space the size of an ordinary living-room once every few million years").[1]

The steady-state model possessed the considerable advantage of being infinite in time; there was no "singular state" when all the matter of the universe was compressed into one compact mass, no "birth" of the universe, as some cosmologists referred to this moment. It possessed the serious disadvantage of violating one of the most fundamental laws of physics: the conservation of matter and energy (because of its hypothesis of the creation of matter). Thus, it became the center of a considerable controversy in many countries. Fortunately from the standpoint of the resolution of the debate, it was a testable hypothesis. Its assumption that the universe was always the same in terms of time could be tested against observations of far-distant galaxies, which are "distant in time"; its assumption that all elements could be synthesized at the present time (the heavy ones presented particular problems) was also open to inquiry; and its rejection of a primary cataclysm could be tested by a search for evidence of that cataclysm. These efforts, and others, have been made in recent years; the over-all result, I believe it would be fair to say, has been a retreat by the advocates of the steady-state theory, although Hoyle and his pupil J. V. Narlikar after 1961 developed a compromise version that remained a candidate.

In order not to spend more time on a description of cosmological models, I will introduce a schematic representation to which I will refer in subsequent discussion of Soviet views. Because of certain degrees of overlap it is quite difficult to reduce the models to distinct categories, but I have attempted to do so, drawing upon the discussions of William Bonnor, G. C. McVittie, Otto Struve, Gerard de Vaucouleurs, and others.[2] An indication of the complexity of the problem can be gained by noting that this simplified categorization includes four variants of the big-bang theory (IIa, IIb, IIc, IIIc) and three variants of the steady-state theory (under VI), not to speak of others.[3]

I. *Static*
   a. Einstein equations of 1915
   b. Einstein (with cosmological term [λ]), 1917

II. *Expanding models without cosmological term* (λ) [4]
   a. Einstein–de Sitter, 1932 ⎫
   b. Cycloidal   ⎬ based on A. A. Friedmann's work, 1922
   c. Hyperbolic   ⎭
   d. Oscillating without singular state

III. *Expanding models with cosmological term* (λ)
   a. Einstein (as modified by Eddington, 1930)
   b. de Sitter, 1917
   c. Eddington (based on Ib)
   d. Lemaître, after 1927
   e. Infinite contraction–infinite expansion

IV. *Expanding and rotating*
   a. O. Heckmann *et al.*, based in part on Gödel, 1949

V. *Kinematic relativity*
   a. Milne, 1935

VI. *Steady State*
   a. Bondi–Gold–Hoyle, 1948 (modification of IIIb)
   b. Electric universe, Lyttleton–Bondi, 1960
   c. Hoyle–Narlikar, 1963

Many discussions of Soviet cosmology that have appeared outside the Soviet Union have concentrated on the most elementary and dogmatic of the sources. Before Stalin's death there was a considerable body of Soviet literature with an extremely simple message: Any interpretation of the universe that could be turned into an argument, however strained, for divine interference was automatically condemned.[5] This condemnation was usually issued without much consideration of the scientific merits of the interpretation or of the possibility that its scientific core could be maintained without the particular theological overtones placed upon it by certain European and American writers. Thus, many prominent non-Soviet astronomers and physicists, such as James Jeans, Arthur Eddington, G. E. Lemaître, F. Hoyle, H. Bondi, T. Gold, O. Struve, C. F. von Weizsäcker, and Bart Bok, were accused, at one time or another, of "idealism," "mysticism," or "popism." It is easy to ridicule these Soviet propaganda pieces, and many of them deserve ridicule, but it should be recognized that a number of the above-named authors—but by no means all—did indeed introduce religious elements into their astronomical writings. In some cases, such as James Jeans's discussions of the "finger of God" that started the planets in their orbits, the references were little more than the results of a colorful style; in other cases, such as Abbé Lemaître's frequent references to the "birth of the universe" just before the beginning of its expansion, the statements probably did have a connection with religious belief.[6] And in still other instances, the statements were simply too strong to be brushed off; such a remark was E. T. Whittaker's "It is simpler to postulate a creation *ex nihilo,* an operation of the Divine Will to constitute Nature from nothingness." [7] Not only Soviet ideologists were disturbed by some of these references; as the British astronomer William Bonnor wrote:

. . . one can well understand the enthusiasm with which some theologians accepted the idea that the universe was created 10,000 million years ago. Here was the vacancy for God which they had been seeking. Archbishop Ussher had been a few years wrong with the date, but he had the right idea when he said that God created the world in 4004 B.C.

Unfortunately, some cosmologists have been sympathetic to this attitude. This seems to me quite reprehensible for the following reason. It is the business of science to offer rational explanations for the events in the real world, and any scientist who calls on God to explain something is falling down on his job. This applies as much to the start of the expansion as to any other event. If the explanation is not forthcoming at once, the scientist must suspend judgment: but if he is worth his salt he will always maintain that a rational explanation will eventually be found. . . . There has been a curious reluctance on the part of the cosmologists I mentioned to do this, and they have preferred to identify the singularity in the equations with God. Now, I argue that scientifically this is unexcusable. . . .[8]

If the use of religious metaphors and even the intentional introduction of religious elements occurred in the writings of some non-Soviet cosmologists, the similar fault of distorting their arguments in the name of militant atheism was even more common in the Soviet Union before the late fifties. When ideologists with little knowledge of mathematics were the authors of these articles, the results were frequently scientifically erroneous. One of the most common arguments was that dialectical materialism required an infinite universe. Historically speaking, the association of spatial infinity with modern science, of course, is a very close one; this association is revealed, for example, in the very title of Koyré's "From the Closed World to the Infinite Universe." [9] Religion was a genuine obstacle at certain moments to the theory of an infinite universe (although it should not be forgotten that to Newton an infinite universe with absolute space and time implied God rather than denied him).[10] Because of this association, many recent opponents of religion, in the Soviet Union and abroad, found the relativistic closed models of the universe of this century uncongenial. An important difference, however, is that the closed models of the twentieth century were four-dimensional, while the finite model of medieval scholastic thought was three-dimensional, bounded by the fixed stars. (The "crystalline spheres" were not always taken literally by scholastic thinkers, but the idea of an enclosed space was.) The basis for linking a finite universe to religion was that of historic association—not an entirely trustworthy foundation. One is tempted to say that there was no more logical necessity for dialectical materialists to demand an infinite universe than there was for medieval theologians to demand a finite one (or, for that matter, for many contemporary non-Soviet astronomers to assume a beginning in time). Accordingly, Soviet cosmologists had every reason to be cautious about criticizing finite models on extrascientific grounds. While a few astronomers and mathematicians were very aware of these reasons for caution, the general ideological antipathy toward finite models was very strong. As late as 1955, a Soviet author commented in an astronomy journal:

> The Marxist-Leninist doctrine of the infinitude of the universe is the *fundamental axiom* at the basis of Soviet cosmology. . . . The denial or abandonment of this thesis . . . leads inevitably to idealism and fideism, i.e., in effect, to the negation of cosmology, and therefore has nothing in common with science.[11]

The question of the "birth" of the universe is more controversial than its configuration, as indicated above in Bonnor's remarks. While everyone should tread extremely lightly where questions of cosmology are concerned, there are a number of serious reasons for not accepting the concept of a beginning of all time unless absolutely necessary. It is, furthermore, quite difficult to think of circumstances that would make such a concept absolutely necessary. The Soviet critics of "big-bang" theories were usually more aware, at least in their writings, of these considerations than their non-Soviet counterparts; they correctly saw that a hypothesis of the birth of the entire universe (and not of merely one of its parts) was linked to a religious view. These Soviet writers frequently squandered their philosophic advantage on this issue, however, by extending their arguments far beyond what was necessary in order to avoid commitment to a concept of the absolute beginning of the universe. Thus, G. A. Kursanov in an article in 1950 questioned those relativistic models that begin with what astronomers call a singular state—that is, a moment when all matter is compressed into a very compact, hot mass of neutrons. According to a version of this model known as the $\alpha \beta \gamma$ theory because its best-known description was written by Alpher, Bethe, and Gamow, all of the heavy elements in the universe were formed in the first half-hour or so after the beginning of expansion.[12]

There are many serious questions that can be asked of this theory, and Kursanov asked some of them: "What led to the formation of this compact mass of neutrons?" "What was the reason for its sudden, unexpected, and terrible explosion?" But rather than contenting himself with asking the questions that astronomers everywhere were posing, Kursanov concluded that

This "theory" yields no articulate answer of any sort to this question and can yield none; it is the same old story in a new form of preaching the interference of incomprehensible, secret forces in the fate of the world, of preaching the same mysticism . . . as the old "pulsating" and similar theories.[13]

Kursanov's criticism here is so broad that it is rather difficult to discuss accurately. To some authors, a pulsating universe, which he also condemned, is a preferable alternative to a big-bang model, with all its difficulties about an "origin" of time. The opposition to incomprehensible or mystical forces is clear enough, and understandable, but some astronomers who support big-bang or pulsating universes are equally opposed to such forces.

The most interesting exploration of some of these issues can be found in the writings of several Soviet scientists of recognized distinction; they will be discussed in the following sections.

O. Iu. Schmidt

One of the leading Soviet writers on planetary cosmogony was O. Iu. Schmidt (1891–1956). Schmidt was originally trained as a mathematician and ultimately became a leader of the Moscow school of algebra; his great popularity in the U.S.S.R. derived largely, however, from his exploits as a polar explorer. He became a very famous man, a hero to a whole generation of Soviet citizens. A member of the Communist Party since 1918, Schmidt held a series of important administrative posts, including director of the State Publishing House, editor of the *Great Soviet Encyclopedia,* and member of the Central Executive Committee of the U.S.S.R. After 1935 he was a full member of the Academy of Sciences. No doubt his most famous exploit was his command of the ship *Cheliuskin* in 1933 and 1934 in an attempt to repeat his complete transit of the Northern Sea Route in 1932 (the first complete transit in a single season). During the 1933–34 trip, Schmidt and his men were trapped in the ice of the Arctic Ocean for months; eventually they had to evacuate their sinking ship onto the ice many kilometers from shore. A spectacular rescue followed, and Schmidt became an international hero.[14]

Schmidt gave lectures in the twenties and thirties on the history and philosophy of science; in his scientific work he spoke proudly of the importance of Marxist philosophy. It is said (and if true, it is matter for congratulation) that when he and his crew were stranded on the Arctic ice, he organized discussions of dialectical materialism in order to cause the men to forget about their plight.[15]

Schmidt is best known to cosmogonists for his theory of the origin of the earth and the planets, published in the form of four lectures in 1949.[16] Since he restricted himself to the solar system, Schmidt did not encounter any of the large-scale problems of theorists of the universe, such as relativity or red-shift, but he nonetheless saw his scheme in terms of a conflict of world views. In his first lecture he wrote: "The history of cosmogony has meaning and is instructive when it is seen as a struggle between idealism and materialism that never ceases at any stage in history." [17] And as we later will explain, Schmidt maintained that his theory of the sun's capture of a gas-dust cloud was supported by a dialectical concept.

Schmidt's views on cosmogony began with a recognition of the continuing importance of the nebular hypotheses of Kant and Laplace. According to these well-known theories (which differ in certain respects), the sun and the planets arose from the gradual condensation of a diffuse mass of material into discrete bodies. Although Kant's and Laplace's hypotheses had enjoyed great popularity in the nineteenth century, by the early twentieth they were seriously challenged on the grounds that they could not account for angular momentum. One of the very odd characteristics of the solar system is the fact that the major planets, which have less than 1/700 of the total mass of the system, nonetheless possess ninety-eight per cent of the angular momentum. On the other hand, the sun, with almost all the mass, has only two per cent of the angular momentum. The consequent dilemma of astronomers was described in 1935 by H. N. Russell: "No one has ever suggested a way in which almost the whole of the angular momentum could have got into so insignificant a fraction of the mass of an isolated system." [18]

After 1900, various kinds of "tidal" theories were developed to try to explain this phenomenon. The essence of the tidal theories was the hypothesis that the sun had been approached by another star so closely (perhaps even a grazing collision occurred) that solar material was strung out in space. This material later formed the planets. In the Chamberlin and Moulton version, the matter was ejected from opposite sides of both the sun and the star in the form of violent tides; in the variant developed by James Jeans and H. Jeffreys, a cigar-shaped stream of matter was strung out between the star and the sun. The cigar shape (thicker in the middle) would account for the large size of the planets Jupiter and Saturn.

Schmidt believed that the popularity of Jeans's theory of planetary cosmogony in the twenties and thirties was connected with social factors. He commented: "The Jeans hypothesis lasted longer than any of the other 20th century hypotheses. The reason for its popularity was not its scientific value (it had none) and not the undoubted talents of its author but because it was the one most acceptable to the idealist, religious philosophy dominating bourgeois society." [19] The connection between Jeans's explanation of the creation of the planets and bourgeois values was, in Schmidt's mind, the emphasis upon the rarity of the events involved and the consequent miraculous aura of the universe, which Jeans exploited. For a sun and a star to approach each other closely enough for the events described by Jeans and other tidalists to occur would be an exceedingly rare phenomenon. It is obvious that scientists would prefer not to rely upon extremely rare events to explain nature; if the rarity of the event approaches uniqueness, the event tends to pass out of the realm of phenomena explainable by scientific laws, which depend upon repetition. Of course, the grazing of two stars would not be unique, given enough time, but to say only that the creation of the earth was very rare, not unique, would cause some astronomers discomfort.[20] These were the years in which the "age" of the universe was being placed by many astronomers at only a few billion years; therefore, planetary systems would indeed be rare. The issue here is what astronomers call "embarrassment of privilege." If the planetary system is very special, then its inhabitant man is also. Ever since the discrediting of the Ptolemaic system, any variety of anthropocentrism has been considered suspect by most scientists. Schmidt looked upon Jeans's theory as, at least in part, a careless, perhaps even intentional reversion of that tradition.

Schmidt believed that in order to explain the origin of the planetary system it was necessary to disregard the tidalist theories and to build on the inadequate but nonetheless promising nebular hypotheses of Kant and Laplace. The central idea of these systems—the formation of planets from diffuse matter—seemed to him more believable than near-collisions between stars.[21] He postulated that the sun in its orbit had passed through a cloud of dust, gas, and other matter. This cloud possessed a momentum of its own. Out of the interaction of the different momenta Schmidt believed that the peculiar distribution found in the solar system could have evolved. As he wrote:

If the Sun, passing through a cloud, or near it, could "capture" part of the material and take it with it, the Sun would be surrounded by a cloud out of which the planets could later be formed. If the cloud originated in this way there is no further difficulty with the distribution of the angular momentum. This momentum would result from a redistribution of the angular momentum of the Galaxy; part of the angular momentum possessed originally by the cloud in respect of the passing Sun would be retained by the part of the cloud captured by the Sun.[22]

So far as philosophic considerations are concerned, the superiority that Schmidt claimed for his theory, at least initially, was the greater credibility of the events involved as a result of their greater probability. Interestingly enough, in the passage immediately following the one quoted above, Schmidt did defend his system on philosophic grounds, but not on the issue of the rareness of events. Perhaps he recognized that the events he described would also be considered very unlikely by many astronomers, but that can only be surmised. The specific philosophic grounds that he chose concerned the dialectic concept of the interconnection of all phenomena, which has already been mentioned in the debate over quantum mechanics (see p. 85). Schmidt continued:

For our explanation of the origin of the solar system we introduce the matter and the forces of the Galaxy. Is this correct? Would it not be more correct to explain the origin of the solar system by the development of the internal forces of the system itself?

The concept of the general interconnection of all phenomena is one of the basic dialectic concepts and is well enough known to all of us. The problem of the relationship existing between the external and internal is solved concretely by materialist dialectics where everything associated with the given phenomena is taken into consideration. . . . It is this circumstance that makes the "capture" hypothesis so tempting despite the fact that there are some difficulties connected with it which we shall discuss later.[23]

This reliance on the interconnection of phenomena for support of a particular thesis in planetary cosmogony was a much weaker argument than Schmidt's original critique of Jeans's theory on the basis of its improbability. Whether a scientist should study a particular realm of activity in isolation is usually a result of a consideration of the influence of the greater outside realm, rather than a simple statement that one should or should not consider it. It is well understood, for example, that every problem concerning the influence of gravity on any specific body in the universe is in actuality a "n-body" problem, and inherently impossible of solution. The scientist, however, decides to what degree he can disregard other bodies. Similarly, the support for Schmidt's above argument that most scientists would consider telling is not that he is willing to consider a larger realm, but that such consideration in this particular instance results in more plausible explanations of the origin of the planetary system. Whether the last half of the preceding sentence is true or not has been the subject of much debate.

Before returning to the essential problem of probability of events, it is necessary to observe that Schmidt's system as described is still incomplete. A mathematician himself, he clearly realized that the sun could not capture a dust-gas cloud in the way described so far. In order for capture to occur, the resultant motion would have to be elliptical; that is, orbits would have to be formed around the sun. In the case of two isolated bodies, however, the resultant motion would be hyperbolic and capture could not occur. In order to achieve the necessary capture, Schmidt introduced the hypothesis that three bodies were involved; in other words, one could imagine a scenario in which the sun enters a dust-gas cloud at the same time that a star is also entering the cloud. Even then the question of the possibility of capture was problematic; it was an important feature of the famous "three-body problem," which has occupied mathematicians for several centuries. It has been demonstrated that no general algebraic solution of the problem is possible, but in specific cases when initial conditions are known, numerical solutions are possible, though they were extremely laborious before the widespread use of computers. In 1947 Schmidt carried out such a numerical solution that convinced him that capture was possible in a three-body situation.[24] This conclusion was supported by G. F. Khil'mi.[25]

There remained the problem of the probability of the events, presumably one of the main advantages of Schmidt's system, philosophically speaking, over that of Jeans. Yet most observers would say that Schmidt's scheme demanded very unlikely occurrences as well. Schmidt pointed out, however, that since it was possible for capture to occur in a three-body situation, so was it possible in a proximate arrangement of any number greater than two, given certain ranges of distances and velocities. Furthermore, his supporters introduced other variants of capture, including the influence of collisions and light pressure.[26] Nonetheless, the central question of the rareness of the birth scene of planetary systems remained a major problem to Schmidt; as he commented:

The inevitable question arises—how often can capture take place, in other words, what is the probability of capture? The passage of a star through a cloud is not a rare phenomenon: it has been estimated that the Sun passes through clouds on about a thirtieth of its way through the Galaxy. For capture in its classic form, however, the simultaneous passage of a second star in close proximity is essential (for other forms of capture this is not essential). The passage of two stars in close proximity is a much rarer phenomenon. It must not be thought that they have to be very close for capture to be effected. According to the Jeans hypothesis, for example, the passage must be quite close to the Sun (a distance of a few solar radii). For the acquisition of dust and gas materials such as the planets are made of, distances 10,000 times greater may be effective and this increases the possibility of stars encountering one another by 10^{12} times. The phenomenon, however, is still a rare one, especially in view of the fact that capture requires certain velocity restrictions.[27]

In the last sentence of the above passage we see the worry that continued to haunt Schmidt. According to his own philosophic beliefs, the system that he had erected was rather awkward, although preferable to the alternatives. The last part of his life was one of interminable illness; confined to his bed with tuberculosis, he struggled to perfect his system. In the last years he turned to the mechanism of capture on the basis of inelastic collisions of particles as the most promising avenue, but the main features of his system remained constant.

It should be noticed that Schmidt was not the only astronomer in the 1940's to turn away from theories that demanded such rare events as stellar collisions to explain the creation of planetary systems. C. F. von Weizsäcker in Germany, W. H. McCrea and Dutch-born G. P. Kuiper in the United States, among others, were carrying out similar searches. Weizsäcker's theory became best known.[28] Frequently termed a theory of "turbulence," Weizsäcker's system postulated a nebula that had formed around the sun at about the same time as the birth of the sun itself. Here already was a clear difference between Weizsäcker's and Schmidt's theories. Weizsäcker believed that systems of convection currents, or vortices, had originated in the nebula; these whirlpools of matter had condensed into planets. The problem of direction of rotation was hopefully solved by thinking of the whirlpools as balls in a ball bearing; the outside of the nebula would rotate in a direction opposite to the inner whirlpools. Stimulating as Weizsäcker's theory was, it ran into many serious problems, a consequence that seems to be inevitable in cosmology and cosmogony. Schmidt was one of the vocal critics of Weizsäcker's theories.

Schmidt's system was the subject of much discussion at the Conference on Problems of the Cosmogony of the Solar System held in Moscow on April 16–19, 1951.[29] This conference occurred when several areas of science in the Soviet Union were under rather heavy ideological pressure; other sessions had been held within the previous three years in biology, medicine, and physiology, and the meeting on resonance chemistry would follow by two months. Schmidt's theory of planetary cosmogony emerged from this conference with a quasi-official status; no doubt his immense popularity and his past assistance to political and economic efforts were important factors in attracting nonscientific support. Nonetheless, Schmidt's theory and his general approach to astronomy were subjected to considerable criticism.

Schmidt repeated at the conference his opinion that the widespread influence of Jeans's theory must be explained by religious factors. He said that after the "clearly materialistic" theories of Kant and Laplace ran into difficulties, religiously inclined astronomers attempted to salvage a cosmogony that would "not contradict the Biblical legend of the creation of the world too strongly." [30] Such an effort, he said, was that of the French Academician Hervé Faye, "a zealous Catholic." [31] All these efforts were, however, a failure, Schmidt continued, and therefore the

Church was prepared—if not to snatch up such a theory—to silently give it the Church's blessing; even though it did not coincide with the biblical myth, it was nonetheless compatible with the position of Christian religion— with the exceptional character of the earth.[32]

Hence, he thought, the reception by Western society of the theories of Jeans.

Schmidt also considered the hypothesis of Weizsäcker, now the "most modish in the West" and one that had been supported by such prominent scientists as the Indian-born S. Chandrasekhar at the University of Chicago and the Dutch D. Ter Haar. Schmidt applauded Weizsäcker's use of thermodynamics and statistical physics, but criticized "the epistemological framework of the author [that is] found in the very first proposition of his hypothesis." [33] Weizsäcker's system of oppositely rotating vortices was "extremely artificial, clearly far-fetched. Weizsäcker does not even conceal his idealistic world view." [34] Turning to the defense of his own theory, Schmidt observed that "only in the Soviet Union does there exist a detailed cosmogonical theory that presents an explanation of all the basic characteristics of the planetary system." [35] Among the Soviet works that he praised was that of V. G. Fesenkov, even though Fesenkov considered Schmidt's cosmogonical theory "premature."

The disagreement between Fesenkov and Schmidt contained a number of interesting elements, including the professional one of the differences in approach of a research physicist and a mathematician. V. G. Fesenkov (1889–   ) was another internationally known scientist, an astrophysicist and astronomer. Elected to the Academy of Sciences of the U.S.S.R. in the same year as Schmidt, he was the author of many works on cosmogony, both popular and technical, and was known for his research on zodiac light, luminescence, corpuscular photogenesis of stars, and solar cosmogony. He served as chairman of the U.S.S.R. Academy of Sciences Committee on Meteorites and director of the Astrophysical Institute of the Academy of Sciences of Kazakhstan. He was also a member of a number of committees of the International Astronomical Union. In 1969, on his eightieth birthday, he received the Order of Lenin in honor of his contributions to the development of astronomy.

At the time of the 1951 conference Fesenkov was quite critical of Schmidt's work. The principal inadequacy that he saw in his colleague's approach was the "narrowness of his cosmogonical conception." Schmidt, he said, examined only the solar system, and within that, only the planets. The sun itself, the major body by far of the system, was outside his purview. The implication here was that Schmidt fell short in terms of his own criterion of studying the interconnection of all phenomena.

Furthermore, Fesenkov thought that Schmidt was not an astronomer, and in a certain sense not even a natural scientist, if one makes the frequently forgotten distinction between natural science and mathematics. The issue here was not Schmidt's mathematical abilities, which Fesenkov fully appreciated, but the fact that, in his opinion,

Academician Schmidt pays attention only to the deductive part of his constructions; the points of departure of his theory are in no way proved, and even its probability is not shown; and this occurs at a time when modern astrophysics provides sufficient material for the construction of a cosmogonical theory on the basis of observational evidence.[36]

If Schmidt had broadened his conception, Fesenkov continued, he would have seen that V. A. Krat and Fesenkov were correct in viewing planetary cosmogony as a part of stellar cosmogony. Thus, Fesenkov felt that the solution of the problem of the formation of the earth must depend on the problem of star formation, rather than being isolated from it.

The editors of *Problems of Philosophy* made several interesting comments on the April 1951 conference. They noted that the conference had "approved" Schmidt's theory, with some reservations. In an oblique reference to Fesenkov's criticisms they observed that doubts about the possibility of working out an independent theory of planetary cosmogony could lead to "agnosticism and slow down the development of science." They also chided the professional philosophers for not taking a more active role in the conference.[37]

Both Schmidt and Fesenkov occasionally supported their views with philosophic arguments, as, indeed, almost all cosmologists in other countries intermittently have done. Perhaps no other field of science is forced to build so much on so little actual evidence. Nonetheless it may be worthwhile to notice the particular arguments that the Soviet cosmologists employed. Schmidt's keen interest in dialectical materialism has already been noted. In an article that appeared in *Problems of Philosophy* in 1952, Fesenkov also discussed the philosophic assumptions of certain theories of modern astronomy. He said that Soviet astronomers had pointed out that stars are in evolution, as illustrated by their gradual conversion of hydrogen into helium and their loss of mass, and that they are being created at the present time in our galaxy. In terms of relative emphasis, he thought the research of non-Soviet astronomers was discernibly different, and for philosophic reasons. Outside the Soviet Union, he thought, the preoccupation of cosmologists was with the term "creation"; as he described the situation:

A completely different situation exists abroad, where there is no effort to correctly generalize the observational evidence and where factual evidence, spontaneously collected at various observatories, as a rule is not generalized and is not used for the solution of the problems of cosmogony and astrophysics. There the dominating question concerning the formation of the stars is the idea of the so-called red-shift, i.e., the systematic expansion of the whole visible universe, which is then connected with its birth and the process of creation.[38]

Fesenkov seems to have changed his position on several important issues around 1951 or 1952. For over twenty years, for example, he had denied the possibility of the formation of the solar system from diffused matter, but in 1953 he wrote that "no one could doubt" that planets arose from some sort of gas-dust environment. He had also earlier doubted the possibility of constructing a theory of planetary cosmogony as long as stellar cosmogony remained so unclear, but he also modified that view. Whether these changes were in part responses to the ideological pressures of 1951–52 and the popularity of the Schmidt thesis or were the result of scientific reassessments seems quite impossible to determine.[39]

V. A. Ambartsumian

Perhaps no great Soviet scientist has made more outspoken statements in favor of dialectical materialism than the astrophysicist Viktor Amazaspovich Ambartsumian (1908–   ). Ambartsumian studied at Pulkovo Observatory under the Russian astronomer A. A. Belopol'skii, then went on to hold many distinguished positions in Leningrad University, the Armenian Academy of Sciences, and the U.S.S.R. Academy of Sciences. He supervised the building of the famous Biurakan Astrophysical Observatory near Erevan. For his works of fundamental importance on stellar astronomy and cosmogony he was honored several times with governmental prizes.[40] He became one of the best-known abroad of all Soviet scientists. His praise of dialectical materialism has been voiced again and again over the years; these affirmations have come when political controls were rather lax as well as when they were tight. We have every reason to believe that they reflect, at root, his own approach to nature. As an example, in 1959 Ambartsumian declared:

The history of the development of human knowledge, each step forward in science and technology, each new scientific discovery, irrefutably attests to the truth and fruitfulness of dialectical materialism, affirms the correctness of the Marxist-Leninist teaching concerning the knowability of the world, the magnitude and transforming power of the human mind, which is penetrating ever deeper into the secrets of nature. At the same time the achievements of science convincingly demonstrate the complete unsoundness of idealism and agnosticism, and the reactionariness of the religious world view.[41]

Two years after the publication of the article in which this statement appeared, Ambartsumian was elected president of the International Astronomical Union in Berkeley, California. Here was an internationally prominent scientist, an honorary or corresponding member of the scientific societies of most of the nations leading in science, an authority in the field of stellar physics, who declared that dialectical materialism was of assistance to his work. Non-Soviet observers usually brushed aside such comments as ornamentation or the result of Party pressures.

At the same time, Ambartsumian was not afraid to reprimand the

Communist Party ideologues when they obstructed his research. He observed, "When we boldly posed certain questions and when science came upon something that was still unexplained, several philosophers tried to hold us back—as if our scientists had fallen prey to idealism!" [42]

Ambartsumian confined his most important work to problems of stellar cosmogony rather than planetary, galactic, or universal cosmogony. Each of these problems—the description of the formation of stars, planets, galaxies, or the universe—presented its own particular problems. Like V. G. Fesenkov, Ambartsumian believed that stellar cosmogony would provide important clues to the other fields and that in the absence of a plausible theory of stellar cosmogony work in the other areas would have to be based on what he termed an excessive degree of speculation. Nonetheless, Ambartsumian's preferences on the larger-scale cosmogony and cosmology can be rather easily discerned by reading his frequent criticisms (particularly before the sixties) of astronomers in Europe and America. His work on stellar cosmogony also contained philosophical elements of general implication; furthermore, despite his reserve, Ambartsumian admitted that the larger cosmological problems were the final and most important problems of astronomy. From these writings we will see that although Ambartsumian considered the construction of a world system premature, he generally preferred, like many Soviet astronomers, a model that was relativistic, inhomogeneous, expanding, and infinite in time. These preferences will be discussed more fully in the following section. They implied a rejection of steady-state models and either a rejection or serious modification of big-bang models.

In the field of stellar cosmogony Ambartsumian was, in his early years, a frequent critic of those "ardent advocates of idealism of the twenties," James Jeans and Arthur Eddington. This criticism extended in several different directions, but one area concerned an argument over the rate of change in the evolution of stars. As a dialectical materialist, Ambartsumian believed that all nature is constantly in evolution; he was suspicious of attempts to postulate even relatively unchanging entities in nature. Jeans and Eddington believed that the mass of stars diminishes over time primarily as a result of the emission of electromagnetic radiation. According to their calculations many hundreds, perhaps thousands, of billions of years would be required before a noticeable change in mass would occur in the average star as a result of this process. But according to the theory of the universe that Eddington and Jeans supported, which was a form of the big-bang class, the age of the universe was only several billion years. Therefore, following Jeans and Eddington, the mass of the stars had changed insignificantly since the birth of the universe; man saw them essentially as they were at the beginning of time.

The Church in the Middle Ages had, of course, favored a view of the universe that went beyond such relatively static celestial bodies to an absolutely unchanging heavenly sphere. One of the more dramatic achievements of Galileo was to show change and irregularity in celestial bodies. Ambartsumian felt that he was continuing that tradition by maintaining that the stars change much more rapidly than Jeans and Eddington believed, and according to a mechanism of which they were supposedly unaware. Speaking of the theories of the two English astronomers, Ambartsumian commented:

Soviet astrophysics long ago opposed these idealistic views, full of internal contradictions and not corresponding with observational data, with their own materialistic point of view, based on facts. The change of mass in the universe, which existed and will exist infinitely, depends primarily on the direct ejection [vybrasyaniem] of matter.[43]

Ambartsumian said that this phenomenon of the ejection of mass by stars leads relatively quickly to significant changes in the physical state of the stars.[44] His colleagues D. A. Martynov, V. A. Krat, and V. G. Fesenkov had all done work in examining the results of the phenomenon; Fesenkov attempted to follow changes in the speed of rotation of the sun on this basis. Thus, said Ambartsumian, "one of the most important results of the work of Soviet astronomers has been the conclusion that the stars themselves change and also change the interstellar environment that surrounds them." [45]

Just as Ambartsumian believed, contrary to some earlier astronomical views, that the stars are perceptibly changing in mass, so he also believed that stars are constantly being born. His theory of the continuous formation of stars at the present stage in the development of the galaxy is now widely known and usually considered a disproof of the belief that all stars in the galaxy were simultaneously formed.[46] This work was also, in Ambartsumian's mind, a confirmation of dialectical materialism. As we will see later, this issue of the evolution of stars bears certain resemblances to the discussions of geology in the early nineteenth century; like Lyell, Ambartsumian believed that the features displayed by nature should be—if at all possible—explained on the basis of processes presently being witnessed in nature.[47]

The exact details of the early life of stars according to Ambartsumian's scheme are necessarily uncertain, as are most such descriptions. The main outlines can, however, be given. Since his description is based on the Hertzsprung-Russell diagram of the relation between the spectral types of the stars and their luminosities, it is necessary to briefly discuss the diagram.

All stars demonstrate dark-line spectra just as the sun does. The absorption lines of these spectra not only tell us the composition of stars, but also allow us to classify them in different groups, with a gradual gradation between types. The standard types are O, B, A, F, G, M, K, R, N, and S.[48] Because of the relative rarity of types O, R, N, and S, we can limit ourselves to a consideration of the six other types. Stars can also be classified in terms of their absolute luminosities, a luminosity of one being equal to that of the sun.

When the stars are placed on a graph in which the abscissa corresponds to spectral type and the ordinate to luminosity, it turns out that they do not fall upon the diagram in a random fashion, but instead form groups, including a diagonal belt known as the main sequence, which includes the great majority of all stars. The resulting Hertzsprung-Russell (H-R) diagram has the following appearance:

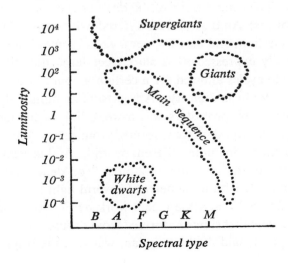

Spectral type

Ambartsumian's description of the life cycle of stars traced their positions on the H-R diagram, and he addressed himself only to the stars in the main sequence. They are formed, he said, in groups ranging from a few dozen to even thousands of members; these loose clusters of young stars are those known to astronomers as "associations." He believed that such associations are being formed out of matter in a prestellar state in our own galaxy at the present time; in other words, the process being described was valid both historically and contemporaneously. These new stars initially appear on the H-R diagram in positions that on the average are above the center line of the main sequence; later they move toward the center line as a result of changes in their states. The cause of these changes, said Ambartsumian, is a "mighty outflow" of matter from the interior of the stars into surrounding space.[49] Thus, young stars lose mass, decline slightly in luminosity, and enter into the main sequence along its entire front. Once they have entered into the main sequence, the state of the stars begins to stabilize; the outflow of matter continues, but at a much slower rate, and those stars that had a significant rotational moment lose it almost entirely. Ambartsumian believed that the time required for the average star to move into the main sequence was several dozen million years. In view of the several (now frequently about fifteen) *billions* of years given as the "age" of our galaxy, such a shift represented an appreciable rate.

According to this scheme, stars are constantly being born, but not "out of nothing," as Ambartsumian said certain non-Soviet astronomers would have it. The exact details of the birth of stars was one of the most difficult problems; Ambartsumian differed with other Soviet astronomers on the question, and his own views varied somewhat with time. His opinion, already expressed, that stars eject large quantities of matter during their life cycles allowed for a certain amount of available matter for additional star formation. But Ambartsumian admitted that the specific character of the protostars was a weak link in his theory. Rather than believing that the stars were formed from diffused matter, as many Soviet astronomers including Fesenkov did, Ambartsumian believed that the protostars are possibly "globules," or dense, dark clouds of spherical form, and with a diameter of several light-months.[50] But Ambartsumian admitted that some recent work by Fesenkov on star chains supported the view that stars are formed from diffused matter.

The main point, said Ambartsumian, was that the origin of the stars was not "from nothing," as he described the theory of the West German physicist P. Jordan.[51] Ambartsumian said that not only did Jordan postulate the spontaneous and causeless appearance of stars, but that this view was incorporated into Lemaître's description of the birth of the universe. Ambartsumian considered the terms "birth" or "age" of the universe, used by Jordan, as careless and inaccurate. Nonetheless, he respected highly much of Jordan's work.

Ambartsumian also disagreed with the theory of Weizsäcker, on the basis of what he considered Weizsäcker's unawareness of the importance of dialectical quantitative-qualitative relationships in nature. The crux of the issue here was that Weizsäcker's explanations of the formation of planets, stars, star associations, and galaxies were essentially the same in that they all relied on the clustering of originally diffused matter as a result of gravitational instabilities. According to Ambartsumian,

Such a standard, unified approach to the solution of the question of the origin of cosmic systems of different structure must be decisively rejected. Galaxies differ from stars and stars from planets not only in the quantity of matter, that is, the mass, but also deeply in the qualitative terms. An enormous quantitative difference in masses leads to deep qualitative differences in the physical nature of these bodies and in their structural features. Therefore, it is not surprising that Weizsäcker stops helplessly in trying to explain the concrete features of different systems and at each step falls into contradiction with the facts.[52]

Rather than describing the birth of stars, Ambartsumian's theory described their lives during the period immediately after birth. And even within that limitation, it accounted only for stars of the main sequence. It either omitted or was unclear on the evolution of such types as white dwarfs and cold giants. Furthermore, the final phases of the stars of the main sequence were not described. Nevertheless, in the field of cosmogony no theorist can claim completeness for his views. Ambartsumian's theories of star formation have justifiably won him the reputation of one of the leading researchers in the field.[53]

The position that Ambartsumian took on the question of the red-shift provides important clues to much of his thinking on cosmology and cosmogony. Contrary to much opinion outside the U.S.S.R., Soviet scientists who at this time interpreted the universe with reference to dialectical materialism did not necessarily question the idea of expansion. In other words, they may have accepted the Doppler-effect interpretation of the red-shift, they may have accepted the conclusion that our observable universe is expanding, but they usually had strong reservations about concepts such as the "creation" or "age" of the universe as a whole. There is an obvious connection between the issues of "expansion" and "age." If astronomers can arrive at a constant speed of expansion, or a variable speed of expansion that can be described mathematically, they can then extrapolate back to a point in time when the universe was concentrated at one infinitesimally small mass; this moment becomes the "birth" of the universe, and the distance in time between that moment and the present, the "age" of the universe. It is such extrapolations that Soviet astronomers such as Ambartsumian warned against.[54] A number of logically legitimate alternatives to "creation theories" of different persuasiveness existed. One could maintain that there was no overwhelming proof of a long-term calculable velocity of expansion. One could maintain that expansion was only one phase in the history of a universe that alternately expands and contracts. Or one could advance the belief that the calculations of the point in time when the present universe began expansion may be perfectly valid without concluding that this same point was the beginning of all time and of all universes.

Ambartsumian tried to steer a middle course between those people who, on the one hand, rejected the interpretation that the red-shift meant that the universe was expanding and those, on the other, who concluded that this expansion permitted an extrapolation in time back to the birth of the entire universe. When he was asked in 1958 if there was any solution for the red-shift other than the Doppler effect, he replied, "No, it is impossible. At any rate, so far no other plausible interpretation has been proposed. Therefore, we must conclude that the system of galaxies and clusters of galaxies surrounding us is expanding. This is one of the most fundamental facts of contemporary science." Similarly, Ambartsumian was asked what his attitude was toward "relativistic cosmology," to which he replied, "Cosmology can only be relativistic." [55] Consequently, any consideration of Ambartsumian's cosmological theories should recognize that the simpler discussions of Soviet views of cosmology stating or implying that Soviet cosmologists rejected both Einsteinian relativity and the concept of an expanding universe were rather gross reductions of a considerably richer debate.

Ambartsumian criticized both the idealistic and mechanistic schools of cosmology. The idealists, he said, played on lack of knowledge and the frustration involved in trying to answer extremely difficult questions by appealing either to epistemological idealism or religion. The mechanists, to the contrary, simplified nature by trying to explain everything on the basis of principles already known, failing to realize the need for new conceptualizations. Ambartsumian then applied this model of the two erroneous schools of cosmology to the problem of the red-shift phenomenon and its implications for the structure of the universe. A number of physicists and astronomers, he said, assumed that the metagalaxy is ideally homogeneous and then further proposed that this type of system fills the entire universe.[56] Taking the red-shift into consideration, they applied Einstein's interpretation of gravitation to the hypothesis of a homogeneous universe; they then concluded that the universe is finite and expanding. (Such a model would be the Einstein–de Sitter one, IIa above.) At this point, said Ambartsumian, the idealist-philosophers and similarly inclined physicists stepped in and drew dramatic conclusions about the creation of the world and about the mysterious force responsible for this creation. This final step was, said Ambartsumian, totally unjustified. It occurred, he said, only because the inadequacy of knowledge about the structure of the metagalaxy left room for "unrestrained extrapolations that led these hypotheses rather far from real science. . . ."[57]

But the opposite extreme, that presented by the mechanists, was no more justified. "Without any kind of experimental evidence they attempted to show that the red-shift is not connected with the Doppler effect but has instead some other cause." Such an attempt was the hypothesis that photons "age" over long periods of time; according to this interpretation, the shift toward the red end of the spectrum would not be a result of a receding velocity, but a change in the nature of light itself. These changes have not previously been witnessed, so the backers of the aging hypothesis proposed, because of insufficient passage of time in laboratory experiments. Ambartsumian believed, as have most other astronomers, that the aging hypothesis was extremely arbitrary; it was an ad hoc explanation developed as a means of avoiding the expansion hypothesis. No evidence for aging of photons outside of the astronomical red-shift existed. Scientists do not resort to explanations that depend upon unknown and unverifiable phenomena if they have at hand another explanation, verified in at least some instances, that will explain equally well the data in hand. Such an existing explanation was the Doppler effect.

Ambartsumian noted that since the original interpretation of the red-shift as a Doppler effect in the second decade of the twentieth century, astronomers had found other data supporting the hypothesis. In 1955, for example, A. E. Lilley and E. F. McCain, of the U.S. Naval Research Laboratory, obtained a red-shift for the line of neutral hydrogen (HI) in radio source Cygnus A. Thus, both optical and radio red-shifts have been observed. As a result of such data Ambartsumian described the efforts of the mechanists and conservatives who would deny the Doppler interpretation as having "suffered a complete failure." [58]

If Ambartsumian's views on the Doppler interpretation were similar to those of the majority of astronomers elsewhere, on what grounds did he criticize many of these same astronomers? The issue on which Ambartsumian concentrated in building his own interpretation, at least until 1958, was that of the homogeneity of the universe. At the time that Hubble formulated the red-shift relation, he assumed that extragalactic nebulae populate space with an approximately constant density. On such an assumption, the velocity-distance relationship was linear out to a distance of 250,000,000 light-years. As more complete and refined data were collected, the question of whether the velocity-distance relationship was linear or curved became pressing. The conclusion has obvious implications for cosmology: A curved relationship could indicate that the expansion of our part of the universe is slowing or accelerating, making extrapolations difficult and calling the "age" of the universe into question. As Ambartsumian commented:

Twenty years ago one could try to justify the hypothesis of the homogeneous density of the universe by saying that in the absence of sufficient evidence on the distribution of distant galaxies the assumption of the homogeneity of the metagalaxy is a natural but very rough approximation; at that time this view found a certain support in Hubble's calculations; but now the situation has radically changed. New evidence concerning the visible spatial distribution of galaxies turns out to be in complete contradiction with the assumption of homogeneity, even in the roughest approximation. It seems to me that if one tried to characterize in two words that understanding of galactic distribution that has begun to form in the last years on the basis of the latest evidence, then the most successful expression would be "extreme inhomogeneity." [59]

In elaborating his view that the universe is inhomogeneous, Ambartsumian acknowledged that the mere fact that clusters and groups of extragalactic nebulae exist is not necessarily evidence that nebulae do not populate space with an approximately constant density. One could maintain that the clusters and groups were only small islands in a broad, general metagalactic field. This general field could be homogeneous, or it could vary according to a continuous gradient. Ambartsumian felt, however, that astronomical research of the early 1950's pointed to the conclusion that the tendency toward the formation of groups and clusters was a basic characteristic of the metagalaxy rather than an exceptional circumstance. Such evidence of inhomogeneity caused a crisis, since the major cosmological models were endangered, a prospect that Ambartsumian found not unpleasant on philosophic grounds. He believed that most of the matter of the metagalaxy was located in clusters and groups. One could still try to save the hypothesis of homogeneity, he observed, by saying that homogeneity was expressed in terms of a mass scale: that is, that not the galaxies themselves were distributed homogeneously in space, but that the centers of their clusters were. But according to Ambartsumian, such an assumption was "completely incorrect." [60] He cited the research begun in 1952 by G. de Vaucouleurs, of the Australian Commonwealth Observatory in Canberra, into the problem of a "local supercluster." This supercluster seemed to consist of a gigantic flattened system composing one of the parts of the metagalaxy. This evidence of inhomogeneity could be supported, thought Ambartsumian, with other recent researches, such as the Palomar Observatory counts of F. Zwicky and his associates, and the Lick Observatory counts of C. D. Shane and C. A. Wirtanen.

The question of the homogeneity or inhomogeneity of the universe remained troublesome to Ambartsumian after this date. The above views were written in the latter part of 1957. During the next year (October 1957–October 1958) new evidence led Ambartsumian to believe that his theory of the inhomogeneity of the universe was somewhat more in question. Before 1957, Ambartsumian had based his conclusion on studies showing that up to dimensions of two hundred million parsecs, the universe seemed inhomogeneous.[61] Thus, at a distance of ninety million parsecs from the earth in the constellation *Coma Berenices* (Berenice's hair) astronomers may observe a cluster of galaxies larger than all other nearer galaxies. If one looks in other directions up to a distance of ninety million parsecs, no other galaxy of such large dimensions can be observed. Therefore, Ambartsumian felt justified in concluding that in order to postulate homogeneity, one would have to speak of volumes with diameters greater than two hundred million parsecs. However, in 1957–58, Ambartsumian received reports from the researches of Zwicky at Palomar that in volumes with diameters in the vicinity of a billion parsecs, it is possible to discern distributions approaching homogeneity. Thus, contrary to Ambartsumian's earlier views, astronomers could speak of an average density of matter at a distance of a billion parsecs approximately equal to the density in the vicinity of our galaxy. Thus, the case for existing cosmological models regained strength.

Nonetheless, Ambartsumian was still willing to defend his characterization of the universe as "extremely inhomogeneous." The universe is inhomogeneous in many more ways than in its distribution of matter, he said; for example, in color of galaxies, electromagnetic radiation, and so on. He called the latter "qualitative inhomogeneity." [62] In his opinion this particular type of inhomogeneity was becoming more and more evident. Thus, he concluded:

Therefore, I venture to add to the observed great inhomogeneity in density and distribution of a number of galaxies the fact of their great qualitative diversity. This diversity is all the more revealed the greater the distances involved.

. . . .

Thus, one of the bases of contemporary simplified models is undermined. A homogeneous universe such as is described in these models does not exist. But there is also a second consideration: All these models apply as a basic postulate the linear dependence of velocity on distance. Unfortunately, we do not have exact enough knowledge of galactic distances to verify this assumption.[63]

Ambartsumian pointed out that only recently astronomers had drastically revised their distance scales; in 1952, Hubble's colleague W. Baade reported in Rome to the International Astronomical Union that the previously accepted distance to the Andromeda nebula was probably too small by a factor of two; in 1958, at Brussels several astronomers proposed enlarging the scale of distance for the more distant galaxies by a factor of five or six over the pre-1952 figure. Each of these revisions obviously affected the distance-velocity relationship. Thus,

Ambartsumian pleaded for caution in computing a universal constant that multiplied times the distance to a certain galaxy supposedly would yield its velocity. And without galactic velocities, extrapolations of the "age" of the universe on the basis of Hubble's law of red-shifts were obviously impossible. Thus, Ambartsumian concluded that the universe is indeed expanding, as established by the red-shift, but that our data is so rough that we can not conclude very much from it. We certainly had not yet established a linear relationship between distance and velocity. In his mind, discussions of the age of the universe were not only philosophically unjustified, but scientifically premature.

Ambartsumian's position was supported at the October 1958 all-Union conference on philosophic problems of natural science by G. I. Naan, an able philosopher of science who has figured earlier in this study. Naan developed the argument in his own way, however. In his opinion, the main issue revolved around the old question of catastrophism. Are we not, he asked, in a position analogous to the one that existed in paleontology in the early nineteenth century, when it seemed impossible to explain the change of forms of the organic world without appealing to a series of natural catastrophes? It is quite true, he continued, that many of the facts of modern astronomy can be incorporated into a catastrophic hypothesis involving the birth of the universe, but there are already sufficient reasons for rejecting such an opinion. Notice, for example, he said, the problem of the origin of the chemical elements. Until recently it seemed almost impossible to explain the cosmic distribution of the same elements without relying upon a scenario that presented the birth of elements several billion years ago in the protostellar state of matter or in the first half-hour of expansion of the universe. Now it seems more clear, said Naan, referring to recent research, that all necessary processes of nucleogenesis could occur in stars of various types. Similarly, said Naan, we should be careful about interpreting the red-shift in such a way that a birth of the universe becomes an essential part of the scheme. He felt this to be true even though, in his opinion, Ambartsumian may have overstated the case in maintaining that the universe displays extreme inhomogeneity.[64]

In the sixties, Ambartsumian wrote more and more on philosophic subjects and on aspects of astronomy with philosophic implications. Although his most recent views were logical outgrowths of his previous ones, his emphases changed somewhat. He very rarely, for example, criticized in recent years non-Soviet astronomers and philosophers for their idealistic viewpoints. It is quite clear that he highly evaluated the work of such men as Jeans, Eddington, Jordan, and Weizsäcker, his early criticisms and continuing reservations notwithstanding. In addition, he spoke in the late sixties much less frequently of the problem of homogeneity or inhomogeneity in the universe. Instead, he concentrated on the problem of the possibility of formulating a single naturalistic picture of the universe and on the problem of astronomical evolution. In recent papers, including one given in 1968 at the Fourteenth International Congress of Philosophy in Vienna, Ambartsumian has refined and developed his views on these subjects.[65]

Ambartsumian has always been opposed to the building of models of the entire universe on the grounds that such efforts are premature. In a lecture in Canberra in 1963 he commented, "The character of these models depends so much on simplifying assumptions that they must be considered far from reality. I personally think that at the current stage it does not even make sense to compare these models with observations in a detailed fashion." [66] In some of his more recent statements he has supported this view with philosophical arguments. The philosophical principle at the bottom of his interpretation is the antireductionist principle of the transition of quantity into quality; that is, he believes that current physical theory is based on such a limited area of observation that the quantitative transition to truly cosmic scales will reveal qualitatively new physical regularities, laws that are as yet unknown. Like the biologist Oparin and many other Marxist scholars, Ambartsumian believed that objective reality consists of different levels of scale, with each level possessing its own physical principles. When astronomers erected models of the universe based on current physical theory, they were passing from a lower scale to a higher one, and their models thus failed to describe physical reality adequately. As he observed,

In our view, the idea that an infinite number of different physical *phenomena* can be explained in terms of a limited number of basic laws and theories is inadequate. Nature is infinite in its diversity and even in the levels of its laws. . . . The higher the level of organization of the system, the greater the significance attached to the relationship and interaction of its elements. Consequently, the system increasingly manifests complex properties, the regularities of which may prove so essential to the system that the elementary laws which individual parts of the system are dependent on begin to play but a subordinate role. In this sense biological systems should be regarded as the outcome of natural synthesis leading to the emergence of new properties. If those properties are compared to the original physico-chemical properties of the elements of those systems, the latter properties prove trivial, and it is simply ridiculous "to reduce" the living organisms to the mere sum total of their constituent elements.[67]

Ambartsumian believed that just as living organisms cannot be reduced to the known principles of physics and chemistry, neither can the universe. He believed that evidence of the inadequacy of contemporary physics to explain the large-scale phenomena of the universe was to be found in the supernovae: "At present there are solid grounds to assume that the processes accounting for those explosions could hardly be explained in terms of existing physical laws. The same is true of the energy sources in quasars, discovered in 1963." [68] Ambartsumian believed that the peculiarities of nature being found in supernovae, quasars, and pulsars called for a revolution in physics, and that for the first time since the days of Copernicus, Brahe, and Kepler physics would be overturned by data from astronomy. But even after that revolution occurred, Ambartsumian implied he would remain skeptical of the universal model-builders, since the universe was "infinite in the levels of its laws." Nature possessed a double-ended infinite diversity; in the microscopic realm, the subatomic particles are infinitely inexhaustible, as Lenin had emphasized, and on the macroscopic level, the universe itself is infinitely inexhaustible.

Another theme in Ambartsumian's recent writings was his belief in the cardinal importance for astronomy of unstable objects. He had, of course, always emphasized unstable systems and unstable stars as keys to the understanding of the universe; such an approach underlay his early study of the evolution of stars, already described. In 1952, he explained his emphasis on unstable objects in terms that were, again, close to the Marxist philosophical description of evolution as arising from conflicting forces:

Why is the study of unstable states so interesting for cosmogony? It is well known that the important mover of every developmental process in nature is contradictions. These contradictions are displayed in an especially clear fashion when the system or body is in an unstable state, when there is a struggle of opposing forces, when the bodies are at turning points in their development. . . . This means that objects in unstable states deserve special attention. In recent years important success has been achieved in the study of unstable systems and stars.[69]

In 1969, reflecting on his own work on the evolution of stars, Ambartsumian commented, "Before the middle of the thirties . . . evolutionary ideas did not play an essential role in astrophysics, although the majority of astrophysicists knew very well that they were dealing with changing, developing objects." [70]

In sum, the strongest element of continuity in Ambartsumian's professional life, from his early emphasis on the birth and evolution of stars to his recent emphasis on such rapidly changing phenomena in the universe as supernovae and quasars, has been the principle of astronomical evolution.

In 1952 the Eighth Congress of the International Astronomical Union was held in Rome; a sizable Soviet delegation attended. A number of the delegates visited the Vatican observatory while in Rome, where Pope Pius XII told them that those astronomers who were penetrating ever more deeply into the universe were at the same time confirming the existence of almighty God. Several of the Soviet delegates reacted angrily to this assertion; if one were to take the comment seriously, such reaction would indeed be appropriate.[71]

One of the views discussed at the Rome congress was that of Pascual Jordan, the German physicist already mentioned who published in 1947 a small book on stellar cosmogony.[72] Jordan's theory was in one respect similar to the steady-state theory; it postulated the creation of matter, but was different in terms of the location of this creation. While the advocates of steady state saw the creation occurring in minute quantities throughout all space, Jordan believed that matter appeared in the exploding stars, or supernovae. B. V. Kukarkin and A. G. Masevich, two of the Soviet delegates, criticized Jordan's views as an example of the type of theory so easily exploited by the Vatican. They also objected to Jordan's descriptions of the "appearance" of the universe "ten seconds after its creation." In the meantime, however, several non-Soviet astronomers, such as H. Bondi and O. Heckmann, had criticized Jordan's view of stellar cosmogony on other grounds. Bondi, one of the developers of the steady-state theory, maintained that Jordan's theory would require the appearance, on the average, of one supernova annually in each galaxy, a prediction not supported by observations.[73] Kukarkin and Masevich, however, had no kinder words for the "thoroughly idealistic and absurd theory of the creation of matter" advanced by Bondi, Hoyle, Lyttleton, and other members of the Cambridge school. Indeed, the two authors presented a rather negative review of most prominent non-Soviet astronomical theories. They spoke positively, however, of the theory of star formation being developed in the Soviet Union by Ambartsumian and others. Soviet work was based, they said, on a recognition of the importance of the dialectical Law of the Unity and Struggle of Opposites, as evidenced in unstable states of stellar bodies:[74]

Each such unstable state shows that as a result of lengthy quantitative changes the system or celestial body is in the stage of transition to a new qualitative state. At the same time, the features of the unstable state characterize not only the nature of this leap, but also its initial and terminal states, i.e., they give a direct indication of the direction of development. Unstable objects in the Universe deserve, consequently, special attention. Practice has shown that this direction of research in cosmogony, which is a direct application of the methodology of dialectical materialism to the study of celestial bodies, is very productive.

In 1954 in the Soviet journal *Nature,* A. S. Arsen'ev praised the article by Kukarkin and Masevich just discussed.[75] Arsen'ev attempted to classify into two categories the "philosophically weak or erroneous" theories circulating outside the Soviet Union. Like Naan, he referred to the nineteenth-century controversy between catastrophists and uniformitarians as an analogy for present debates in astronomy. "Catastrophism" is, of course, a term referring to the stormy debates in the eighteenth and early nineteenth centuries among such natural scientists as Cuvier, Werner, Hutton, and Lyell. The catastrophists relied on unusual and violent events, unlike any to be presently witnessed in the world, that might explain such phenomena as the presence of marine fossils on mountain tops, or the existence of mountains themselves, with their radically tilted strata and dizzying heights.[76] Arsen'ev considered the catastrophists of modern cosmology to be Jeans, Jeffreys, Lyttleton, and Hoyle. To him, Jeans's theory was a form of catastrophism in its reliance on rare stellar conjunctions; Jeffreys's theory was even more catastrophic in demanding an actual grazing collision of the stars.

Lyttleton's suggestion in 1941 that the sun had once been a part of a trinary star system in which two components combined only to fly apart again, forming the planets in the process, was in Arsen'ev's mind a particularly far-fetched form of catastrophism. And the last form was that of Hoyle, whom Arsen'ev considered a symbol of the "progressive degradation" of the Cambridge school. To Arsen'ev, Hoyle's 1946 theory of the association of the sun with the explosion of a supernova, which ejected matter around the sun forming the planets, was yet another variety of catastrophism.[77]

The second category of faulty cosmogonical theories which Arsen'ev saw was that of "mechanistic transfer." These were the theories in which the authors, according to the Soviet critic, selected a certain group of physical processes as those that determined celestial evolution as a whole; these processes were then allegedly used to solve all problems of cosmogony. Among the authors whom Arsen'ev placed in this category were H. Alfvén, Weizsäcker, B. Lindblad, and F. L. Whipple. The Swedish astronomer Lindblad is probably best known for his mathematical studies of galactic rotation; he also, however, discussed a method by which a gas cloud could be converted into very small meteorites by condensation; the meteorites would then lose their kinetic energy through collisions and be attracted together, forming the sun. Those meteorites that lost less kinetic energy would orbit around the center, eventually forming the planets. The American astronomer Whipple investigated the possibility of the formation of the solar system by the condensation of dust globules. The Swedish physicist Alfvén studied electromagnetic forces as possible formative influences in the creation of solar systems. All of these attempts (some of which were rather similar to that of Schmidt) Arsen'ev considered methodologically faulty, however interesting and helpful they might be in certain ways:

Without question all these processes may play a great, even leading role in individual stages of the development of some celestial objects, but to try to explain with the aid of one and the same group of processes the origin and development of objects that are in qualitatively different states of matter would mean to contradict objective reality. In step with qualitative changes in the state of matter there also occur changes in the forms of its movement; other processes and regularities begin to play a predominant role in the evolution of the object.[78]

The question of catastrophism was explored in 1956 in a more philosophic vein by O. S. Gevorkian in a long article summarizing most of the issues in the cosmogonical controversy.[79] To Gevorkian, a philosopher, the key to the conflict between catastrophism and uniformitarianism was the dialectical Law of the Transformation of Quantity into Quality. In the history of the earth, or of the celestial bodies, periods of long and gradual evolution are interrupted by the emergence of new qualitative relationships. The appearance of new qualities (a qualitative leap) can occur, thought Gevorkian, in two different ways: a burst or explosion (*vzryv*), characterized by the simultaneous destruction of the old and the appearance of the new; or a gradual accumulation of the new elements of a new quality, accompanied by the dying out of the elements of the old quality.

Gevorkian believed that dialectical materialism seen from this standpoint could, in a sense, unite the catastrophic and uniformitarian views. Depending upon the particular circumstances, astronomers and geologists could call upon either lengthy gradual processes or sudden bursts to explain the universe. Certainly "catastrophes" have occurred in natural history, Gevorkian noted, if one means by that term such things as great ice ages, collisions of celestial bodies, volcanic eruptions, stellar explosions, and so on. Nonetheless, he believed that reliance upon catastrophes for scientific explanations was dubious; it was a step that should be taken only when it was particularly well supported by existing scientific evidence. In the absence of such evidence, it was far better to look for the causes of things in long historical evolutions.

In a book that appeared in the Soviet Union in 1956, V. I. Sviderskii seemed to reject all cosmological models that included concepts of a universe finite in space or time.[80] His criticism of general relativity in cosmology was based on the belief, somewhat similar to Fock's but much more simply expressed, that cosmologists had unjustifiably extrapolated space-time from a limited area to the entire universe. Sviderskii believed the universe as a whole to be qualitatively different from the limited areas in which the few observations supporting general relativity had been made (perturbation of the motion of the perihelion of Mercury; displacement of light rays passing through the sun's gravitational field). Sviderskii's criticism of relativistic models was directed most pointedly against expanding models (categories II and III above), although one can assume that the steady-state model, with its creation of matter, would have been equally objectionable to him. Sviderskii did not pose an alternative theory. Indeed, the person who rejected both relativistic and steady-state cosmology had very little else available at this time. The best case that could be made for Sviderskii's position was that occasionally a scientist should recognize the unsuitability of any existing explanation for a certain aspect of nature. Sviderskii, however, did not help his case by commenting that "modern cosmology in the framework of relativity theory (Einstein's model, Lemaître's model, and so forth) is coming to clearly antiscientific, popish conclusions concerning the finitude of the universe in space and time." [81] This equal imputation of religious or antiscientific grounds to the work of prominent scientists of such differing religious beliefs as the priest Lemaître and the agnostic Einstein was one of the characteristics of some Soviet cosmological writings under Stalin, but one could have hoped that by 1956 the thaw already evident in many other fields would have occurred in cosmology as well.

A refreshingly new tone was struck by S. T. Meliukhin in his book *The Problem of the Finite and the Infinite,* published in Moscow in 1958.[82] Meliukhin's work was a transition, a bridge between the older orthodoxy and a new readiness, even eagerness, on the part of some Soviet scholars to combine dialectical materialism, factual discussions of recent astronomical evidence, and conditional tolerance of relativistic models of the universe. Indeed, Einstein began to emerge in a very favorable light, a defender, unwittingly, of dialectical materialism. This view of Einstein would in later years gain much strength.[83] Nonetheless, Meliukhin's interpretation of the universe was not a simple recognition of viewpoints previously considered inadmissible, but something of an independent statement.

Meliukhin organized the most interesting part of his discussion around the problems posed by Olbers's and Seeliger's paradoxes. It will be reasonable, therefore, to briefly consider these noted issues in astronomy.

In contrast to the finite universe of the Middle Ages, the Newtonian universe was considered to be made up of an infinitely large number of bodies in infinite Euclidean space. As Newton remarked, if a finite amount of matter were located in infinite space, the force of gravity would result in a tendency for all the matter to concentrate in one mass.

By supposing that the number of stars and other celestial bodies was infinite, Newton avoided this problem, since an infinitely large set of bodies has no center. His view of an infinite universe became the standard interpretation of the late eighteenth and nineteenth centuries.

In 1826 H. W. M. Olbers pointed to a problem within the Newtonian universe that has come to be known as Olbers's paradox: If the total number of stars were infinite, then a terrestrial observer should see a sky of blinding brightness, a sheet of solid light. Since intervening stars would block those farther away from the earth, Olbers thought that the level of brightness should be equal to that of the sun in each direction, rather than infinitely bright. Attempts to solve this paradox occupied the efforts of a number of astronomers of the past century; even yet it is of some interest, although the assumption of an expanding universe can account for the phenomenon. The point is not that there is no path out of the paradox (it can be avoided by assuming that the brightness of stars diminishes with distance from the earth, or by attributing certain types of relative motion to the stars, or by assuming that the universe changes in certain ways over periods of time), but that any one of the assumptions necessary for eliminating the paradox was, until Hubble's work on the red-shift, of a rather marked ad hoc nature. In other words, the assumption would be introduced for this reason alone, without further supporting evidence. Each of the assumptions, furthermore, would have radical cosmological implications. Olbers himself believed that he could solve the puzzle by assuming the existence of dust clouds between the stars and the earth, which would obstruct the passage of light. We now know that Olbers's hypothesis could not provide the answer, since the dust would absorb energy from the stars until it eventually became equally incandescent.

The other paradox to which Meliukhin pointed as a preface to his attempt to provide a coalescence of dialectical materialism and modern cosmology was that of H. Seeliger in 1895. Seeliger maintained that if infinite matter were indeed distributed uniformly through infinite space, as Newton believed, the intensity of the gravitational field resulting from the infinite mass of the universe would also be infinite. Since no such gravitational field exists, the Newtonian assumptions must be incorrect. Seeliger attempted to solve the paradox by introducing to Newton's gravitational law a modification that would have perceptible effects only when very great distances are involved.

Upon reflection, it becomes clear that both Olbers's and Seeliger's paradoxes are merely different expressions of the problem of conceptualizing infinity, as are many of the major problems of cosmology. Yet this does not explain them away, since the infinitude of the universe was considered essential to the Newtonian universe.

Meliukhin realized that since the 1920's an escape from Olbers's paradox was available in the form of theories of the expanding universe (the relative motion away from the earth would result in a weakening effect on the stellar light), but he was reluctant to accept expansion as a phenomenon of the universe as a whole even though he was prepared, as we shall see, to accept it as a phenomenon within limited areas. Furthermore, the expansion theory, even as pure conjecture, did not satisfy him, since he believed it would solve Olbers's paradox only in terms of visibility; the wave length of the electromagnetic radiation reaching the earth from the stars would be shifted from the band of light to that of radio waves as a result of the outward expansion, but the paradox would, Meliukhin thought, remain. (The problem of the threshold of measurability of radio waves would be relevant to this line of attack, but Meliukhin did not pursue it.)

Instead of explaining the paradoxes on the basis of a postulated expanding universe, Meliukhin posed other possibilities: the Lambert-Charlier model of a hierarchical universe (to be explained below) and the possibility of an interaction of the electromagnetic and gravitational fields with cosmic matter. As we will see, the latter view was the one that Meliukhin preferred.

The Lambert-Charlier model, first suggested in the eighteenth century, consists of a universe organized in terms of systems or clusters of the first order, second order, and so on indefinitely, each succeeding system being of larger dimensions than the previous one. Thus, there would be galaxies, supergalaxies, and super-supergalaxies, ad infinitum. The Swedish astronomer C. V. L. Charlier showed how such a hierarchical model could, within the context of classical theory, avoid the problems of the Olbers and Seeliger paradoxes. Meliukhin, however, pointed out that in a hierarchical universe of the Charlier type an extrapolation to infinity would yield an average density of matter in the universe equal to zero, since the average density of matter at each higher level is lower than the preceding one; such "abolition" (*uprazdnenie*) of matter was to him unacceptable on philosophic grounds:

"From this point of view even the very concept of space becomes inconceivable, since space does not have an existence independently of matter, it expresses the extension of matter. . . ." [84] Thus, Meliukhin, like Aristotle and many subsequent thinkers, denied the existence of a void. Furthermore, Meliukhin observed, all those persons who have proposed a hierarchical universe have merely assumed its existence without asking how it came into being. According to dialectical materialism, no state of matter can be maintained indefinitely without change, since no "material system, no matter how great its dimensions, can be eternal. It arose historically from other forms of matter." [85]

Meliukhin had turned away from two possible exits from the Olbers and Seeliger paradoxes: the concept of a universe that, as a whole, expands, and a hierarchical universe. What solution did he propose? He considered the most promising avenue of investigation to be the conversion of quanta of electromagnetic and gravitational fields into "other forms of matter." Accepting the equivalence of matter and energy inherent in relativity theory, Meliukhin believed that in both paradoxes the problem of excesses of electromagnetic or gravitational energy could be solved by the absorption of such energy "accompanying its transition into matter." Modern field theory, he observed, describes both gravitational and electromagnetic fields as specific forms of matter. No violation of the conservation laws is involved in posing the transformation of gravitational energy into matter, Meliukhin noted; such conversions in the reverse direction are evident in the conversion of the mass of stars into radiation. Therefore, Meliukhin believed that he had posed a possible solution to the paradoxes.

This candidate for solution of the problem has not enjoyed much popularity in the Soviet Union, particularly since the expansion of the metagalaxy, supported by other evidence, solves the same problem. Furthermore, in order to accept Meliukhin's hypothesis one would have to reconsider the Second Law of Thermodynamics, which is usually interpreted as saying that electromagnetic radiations such as heat and light are at the bottom of an irreversible ladder. However, Meliukhin would probably not be dismayed by such a consideration, since "heat death" (*wärmetod*) has been criticized by dialectical materialists on other grounds.[86]

Meliukhin criticized those relativistic models of the universe that included reference to the birth of the universe, but in contrast to a number of previous Soviet authors, he spoke positively of certain aspects of some relativistic models:

In relativistic cosmology there are many rational points and profound propositions that must be utilized and further developed. . . . [E]ven the very idea of a positive curvature of space deserves attention, for the possibility is not excluded that in the infinite universe there exist areas with a density of matter that is in keeping with a positive curvature of space.[87]

Not only did Meliukhin find much that was rational in relativity theory, but he also believed that relativity theory supported dialectical materialism in stating that the characteristics of space-time are determined by the amount of matter within the continuum. Thus, "as incontrovertibly follows from the most important principles of the theory of relativity and dialectical materialism, space and time are forms of the existence of matter and without matter cannot have an independent existence." [88]

In view of his acceptance of general relativity theory, one is tempted to ask why Meliukhin did not rely upon one of the expanding relativistic models (categories II, III, and IV above) to explain the Olbers and Seeliger paradoxes, rather than introduce an otherwise unnecessary hypothesis. The answer to this question seems to be that this would have called for too great an area of expansion; a belief in a universe that expanded as a whole was to him still "antiscientific, contributing to the strengthening of fideism." [89] Nonetheless, Meliukhin did not deny the expansion of the observable portion of the universe: "All the evidence points to the fact that our area of the universe is apparently in a state of expansion—regardless of its causes." [90]

If many Soviet scholars were willing to accept the concept of the expansion of the part of the universe that we can observe yet were not willing to accept an absolute beginning in time, one type of cosmological model among those being proposed was a logical candidate: a pulsating one. The reluctance of Soviet philosophers and astronomers to embrace such a model has been rather strange, considering the philosophic advantages that it possesses from their point of view; it is infinite in time and does not include a concept of the creation of matter such as the steady-state model supposes. It might, however, contradict the principle of the over-all evolution of matter, contained in dialectical materialism. Even the relatively unphilosophical Soviet astronomer I. S.

Shklovskii would, at a later time, find the pulsating universe philosophically flawed; he commented, "The simple repetition of cycles in essence excludes the development of the universe as a whole; it therefore seems philosophically inadmissible. Further, if the universe at some time exploded and began to expand, would it not be simpler to believe that this process occurred just once?" [91] Much could be said of this statement of Shklovskii's; it is quite possible, for example, to pose a pulsating universe without the absolute reduction of all matter to its most primitive state (type IId), as William Bonnor in particular has suggested. Some concept of evolution through successive cycles could then be rescued, although at this point such evolution would be conjecture.

With the increasing acceptance of relativisitic models in Soviet cosmology in the late fifties, the main issues shifted from criticism of those models as a group to a discussion of particular types within the group. In particular, the question of the type of curvature of space-time within the relativistic framework became pressing. This issue was usually seen, as already indicated, in terms of the *observable* portion of the universe. The problem of a model for the entire universe was, with some exceptions,[92] considered inappropriate for scientists. As A. S. Arsen'ev commented in an issue of *Problems of Philosophy* in 1958, "The natural sciences cannot answer the question, Is the universe infinite or finite? This problem is decided by philosophy. The materialist philosophy comes to the conclusion that the universe is infinite in time and space." [93]

The philosopher G. I. Naan, now familiar to these debates, said in 1959 that the question of curvature in cosmological models was entirely open; he did not believe that even if scientists agreed on a model of positive curvature, this decision would necessarily lead to a concept of a finite universe: ". . . contrary to a very widely held opinion, it would in no way prove the spatial finiteness of the universe. One cannot judge the properties of space as a whole by the metric. For an evaluation of these properties one must know the character of the interrelatedness of space. And this is the most complicated question in cosmology. We simply have no evidence to help us answer this question." [94]

The degree to which discussion of various cosmological theories broadened by 1962 was clearly indicated at a conference held at Kiev in December of that year. Called to inquire into the philosophic aspects of the physics of elementary particles and fields, the conference was attended by approximately three hundred philosophers and scientists.[95] Although cosmology was not the major concern of the meeting, reports by P. S. Dyshlevyi and P. K. Kobushkin on, respectively, the general theory of relativity and relativistic cosmology attracted considerable attention. Kobushkin, who agreed with Dyshlevyi's definition of space and time ("space and time are essentially the totality of definite properties and relations of material objects and their states"),[96] presented the report that was most pertinent to the cosmological issues discussed so far.[97]

Relying on the work of G. I. Naan, V. A. Ambartsumian, V. A. Fock, and Iu. B. Rumer, Kobushkin sketched out an approach according to which it would be possible for him to accept "closed" and expanding relativistic cosmological models, including the much-maligned Lemaître one, without contradicting the principles of dialectical materialism. The operation involved the addition of several interesting amendments, or possible amendments, to the existing models.[98]

In order to arrive at his goal of describing the metagalaxy as a "quasi-closed system immersed in the 'background matter' of an infinite Universe," Kobushkin pointed out that certain hypothetical stars, those having densities similar to the white dwarfs and masses typical of the super-giants, would have, on the basis of the solutions of the general relativity equations obtained by Oppenheimer and Volkoff, the following strange characteristics: Material particles and light quanta that were ejected would go out only to a certain boundary, or "horizon," and then would return to the star.[99] The importance of this phenomenon was the following:

This means that, from within, such a star may be considered a completely closed system with respect to its energetic interaction with the external world by means of such "carriers" of interaction as the usual material particles and light quanta.

Therefore the metric of space-time near such stars is identical to that of spatial closedness. . . . Nevertheless, such a star is not an absolutely closed system; in actuality an interaction with other stars occurs by means of gravitons; this interaction with other stars is formally expressed in the . . . superposition of their gravitational fields. Moreover, such closedness of stars of this type is characteristic only for static fields of spherical symmetry, considered up to that moment of time when the surface of the star, during expansion, touches the "horizon" of [mathematical] solution. As soon as the physical surface of the star touches the "horizon" of solution, the metric of the field is "broken," so to speak, and it then becomes possible for there to be an energetic interaction of the star with the external world by means of normal material particles and light quanta. Henceforth, we will call the solution of this type of equations in general relativity theory *semiclosed solutions*.[100]

Kobushkin believed that the possibility of such a solution of relativity equations for a star posed the further possibility of a similar solution for the metagalaxy. However, he realized that one of the difficulties with the analogy was that in the case of the hypothetical star cited above, the system would be closed in terms of normal material particles and light quanta, but not in terms of gravitational interaction, while the large cosmological models were considered closed in gravitational terms as well. However, Kobushkin questioned whether the closed cosmological models were, indeed, absolutely "closed." He posed the possibility of the existence of a new kind of particle (called "background particles," or following Rumer, "fundamentons") for which no "world horizon" would exist in the current closed cosmological models, just as there does not exist a "horizon" for the gravitons in the case of the massive superdense stars already described. The introduction of particles with such "strange properties, at first glance" would be "entirely consistent with the dialectical materialist thesis of the inexhaustibility of the different forms and properties of moving matter." [101] By postulating the existence of these new particles, he could further suppose the interaction of the gravitationally closed models of Einstein, de Sitter, and Lemaître with other parts of the infinite universe.

It might be appropriate to note, at this point, the inaccuracy of the view, expressed from time to time, that Soviet cosmologists could not accept big-bang or expanding cosmological models. Kobushkin in the above passage considered a big-bang and expanding model the most satisfactory of the existing models—with an important amendment about the extent to which the model is identical with the entire universe.[102]

One of the most interesting books on cosmology to appear in the Soviet Union in the sixties was the result of yet another conference of physicists and philosophers held in Kiev, also in 1964, and entitled *Philosophic Problems of the Theory of Gravitation of Einstein and of Relativistic Cosmology*. (Note "theory of gravitation" rather than "theory of relativity.") On first reading the book seemed to be openly revisionist; its suggestions for changes in cosmological conceptions went beyond anything promoted in the Soviet Union since the end of World War II. Philosophically speaking, however, the book represented something of a return to an older tradition in Soviet Marxism, that of centering on the philosophic "categories" as the area of greatest flexibility within dialectical materialism with respect to the advance of science. An effort to clear the conference of the revisionist label was made by P. V. Kopnin with an opening quotation from Lenin, but the effect, combined with what followed, tended to justify necessary revisionism:

The revision of the "forms" of materialism of Engels, the revision of his *naturphilosophie* views, not only is not "revisionist" in the accepted sense of the word, but, on the contrary, is necessarily demanded by Marxism.[103]

Kopnin believed that it was time to consider a revision of certain Marxist categories, particularly those of the "finite and infinite," in order to bring Marxism in step with science. He did not doubt the possibility, nor the legitimacy, of making revisions to the categories of dialectical materialism. He pointed out that "contradictions between the content of the philosophic categories and new scientific evidence are connected only with great, epochal discoveries," but recognized that the preceding fact does not "in principle exclude the possibility of such contradictions." [104] And the changes of the categories could go beyond "modification" of existing ones:

Since philosophy is a science and not an item of faith, the development of its categories is obviously subject to the general dialectical laws that it itself establishes; this evolution includes both change and clarification of the content of previous concepts and also the birth of new and the death of old concepts.[105]

This effort to modernize dialectical materialism, Kopnin remarked, must go beyond Engels to Lenin himself, who, despite his genius, by no means "did everything" in the direction of bringing dialectical materialism in line with modern science. Each man and each epoch has its own characteristics, and just as Engels and Lenin in their times attempted to incorporate the latest science into the Marxist world view, so must contemporary Marxist-Leninists. Kopnin observed that a revolution had occurred in science since the early twentieth century and in light of this it was only logical, according to Marxism, that such concepts as "finite" and "infinite" must be re-examined:

These categories were formulated a very long time ago, at the dawn of the development of scientific knowledge. Their content was defined in the period when natural science was hardly developed (antiquity, the Renaissance), when the mature concepts of astronomy and physics did not exist, and the only geometry was Euclidean.[106]

Kopnin went on to pose the possibility that the categories "infinite" and "finite" meant nothing other than the affirmation that matter can be neither created nor destroyed, but can, instead, be infinitely transformed into different forms. Such an equation of infinity with the principle of the conservation of matter had a considerable implication for the discussions of cosmological models. Kopnin observed that if dialectical materialists defined "infinity" in such a way that a certain type of metric of the universe was required, they would be repeating the old mistake of "dictating" to science rather than drawing upon it. He believed that dialectical materialists need not bind cosmology to specific geometric characteristics.[107]

Several other authors extended this argument to the point of maintaining, contrary to Soviet cosmology of the period immediately after the war, that dialectical materialism *requires* the finitude of man's cosmological models. In the opinion of Sviderskii, this requirement flowed from the dialectical Law of the Transformation of Quantity into Quality. If one grants spatial extension to the limit of infinity, Sviderskii observed, such an enormous accumulation of quantity will surely result in a change in quality, namely a change in space-time structures. To maintain the opposite would be to "absolutize" a specific state of matter (its space-time metric); dialectical materialism is based, he said, on the change of the states of matter in accordance with the dialectical laws.[108] Sviderskii's views on the implications of this dialectical law were by no means accepted by all the other participants at the conference (S. T. Meliukhin and G. I. Naan, in particular, objected to them), but his readiness to accept relativistic models, including those with Riemannian space-time, was common to most of the speakers at the conference, philosophers and scientists alike.

This conference was a further development of the ideas expressed at the 1962 meeting; together they marked a new stage in the development of Soviet cosmological discussions. No longer were Soviet cosmologists engaged primarily in polemics (usually unheard by the opposition) with non-Soviet writers. Very little time was spent at these gatherings in the old activity of sparring with the "idealistic views of bourgeois scientists." Instead, the dialectical materialist philosophers and scientists arrived at a point where their primary disagreements were among themselves rather than with scientists and philosophers living in other social systems. At the same time, among their arguments could be found, as before, consistent lines of thought, agreement on certain basic issues, beginning, of course, with a commitment to the objective reality and materiality of nature.

In cosmology, the common themes that could be seen in the 1964 conference concerned, in particular, the belief that one should not refer to the metagalaxy as the universe as a whole. All man's cosmological evidence stems, they said, from only a small part of the universe, and extrapolations from this portion to the entire universe are dubious speculations. The implication of this viewpoint was to cast into doubt the "cosmological principle," the belief in the homogeneity of the universe, an assumption upon which almost all existing models are built. Rejection of these models is not necessary, however, the Soviet authors usually maintained, since they may be perfectly adequate and helpful when applied to the metagalaxy. A further view on which there was agreement was that of the universal "connectedness" of the universe. In other words, one should not postulate the existence of "completely closed finite systems" within the universe—not for the metagalaxy nor for other systems. To do so would deny the integration of nature into a unitary whole. A final agreement—although somewhat less general than those just discussed—was the belief that a pulsating or oscillating metagalaxy would not, by itself, be a satisfactory exit from the cosmological problem since the existing forms of such models are based on endless identical repetition, in which stages are indistinguishable one from another. Such a hypothesis would contradict the principle of the *evolution* of nature, casting man's scientific explanation of the universe back into an entirely static mold, similar to the Aristotelian model as interpreted by the scholastic philosophers. Presumably a pulsating model that somehow preserved an evolving continuity between stages would be consonant with the dialectical materialist interpretation as expressed at the conference. This latter suggestion was an example of the inner flexibility of Soviet cosmology in the sixties. Almost any model under consideration by scientists anywhere in the world could be, with appropriate adjustments, fitted to the above philosophic requirements, although some more easily than others (the steady-state model would be rather difficult). This inner flexibility did not, however, reduce Soviet philosophy of science to the level of triviality, as some of its critics would maintain; the preferences of dialectical materialists concerning cosmology were still clear, even if they no longer possessed require- ments. Furthermore, the same lack of requirements imposed on nature is characteristic of many other philosophic systems. A person might, therefore, be tempted to throw out philosophy and rely on "hard facts." The catch is that then we would almost surely not have something we could call science, and certainly not cosmology.

The degree to which one should rely on facts as opposed to philo- sophic considerations in constructing cosmological models was a sub- ject of controversy in the Soviet Union, as elsewhere. An example of a person who veered to the side of philosophy at the 1964 conference, relatively speaking, was Sviderskii, while Naan spoke for more attention to science. Sviderskii chose to make statements such as:

A consideration of the problem of the finite and infinite obviously attests to the fact that these concepts can be clarified only by means of philosophy; the very problem is in the competence first of all of philosophy, and not cos- mology, as some people maintain.[109]

Naan looked at things from a different viewpoint:

The position of dialectical materialism on the infinity of the universe is not a *demand,* but is a *conclusion* from the facts of natural science. The content and form of this conclusion constantly change. At first man established that the universe is infinite in the sense of practical infinity. Then it became clear that it is infinite in a deeper sense, in the sense of unlimited spatial extension. Now we say it is infinite in space-time (but not at all in space and time!) in a still more profound, metrical sense. Topological and still more complex aspects of the problem force us once again to change the very way in which the issue is posed in dialectical materialism.[110]

Naan's emphasis on the determination of philosophy by science might seem to be fundamentally antiphilosophical in intent. Yet Naan, a phi- losopher himself, further maintained that philosophy can make "very significant contributions" on the issue of cosmological infinity, which was to him a "problem where the boundaries of mathematics, natural science and philosophy meet." [111] Naan clearly stated that his views were influenced by dialectical materialism: His belief in the validity of the "infinity" of the universe, according to a new definition, and his hope to construct "quasi-closed cosmological models" were both results, at least in part, he maintained, of his dialectical materialist viewpoint. With Naan we meet the issue that we have met with Soviet scientists such as Fock, Ambartsumian, Schmidt, Blokhintsev, and others. The degree to which dialectical materialism has been important to them is not something that the outside observer, particularly the observer who comes from a society that scoffs at dialectical materialism, can easily assess. In principle, it is no more difficult to believe that certain Soviet scientists have occasionally been affected in their researches by dialectical materialism than it is to believe that other significant scientists, such as Kepler, Newton, Poincaré, and Heisenberg, have been affected in their work by philosophic preferences.

Naan's exposition of the relation of gravitation and infinity at the 1964 conference deserves additional attention. He began by taking note of the criticism of him by Sviderskii. According to the latter, several scientists and philosophers (E. Kol'man, A. L. Zel'manov, Ia. B. Zel'dovich, and Naan) were "attempting to make the conclusions of modern relativistic cosmology, especially the Friedmann model, agree with the requirements of dialectical materialism concerning the infinity of the universe in space and time," [112] but had "accomplished" this at the cost of "absolutizing gravitation," a price that in Sviderskii's opinion was unacceptable from the standpoint of dialectics. Sviderskii hoped to accomplish the same goal, but paying a different price: He would demote the gravitational field to the status of the electromagnetic field, to which certain kinds of particles are neutral. Sviderskii's approach, then, was similar to Kobushkin's, and indeed, Kobushkin had referred favorably to Sviderskii's 1956 analysis in his paper at the 1962 conference.

Naan disagreed strongly with this viewpoint. In contrast to the electromagnetic and nuclear fields, the gravitational field was, in his opinion, a *universal* form of the mutual influence of the forms of matter. This conclusion followed from the following analysis: Physicists had established the equivalence, to extremely exact degrees of measurement, of gravitational mass and inertial mass. Relativity physics had, furthermore, established the equivalence of inertial mass and energy. There-fore, maintaining the existence of nongravitating forms of matter, in the fashion of Sviderskii or Kobushkin, would mean the existence of forms of matter "that are deprived of energy, do not transmit any kind of information, do not take part in any kind of interactions, the existence of which is not evident anywhere or by any means. . . ." [113] To Naan, a postulation of this type violated modern physics "more strongly than the hypotheses of Milne, Jordan, and Bondi." He continued that Svider-skii's statement that science must not make an absolute of gravitation was a vestige of the recent past, when philosophers attempted to state flatly what nature could and could not be like: "The spirit of dialectical materialism is alien to such attempts; it takes nature just as it is." [114]

Naan hoped to achieve a model of a "quasi-closed metagalaxy" just as did Sviderskii, but by means that he considered more consistent with the scientific evidence. This model was based on the researches of Ein-stein and E. G. Straus on the gravitational fields of point particles, the work of I. D. Novikov on the metagalaxy as an "anticollapsing" system, and the Lambert-Charlier hierarchical model. Einstein and Straus in 1945 had sketched out the beginnings of a "vacuole" model in which point particles in the cosmological substratum are surrounded by ex-panding spherical vacuums. [115] They showed that the field created by such a point particle does not depend on the field created by the sur-rounding substratum. I. D. Novikov published in 1962 an article in which he maintained that the metagalaxy could be looked upon as a vacuole in a supermetagalactical expanding substratum. Such a meta-galaxy would be an "anticollapsing system." [116]

Naan then combined this possibility with the Lambert-Charlier model, with the exception that it was made up of a hierarchy of relativis-tic "closed" models, rather than the three-dimensional Euclidean ones of the original Lambert-Charlier model. Naan postulated the existence of a whole series of expanding "bubbles" of vastly different scales. In this way he would combine "great homogeneity" with "extreme inho-mogeneity" in a "dialectical unity." [117] The resulting model for our metagalaxy would be that of a vacuole immersed in the substratum that had a "closed" space-time "carcass" whose spatial intersection could be either finite or infinite. The borders of the vacuole would be a Schwarz-schild shell, [118] which would not be an "insurmountable barrier," but on the contrary, a unidirectional barrier: It would permit signals to pass inward if the system were collapsing, outward if the system were expanding.

Thus, Naan had constructed a model called, like Kobushkin's, "quasi-closed," but resting on different principles. So far as the metagalaxy is concerned, it could be a relativistic model, finite in Riemannian terms, but nonetheless infinite in the sense that it existed within a larger system, which, in turn, existed within a still larger system. It was "infinite," said Naan, in a space-time sense (but not in terms of space and time taken separately). It was a part of unitary nature in that it could, in principle, transmit or receive signals to or from the "outside" (but not simultaneously). It was also, one might add, exceedingly complicated.

A. L. Zel'manov

One of the most interesting Soviet writers on cosmogony at the present time is Abram Leonidovich Zel'manov (1913–   ), a theoretical astronomer at the Shternberg Astronomical Institute of Moscow University. A student of V. G. Fesenkov, who was previously discussed, Zel'manov from an early age has been interested in the application of the general theory of relativity to astronomy. In his attitude toward model-building he has an extremely eclectic approach, granting the possibility of many types of cosmogonical models for different areas of the universe.[119] He has stoutly resisted any attempt to rule out a priori either "closed" or "open" models. He believes that outside the Soviet Union theoretical astronomers have been too willing to assume that the universe is homogeneous and closed.

Zel'manov, like many of his contemporaries, has a strong interest in dialectical materialism. He wrote in 1969 that "Dialectical materialism has been and remains the only system of philosophical views that is at one and the same time logically consistent in a philosophic sense and harmonious with all of human practice." [120] Similar to Ambartsumian, he speaks of "qualitatively different" areas of the universe, pointing out that different physical forces predominate on different levels of being. Thus, he observed, the most influential kinds of forces on the microscopic level are nongravitational (the so-called "strong," electromagnetic, and "weak" ones) while on the cosmic level gravitational force is predominant. These different levels illustrate, he maintained in 1955, "the dialectical materialist position concerning the inexhaustibility of matter and the infinite multiformity of nature." [121]

One reason for his permissiveness on the question of model-building for the metagalaxy was his belief that within the framework of modern physics the question of infinity in traditional terms is "almost trivial." [122] In order to choose a model, he observed in 1959, it is necessary to accept a congruence relation. In other words, in order to construct a model of the universe (or of any surface or volume), it is necessary to accept a convention about what constitutes a unit length in different places, at different times, or in different directions at the same time. By choosing different congruence relations, it would be possible to construct models of an unlimited variety of curvatures.

Zel'manov was critical of both those more orthodox dialectical materialists who resist certain cosmogonical models and also those astronomers, frequently non-Soviet ones, who uncritically accept the cosmological principle, upon which all popular relativistic models rely. In a 1964 article Zel'manov said that rather than assume a homogeneous, isotropic universe, one should notice the possibility for an inhomogeneous, anisotropic one.[123] The cosmological principle dictates, he said, that whatever the curvature of space (positive, negative, zero) it must remain a constant, since the curvature results from the amount, distribution and movement of matter; if one assumes a universe that is uniform everywhere, the resulting curvature, whatever its sign, would be a constant.

Such an assumption of constant curvature was, to Zel'manov, a gross simplification; it forced the concepts of "closed" and "infinite" to be mutually exclusive. This exclusivity was true even though the Einsteinian theory of gravitation did not, by itself, give a unique answer to the question of the finiteness of the universe. The Einsteinian theory could be retained without forcing commitment on the question of infinity if cosmologists would not insist on unnecessary assumptions: "It is not necessary for cosmologists to supplement Einstein's theory of gravitation with any kind of simplifying assumptions such as those of homogeneity and isotropy." [124]

By posing an inhomogeneous, anisotropic universe, Zel'manov could provide many types of local space-time continuums, including both closed and infinite ones. Furthermore, the fact that the space (considered separately) in a space-time continuum may be infinite does not mean that the continuum as a whole fills the entire universe. As Zel'manov wrote:

A space-time world, infinite in time and space, possibly does not occupy the whole universe: It may be part of another space-time world, spatially finite or infinite. The space-time world occupying the whole universe, on the other hand, may not be infinite in space and, moreover, may contain spatially infinite world areas. . . . It seems that the question concerning the actual situation, the one interesting us, remains open, and a consideration of homogeneous isotropic models, empty or not, will not give us an answer. But one can hardly expect that the properties of actual space will be simpler than that in simplified models.[125]

Zel'manov envisioned a new stage in physical theory, building upon relativity but going beyond it in the sense that relativity physics went beyond classical physics, introducing necessary corrections in the realm of great distances and velocities.[126] What were the ways in which Zel'manov thought that Einstein's theory of gravitation might be modified to affect cosmology? First, he noted that the cosmological models within the framework of Einsteinian relativity "exclude each other"; in other words, although all these models are relativistic, they describe "different universes." Since there can be, by definition, only one universe, all but one of the models should be false. Yet Zel'manov felt there were advantages in more than one. He seemed to be hoping for a sort of "complementarity" in cosmology that would allow the possibility of combining mutually exclusive explanations, although he did not use the term. The different models would not be mutually exclusive, Zel'manov maintained, if each of them had its own congruence relation, its own space-time metric. Thus, he returned to the importance of congruence relations, saying that by utilizing various relations, different models could give different descriptions of one and the same universe and its various parts. This would involve giving up the concept of an "ideal standard" of length and time. Such a modification would have other implications:

This [view] includes not only a change in the understanding of gravitational interaction, but also a change in the very concept of finiteness or infinity in space and time: This [new] finiteness or infinity can not be considered metric, i.e., as a finiteness or infinity of cubic meters or parsecs or of the number of seconds or years.[127]

Thus, Zel'manov sketched out a model embracing many submodels. Without going further into the details of an exceedingly speculative subject, one can observe that many scientists would consider adopting the complex model that Zel'manov proposed only if there were reasons for that adoption other than the avoidance of a scientifically acceptable, considerably simpler, but philosophically unpleasant alternative. Quite a few scientists would consider existing models of constant curvature preferable in terms of scientific acceptability and structural simplicity, whatever the philosophic implications. Yet this observation does not belittle Zel'manov's major goal; the extrapolations of systems built on the basis of observable evidence to the entire universe (as all existing models are) are, as Zel'manov observed, dubious operations. And the philosophic obstacles involved in accepting a closed universe are troublesome to many scientists in many parts of the world.

In some of his most recent publications Zel'manov has attempted to integrate his views on cosmology and cosmogony with a conception of all physical knowledge; in his opinion there exists in nature a "structural-evolutionary ladder" extending from the subatomic level to the universe.[128] This material, multiform ladder contains qualitatively distinct levels, but constitutes an interconnected whole. Its most distinguishing characteristic is its incredible variety. Indeed, Zel'manov recommended that scientists accept as a "methodological principle" the view that nature contains the entire multiformity of conditions and phenomena that currently accepted fundamental physical theories would permit. Hence, Zel'manov would imagine, heuristically, the presence somewhere in nature of all the forms of matter and all the cosmogonical models permitted by existing physical theory.[129] As physical theory changes over time, this hypothesized infinite reservoir of models would change in step, but Zel'manov saw no reason to rule out any model in advance.

In the late 1960's Soviet writing on the philosophic problems of cosmology and cosmogony continued to improve in quality. In 1969 there was published in Moscow an interesting volume of eighteen articles entitled *Infinity and the Universe*.[130] In this volume and in other articles an impressive effort was being made to reach a philosophic understanding of the structure and evolution of the universe. A strong link existed between several leading Soviet astrophysicists and philosophers. Among the scholars doing important work in 1969 and 1970 in this field were A. L. Zel'manov, V. A. Ambartsumian, G. I. Naan, V. V. Kaziutinskii, and E. M. Chudinov.[131] The first two, both of whom have already been mentioned, are distinguished natural scientists; the latter three are able philosophers of science. Ambartsumian and Kaziutinskii have published works jointly in an effort to blend the views of a professional philosopher of science and an astronomer.[132]

These writers all make careful distinctions between science and the philosophic interpretation of science. They maintain that they are ready to accept completely the evidence of science, no matter how upsetting it may be to previous conceptions, but say that they are convinced that this evidence can not only always be accommodated with dialectical materialism, but illuminated by it. Furthermore, just as they see no threat to dialectical materialism in science, they also see no threat to science in dialectical materialism. Kaziutinskii quoted approvingly the words of the Soviet physicist V. L. Ginzburg that dialectical materialism "does not and cannot place a taboo on any model of the universe." [133] Kaziutinskii believed that even Lemaître's hypothesis of an original exploding atom as the beginning of the universe could be accommodated with dialectical materialism after a few terminological clarifications had been made:

We should not think that the idea of an explosion of a dense or superdense "primeval atom" is in itself idealistic. If it turns out that the metagalaxy (but not the whole universe!) formed in the way that Lemaître proposed, this would only mean that nature is more "odd" than it seemed earlier, and that it has posed for us still one more difficult question, the answer to which will be found, however, within the framework of natural science.[134]

Dialectical materialists have become so flexible on questions of cosmology and cosmogony that one might suppose that their philosophy has become irrelevant to their approach to nature. This conclusion would not be quite correct, however. They still seek to retain a concept of infinity, often in terms of time for the universe as a whole, but always, as a minimum, in terms of the "inexhaustibility" of matter as man studies it ever more carefully.[135] They differentiate between the words "infinite" and "boundless," pointing out that closed space-time possesses no boundaries "beyond which there must exist something non-spatial." [136] Quite a few of them continue to prefer inhomogeneous, anisotropic models of the universe, finding in them the richness and infinitude that they seek in material reality. And they continue to make a distinction between the observable portion of the universe and the universe as a whole.

In the last few years considerable excitement has been caused among cosmologists by the discovery of a cosmic microwave background that has been interpreted by some scientists as the remnant of the primeval fireball in which the universe began. This development has added considerable strength to the arguments of those cosmologists who favor big-bang models of the universe. Coming as it has at a time when the steady-state theorists were retreating on other grounds, it is fair to say that there is a long-term and significant tendency among cosmologists toward some version of a big-bang theory. The discoveries of quasars and pulsars have also added evidence to a field in which new observational evidence is often difficult to obtain. We can expect detailed philosophical interpretations of this evidence in the Soviet Union in the near future. It should be clear from the foregoing discussion that dialectical materialists have a number of potential avenues of interpretation, all of which could take this evidence into account. The distinction between "universe" and "metagalaxy" remains significant. And the cosmologists outside the Soviet Union are by no means agreed on the meaning of the new cosmic microwave background. A few theorists have attempted to explain it without reference to a primeval fireball.[137] Indeed, we cannot divide cosmologists into "Western" and "Soviet" groups, since the various trends of interpretation are supported by scientists in all countries where such work is done. An American astronomer wrote in a recent article summarizing research on the microwave background that "Many ideas first considered on a paper napkin at Cal Tech, on a blackboard in Russia, or over lunch here at Princeton have found their way into this paper." [138] But we can, nonetheless, observe that a tradition of interpreting astronomy within the framework of dialectical materialism still exists in the Soviet Union.

Soviet cosmologists had demonstrated by the middle of the 1960's a rather striking ability to fit cosmological models into the system of dialectical materialism. These efforts to solve the cosmological problem, to find a model of the metagalaxy that would not violate certain philosophic principles, were not so dissimilar, in essence, from the efforts of many non-Soviet philosophers or scientists. When Michael Scriven, a philosopher in the United States, spoke of the "phases in the Universe prior to the present 'expansion,'" or William Bonnor, a British theoretical astronomer, spoke of his "preference for the cycloidal model," both were significantly influenced by nonempirical considerations.[139] They both were seeking a model in which a certain concept of infinity could be retained. Soviet cosmologists have experienced similar desires.

VI   Genetics

To many persons the phrase "Marxist ideology and science" will bring to mind one word—"Lysenko." Of all the issues discussed in this volume, the "Lysenko affair" is best known outside the Soviet Union. It is frequently considered the most important of the various controversies concerning dialectical materialism and the natural sciences. It has been discussed in hundreds of articles and dozens of books.

How ironic it is, then, that the Lysenko affair had less to do with dialectical materialism as Marx, Engels, Plekhanov, and Lenin knew it than any of the other controversies considered in this study. The interpretations advanced by Lysenko arose neither among Marxist biologists nor established Marxist philosophers.[1]

Compared with the other scientific issues involving dialectical materialism, the Lysenko controversy was unique in still other ways. Intellectually it is far less interesting than the other discussions. A person may experience at moments a certain fascination in watching in detail through historical sources the suppression of a science, but this reaction surely issues only from either a dramatic sense of tragedy or a desire to know how to avoid such occurrences in the future. The Lysenko episode was a chapter in the history of pseudoscience rather than the history of science.

A number of authors have maintained that one of the most important reasons for the rise of Lysenko was the existence in pre-Revolutionary Russia of an unusual school in biology.[2] Some would tie the birth of this movement to Marx and Engels, while others would look to the populist writers such as Pisarev and Chernyshevskii.[3] It is quite true that frequent support for the concept of the inheritance of acquired characters or criticism of the early ideas of genetics can be found in the works of pre-Revolutionary Russian writers, often of leftist persuasion. But such writings can be found in other countries as well. The last half of the nineteenth century was the great age of biological controversy in western Europe, and those discussions found their reflections in Russia. Leftist writers everywhere objected to the "heartlessness" of biological theories from Darwin onward. The views on biology of populist writers in Russia such as Pisarev, Nozhin, and Chernyshevskii were rather diverse; the belief in the inheritance of acquired characters was a part of nineteenth-century biology, not a characteristic of Marxism or populism.[4] When Marxism was introduced in Russia, its early leaders, such as Plekhanov and Lenin, did not select biology for special attention; indeed, if any field of science was proposed as a candidate for ideological concerns by the founders of Russian Marxism, it was physics.

In Russia at the time of the Revolution there were some older biologists, such as K. A. Timiriazev, who were not able to accept the new field of genetics, but such biologists also existed in other countries. As we will see, some of the greatest men in the history of genetics also wrestled in the first decades of the century with what seemed to them the troublesome implications of genetics. Russia was not unique in this respect; by the late twenties it was distinguished, on the contrary, by the degree to which the new genetics was flourishing.[5] Soviet Russia by the end of the twenties was a center of outstanding genetics research, entirely in step with the new trends and in some respects leading the way.

More interesting from the standpoint of later events is the person of I. V. Michurin (1855–1935), a horticulturist whose name was to become the label for Lysenko's particular type of biology.[6] Michurin has often been described as a Russian Luther Burbank, and there is much to be said for that description, despite Michurin's occasional criticism of Burbank.[7] Like Burbank, Michurin was a practical plant breeder and an exceptionally gifted selectionist and creator of hybrids. Also like Burbank—and most selectionists before the proliferation of modern genetical concepts—Michurin believed the environment exercised an important hereditary influence on organisms. This influence was particularly strong, he thought, at certain moments in the organism's life cycle or on certain types of organisms, such as hybrid seedlings. Furthermore,

Michurin disputed, at least at one period of his life, the Mendelian laws of inheritance, which he felt were valid only under certain environmental conditions. Another of his beliefs was in graft hybridization; according to his "mentor" theory, the genetic constitution of the stock of a grafted plant could be influenced by the scion. And yet another of his theories concerned the phenomenon of dominance in inheritance; he thought that dominant characters were those that gave its organism advantages in local conditions.[8]

In all of the above theories Michurin prefigured in important ways the views of Lysenko. Despite this marked degree of resemblance, however, the fact remains that Lysenko manipulated Michurin more than he drew sustenance from him. Determining the exact correspondence of Lysenko's views with Michurin's has been complicated by the fact that for thirty years most books and articles published in the Soviet Union portrayed the positions of the men as identical. It is only within the last few years that articles and books distinguishing the two men have begun to appear.[9]

Michurin never made a claim to a great biological system—as Lysenko did in his name. He also did not emphasize the influence of environment in inheritance to the exclusion of the internal hereditary constitution of the organism. And in the final part of his life he began to recognize the validity of Mendelism, stating that several of his experiments designed to disprove Mendel's laws had actually affirmed them.[10]

Rather than looking primarily to pre-Revolutionary ideology or to Russian selectionists for the most important reasons for the rise of Lysenko, it is necessary to consider the history of Lysenko's early activities against the background of the economic and political events of the late twenties and early thirties.

Trofim Denisovich Lysenko was born in 1898 in the Ukraine near the city of Poltava, where he grew up in a peasant family. He received an education as a practical agronomist at the Horticultural Institute of Poltava, later continued his studies and research at several different locations in the Ukraine, and after 1925 began to investigate the vegetative periods of agricultural plants at the Gandzha (now Kirovabad) Plant-Breeding Station in Azerbaidzhan.[11]

Between 1923 and 1951 Lysenko published approximately 350 different items, although a great many of these are repetitions.[12] The first publication in 1923 concerned sugar-beet grafting; this was followed by another 1923 article on tomato breeding. Then for five years he published nothing. It was during this time that he began to work on the effects of temperature on plants at different points in their life cycles, a topic that led him to his well-known concepts of vernalization and the phasic development of plants.

In Azerbaidzhan Lysenko was confronted with the very practical problem that the leguminous plants needed for fodder and for plowing under as green manure require considerable amounts of water for growth. Azerbaidzhan is an area of marginal rainfall for many crops, but irrigation provides additional water in moderate amounts. However, the main crop of the area, cotton, required all the water in the summer. Therefore, unless a way could be found to grow the legumes in the period from late fall to early spring, when sufficient water was present, a solution to the problem did not seem apparent. The possibility of growing the legumes in the winter was worth considering since Azerbaidzhan, located in the Southern Caucasus, enjoys a mild climate. Nonetheless, temperatures below freezing are encountered in the winter, although usually only for a few days.

Lysenko decided to grow hardy legumes during the winter season. By choosing early ripeners and planting in the fall, he hoped the plants would reach maturity before the coldest days arrived. Although this goal was fulfilled "not badly" according to Lysenko, the phenomenon that he now centered on was a side effect, drawn to his attention by his knowledge of the performance of the same plants in his native Ukraine.[13] Lysenko maintained that some of the peas that in the Ukraine were early-ripeners became late-ripeners in Azerbaidzhan. He decided that the reason for this change in vegetative period was the "unsuitability of the environment" for the development of the pea. The whole process of the growth of the pea was, as it were, "slowed down" in these unfamiliar conditions; therefore, the peas either did not reach maturity or did so very late. The same "slowing down" concept seemed to Lysenko also to be a good explanation of the difference between winter and spring varieties of certain cereals, such as wheat. A winter variety of wheat that is—contrary to normal practice—planted in the spring finds itself in "unfamiliar conditions," its growth is slow, and it fails to reach maturity.

On the basis of this kind of analysis Lysenko came to the conclusion that the most important factor in determining the length of time between seed germination and maturity in a plant is not the genetic constitution of the plant, but the conditions under which that plant is cultivated. The underlying theme here is, of course, that of plasticity of the life cycle, although still only with reference to the limited character of length of vegetative period. As Lysenko summed up the results of these early experiments:

We observed that *plants of the same variety, when grown under different conditions,* may, depending upon these conditions, be winter, early-spring or late-spring plants, and the behaviour of different varieties, when grown under the same *definite* conditions, may be different. Some varieties of wheat may behave like winter wheat, others like late-spring varieties, and others again like early-spring varieties. From all the material we obtained in our experiments in 1927, we arrived at the conclusion that the *length of the vegetative period of plants from the sowing of the seeds to the ripening of the new seeds depends upon the interaction between the plant organisms and environmental conditions.* By changing external conditions it is possible to change the behavior of different plants of the same variety.[14]

Lysenko and his co-workers in Kirovabad then attempted to determine the causes of this variability of the length of the vegetative period. They decided that the critical factor was the temperature immediately after sowing. The reason winter wheat could not reach maturity if sown in the spring, they decided, was that the temperature immediately after sowing was too high. This excessive heat, said Lysenko, prevented the plant from passing through the first stage of its development.

Could anything be done about this? The prospect of shortening the period of growth of cereals was a very attractive one, particularly in those parts of Russia where the winter was so severe that wheat frequently died then. But one could hardly hope to control, on a practical basis, the temperature on the field once the plants had sprouted. Fortunately from the standpoint of manipulation of the growing period, it was found, as Lysenko stated it, that "plants may pass through this phase of development even when still in the seed state, i.e., when the embryo has just begun to grow and has not yet broken through the seed integument." [15]

Therefore Lysenko thought it was possible to influence the length of the vegetative periods of plants by controlling the temperature of seeds before planting. Lysenko tried to work out an algebraic law to express this relationship. In an article that he published in 1928 entitled "The Influence of the Thermal Factor on the Duration of the Phases of Development of Plants," [16] Lysenko presented the formula

$$n = \frac{A_1}{B_1 - t_0}$$

by which n number of days of cooking could be computed to achieve the necessary preconditioning of seeds (B_1 equals the maximum temperature that can occur "without the preconditioning"; A_1 equals the sum of degree-days necessary for completion of the phase; and t_0 equals the average daily temperature).

This 1928 article is the only one I know in which Lysenko attempted to use mathematical methods—however simple—in his research. And this venture was soon severely criticized. A. L. Shatskii chastised Lysenko in a subsequent article for his "gross error" in trying to reduce relationships to a "physical truth" that can at best be described statistically. Shatskii also criticized Lysenko for believing that he could isolate the influence of the thermal factor alone when there were so many other factors that were also pertinent, such as light, humidity, soil moisture, and so forth.[17]

In later years Lysenko was extremely antipathetic to all attempts to describe biological laws mathematically. Although the reasons for this dislike of mathematics are unknown, it is possible that at least part of the explanation is that while a young man, he was submitted to embarrassing criticism in an area where he felt, at best, insecure. His frustration in the face of mathematics was commented upon at later times by a number of writers.[18] The 1928 article may represent an attempt by Lysenko to join academic biology; it was followed by a rebuff.

Lysenko continued, however, to expound his views on the importance of temperature in determining the development of plants. In January 1929 he reported on the Azerbaidzhan researches at the All-Union Genetics Congress in Leningrad. The paper was only one of more than three hundred presented and attracted no particular attention. At this time the exciting developments in the field of biology and genetics in the U.S.S.R. were coming from such scientists as Iu. A. Filipchenko, director of the Bureau of Eugenics of the Academy of Sciences, and Nikolai Vavilov, who in the same year became president of the new All-Union

Academy of Agricultural Sciences. Filipchenko and Vavilov were in a completely different circle than Lysenko, that of the academicians thoroughly trained in the neo-Mendelian genetics that emerged in the first decades of this century. Much more will be heard of Vavilov, who became the opponent of Lysenko in the great genetics debates of the late thirties.

After the Leningrad congress Lysenko decided to apply his new theory concerning the importance of temperature in plant growth to practical problems of agriculture. The term "vernalization" was utilized in 1929 in connection with an experiment in the Ukraine on the farm of Lysenko's father, D. N. Lysenko. In order to successfully sow a winter wheat in the spring, the workers buried sacks of germinating grain in snowbanks for a number of days before planting. This process of applying moisture and coolness to the grain became known as vernalization. In later years the mechanics were modified, but the principle remained the same. The grain was then planted, and later in 1929 the announcement was made in the press "of the full and uniform earing of winter wheat sown in the spring under practical farming conditions in the Ukraine." [19] This was only the first of the public claims made by Lysenko that I will attempt to evaluate in the following pages.

Within a few years, and for reasons shortly to be more carefully examined, the term "vernalization" became one of the best known in Russia. Lysenko became a hero of socialist agriculture and a mighty spokesman of agronomic science. He was transferred to the Ukrainian Institute of Selection and Genetics at Odessa, where the government established a special laboratory for the study of vernalization. Between 1930 and 1936 Lysenko published dozens of articles and pamphlets detailing the methods of vernalization, which was soon extended to include specific treatments of cotton, corn, millet, sugar beets, sorghum, barley, soya, potatoes, vetch, and various other grains, tubers, and fruits. On July 9, 1931, the U.S.S.R. Commissariat of Agriculture issued a resolution establishing a journal, the *Vernalization Bulletin,* for the purpose of popularizing the researches of Lysenko's Odessa laboratory and for issuing instructions for the vernalization of crops. The thirty-four-year-old Lysenko now had a journal; in different forms, it would be one of his main sources of strength for the next thirty-five years.[20] In 1935, after a hiatus, it was revived under the name *Vernalization,* and in 1946 it became *Agrobiology,* the increasingly general title growing in

step with Lysenko's increasingly general biological conceptions and ambitions. The first issue gave pathetically simple directions to the peasants
concerning the means of accomplishing vernalization, carefully citing
inventories of all necessary equipment: buckets, shovels, barrels,
scales, thermometers.[21] Here was a novel method of agriculture that
could be applied with only the simplest tools and yet that in its scale of
operation seemed suited for large collective farms. Its primary requirement was labor. But that was one commodity the predominantly rural
Soviet Union could supply, provided the peasants would co-operate. By
1935 Lysenko announced that the vernalization of spring cereals alone
in the Soviet Union had been carried out on forty thousand collective
and state farms and on a total area of 2,100,000 hectares (5,187,000
acres).

The historian of this process is immediately confronted with two
basic questions: (1) How valuable was vernalization? (2) If its value
was slight—as will be maintained—why did the government and Party
support it?

A definitive answer to the first question will probably never be given,
as a result of the extremely inaccurate records kept of the vernalization
trials and the methodological errors involved. The most obvious
methodological error in these trials was the almost total absence of control groups. But an attempt to judge the value of vernalization, based
both on non-Soviet and Soviet accounts, can be made.

First, it should be readily granted that the treatment of seeds before
or after germination does permit, under some conditions, the shortening
of the vegetative period and growing of winter varieties of grains during
the summer. This technique was known in the United States as early as
1854 and was also the subject of research in Germany by G. Gassner
shortly before the end of World War I. (Lysenko was aware of Gassner's work and credited him in his writings.) And the fact that seeds of
various kinds of plants require certain conditioning periods, during
which temperature and moisture are critical factors, has been a commonplace in the field of plant propagation for decades. The actual
processes that take place within seeds before germination are extremely
complex and are even yet far from being fully understood, not to speak
of the state of knowledge in the twenties. These processes involve complex biochemical and physical changes, including natural inhibitors and
hormone balances. In an effort to manipulate these processes, research-

ers have not only controlled the temperature and humidity of the seeds,
but have alternated such changes in complex patterns, scraped the seeds
with sandpaper, and even treated them with acid solutions in order to
render the seed coat (testa) more permeable. The refrigeration and
moistening of seeds preparatory to planting is generally known as cold
stratification, and the term "afterripening" is used to describe the complex processes that occur in the testa or endosperm before the plant
develops normally.[22]

But not every potentially useful technique that works under laboratory
conditions can be economically employed; the opinion of researchers outside the Soviet Union generally was that such techniques as vernalization involve greater losses than gains. There were a formidable
number of reasons for remaining skeptical of most mass pretreatments
of seeds, particularly in primitive areas. First of all, in the unmecha-
nized conditions of Soviet agriculture in the early thirties it was an extremely labor-intensive operation. The spreading of such seeds on the
ground or in trays, the application of water at controlled temperatures
for what amounted in many cases to weeks, the necessity to provide huts
and buildings for the protection of the seeds during soaking—all this
required the expenditure of enormous amounts of labor. Furthermore,
the process of vernalization was an ideal situation for the spread of certain fungi and plant diseases. The losses from such diseases must have
been considerable. And lastly, in the conditions of Soviet farms, where
there was often no electricity and no refrigerating equipment, it must
have been nearly impossible to keep the seeds in uniform conditions
over long periods of time. Sometimes the seeds became too hot, too
cold, too wet, too dry. Some seeds germinated too rapidly, some too
slowly, some not at all. But perhaps these very losses also provided ex-
cuses; if vernalization was not a success on a particular farm, the failure
could easily be blamed on the conditions, not the process of vernalization.

Another fact to consider in judging the vernalization program is that
Lysenko used the term in an exceedingly loose way; it covered almost
anything that was done to seeds or tubers before planting. Non-Soviet
scholars who have written about Lysenko's vernalization have usually
concentrated on the more spectacular attempts, such as the "conversion" of winter into spring wheat. The "vernalization" of potatoes promoted by Lysenko included the sprouting of the tubers before planting

—a practice that practically every gardener in potato regions is aware of. Eric Ashby commented that some of the methods advocated under the rubric of vernalization were nothing more than ordinary germination tests (although these tests may have been urged as a face-saving device after the more radical vernalization measures failed).[23] And many of the crops that were grown with vernalization techniques might well have succeeded without them. In the absence of control plots it was absolutely impossible to determine to what degree vernalization contributed to the harvest.

The last point needs some elaboration. Many experiments with vernalization worked both ways. Lysenko frequently presented his evidence in terms of yields in a certain season with both vernalized and unvernalized plantings of the same crops. While the comparisons were not rigorous enough to serve as controlled samples, they do point out one significant fact: Vernalization was only very rarely used as an attempt to make possible the previously impossible—growing crops that had never been grown before in the region because of the climate. Rather, it was usually directed toward making traditional crops ripen earlier, or to grow a grain that because of the length of its growing season could only occasionally be successfully harvested by traditional methods in a certain region before frost. These are the kinds of experiments in which the evidence can be manipulated very easily, or where even sloppiness in record-keeping can conceal results from an honest researcher. A two- or three-day difference in date of ripening of a grain is a very inconsiderable period, subject to many different kinds of interpretation. A little enthusiasm in claiming victories for vernalization would go a long way in conditions of inaccurate records, uneven controls, variable agronomic conditions, impatience about verification, willingness to discount contradictory evidence on the basis of peasant methods, and impure plant varieties.

The more spectacular of Lysenko's vernalization claims can probably be accounted for by the impurity of Russian plant varieties and by Lysenko's extremely small samples. The best known of his examples of the conversion of winter wheat into spring wheat is the case of the *Kooperatorka* winter wheat.[24] Lysenko himself called it in 1937 "our most prolonged experiment at the present time." (This was at a time when vernalization had already become the subject of an enormous publicity campaign.) On March 3, 1935, Lysenko sowed this variety of winter wheat in a greenhouse that was kept until the end of April at a very cool temperature, 10 to 15 degrees centigrade. After the vernalization treatment the temperature was raised. Originally there were two (!) *Kooperatorka* plants, but one perished, Lysenko said, as a result of pests. The sole surviving plant matured on September 9, proving to Lysenko that vernalization had worked, since *Kooperatorka* normally matures in the spring. Grain was then taken from the plant and immediately sown, again in a greenhouse, where it eared as an F_2 generation at the end of January. Then on March 28, 1936, the third generation was sown, producing seed in August 1936. Hereafter the grain acted as a spring variety, and Lysenko maintained that its habit had been converted.

All that can be concluded from such an experiment is that Lysenko's methods were incredibly lacking in rigor. The ridiculousness of basing scientific conclusions on a sample of two need not be emphasized. The *Kooperatorka* was probably heterozygous; the one plant that survived could well have been an aberrant form. Even had several plants survived, a selection out of the variations would naturally occur. If one is attempting to convert a winter wheat into a spring wheat, and one sows in the spring, one will be likely to gather in the fall only the grains from those plants that did in fact mature. The effects of selection could be avoided, or rather determined, only by using a variety of known purity, coupled with careful statistical studies of many plants over a number of generations, including statistics for those plants that did not mature, and including large control groups of nonvernalized plants. (Such attempts to duplicate Lysenko's results were soon made outside the Soviet Union and did not succeed.)[25]

Despite the inaccuracy of Lysenko's methods as so far described, he has still not emerged as the dictator in biology that he later became. Vernalization is a perfectly respectable field in agronomy, and despite all the inaccuracy of his methods, some genuine contribution should be granted Lysenko in this area. He may not have been the original developer of the field, but he organized greater efforts in this sort of activity and attracted more attention to it than any predecessor. Many farmers and selectionists around the world have performed experiments without proper controls and have claimed results that other people could not duplicate. Why did not Lysenko remain a somewhat eccentric agronomist or selectionist, hoping for recognition by the academic biologists, working feverishly within the narrow confines of his method? And how did the cause of Lysenko become connected with that of dialectical materialism? There was no such effort in his early publications. And why, if the value of vernalization was at best dubious, did the government support him?

In order to attempt to answer these questions, it is necessary to turn from agronomy to politics. The essential clues to the Lysenko affair lie not in theoretical biology, not in Marxist philosophy, nor even in practical agronomy, but in the political, economic, and cultural environment of the Soviet Union in the late 1920's and early 1930's.

During most of the 1920's political and economic controls were rather lax, at least compared with what occurred later. The Communist Party would not, it is true, tolerate competing organized political groups; the Soviet Union was even then an authoritarian state, and the state security organs dealt summarily with persons suspected of active political opposition to Soviet power. But for the average Soviet citizen who accepted or was resigned to Bolshevik rule, the state was not seen as a threat. The workers had lost the possibility of actually controlling the factories, as some in the early twenties had wished to do, but the regime was partial to the workers as a class, and the industrialization program had not yet attained the strained tempo of the later five-year plans. The peasants were more prosperous than either before the Revolution of 1917 or after the collectivization program beginning in 1929. They had occupied most of the arable land which had belonged before the Revolution to the church, nobility, or crown, and the loose regulations on trade permitted them to profit from the sales of their produce. The academic intelligentsia, still overwhelmingly pre-Revolutionary in educational background and attitudes, was more uneasy than either the proletariat or the peasantry, but still tried to maintain something of its pre-Revolutionary mode of life.

All of this was changing by 1929, the year that Stalin called the Great Break.[26] The first five-year plan, launched in 1928, was marked by the nationalization of virtually all industry and the beginning of a frenetic pace of industrialization. The wrench of rapid industrialization was felt by every Soviet citizen. In late 1929 the peasants were swept into a collectivization program that within a few months reorganized the entire countryside into massive state or collective farms. Many of the peasants resisted this program bitterly, destroying their crops and animals when all other opposition failed. Stalin is supposed to have told Winston Churchill at Yalta that the collectivization program was more difficult for the Soviet Union than the later battle of Stalingrad. The academic profession also suffered the trauma of those years; re-elections of the members of the faculties of the universities resulted in the forcible installation of Communist professors. Members of the intelligentsia were exhorted to work for the success of the industrialization and collectivization programs.

Such, in the briefest scope, was the political and economic background impinging on intellectual life in the 1930's. The "second revolution" of those years was intended to construct socialism. Soviet socialism would involve new forms of organization of industry and agriculture that were assumed to be inherently superior. The new form in industry was based on state ownership and control of the means of production, a principle that involved a loss primarily by the previous owners or managers of industry, not the workers themselves. The new form in agriculture, however, was very different in its effects. All but the poorest peasants were deprived of their possessions and of control over land that they considered their own. The result of this deprivation was opposition by the peasantry to the government and a consequent agricultural crisis. Many peasants were deliberately withholding or destroying their produce. The survival of the Soviet regime in the early thirties was directly connected with its success in dealing with this agricultural crisis.

One of the many desperate needs of the Soviet government at this time was for politically committed agricultural specialists. The professional biologists in the universities and research institutes were ill suited for this role, both in terms of their politics and of their interests. The best of them were involved in theoretical questions that only later would have great economic benefit;[27] the twenties were the years of the fruit fly *Drosophila melanogaster,* not the years of hybrid corn, although a direct connection between the two types of genetic research showed itself dramatically in later years. Hybrid corn's day came primarily in the forties, and it would be only one of the practical triumphs issuing from the science of genetics.[28] Furthermore, the professional biologists, like many leading Soviet scientists of this time, were frequently from bourgeois families. Often educated abroad and almost always aware of foreign developments, at least in their fields, they were members of that class falling under suspicion in the early thirties. It would require only a little imagination to convert their disinterest in agriculture into purposeful "wrecking" of the socialist economy, or their interest in eugenics into sympathy with fascist theories of racism, or their emphasis on the relative immutability of the gene into an attempted rescue of the biological fixity favored in earlier times by the Church.

Lysenko, on the other hand, was seen by many Soviet bureaucrats as a precious commodity.[29] Of peasant family background, he was committed to the cause of the Soviet regime, and instead of trying to avoid the tasks of practical agriculture, he placed all his limited talents at its disposal. Whatever the Party and government officials urged in the way of agricultural programs, Lysenko supported. In later years his shift of attention to support whatever the Party called for became a studied maneuver. After World War II Stalin said he would "transform nature" through the planting of shelter belts, and Lysenko came up with a plan for the nest-planting of trees; after Stalin's death his successor, Malenkov, called for an increase of crops in the nonblack-earth belt, and Lysenko produced a suggested method for fertilizing this kind of land; Khrushchev in turn became entranced with growing corn after visiting the United States, and Lysenko, swallowing his pride as he accepted this product of modern genetics, promoted the square-cluster method of planting it; later, Khrushchev called for the U.S.S.R. to overtake the U.S. in milk and butter production, and Lysenko shifted his attentions to the breeding of cows with high-butterfat milk.

In the early and mid-thirties Lysenko built up his strength by urging vernalization on the collectivized farms. Completely aside from its dubious practical value, vernalization had a significant psychological value. The primary question of the times was not so much whether vernalization would work as whether the peasants would work. Still alienated by the collectivization program, the peasants found difficulty seeing very much "new" about "socialist agriculture" except the fact of dispossession. Lysenko and his followers introduced a great deal that was new. They organized the peasants weeks before spring plowing and planting normally began, in the historically "slack period" for the countryside, and had them preparing seed. Lysenko and his assistants not only saw to it that the seed was prepared, but that it was, in fact, planted, no mean feat at that time. They soon developed other plans that involved the peasants in projects they had never before witnessed; if they were not soaking seeds in cold water, they were planting potatoes in the middle of the summer, or plucking leaves from cotton plants, or removing the anthers from spikes of wheat, or artificially pollinating corn.[30] These are only a few of Lysenko's projects. The intrinsic value of them is doubtful —today the Soviet government does not promote a single one.[31] Yet in their time they were genuinely valuable to the Soviet regime, though for reasons that have very little to do with principles of agronomy. Every peasant who participated in these projects was enrolling in the great Soviet experiment; a peasant who vernalized wheat had already clearly graduated from the stage when he destroyed his wheat so that the Soviet government would not receive it.[32] Every one of Lysenko's projects was surrounded with the rhetoric of socialist agriculture, and those who liked his projects committed themselves to that cause. A novel action in the service of a cause represents an important psychological transition. One is tempted to say that the important thing about Lysenko's proposals was that they did not do too much harm, rather than that they did a great deal of good.[33] Some of the later ones did cause much damage, but only after his strength was already very great.

After Lysenko moved from Azerbaidzhan to Odessa in 1930, he met I. I. Prezent—in contrast to Lysenko, a member of the Communist Party and a graduate of Leningrad University. Prezent had once thought that Mendelian genetics was a confirmation of dialectical materialism, but he later "diverged from the formal geneticists on the most cardinal questions." [34] Unfortunately, very little is known about the causes of that change of opinion, so fateful to Soviet genetics. The economic and social issues already referred to must have played a role. Prezent is frequently described, both in and out of the Soviet Union, as the ideologue who was primarily responsible for systematically formulating Lysenko's views and for attempting to integrate them with dialectical materialism.[35] To ascertain the relative contributions of Lysenko and Prezent to the full system of Michurinist biology is an impossible task since they worked closely together and published several important works as co-authors. It is quite possible that once alerted by Prezent to the ideological possibilities of his biological views, Lysenko was as active as Prezent in expanding the system. But the fact remains that not until Prezent became his collaborator did Lysenko make an attempt either to connect his biological views with Marxism or to oppose classical genetics. The joint publication in 1935 by Lysenko and Prezent of "Plant Breeding and the Theory of Phasic Development of Plants" marks an entirely new stage in the development of Lysenko's career. It was his first publication in which he reached beyond agronomic techniques to a theoretical conception of plant breeding science, and it also was his first publication in which he submitted classical genetics to substantial criticism. The theoretical tenets of this publication will be considered in some detail in the section of this chapter that concerns Lysenko's biological system. At this point it is necessary only to notice several alterations in Lysenko's approach. Now Lysenko was beginning to think in terms of a polarity between socialist science and bourgeois science:

The Party and the government have set our plant-breeding science the task of creating new varieties of plants at the shortest date. . . . Nevertheless, the science of plant breeding continues to lag behind and there is no guarantee that this socialist task will be carried out within the appointed time.

We are convinced that the root of this evil lies in the critical state of plant biology that we inherited from methodologically bourgeois science.[36]

The tone of this publication differed sharply from Lysenko's earlier, pedestrian publications on vernalization. His ambitions had grown enormously: "We must fight uncompromisingly for the reconstruction of genetic plant-breeding theory, for the building of our own genetic plant-breeding theory on the basis of the materialist principles of *development,* which actually reflect the dialectics of heredity. Only by consciously building such a theory can the breeding process be provided with real guidance suitable for the requirements of socialist production." [37] Here we see that Lysenko had found a new vocabulary, based on "materialism" and "dialectics." How meaningful these references could be made remained to be seen.

Criticism of academic biologists in the Soviet Union was not totally new in 1935; it actually began at the end of the twenties, but these earlier censures should probably be seen as a part of the general suspicion of the bourgeois specialists, whatever their fields, rather than a specific attempt to displace classical genetics with a rival theory. Sometime before 1935 the various critical tendencies began to come together. Other rivulets of criticism joined the growing stream of disapproval of classical genetics in those years; the sources of these negative judgments were quite diverse. The relatively uneducated selectionists and a few of the older biologists had their own reasons for opposing modern genetic theories—reasons that had effects in other countries as well, including the United States. And the rise of fascism in Germany, supported by several prominent geneticists in Germany (and, of course, opposed by some), added a certain urgency to the growing controversy.[38]

Genetics had been seen by a number of its notable proponents as a key to radical social reform, a natural ally of Soviet socialism, rather than its opponent. A prominent geneticist in the twenties was Iurii A. Filipchenko, director of the Bureau of Eugenics of the Academy of Sciences. Filipchenko was concerned for the fate of the Russian intellectual elite, which he thought was not reproducing itself; he considered the dissemination of marriage advice to be one of the responsibilities of his bureau, and he hoped thereby to strengthen the genetic position of Russian scholars.[39]

The possibility of a Soviet sponsorship of eugenics for the cultivation of talent may seem remote in view of the later opposition to genetics as a whole, but the twenties were a period when many things seemed possible. Although Filipchenko backed away from radical eugenic proposals, other writers in this period spoke of how the dissolution of bourgeois family relations would permit couples to choose sperm donors of great intellectual ability who would provide for "1,000 or even 10,000 children." [40]

Nikolai Vavilov, the most prominent of Soviet geneticists, was also clearly attracted by the possibility of a union between the Soviet state and genetics, although on different grounds. This commitment to an alliance of socialism and science is frequently forgotten by non-Soviet observers who know only of his subsequent martyrdom. Born in a wealthy merchant family in 1887, educated in England under William Bateson, one of the leaders of neo-Mendelism, Vavilov returned to Russia at the beginning of World War I. After the Revolution he became a leading administrator of Soviet science.[41] His most important work, *The Centers of Origin of Cultivated Plants,* published in 1926, developed the theory that the greatest genetic divergence in cultivated plant species could be found near the origins of these species. This conclusion led him to expeditions to many remote places. His other major theoretical work, "The Law of Homologous Series in Variation," first published in 1920, was based on the belief that related species tend to vary genetically in similar ways. He later criticized this work for regarding the gene as too stable.[42]

Vavilov's real importance lay, not in his theoretical work, but in his collection of plant specimens from all over the world and his administration of a network of research institutions devoted both to theoretical genetics and the improvement of agriculture. He believed that the two goals could be reached best in Russia, under a socialist government.[43]

Among the foreign geneticists attracted to Moscow by the prospect of a union of socialism and genetics was the American future-Nobelist H. J. Muller, who came in 1933 expecting to find a place where he would not suffer for his communist sympathies. An earlier visit to the U.S.S.R. had had a great impact on Muller and on Soviet genetics.[44] Muller had from his early youth been committed to socialism and to the control by man of his own genetic future. In his unpublished autobiographical notes, written about 1936, he commented that after being shown fossil horses' feet at the age of eight, "the idea never left the back of my head, that if this could happen in nature, men should eventually be able to control the process, even in themselves, so as greatly to improve upon their own natures. In 1906 I began a lasting friendship with Edgar Altenburg, then a classmate. . . . He and I argued out vehemently and to the bitter end all questions of principle on which we differed, and thus he succeeded in converting me both to atheism . . . and to the cause of social revolution." [45]

How ironic that not Lysenko, but his opponent Muller, should be the true revolutionist, the scientist committed to basic social change and the modification of man's heredity through eugenics! (See p. 236 for a discussion of Soviet opposition to eugenics.) In 1935 Muller published a book, *Out of the Night,* in which he stated that only in a society where class differences had been abolished could eugenics be properly implemented. In the preface to this book, the only one of which he was sole author, Muller commented, "our airy imaginings concerning the future possibilities of cooperative activity on a grand scale are brought down to earth and given substance when we turn to the great and solid actualities of collective achievement which are becoming increasingly evident in that one section of the world—the Soviet Union—in which the fundamental changes in the economic basis have already been established." [46] In the Soviet Union Muller tried to promote his book, but was rebuffed.[47] As Lysenkoism grew in strength, Muller became a firm anti-Stalinist; he made the struggle against Lysenkoism one of the two major campaigns of his life, the other being his fight against radiation hazards.

But there is no evidence that Muller's opposition to Stalinism resulted in a change of heart toward socialism. His colleague T. M. Sonneborn of Indiana University wrote of him that "his disillusionment with Stalinism left completely unchanged his conviction that a socialistic economy was necessary for effective and wise control of human evolution." [48]

The first known attack upon Vavilov and his Institute of Plant Industry came in an article in 1931 by A. Kol', who called the institute "alien" and "hostile"; he criticized it for devoting its attention to the morphology and classification of plants rather than their economic significance.[49] This attack, though serious, was typical of criticisms leveled at theoretical institutes in those days, including many in fields outside biology. Vavilov attempted to answer the charges by pointing to the many varieties of plants (potatoes, corn, wheat) found by his institute around the world, which might eventually help the Soviet economy.[50] He stressed how deeply his institute felt its responsibility to socialist construction. But the disadvantage of the theorist in defending his science was clearly revealed by the editor's note to the exchange between Kol' and Vavilov, which commented that despite Vavilov's reply, Kol' was correct in noting many deficiencies in Vavilov's institute. The source of these shortcomings, said the editor, was that the

orientation toward the "needs of tomorrow" about which Academician Vavilov writes turns out to be for many partisans of "pure science" a convenient cloak for ignoring the needs of *today* in bringing about a socialist reconstruction of agriculture.[51]

No attempt will be made here to follow the entire sorry story of the growing campaign in the thirties against Vavilov and the classical geneticists, a campaign that Lysenko had clearly joined by 1935. That series of episodes can be best followed in the careful studies of David Joravsky. The important fact is that although Lysenko may have been the architect of a great deal that was done in his name, no aspiring promoter of a peculiar scientific system ever fell into a more personally fortunate (and historically tragic) situation. The relationship between Lysenko and his environment was one of mutual corruption. As C. D. Darlington commented:

His modest proposals were received with such willing faith that he found himself carried along on the crest of a wave of disciplined enthusiasm, a wave of such magnitude as only totalitarian machinery can propagate. The whole world was overwhelmed by its success. Even Lysenko must have been surprised at an achievement which gave him an eminence shared only by the Dnieper dam. . . .[52]

Just who the early promoters of Lysenko within the official bureaucracy were is difficult to determine. Lysenko himself gave a great deal of credit to Ia. A. Iakovlev, who after December 1929 was people's commissar of agriculture for the U.S.S.R. Iakovlev was in an instrumental position to assist Lysenko; that he had previously been assistant commissar of the Workers' and Peasants' Inspectorate, an influential apparatus for checking on the performance of the governmental organs, is an additional sign of his significance. In 1935 Lysenko commented that if Iakovlev had not snatched up the method of vernalization in 1930, "vernalization today would not be in the same shape as it presently is." [53] Professor Joravsky has also added P. P. Postyshev, M. A. Chernov, and K. Ia. Bauman as early important supporters of Lysenko. But since all three men disappeared in the purges in the late thirties, as Joravsky notes, they were obviously not indispensable to the agronomist. That Lysenko also had the support of Stalin after 1935, at least intermittently, there is no doubt. In February of that year at the Second All-Union Congress of Collective Farmers and Shock-Workers, Lysenko presented a speech entitled "Vernalization Means Millions of Pounds of Additional Harvest," in which he called for the mobilization of the peasant masses in the vernalization campaign. At the same time, Lysenko apologized for his lack of ability as a speaker, saying he was only a "vernalizer," not an orator or a writer. At this point Stalin broke into the speech crying, "Bravo, Comrade Lysenko, bravo!" [54]

The support that Lysenko won from Stalin was doubtless very important in his continued rise. But it is difficult to find the reason for this sympathy in Stalin's theoretical writings. Some authors have maintained that Stalin was from a very early date committed to neo-Lamarckism; in support of this, frequent references are made to Stalin's "Anarchism or Socialism?" published in 1906. This argument becomes less convincing upon examination; only one phrase of "Anarchism or Socialism?" refers to biology, and it may not be significant.[55] Stalin's occasional praise of Lysenko was no guarantee of permanent favor; his praise for other prominent Soviet citizens was sometimes followed by their imprisonment. Rather than enjoying an assured place, it appears that Lysenko struggled constantly, along with many others, to maintain himself under Stalin.

In 1935 a steady stream of pro-Lysenko propaganda flowed in the meetings of agriculturists, in the popular press, and, increasingly, in journals. Lysenko was by this time receiving significant support from the official bureaucracy. Vavilov was replaced as president of the Lenin Academy of Agricultural Sciences by A. I. Muralov, who tried to compromise between classical genetics and Lysenkoism. In 1936 a "socialist competition" was conducted between Vavilov's Institute of Plant Industry and Lysenko's Odessa Selection-Genetics Institute. The results are unknown, but with the emphasis placed on quick results and declarations of plan-fulfillment, it is not difficult to guess.[56]

In December 1936 a great conference was held to discuss the issue of what Lysenko now called "the two trends in genetics." The conference came as a replacement for the Seventh International Congress of Genetics, scheduled to be held in Moscow, but cancelled by the Soviet authorities. The edited record of this conference, later withdrawn from circulation by the Soviet government, is one of the most interesting sources for the history of the Lysenko affair. Appropriately entitled "Controversial Questions of Genetics and Selection," it is, despite the editing, by no means a document of pro-Lysenko propaganda.[57] The speeches are so diverse in view that no classification system would be accurate. In order to give some sort of idea of the alignment of forces, however, I have categorized (somewhat arbitrarily, since a spectrum of opinion is involved) the forty-six speakers as seventeen anti-Lysenko, nineteen pro-Lysenko, and ten unclear in their stated opinions (which, of course, may not reflect their inner opinions). The roster of speakers included many of the major participants in the long struggle over Lysenko, including Vavilov, Lysenko, Dubinin, Ol'shanskii, and Prezent. Many of the opinions expressed were sharp. The theoretical aspects of the discussion will be taken up in the second section of this chapter, but some comments are appropriate at this point.

One of the most outspoken of the speakers was A. S. Serebrovskii, who said that although he agreed with the need to establish scientific research on a new socialist basis, he was horrified by the monstrous form this campaign was taking:

Under the supposedly revolutionary slogans "For a truly Soviet genetics," "Against bourgeois genetics," "For an undistorted Darwin," and so forth, we have a fierce attack on the greatest achievements of the twentieth century, we have an attempt to throw us backward a half-century.[58]

A similar portrayal of possible disaster was made by N. P. Dubinin, who three decades later would be one of the leaders in the reconstruction of Soviet genetics:

It is not necessary to play hide-and-seek; it is essential to say outright that if the view triumphs in theoretical genetics that Academician T. D. Lysenko says is best represented by I. I. Prezent, that will mean that modern genetics will be completely destroyed. (Voice from the hall: How's that for pessimism!)

No, this is not pessimism. I wish to pose the question sharply only because the topic of discussion concerns the most cardinal issues of our science.[59]

One of the most poignant moments of the conference came when the American H. J. Muller began his rebuttal to the followers of Lysenko by quoting from a letter he had just received from the English geneticist J. B. S. Haldane, who wrote that he had dropped his laboratory work in order to go to Madrid to participate in the defense of that city against Franco's forces. Muller and Haldane were members of that generation and of that group of intellectuals who saw the relationship between Marxism and science as a natural alliance, mutually beneficial. They were international in their politics, humanitarian in their ethics, and motivated by the principles of intelligence and co-operation.[60] Muller said that by encouraging Lysenkoism, the Soviet Union, which had long represented to him the march of progress, was turning its back on its own ideals. Lysenkoism was not Marxism, he suggested, but its opposite. He criticized the supporters of Lysenkoism from within the framework of Marxism.[61] The Lysenkoites, not the geneticists, were guilty of "idealism" and "Machism":

Only three kinds of people can at the present time speak of the gene as something unreal, as only a kind of "notion." These are, first, confirmed idealists; second, "Machist" biologists for whom exist only sensations about an organism, i.e., its external appearance or phenotype; some of these biologists at the present time are hiding behind the screen of a falsely interpreted dialectical materialism. And finally, the third category of such people is those simple minds who do not understand the subject of discussion.

The gene is a conception of the same type as man, earth, stone, molecule, or atom.[62]

But Muller came under considerable criticism at the conference for his comment that the gene was so stable that the "period between two successive mutations is on the order of several hundred or even thousands of years." The problem of the mutability of the gene was one of the three major controversies of the meeting; the others were the mechanism of change of heredity (influence of environment; role of chance) and the practical usefulness of the two main trends in Soviet biology.

Another great conference of Soviet biology was held October 7–14, 1939. A significant difference of this conference from previous ones was that it was organized and controlled by philosophers, the members of the editorial board of the theoretical journal *Under the Banner of Marxism*.[63] Many of the philosophers had by this time begun to grant Lysenko's claims that he represented the ideologically correct attitude toward genetics, although earlier they had resisted this conclusion. The incomplete record of the conference published in *Under the Banner of Marxism* indicates that there were fifty-three speakers, a number of whom were participants at the 1936 meeting. By the same simplified classification scheme used for the earlier meeting, I would term twenty-nine as "favoring" Lysenko, twenty-three as "opposing" him in terms of their public statements. Thus, although the result of the conference was again something of a victory for Lysenko, the opposition was at this date still strong. A crude sort of compromise that granted the continued right of the classical geneticists to express their opinions was being observed. Vavilov pointed to the growing use of hybrid corn in the United States as a direct result of genetics research.[64]

By this time the tone of the Lysenkoites had become blatantly aggressive;[65] they demanded changes in school curricula and research programs. V. K. Milovanov commented, "Until the present time departments of genetics have continued to exist: we should have liquidated them long ago." [66] Lysenko had earlier said that Mendelism should be expelled from the universities.[67] Prezent was now working with the Commissariat of Education in order to revise the biology courses of the grade schools; as a result, the teachers and pupils were "completely disoriented on biological questions." [68] The belligerence was apparent in the way in which Lysenko described himself and his opponents. He appropriated the word "geneticists" for his followers; his opponents were "Mendelists." Only the Mendelists were grouping together; Lysenko refused even to admit that he had a "school." Instead, he stood for the broad science of biology, loyal to Darwin and Marx, while his opponents succumbed to antiscientific and clerical views. The *rapporteur* of the conference, V. Kolbanovskii, hardly neutral, called Lysenko's theories "progressive" and "innovative." P. F. Iudin, the philosopher who closed the conference, commented that there was no need to adopt a resolution since he was certain that all institutions and individuals connected with biology would support the opinions of the editors of the philosophy journal. He called upon the academic geneticists to reject that "rubbish and slag that have accumulated in your science." [69]

In 1940 Nikolai Vavilov was arrested and subsequently died in a prison camp.[70] The disappearance of the leader of the academic geneticists, a man whose talents were recognized even by his opponents, meant that no scientist was immune. With Vavilov gone, many of the neo-Mendelian geneticists became silent. Some sought work elsewhere, in less controversial fields. Others continued research in genetics, but on a more limited scale than previously.

The culmination of the genetics controversy in the Soviet Union came at the 1948 session of the Lenin Academy of Agricultural Sciences, when genetics as known in the rest of the world was prohibited. The background of this conference is still not clear; it seems to have been preceded not by growing support for Lysenko, as one would imagine, but by growing criticism. A Soviet biologist who wrote a history of the Lysenko affair commented that by late 1947 Lysenko's political standing was much lower than before the war.[71] For evidence to support my opinion that Iurii and Andrei Zhdanov were among the most influential critics of Lysenko, see Appendix I of this study.

The sad story of the 1948 genetics conference has been told outside the Soviet Union many times; the proceedings of the conference are available in English, unlike the records of the earlier meetings.[72] Of the fifty-six speakers, only six or seven defended genetics as it was known elsewhere, and of these the most important were later forced publicly to recant. Lysenko revealed in his final remarks that the Central Committee of the Communist Party had examined and approved his report. Evidently he knew of this all through the conference while some of his opponents, ignorant of the prior Party decision, seriously implicated themselves by resisting Lysenko. At the moment the Party decision was announced, the entire conference arose to give an ovation in honor of Stalin. The participants sent the Soviet leader a letter of gratitude for his support of "progressive Michurinist biological science," the "most advanced agricultural science in the world."

In the months following the 1948 conference, research and teaching in standard genetics was suppressed in the Soviet Union. The ban remained until after Stalin's death in 1953. The recovery that occurred during the years after Stalin's passing was painful and fitful.

Lysenko's Biological System

By 1948 all the major components of Lysenko's biological system had been developed. Lysenko's views on biological development were contained in a vague doctrine, the Theory of Nutrients. The word "nutrient" (*pishcha*), used in a very broad sense, seemed to include for him such environmental conditions as sunlight, temperature, and humidity as well as chemical elements in the soil, or organic food, or gases present in the atmosphere.[73] The Theory of Nutrients was, then, a general theory of ecology. To Lysenko any approach to the problem of heredity must start with a consideration of the relationship between an organism and its environment, and the environment in the final analysis determines heredity, although through intermediate mechanisms in such a way that each organism possesses a certain hereditary stability at any point in time.

The most important influence on Lysenko in the development of his Theory of Nutrients was his work at the end of the twenties and the beginning of the thirties on the effects of temperature on plants. Lysenko came to the conclusion that the ecological relationship between an organism and its environment could be divided into separate phases or periods during which the requirements of the organism differ sharply. Hence, his views were sometimes broadly labeled the Theory of Phasic Development of Plants, although the Theory of Nutrients is a more comprehensive title, describing both plants and animals, both the phasic development and other ramifications of his views.

As an example of the different phases that plants pass through, Lysenko pointed to the requirements of many cereals, including both spring and winter varieties, for lower temperatures near the beginning of their growing periods than at the end. Whether the grain is normally planted in the fall or in the spring, the period immediately before and after germination is much colder than the period immediately before and during maturity. This relative coldness is not only a circumstance of the climate but is, at the present point in the evolution of many of these cereals, a necessity for their full life cycle. Lysenko noted that many varieties of winter wheat will not reach maturity if grown under uniform conditions at a relatively high temperature (10 to 12 degrees centigrade).[74]

Lysenko did not see the vernalization phase as necessary for cereals only; all plants pass through different stages, he thought, and for many of them the vernalization phase is the first. Cotton, for example, needs very warm temperatures in its first stage, relatively lower ones in its last, or boll-ripening, period. Even though the cotton plant, therefore, has requirements just the opposite of wheat's—relative heat in the beginning, relative coolness at the end—the first stage of the cotton plant can also be called its vernalization phase, Lysenko thought, a period in which temperature is of fundamental importance. Unless a plant, whether it be wheat or cotton, has passed through this vernalization phase, it cannot reach maturity—that is, bear fruit. From this observation Lysenko proceeded to a generalization:

The changes in the environmental conditions which developing plants need indicate that the development of annual seed plants from germination of the seed to the ripening of the new seeds is itself not uniform in type, not uniform in quality. *The development of plants consists of separate qualitatively different stages, or phases.* To pass through the different phases of development, the plants require different external conditions (different nutriment, light, temperature, etc.). Phases are definite *necessary* stages in the plant's development, and serve as the basis of the development of all the plant's *particular forms*—its organs and characters. The different organs and characters can develop only at definite phases.[75]

The vernalization phase is only one of several different stages that plants supposedly must pass through in order to be able to bear fruit. But Lysenko never gave a coherent description of just what these other stages were. He did maintain that for many cereals the stage immediately succeeding vernalization—in which temperature is so important

—is the photo phase, during which duration of daylight becomes critical. But while in each of the two phases described Lysenko indicated that one factor becomes critical to the development of the organism, he also emphasized that these factors *alone* are not sufficient to guarantee correct development. Each phase should be seen as a complex of factors necessary for the organism. Here, as in many other cases, Lysenko was unclear on just how one differentiates between the phases, since in every phase both temperature and light are among the complex of factors affecting growth. He tried to overcome this difficulty of defining the phases by maintaining that under normal conditions all the other factors except the crucial one are present in correct measure to enable the organism to successfully pass through the phase. As Lysenko described the vernalization phase:

If it were possible in practical farming artificially to regulate the temperature over wide field spaces, the thermal factor would be the chief factor in the spring sowing of winter plants. It would only be necessary to regulate it to suit the plants' requirements for going through the vernalization phase. All the other factors of the set required for vernalizing cereals under the spring field conditions prevailing in our districts are always present in the necessary doses.[76]

Here is a typical example of Lysenko's inexact approach to biology. It is clear that he was attempting to develop a general law of the development of organisms. This law is based on the concept of phases through which all plants must pass. A priori there is, of course, nothing wrong with such an attempt; its success depends on the rigor, usefulness, and universality of the stages that are defined. In order to give his law consistency and rigor, Lysenko had to define the phases through which the plant passes. He endeavored to do this by identifying certain factors, such as temperature, that during certain periods of development are "critical." But then he seriously limited the generality of his definition, and hence of his law of phasic development, by saying that in *our* districts all factors other than the critical one are always present in the necessary doses. The implication of the term "our" is that Lysenko's "critical factors" might not be "critical" at all in other districts, under other conditions. Might there be areas where no one factor can be selected as "critical"? Or is there *any* area where *one* factor is clearly critical? And the phases through which the plant must pass after vernalization were not even discussed to the inexact degree that vernalization was.

Vernalization itself is a perfectly legitimate topic of investigation in plant science; Lysenko's major errors were not in the subject of study he undertook but the methods he used and conclusions he reached. A perusal of the scientific literature reveals a vast amount of evidence on vernalization, some of it obtained in the same years during which Lysenko was working.[77]

In his "Plant Breeding and the Theory of Phasic Development of Plants," published in 1935 in collaboration with Prezent, Lysenko began reaching beyond simple studies of vernalization to a general theory of heredity.[78] His primary complaint in 1935 against classical genetics seems to have been that the geneticists could not predict which characters would be dominant in hybridization and worked primarily by means of making many thousands of combinations. Lysenko's impatience—linked with the impatience of the government in its hopes for rapid economic expansion—drove him to the hope for short cuts.[79] He believed that dominance was dependent on environmental conditions: "We maintain that in all cases when a hybrid plant is given really different conditions of existence for its development this causes corresponding changes in dominance: *the dominant character will be the one that has more favorable conditions for adapting itself to its development.*" [80]

These views were ramified in succeeding years. The most complete statement of Lysenko's theoretical views was contained in his "Heredity and its Variability," first published in 1943. It is to this source we must turn in an effort to give a fuller statement of Lysenko's system.

Lysenko denied the distinction between phenotype and genotype[81] even over the distance of one generation. He observed that *"all the properties, including heredity, the nature, of an organism, arise de novo to the same degree to which the body of that organism (for example, a plant) is built de novo in the new generation."* [82] The obliteration of this separation lay at the bottom of much of Lysenko's writings.

Heredity was defined by Lysenko as *"the property of a living body to require definite conditions for its life, its development and to react definitely to various conditions."* [83] Lysenko, then, described heredity in terms of the relationship of an organism to its environment rather than in the traditional sense of the transmission of characters from ancestor to descendant. But he confused his definition by adding that "the nature of the living body" and "the heredity of the living body" are nearly alike. Just what the "nature of the living body" consisted of was left unsaid beyond returning to the already cited statement concerning the requirements for "definite conditions of life."

The heredity of a living body, according to Lysenko, was built up from the conditions of the external environment over many generations, and each alteration of these conditions led to a change in heredity. This process he called the "assimilation of external conditions." Once assimilated, these conditions become internal—that is, a part of the nature, or heredity, of the organism: "The external conditions, being included within, assimilated by the living body . . . become particles [*chastitsami*] of the living body, and for their growth and development they in turn demand that food and those conditions of the external environment, such as they were themselves in the past." [84] In the last part of this sentence Lysenko referred to the part of his biological system that avoided a totally arbitrary plasticity of organisms. The mechanics of the transition from "external conditions" (temperature, moisture, nutriments, and so on) to "internal particles" was, to say the least, unclear, but Lysenko did achieve in this way a concept of material carriers of heredity. These "internal particles" may seem at first glance the same as genes, but it is clear from Lysenko's description and later comments that they are not. Rather than being unchanging, or relatively unchanging, hereditary factors passed from ancestors to progeny, they are internalized environmental conditions. [85]

Despite the distinction, Lysenko's particles did perform the function of providing—under certain conditions—heredity in the customary sense. This heredity he described as the conservative tendency of any organism in its relationship to its environment. If an organism exists in external surroundings similar to those of its parents, then it will display characters similar to its parents'. If, however, the organism is placed in an environment different from that of its ancestors, its course of development will be different. Assuming that the organism manages to survive, it will be forced, Lysenko thought, to assimilate the different external conditions of its new environment. This assimilation leads to a different heredity, which in several generations may become "fixed" in the same way in which a different heredity had been fixed in the earlier environment. In the intermediate, or transition, period, the heredity of the organism is "shattered," and therefore unusually plastic.

Lysenko believed that there existed three different ways in which one could "shatter," or remove the hereditary stability of, an organism. One could place the organism in different external environments, as already described. This method was much more effective at certain stages (for instance, vernalization) of the development process than others, he thought. One could graft a variety of a plant onto another, thereby "liquidating the conservatism" of both stock and scion. Or finally, one could cross forms differing markedly in habitat or origin. Each of these methods was attempted in Lysenko's experiments.

Organisms that were in the shattered, or destabilized, state were, Lysenko thought, particularly useful from the standpoint of manipulation. One could, within certain limits, give them new heredities by placing them in environments of carefully specified (and desired) conditions.[86] In several generations the organism's heredity would stabilize to the point that the organism would henceforth "demand," or as a minimum, "prefer," that environment.

Although Lysenko referred to hereditary particles, he was extremely indefinite on the location and function of these particles. His concept of them certainly did not involve "particulate heredity" in the usual sense of nonblending hereditary factors, nor did it permit a conceivable separation of the particles from the rest of the organism. He observed that *any living body part, and even a droplet (if the body is liquid) possesses the property of heredity, i.e., the property of demanding relatively determined conditions for its life, growth, and development.*[87] This view reminds one of Darwin's theory of pangenesis, with Lysenko's "particles" being Darwin's "gemmules," which were supposedly given off by every cell or unit of the body. This theory of Darwin's has, of course, been discarded in the light of modern genetics. One might add that in Darwin's time, the theory explained phenomena that otherwise could not be explained; Darwin was, further, aware of its speculative character, and labeled it "provisional." Lysenko's theory, on the other hand, inadequately and incorrectly accounted for phenomena that were better explained by another existing theory. Thus, even though Darwin's and Lysenko's theories in this particular instance were very similar, the historian of science would easily conclude that Darwin's effort was innovative and useful, even if tenuous, while Lysenko's was essentially retrogressive.[88]

Lysenko's view of the possible types of inheritance included the case of particulate, or mutually exclusive, inheritance, but went far beyond it. His system was borrowed largely from Timiriazev, who in turn had been influenced by earlier biologists. Here again, Timiriazev's scheme was, at the turn of the century, fairly plausible. By the time Lysenko espoused it, genetics had created a far superior scheme, which Lysenko never mastered. Timiriazev's and Lysenko's categories of inheritance can best be described in terms of a diagram given in Hudson and Richens's careful study; the same scheme was described by Lysenko in his *Heredity and its Variability:* [89]

Simple inheritance
  (one parent
  involved)

Complex inheritance
  (two parents
  involved)

Mixed inheritance
  (mosaics of parental
  characters)

Blending inheritance
  (blends of parental
  characters)

Mutually exclusive in-
  heritance (complete
  dominance of one or
  the other parent)

Millardetism
  (F_2 generation
  not segregating)

Mendelism
  (F_2 generation
  segregating)

Examples of simple inheritance, in which only one parent is involved, would include all types of asexual and vegetative reproduction (self-pollination in plants such as wheat, propagation from tubers or cuttings, and so forth) and parthenogenesis.

Complex inheritance involves two parents, and according to Lysenko, this *"double heredity gives rise to a greater viability of the organisms, and to their greater adaptation to varying living conditions."* [90] Lysenko, therefore, felt that the offspring of two parents possessed, in potential, all the characters of both parents, and he looked with disfavor upon inbreeding, or self-fertilization, which led, he thought, to a narrowing of the potentialities of the organism.[91] In the case of double heredity with unrelated parents, the characters that would actually be displayed depended on, first, the environment in which the organism was placed, and second, the unique properties of the particular organism involved. The interaction of the environment and these unique properties led to the "types" of complex inheritance: mixed, blending, and mutually exclusive.

Mixed heredity was, to Lysenko, represented by progeny that displayed clear (unblended) characters of both parents in different parts of their bodies; examples would be variegated flowers, piebald animals, and grafts of the type known to geneticists as chimeras (mosaic patterns of genetically distinct cells formed by artificial grafting of two different plants). Lysenko gave a number of examples of mixed heredity, the best known of which was the supposed graft hybridization of tomato plants by Avakian and Iastreb, in which the coloration of the fruit of the scion was reportedly influenced by the stock. This experiment was investigated by Hudson and Richens, who concluded that it was of doubtful validity.[92] If the tomato plants were heterozygous and if stray cross-pollination occurred, the results could be explained in terms of standard genetics. Whether graft hybridization ever occurs has been a hotly disputed question in biology, but Lysenko's failure to use proper experimental controls eliminated him as a reliable participant in the debate.[93]

Blending inheritance was, to Lysenko, the merging of characters in the hybrid in such a way that they were intermediate between those of the parents. Many cases of such inheritance are known. It is obvious, for example, that the progeny of marriages between humans of distinctly different skin color are frequently intermediate in color, and a whole spectrum of intermediate forms may occur with no clear relationship to the Mendelian ratios. The major difference between Lysenko's interpretation of this continuous variation and that of modern geneticists is that the latter see continuous variation as the result of a series of independent genes that are cumulative in effect, but each of which still functions discretely, while Lysenko spoke simply in terms of blending.[94]

"Mutually exclusive inheritance" was the term used by Lysenko to cover the phenomenon of complete dominance. Lysenko did not see dominance in the customary terms of the mechanism of allelic pairs, only one of which in the hybrid form is expressed in the phenotype, but instead in terms of the relationship of the organism to the environment.

There were no dominant and recessive genes, he thought, but only "concealed internal potentialities" that may or may not "find the conditions necessary for their development." He wrote, "In a certain sense, all properties and characters may be regarded as concealed, recessive. . . ." Those characters would be expressed that found the "proper conditions." Lysenko felt that this theory provided a better means for manipulating heredity, since there was not inevitable dominance of one character over another and characters could be altered by man.

Lysenko saw two different types of mutually exclusive inheritance, which he called "Millardetism" and "Mendelism," or "so-called Mendelism." Millardetism, named after the French botanist, was used to describe hybrids that in subsequent generations supposedly never display segregation. The dominance that was displayed in the F_1 generation continues, Lysenko reported, in all other generations. Lysenko maintained that there was nothing surprising in this, since his general theory of the expression of characters rested on the relationship of the organism to the environment; therefore, the correct environment would always cause the appearance of the appropriate character. Lysenko's followers cited a number of experiments that allegedly supported this conclusion. There is nothing in classical genetics to explain these particular cases, although it is not difficult to imagine errors that might lead one to such a conclusion.[95] Lysenko's results were not verified abroad. But as in many other of Lysenko's claims, refutation of his position would require a universal statement ("Hybrids *always* segregate in subsequent generations"), which in theory can never be established, since one exception would destroy it.

"So-called Mendelism," the last of Lysenko's types of inheritance, refers to hybrids that do segregate in F_2 and subsequent generations. Lysenko considered them isolated cases and insisted, like Timiriazev, that Mendel did not actually discover this type of inheritance. The Mendelian laws themselves were "scholastic" and "barren," did not reflect the importance of the environment, and did not permit the prediction of the appearance of characters before making empirical tests for each type of organism.

So far nothing has been said concerning Lamarckism or the inheritance of acquired characters, two topics that are usually mentioned early in any discussion of Lysenko. It should be clear by now that Lysenko believed in the inheritance of acquired characters. The "internal-izing" of environmental conditions, which he considered the means by which the heredity of any type of organism is acquired, is obviously a type of such inheritance. And Lysenko himself stated his position un-equivocally:

A materialistic theory of the development of living nature is unthinkable without a recognition of the necessity of the inheritance of individual differ-ences by the organism in definite conditions of its life; it is unthinkable with-out a recognition of the inheritance of acquired characters.[96]

This is a clear case of Lysenko's appropriating Marxist philosophy to serve his own dated biological theories. There is nothing in systematic dialectical materialism that requires belief in inheritance of acquired characters. Materialism as a theory of knowledge and a view of nature does not even come close to including such a principle. Since Lysenko's mentors were all representatives of old biology, it is not surprising that he subscribed to the theory. Belief in the inheritance of acquired charac-ters "solaced most of the biologists of the nineteenth century," as a prominent geneticist of the twentieth century observed.[97] Thus Lysenko could cite Darwin as well as Timiriazev and Michurin in support of the view.[98] One might note parenthetically that the surprising aspect of Dar-win's attitude toward the inheritance of acquired characters was not that he believed in it (which he did), but that he relied so little on it for his great theory. That Marx and Engels also accepted it illustrates only that they were aware of the biology of their time.

Whether Lysenko was a Lamarckist is a more difficult question. The very term "Lamarckism" has been so devalued through wide currency that it probably should be discarded.[99] Lamarck did many things other than teach that acquired characters can be inherited, but he is now usu-ally remembered only for that, although most other biologists of his cen-tury believed it also (after Darwin). Furthermore, he believed that only use and disuse and the effort of organisms to improve themselves had effects on heredity, not the "conditions of the environment" that Ly-senko emphasized. Lysenko never seemed to consider use and disuse or self-improvement important, although a few of his enthusiastic follow-ers did.[100] And more interestingly, Lysenko denied that he was a Lamarckist.[101] Lamarck was a typical eighteenth-century materialist, Lysenko maintained, incapable of thinking "dialectically." There also seem to be no equivalents in Lamarckism to Lysenko's theory of shattering heredity, or his theory that heredity is a metabolic process. Therefore, genuine distinctions between Lamarckism and Lysenkoism do exist. Nonetheless, the two systems are similar in that both contain the principle of the inheritance of acquired characters. Some of the Soviet geneticists who later displaced Lysenko described his system as a "naïve Lamarckist view." [102]

There are other aspects of Lamarck's thought that resemble Lysenko's, but an evaluation of these similarities involves one in the very difficult problem of interpreting Lamarck. There is much debate among historians of science over whether Lamarck should be seen as one of the first of the evolutionists or the last of an earlier breed, the romantic scientists. Usually Lamarck is described as an eccentric, perverse, frequently wrong, but nonetheless brilliant precursor of Darwin. But there are some exceptions to this view. Professor Charles Gillispie of Princeton wrote:

. . . Lamarck's theory of evolution was the last attempt to make a science out of the instinct, as old as Heraclitus and deeply hostile to Aristotelian formalization, that the world is flux and process, and that science is to study, not the configuration of matter, not the categories of form, but the manifestations of that activity which is ontologically fundamental, as bodies in motion and species of being are not.[103]

According to Gillispie, it was not accidental that Lamarck achieved his position in the wake of the French Revolution. Gillispie believes that Lamarck belonged to the same radical, democratic, antirational camp as Diderot and Marat. These people, says Gillispie, were rebelling against the cold rationalism of Newtonian science, with its explanations of the "how" of things, with its emphasis on cold mathematics. They believed, says Gillispie, that "to describe is not the same thing as to explain. . . . [T]o analyze and to quantify is to denature."

It would not take very much work of the imagination to put Lysenko in this same romantic camp, responding to the same stimuli as Lamarck, if not to his ideas. Lysenko's hostility to mathematics has already been noted. He was also in a post-Revolutionary society. Gillispie observed, "It is no accident that the *Jardin des Plantes* was the one scientific institution to flourish in the radical democratic phase of the French Revolution, which struck down all others." [104] One might stress that Lysenko similarly flourished after the Russian Revolution. Lysenko believed in the inheritance of acquired characters, and after being philosophically educated by Prezent, subscribed to the view that all the world is in flux, as did Lamarck. But tempting as such a correspondence between Lamarck and Lysenko is in a number of ways, one should notice that it conceals as well as reveals. First of all, Lamarck himself is not entirely explained in this interpretation. We know that he was critical of the excesses of the French Revolution.[105] Although there were aspects of Lamarck's thought that were anachronistic, other aspects, particularly those relating to evolution, were based at least partially on the scientific evidence of his day. Lamarck was *both* a predecessor of Darwin and one of the last of the romantic scientists; he was much more of an intellectual than Lysenko ever thought of being. It seems quite certain that Lysenko will never be regarded as a predecessor of anyone of scientific importance, even if accepted views on the inheritance of acquired characters should greatly change. Furthermore, if one is to tie Lysenko to Lamarck because of his similar commitment to a philosophy of flux, in Heraclitus's sense, what is one to do with the classical geneticist H. J. Muller, Lysenko's opponent, who subscribed to philosophical Marxism on much more genuine grounds? And lastly, Lysenko based his interpretation of nature on Darwinism, which romantics of the late nineteenth century found heartless. Therefore, one is left with the impression that although there are genuine similarities between Lamarck and Lysenko—both in terms of their systems and their historical situations —there are also very real differences.

The discussion above has included mention of Lysenko's Theory of Nutrients, his concept of heredity, and his view of the mechanism of heredity. Many of the issues over which Lysenko quarreled with classical geneticists, such as the genetics of earliness, pollen fertilization,[106] the deterioration of pure lines, rejuvenation, and graft hybridization can be understood within the framework of the system so far described. The missing element in the discussion so far is the philosophical ingredient. In what way was this system connected with Marxist philosophy, particularly in view of its clear basis in the thought of people unschooled in Marxism, such as Darwin, Timiriazev, Michurin, and the pre-1930 Lysenko? We have noted that the genetics controversy seems farther from dialectical materialism than any of the other issues in this study. Nevertheless, manfully struggling and aided by a few eager ideologists, Lysenko was able to drag several of the issues of genetics into the realm of philosophy. The most important of these were: (1) the question of the mutability of the gene; (2) the question of the isolation of the genotype; (3) the question of the union of theory and practice in genetics; (4) the question of probability and causation.

The question of the mutability of the gene was a serious one, and one that occupied the attention of many of the best biologists of the early twentieth century. One is tempted to say that the closest the Lysenko affair came to a legitimate intellectual issue was its concern with the integrity of the gene—tempted, but not compelled, for the questions Lysenko asked had been rather fully answered a decade or two earlier. But in the first years of the century the problem worried many geneticists.

The issue has definite philosophic and religious implications, indirect perhaps, but real enough to many thinkers, including the geneticists. At the bottom of the discussion there existed a tension between two opposite, but not necessarily incompatible, tendencies, that of heredity and that of evolution. Heredity is a conservative force that tends to preserve similarities. Evolution is a process that depends upon differences. If heredity conserved perfectly, there could be no evolution.[107] The striking characteristic of the gene (named in 1909 by Johannsen), as it seemed to the early geneticists, was its stability over many generations. It seemed a threat to the common-sense (and dialectical materialist) notion that everything changes, and to the scientific concept of evolution.

It is sometimes forgotten by people outside the Soviet Union interested in the Lysenko affair—and it was totally ignored in the Soviet Union—that several of the men who created the science of genetics had great difficulty in accepting the concept of the extremely stable and constant gene. It seemed reminiscent of the fixity of the species favored in past centuries by the Church. T. H. Morgan was openly anticlerical in his views, and Muller accepted, as did most men of scientific bent (including the Marxists), the inevitability of change.[108] A. H. Sturtevant, another of Morgan's students, commented:

Do new genes in fact arise, or is all genetic variability due to recombination of preexisting genes? This question was seriously discussed—though the alternative to mutation seems to be an initial divine creation of all existing genes.[109]

But more important than religious of philosophical considerations in causing the early geneticists to be skeptical of the concept of a stable gene was the impact of evolutionary theory. As L. C. Dunn remarked:

The idea that the elements of heredity are highly stable and not subject to fluctuating variability was repugnant to many biologists. These included for a time William Bateson, W. E. Castle, T. H. Morgan, and others who helped to build the new science. There had been a natural growth in nine-teenth-century biology of faith in the opposite assumption: namely, that bio-logical forms and properties were inevitably subject to variation. The closer the biologist had been to Darwin's ideas and evidence on variation as the con-dition of evolutionary change, the more firmly did he hold this faith.

W. E. Castle was a conspicuous example of those who held the view that genes must be modifiable by selection. It was shared by many others to whom the inviolability of the gene to change from its genotypic environment in the heterozygous state seemed like arbitrary dogma. Castle cured himself of disbelief in the integrity of the gene the hard way—by 15 years of ardu-ous experimentation.[110]

Lysenko and his followers did not have the benefit of those fifteen years, nor would they consider seriously the published reports of the classical geneticists that had led them to change their opinions. Instead, the Lysenkoites raised the issue of mutability as evidence of the "ideal-ism" of formal genetics. And here they were able to find some support in dialectical materialism, which, like the philosophy of Heraclitus, in-cludes the principle of universal change. At the 1937 conference Prezent attacked H. J. Muller for his remark that "the gene is so stable that the period between two successive mutations is on the order of sev-eral hundreds or even thousands of years." [111] But by this time Prezent was already striking out against a straw man; the nature of mutations had been investigated rather thoroughly, and the importance of the cu-mulative effects of mutations to evolution was well known. When a per-son considers that one organism is thought to contain thousands of genes, one change even several hundred years in each gene could result in an appreciable rate of change. Biological evolution is built on the concept of great changes resulting from minute variations occurring over vast periods of time. As Vavilov commented at the same confer-ence, "None of the modern geneticists and selectionists believes in the immutability of genes. Essentially, genetics has the right to existence as a science and is attractive to us precisely because it is the science of the change of the hereditary nature of organisms. . . ." [112]

It becomes clear that the relative stability of the gene is not a serious obstacle to dialectical materialism. The rates of change in nature vary enormously; however slow the rate of change of the genotype may seem to the person eager for such change, it is obviously rather rapid when seen on an epochal time scale. Both the Lysenkoites and the formal geneticists took evolution for granted, and evolution is based on truly striking changes in heredity. Furthermore, dialectical materialists were quite willing to accept the existence in nature of matter that changes much more slowly than even the most conservative estimates of the changes in the genotype. The modifications in the interior structure of many rocks are much slower, and the geological clock in general runs more slowly than the biological one. No one has suggested for this reason that geology is undialectical. The commitment of dialectical materialism is that there be change, not that the change occur at a certain rate.

The question of the isolation of the genotype is similar in some ways to the issue of the mutability of the gene. The separation of the genotype from the phenotype was admitted exaggerated by Weismann, but this exaggeration was probably a necessary, or at least an understandable, step in order to throw off nineteenth-century concepts that attributed the property of heredity to all parts of the body instead of to discrete units within the body. When the full meaning of the germ-plasm theory had permeated biological thought, a great change in the concept of heredity resulted. While earlier the body, or soma, had been considered the carrier of heredity, the body was now seen as a temporal husk containing within in it an unbroken series of germ cells. In this view the soma was drastically demoted in status.

Early discussions of the germ plasm put great emphasis upon its isolation from the soma (the body of the organism, excluding the germ cells). Not until 1927, when Muller showed that mutations could be induced by radiation, did it seem possible to affect the genes by any environmental action. Before that time the gene seemed to be unapproachable by external stimuli. This question of the penetrability of the boundary between the gene and the soma became ideologically charged in the Soviet Union. According to the Stalinist version of dialectical materialism, there are no impassable barriers in nature; the short history of the Communist Party (not published until 1938, but indicative of official thought), which Stalin himself supervised, stated that "not a single phenomenon in nature can be understood if it is considered in isolation, disconnected from the surrounding phenomena." [113]

The statement of Lysenko and his followers that formal genetics postulated an entirely isolated genotype was a false one, based on obsolescent theories. Muller himself, known among geneticists precisely because he had disproved this isolation, was not able to establish his point among ideologists who did not wish to listen. Lysenko continued to insist that Mendelism was based on "an immortal hereditary substance, independent of the qualitative features attending the development of the living body, directing the mortal body, but not produced by the latter." To Lysenko, this was "Weismann's frankly idealistic, essentially mystical conception, which he disguised as 'Neo-Darwinism,'" and which still governed modern genetics.[114]

While geneticists had proved by 1927 that genes could be influenced by external stimuli, they could not obtain specifically desired changes in this way.[115] This uncontrollability of induced mutations was a major issue in the third ideological issue of the Lysenko affair, the question of the union of theory and practice. Michurin and his followers emphasized that every experimenter with plants should be a conscious transformer of nature. The formal geneticists, however, emphasized not only the extreme stability of the gene, but also the undirected character of those mutations that did occur. Thus, the Lysenkoites were able to portray the formal geneticists as having nothing of immediate value to the Soviet economy, while they, with their close ties to the soil and their commitments to socialized agriculture, were working constantly to strengthen the Soviet state. Lysenko was, in effect, constantly turning to the theoretical biologists with the query, "What have *you* done for Soviet agriculture?" [116] Michurin, Williams, Lysenko, and their disciples were among the few agricultural specialists who tried to do something immediately for Soviet agriculture. Speaking the same language as the peasants, they built up a strong set of supporters. Vavilov, it is true, was also deeply committed to the improvement of practical agriculture, but he suffered from the disadvantages of his bourgeois background and from his unwillingness to promise more than he could reasonably expect to produce. Vavilov knew well how many difficulties still faced geneticists who were seeking to control heredity. He was forced, therefore, into the position of being less optimistic than the exuberant Lysenko, who recited Michurin's words to the Soviet public: "It is possible, with man's intervention, *to force* any form of animal or plant *to change more quickly and in a direction desirable to man*. There opens before man a broad field of activity of the greatest value to him." [117]

One reason the geneticists were not able to defend themselves successfully against the charge that they were guilty of divorcing theory from practice was the state of genetics as a science in the thirties. Genetics had not yet produced its great practical triumph of hybrid corn, eventually to be so highly valued in both the United States and the Soviet Union. But genetics was clearly not the only such science; astronomy, descriptive zoology and botany, mathematics, and theoretical physics all contained areas of research that were farther from application than genetics. The real reason for singling out genetics, as already indicated, was the agricultural crisis in the Soviet Union and the existence of a group of old biologists and selectionists who were opposed to the new genetics. Thus, although the principle of the unity of theory and practice comes, along with the problem of mutability, as close as any ideological tenets to the issues in the Lysenko affair, it cannot by itself explain the controversy.

The principle of the unity of theory and practice is based on an unstated concept of time: Any theoretical development in science *should* be quickly applied, said the dialectical materialists, but how quickly was not specified. Obviously the application of a theory cannot in every case be simultaneous with its development. Premature application could result in great wastes. Therefore, the whole question of applying theory becomes subject to discussion. In a rational atmosphere, this discussion would revolve around criteria such as completeness of the theory, expenses and risks involved in attempts to apply it, and gains to be obtained from the application. From the standpoint of such criteria the Soviet geneticists of the thirties were not noticeably guilty of divorcing theory from practice. Indeed, Vavilov was devoted to the union of theory and practice in the best Marxist sense: He wished to combine the highest scientific principles with a commitment to the betterment of society.

The last ideological issue in the Lysenko affair was the question of probability and causation. A certain similarity existed here between the genetics controversy and the one over quantum mechanics. Certain writers outside the Soviet Union, such as Erwin Schrödinger, maintained that the undirected character of induced mutations obtained by radiation is connected with the indeterminism of quantum mechanics.[118] Some have advanced the theory that a mutation is similar to a molecular quantum jump.[119] The reason for the necessity of approaching genetics from the standpoint of probability, say these analysts, is essentially the same as the reason for using probability statistics in quantum mechanics. Thus, all the issues involving "denial of causality" that arose in quantum mechanics also arose in genetics—coupled, moreover, with the deep resentment of mathematics long evident in Lysenkoism.

Lysenko commented on this issue in his speech at the 1948 biological conference:

In general, living nature appears to the Morganists as a medley of fortuitous, isolated phenomena, without any necessary connections and subject to no laws. Chance reigns supreme.

Unable to reveal the laws of living nature, the Morganists have to resort to the theory of probabilities, and, since they fail to grasp the concrete content of biological processes, they reduce biological science to mere statistics. . . . With such a science it is impossible to plan, to work toward a definite goal; it rules out scientific prediction. . . .

. . . .

We must firmly remember that *science is the enemy of chance.*[120]

It is not necessary to review here the various interpretations that have been given to probability and causality by Soviet dialectical materialists; the main issues were described in the chapter on quantum mechanics.[121] While the problem of determinism in quantum mechanics still has its philosophically controversial aspects in the Soviet Union (as elsewhere), since the downfall of Lysenko it is no longer significant in biology.[122]

Contrary to many non-Soviet speculations, the inheritance of acquired characters was not upheld in the Soviet Union because of its implications for man. A number of observers of the Soviet Union have assumed that this theory held sway there because of its relevance to the desire to "build a new Soviet man." If Soviet leaders believed that characters acquired in a man's lifetime can be inherited, so the analysis went, then they would believe that a unique Soviet individual would emerge all the more quickly.[123] That this interpretation should play an important role in the U.S.S.R. seems almost predictable in view of Lysenko's belief that one of the advantages of Michurinism was that through knowledge of its principles scientists could control heredity, while the Mendelian approach to genetics was allegedly sterile. The logical extension of Lysenko's views would have been the employment of a "Michurinist eugenics" far more industriously than the Germans applied formal genetics. But this extension never occurred. Discussion of eugenics in the Soviet Union became impossible in the early thirties, because of the international situation. The eugenicist views of Nazi Germany undoubtedly played a large role in discrediting efforts in the Soviet Union to explain the emergence of superior individuals on the basis of biological theories.[124]

Lysenkoism Since 1948

The story of Lysenkoism after the historic 1948 session on biology is largely one of attempts by the biologists to displace Lysenko as the tyrant of their profession, while Lysenko skillfully shifted his emphasis from one nostrum to another—from the cluster-planting of trees, to the use of specified fertilizer mixes, to the square-cluster-planting of corn, to his methods of breeding cows for milk with a high butterfat content. At several moments in the 1950's criticism of Lysenko reached crescendos that seemed to indicate his inevitable demise, but each time he appears to have been rescued by highly placed individuals. Lysenko's resilience, his ability to take advantage of political situations and to curry favor, stood him in good stead. By this time, he was supported by an army of followers in the educational and agricultural establishments, men whose careers were intimately connected with Lysenko's school.

The first new endeavor for Lysenko after 1948 concerned Stalin's grandiose plan for the planting of forest shelter belts to control erosion and to combat dry winds in the steppe regions of the Soviet Union. This plan, heralded as "the transformation of nature," was adopted in October 1948 and expanded during late 1948 and 1949 to encompass eight large shelter belts with a total length of 5,320 kilometers and an area of 117,900 hectares.[125] The area where the belts were planned was extremely dry and unsuited for trees; as the minister of forestry commented, "The history of forestry does not know any examples of forest planting in such an environment." [126]

Lysenko suggested planting the trees in clusters or nests, on the theory that competition exists only between different species in the organic world, not within species.[127] He had suggested cluster-planting before for other plants.[128] Lysenko believed that in nature the life of every individual is subordinate to the welfare of its species. He maintained that while there is no intraspecies competition, there is intense competition between different species of the same botanical or zoological genus. Thus, giving the members of one species a numerical advantage helped them in their struggle with others. This position, similar to the "mutual aid" of Kropotkin, Chernyshevskii, and other nineteenth-century thinkers who found the principle of the survival of the fittest repugnant and hoped to replace it with the principle of co-operation, involved Lysenko in a difference with Darwin.[129] Although Lysenko usually defended Darwinism, especially against attacks based on theories of genetic stability, the "creative Darwinism" that he espoused was different in several important ways from Darwin's views. As already noted, he placed greater emphasis on the inheritance of acquired characters than did Darwin. He also differed with Darwin on the question of competition, believing that since Darwin ignored "qualitative leaps" in evolution, he "resorted to the reactionary pseudodoctrine of Malthus about intraspecific struggle . . . in order to explain the gaps between species." [130] But in reality, said Lysenko, such views sprang not from biology, but from capitalist economic and social views based on competition among antagonistic classes. Lysenko insisted that "there is not and cannot be a class society in any species of plant or animal. Therefore, there also is not and cannot be class struggle, although it has been termed intraspecies competition in biology." [131]

From all available evidence the shelter-belt plan was a failure. Shortly after Stalin's death in 1953 discussion of the project disappeared from Soviet publications. The viewpoint that intraspecies competition does not exist is so obviously false that it hardly needs to be considered. Any person who has witnessed the thinning out of forests or plants in congested clumps can give graphic evidence of the competition for food, water, and light that occurs within a species. The word "competition" here should not be understood in an anthropomorphic sense, but this cautionary note is equally valid, of course, with reference to interspecific competition. Lysenko himself recognized the phenomenon of thinning, but he refused to call it competition.[132] That competition exists within species does not deny the numerous examples of co-operation that can also be found.

The eventual fate of the afforestation project was clarified by an announcement that appeared in 1955 in a Soviet biological journal:

> T. D. Lysenko, contending that intraspecific competition does not exist in the organic world, proposed a method of planting trees in clusters. V. Ia. Koldanov has summed up the results of five years of using this method and has shown that it was erroneous in its very basis. Cluster plantings of trees have caused tremendous losses to the state and have threatened to discredit the idea of erosion-control forestation. T. D. Lysenko's method was refuted by the All-Union Conference on Erosion-Control Forestation held in Moscow in November 1954.[133]

Although it is frequently said that the possibility of mounting a serious attack on Lysenko after 1948 became possible only subsequent to Stalin's death, significant published criticism appeared shortly before the Soviet leader's demise on March 5, 1953. Beginning late in 1952, the publications *Botanical Journal* and the *Bulletin of the Moscow Society of Experimenters of Nature,* both under the editorship of V. N. Sukhachev, carried a long discussion of Lysenko's views, including both support and criticism.[134] The controversy eventually spilled over into other journals and even the popular press. It may not be merely coincidental that both publications that initiated the criticism were the organs of scientific societies, which, as descendants of private, voluntary associations, still preserved a greater sense of independence than the official scientific organizations of the Soviet Union.[135]

The *Botanical Journal,* in particular, organized a rather thorough discussion of Lysenko's opinions on species formation and examined in detail a number of claims promoted by followers of Lysenko concerning species transformation. In an article[136] that appeared in the November–December 1953 issue, A. A. Rukhkian revealed as a fraud the case of a hornbeam tree changing into a hazelnut, which had been reported by S. K. Karapetian in Lysenko's journal *Agrobiology* in 1952 and also in a publication of the Armenian Academy of Sciences. The branch of the hornbeam that had supposedly changed into a hazelnut was actually grafted into the fork of the hornbeam; Rukhkian even turned up a man who admitted making the graft in 1923. Rukhkian went to great pains to point out that although this case discredited Karapetian, it did not necessarily prove that all Lysenko's theories about species formation were wrong. Lysenko, however, destroyed the graceful exit being provided by Rukhkian; he heard of the critical article before it appeared and sent the editors a letter suggesting that its publication be postponed; furthermore, he affirmed his view that Karapetian's original article was completely correct and that the hornbeam had indeed changed into a hazelnut. The editors thereupon published both Rukhkian's article and Lysenko's letter. The article included photographs showing clear evidence of a graft. The result was the elimination of one of Lysenko's important pieces of evidence, and a severe blow to his standing. His integrity was now definitely in question. The editors also indicated their belief that the other cases of species transformation reported by Lysenko and his followers were easily explained on the basis of selection, grafting, or damage due to fungus (teratological changes).

This was only the beginning of a wave of criticism against Lysenko. In the next two years *Botanical Journal* received over fifty manuscripts analyzing some of Lysenko's claims, most of which could not be printed because of lack of space.[137] V. N. Sukhachev and N. D. Ivanov ridiculed Lysenko and his philosopher-defender A. A. Rubashevskii for the belief that intraspecific competition does not exist.[138] They added, "Yes, we disagree with Lysenko's 'new' teaching and his interpretation of natural selection, and we have been saying this forthrightly." At the same time they protested against being called anti-Marxists or anti-Michurinists. (The refusal to discard the Michurin label marked almost all Soviet criticism of Lysenko, including the most recent ones.) A detailed study by a special commission of the Latvian Republic's Academy of Sciences of an alleged pine tree with fir branches growing near Riga concluded, as in the case of the hornbeam, that the phenomenon was a graft.[139] S. S. Khokhlov and V. V. Skripchinskii examined Lysenko's claims concerning the conversion of spring wheats into winter forms, and of soft wheats into hard ones. Khokhlov concluded that the "engendering" of soft wheats from hard ones was the result of hybridization and selection.[140] Skripchinskii's conclusions were similar, and he went on to question the concept of the inheritance of acquired characters.[141] I. I. Puzanov charged that Lysenko was not so much promoting the views of late-nineteenth-century biologists as he was the "naïve transformist beliefs that were widespread in the biology of antiquity and the Middle Ages and that survived to some extent up to the first half of the nineteenth century." [142] S. S. Shelkovnikov maintained that Lysenko's arguments against Malthusianism and intraspecific competition were "based on equating the laws of development of nature and society, an equation that Marxism long ago condemned." [143] V. Sokolov, reporting in *Izvestiia* on his visit to the United States and Canada as a member of a Soviet farm delegation, praised hybrid corn, based on inbreeding and heterosis, two techniques condemned in past years by Lysenko. Sokolov commented, "In the U.S.A. and Canada we did not find a single farmer who planted corn with his own seed. . . . Our trip to the U.S.A. and Canada has convinced us that we are poorly acquainted with scientific achievements in foreign countries." [144] Running through all the criticism was the hope and demand for more freedom in the sciences. Two authors writing in the *Literary Gazette* observed that "the situation that has arisen in areas of such sciences as genetics and agronomy must be recognized as abnormal." [145] They called for the coexistence of differing schools in science. Two other authors, writing in the *Journal of General Biology,* observed, "The time of suppression of criticism in biology has passed. . . ." [146] In its summary of the long debate on Lysenko's view of species formation the editors of the *Botanical Journal* observed, "It has now been conclusively demonstrated that the entire concept is factually unsound, theoretically and methodically erroneous, and devoid of practical value." Furthermore, they observed, "not a single halfway convincing argument was conducted in 1954 or a single strictly scientific argument advanced in support of T. D. Lysenko's views. . . ." [147] A fairly harmless replacement for Lysenko as an idol of Soviet agriculture seemed to emerge in T. S. Maltsev, an experienced soil cultivator.[148]

The Soviet biologist Medvedev later wrote that by the end of 1955 more than three hundred persons had signed a petition requesting Lysenko's removal from the post of the president of the Lenin Academy of Agricultural Sciences.[149] In later months, during 1956 and 1957, the stream of criticism grew and seemed to many people to be irreversible. When Lysenko stepped down from the presidency of the Academy in April 1956, newspapers in countries throughout the world greeted the overdue downfall of a charlatan of biology.

Astoundingly and seemingly inexplicably, this phoenix rose to blight Soviet biology for yet another eight years. This phenomenon is even more striking than Lysenko's original ascent. By the 1950's the Soviet Union was already a modern state, dependent on sophisticated scientists and specialists of almost infinite variety, not the striving nation of the thirties, concentrating on coal, iron, and grain. In the same year in which Lysenko's new strength became discernible, the Soviet Union launched the world's first artificial satellite.

The rebirth of Lysenko in the late fifties seems to be most closely connected with the personal favor of Nikita Khrushchev, curried assiduously by the agronomist. Skillfully Lysenko maneuvered to stay a step ahead of his critics. At the time his views of species formation were being demolished in the biology journals, he was elsewhere pushing the use of organic-mineral fertilizer mixtures.[150] The Soviet fertilizer industry was not sufficiently developed to provide the large quantities of mineral fertilizers needed by agriculture. In the 1950's a desperate effort was being made to expand the fertilizer industry, but output remained insufficient. Lysenko came forward with a plan for mixing artificial and natural fertilizers, thus stretching the available supplies. His fertilizers were manure-earth composts enriched with various mineral fertilizers. This plan, of no theoretical significance to biology, had considerable appeal to the practical Khrushchev. Lysenko applied his method on his experimental farm on Lenin Hills near the city of Moscow. We know now, from a thorough investigation carried out by the Academy of Sciences in 1965, that a large part of Lysenko's considerable success with this method came, not from any genuine innovation in fertilizer techniques, but simply from his farm's very privileged position relative to other farms. Located near the capital city, in constant touch with the agricultural bureaucracy that was controlled by his followers, Lysenko received the best and fullest support in various kinds of agricultural machinery, fertilizers, and other supplies. This extraordinary position, coupled with Lysenko's undisputed talents as a practical agronomist, resulted in his farm being among the several outstanding ones of the region in terms of crop production.

In 1954 Khrushchev paid a visit to Lysenko at his experimental farm; in a later speech the Soviet Premier described the visit in his typically colorful fashion:

Three years ago I visited Lenin Hills. Comrade Lysenko showed me the fields on which he conducted experiments with organic-mineral fertilizer mixtures. We walked around the fields a great deal. I saw the striking results, and I saw how the organic-mineral mixes influenced the crops. Right at that moment I asked Trofim Denisovich [Lysenko] and Comrade Kapitonov, secretary of the Moscow Province Party Committee, to call in the agronomists of the Moscow area and advise them to try this new method of fertilizing fields. I did not hear that they objected to this in the Moscow area. All the collective farms of the Moscow area who fertilized their fields by this method achieved good results. . . . Just why, then, do some scientists object to the method proposed by T. D. Lysenko? I don't know what's going on here. I believe theoretical and scientific arguments should be decided in the fields.[151]

Lysenko had found a new protector at the highest level of the government and the Party, and he moved to support Khrushchev's agricultural policies, even when they contradicted his own earlier positions. Khrushchev, like many other Soviet specialists in agriculture, had been deeply impressed by the use of hybrid corn in North America and promoted its use in the virgin lands being opened according to his policy. Lysenko swallowed his earlier objections to hybrid corn and announced a square-cluster method of planting it that would "accumulate and preserve moisture" in the dangerously dry soil of the virgin lands.[152] Even when Khrushchev spoke critically of the state of agricultural science in the U.S.S.R., a remark that reflected as fully on Lysenko as on anyone, the agronomist accepted the criticism fully and promised better results. At the Twentieth Party Congress in 1956 Lysenko announced, "I am in full agreement with the evaluation of seed-growing and the ways of improving it that Comrades N. S. Khrushchev and N. A. Bulganin set forth in their reports. Production of high-yielding hybrid corn seed is particularly important." [153] At the close of his speech he abjectly remarked, "Permit me to express deep gratitude for inviting me, a non-Party scientist, an agronomist-biologist, to the Congress. . . ." Lysenko's campaign to ingratiate himself with the leader of the Party received new impetus in May 1957, when Khrushchev called for the U.S.S.R. to overtake the U.S. in per capita milk output; in July, Lysenko announced a grand plan for raising milk yields, which he had developed at his Lenin Hills farm.[154] This was to be his last ploy, and one that would end calamitously, not only for his personal standing, but for a portion of the Soviet dairy industry.

As a result of Lysenko's success in gaining Khrushchev's favor, by late 1958 he was coming back strongly. On September 29 *Pravda* announced the awarding of the Order of Lenin to Lysenko on his sixtieth birthday for his great services to the development of agricultural science and his practical assistance to production.[155] On December 14 *Pravda* carried a laudation of Lysenko and an attack on the *Botanical Journal* and the *Bulletin of the Moscow Society of Experimenters of Nature* for their articles criticizing Lysenko. *Pravda* observed, "No other place in the world has achieved results comparable to those obtained at the Lenin Hills Experimental Farm." [156] It lamented the fact that "the reactionary press [abroad] is gleefully trying to prove that Michurinist biology has reached the end of the line in the U.S.S.R." And it called for a consideration of the replacement of the editors of the journals that continued to criticize Lysenko. This request was fulfilled on January 21, 1959, when *Pravda* announced that in response to its criticism the Presidium of the Academy of Sciences had replaced V. N. Sukhachev as editor of both the *Botanical Journal* and the *Bulletin of the Moscow Society of Experimenters of Nature*. Another struggle against Lysenko had ended unsuccessfully. The result had largely been determined by Khrushchev's support for Lysenko's theoretically insignificant innovations in farming,[157] but it resulted in the continuing influence of Lysenkoism throughout theoretical biology, although not to its earlier extent. In 1961 Lysenko returned to his post as president of the Lenin Academy of Agricultural Sciences.[158] The stamina of Lysenkoism seemed incredible, not only to non-Soviet observers, but also to many discouraged Soviet biologists.

In the 1950's and early 1960's genetics research was conducted in the U.S.S.R. under various subterfuges. Such work was protected, particularly, by certain influential physicists such as I. V. Kurchatov (1903–60), who were able to promote genetics research because of the link between mutations and the use of radioactive materials. Later these centers, such as the Institute of Theoretical Physics and the Institute of Biophysics, were able to play a significant role in the resuscitation of full-scale genetics research.

Just as genetics could hide behind prestigious individuals such as the leading theoretical physicists, so also could it seek shelter under the cover of new and glamorous fields. Perhaps the most striking example of this combination of genuine scholarship and artifice was the link between cybernetics and genetics in the years between 1958 and 1965.[159] In the separate chapter on cybernetics in this book, I have discussed in some detail the way in which after 1958 cybernetics was enthusiastically promoted in the Soviet Union. The possibility of linking genetics to this new field was translated into a reality by Soviet scientists eager to overcome the effects of Lysenkoism. By assuming the label of cybernetics, genetics was able to gain access to publications, institutions, and scholarly discussions.

A link between the genetic code and information theory had been seen since early days, both in the Soviet Union and abroad. In his prescient essay of 1944 entitled *What is Life?* Erwin Schrödinger described life as a struggle by an organism against decay (maximum entropy) by means of feeding on information (negative entropy) from its environment.[160] The genes (which Schrödinger described as aperiodic crystals) were described as centers of information acting as reservoirs of negative entropy.[161] In such a description the analysis of genetics seemed quite possible from the standpoint of information theory and cybernetics.

After the outburst of cybernetics research in the Soviet Union in 1958, articles and books on genetics phrased in cybernetics terminology began to appear. Among the authors were I. I. Shmal'gauzen (Schmalhausen) and N. V. Timofeev-Ressovskii, prominent geneticists who suffered much from Lysenkoism, A. A. Liapunov, Zh. A. Medvedev, and K. S. Trincher.[162] Liapunov criticized Michurinist biology from the standpoint of cybernetics and in co-operation with another author termed a gene "the portion of hereditary information and also its encoded material carrier." In the very first issue of the theoretical journal *Problems of Cybernetics* the editor, Liapunov, observed that genetics furnished "another example of a biological science touching on the study of 'control systems.' "

During the early 1960's Lysenko's primary claim for continued preeminence in agricultural biology came from his attempt to raise milk production in the Soviet Union both in terms of over-all production and of butterfat content. The method that Lysenko utilized was the crossbreeding of purebred Jersey bulls, obtained at high cost from western Europe, with other breeds such as East Frisian, Kostroma, and Kholmogory.

Crossbreeding for the purposes of dairy farming is a very old method, but one that carries considerable risks. The goal, of course, is to obtain progeny with the best characteristics of both parent breeds. Jersey cows are known for the remarkably high butterfat content of their milk (often five to six per cent), the result in large part of over 250 years of careful breeding; the total yield of Jersey cows, however, is significantly lower than that of many other breeds. Therefore, a logical crossbreeding would be between the Jersey breed and another, such as Holstein-Frisian, that is distinguished by its quantitative milk-producing ability, but that gives milk with a rather low butterfat content (usually three to four per cent). The dangers or disadvantages of crossbreeding are potential loss of controls and decline of desirable characteristics. In the hands of skillful and educated specialists in genetics and animal husbandry such breeding can have very useful and profitable effects. Artificial insemination has greatly increased crossbreeding possibilities. Careful controls are the key to success in this field. If a mating between a member of a purebred line and one of unknown heredity occurs, the progeny may be valuable in terms of individual qualities, such as milk yield, but their value in terms of breeding is low; if such progeny are used for breeding purposes, the value of pedigreed herds can be quickly destroyed. Furthermore, several of the most important characters of dairy cows seem to be cases of blending inheritance—that is, tied to multiple genes; therefore, a mating between a bull from a breed that has cows with high-butterfat milk and a cow from one with low-butterfat milk usually results in progeny of intermediate butterfat capabilities. Matings in subsequent generations with low-butterfat lines will result in a gradual decline in butterfat content until the contribution of the ancestor of high-butterfat capabilities will be negligible. This absence of dominance in certain valued characters greatly complicates the task of cattle breeders.

Lysenko announced that he had found a method of providing bulls for breeding purposes whose progeny would have high-butterfat capabilities, and whose descendants in subsequent generations would continue to possess this character in an undiluted fashion. Starting with purebred Jersey bulls, he produced crossbreeds, sometimes with pedigrees as low as one-eighth Jersey, that supposedly would sire cows with the simultaneous capabilities of high butterfat and high yield. Furthermore, these qualities, said Lysenko, would not decline in subsequent generations.

The method that Lysenko used was based on his Law of the Life of Biological Species, a very vague concept with connections to his earlier views on shattered and stabilized heredity.[163] By crossbreeding purebred Jersey bulls with cows of regular farm herds that possessed the quality of large milk yield, Lysenko knew that he could produce a first generation with reasonably high merits in both quantity and quality. Lysenko departed from the normal doctrines of cattle breeding, however, in advancing the view that the hereditary qualities of this generation could be "fixed" if certain precautions were taken; these included insuring that the cows were of large stature and that they were fed copiously during gestation. This procedure, said Lysenko, would force the embryo to develop with the butterfat capabilities of the "small breed." If the cow during gestation were poorly fed, the calves would supposedly take after the larger parent.[164] If this stabilization process were followed, subsequent generations would not need to be given special care in feeding. The bulls in this line could be used freely without fear of decline in milk yield or quality.

The bulls from Lysenko's farm were widely sold to collective and state farms in the Soviet Union; the Ministry of Agriculture issued directives recommending such purchases and giving the Lenin Hills farm enviable financial advantages in cattle breeding.[165]

Lysenko continued to project himself as a servant of the Soviet economy. At a plenary session of the Central Committee in January 1961 Lysenko asked:

Is there a risk in my proposal to raise sharply the butterfat content of the milk of the collective and state farm dairy herd in five to seven years? I reply: Yes, there is. But the risk is exclusively to my reputation as a scientist. As for the country and the collective and state farms, they will not suffer a loss even if my scientific proposals fail.[166]

And Khrushchev agreed with Lysenko, crying, "What is there to be afraid of?" [167] On several occasions the Premier returned to Lenin Hills, commenting, "When I want to find out about agriculture in the non-black-earth zone, I go to T. D. Lysenko at Lenin Hills." [168] Thus, the agronomist managed to maintain himself on the basis of political support.[169]

But even before Khrushchev's ouster, there were many signs that Lysenko's case was becoming increasingly desperate. The science of biology continued to advance in other countries, and even an unlimited number of agricultural stratagems by Lysenko could not have offset the publicity that genetics was attracting.[170] Lysenko seems to have tried to meet the challenge of genetics in the raising of chickens, a field revolutionized outside the Soviet Union in the years after the war; his farm attempted to improve egg production, but silently abandoned the attempt after a few years.[171] Rumors that all was not well on the Lenin Hills farm began to circulate among agronomists and even government officials. The biologists, meanwhile, continued quietly to revive their discipline, waiting for the final discrediting of Lysenko.

The word that Lysenko's farm was experiencing difficulties opened a new avenue of criticism. In the past Lysenko had usually been attacked for the poverty of his theoretical views. Almost all his academic critics, from Vavilov onward, had granted his talents as a practical farmer. They had hoped for a *modus vivendi* that would allow them to determine theory and, if need be, permit Lysenko to have his experimental plots so long as he did not try to compete in the area of theory.[172] Now, however, the possibility of destroying the real base of his power—his reputation as a servant of practical agriculture—became apparent.

The transfer in 1956 of Lysenko's farm from the jurisdiction of the Lenin Academy of Agricultural Sciences to the all-Union Academy of Sciences was a helpful step in bringing Lysenko under the scrutiny of his critics. His farm was brought even closer to the academic strongholds by the reforms of the all-Union Academy in 1961 and 1963, which were promoted strongly by N. N. Semenov, the Nobel-Prize-winning chemist who was an opponent of the agronomist.[173] The outspoken Soviet physicist A. D. Sakharov called on Soviet scientists to vote against the Lysenkoite N. Nuzhdin at Academy of Science elections, and Nuzhdin was overwhelmingly defeated.[174] But Lysenko continued to resist inspection; he attempted to approve all information issuing from his farm, confident of political support for his general conduct. And indeed, a Soviet biologist even reported that in July 1962 an investigative committee of the Academy of Sciences had voted a censure of Lysenko's Institute of Genetics only to be overruled by political authority.[175]

Nikita Khrushchev's downfall on October 15, 1964, removed the most important obstacle in the way of a restitution of normal biology in the Soviet Union. In the following weeks articles critical of Lysenko and his views appeared in the popular press.[176] One author revealed the disastrous effect the Lysenko affair had had on high-school textbooks on biology; in the standard text for the ninth year "you would seek in vain a summary of the laws of heredity or a description of the role of the cell nucleus and the chromosome in heredity." [177] An article that was later cited by officials of the Academy of Sciences as being very important in bringing a full-scale investigation of Lysenko appeared in the *Literary Gazette* on January 23, 1965.[178] The author disputed with figures and specific cases the claims of the managers of the Lenin Hills farm to be producing bulls with the property of propagating indefinite numbers of generations of cows with high-butterfat milk. A few days later the Presidium of the Academy of Sciences of the U.S.S.R. created an eight-man committee headed by A. I. Tulupnikov to conduct a thorough investigation of Lysenko's farm. The committee spent over five weeks going over the records of the farm, examining crops and cattle, and checking on the breeding success of bulls sold to other farms. The detailed data, in the form of budgetary balances, crop yields, fertilizer usage, milk and egg output, purchase and sale data on cattle, and breeding records, permitted for the first time in the history of the Lysenko affair an objective and authentic analysis of the agronomist's claims. On September 2, 1965, the reports were presented to a joint meeting of the Presidium of the all-Union Academy of Sciences, the Collegium of the Ministry of Agriculture, and the Presidium of the Lenin Academy of Agricultural Sciences. The importance of this meeting was indicated by the fact that it was chaired by M. V. Keldysh, president of the Academy of Sciences, and a whole issue of the major journal of the Academy was devoted to the final report.[179]

The committee concluded that although the farm did produce a profit and gave high yields, these characteristics could be explained by its extremely favorable position compared with other farms. With approximately 1,260 acres of arable land, the farm possessed, for example, ten to fifteen tractors, eleven automobiles, two bulldozers, two excavators, and two combines. It was practically freed from the obligation to provide grain to the government. On a proportional basis it received several times more investment funds and electrical energy than neighboring farms. The fact that the farm stood out in comparison with many of its competitors was, in the opinion of the investigators, entirely unremarkable.

The heart of the report, however, referred to Lysenko's vaunted breeding methods. During the previous ten years the average yield of milk per cow had dropped from 6785 to 4453 kilograms. No evidence was found to support Lysenko's contention that the descendants of his bulls would have high-butterfat milk through indefinite numbers of generations. On the contrary, a nearly direct relationship was found between the per cent of butterfat and the degree of kinship to the original Jersey bulls:[180]

Degree of Jersey stock	Butterfat content of milk
Pure	5.87
¾	5.46
½	5.01
⅜	4.66
⁹⁄₁₆ [⁵⁄₁₆?]	4.74
¼	4.53
³⁄₁₆	4.50

Furthermore, Lysenko had indiscriminately sold his low-pedigree bulls around the country, where they had ruined herds of higher purity. Some of these bulls and their offspring had to be dispatched to slaughterhouses while still in their prime. Repairing the damage to pedigreed herds in Moldavia alone, said one speaker, would require decades.[181] Thus, Lysenko's claim that only his reputation would suffer if his theories were refuted was disproved, along with the theories themselves. If Lysenko's methods were carried out fully, the result for the country would be, said one of the inspectors, equal to a "natural calamity." "How much milk, meat, leather, and livestock we would lose!" he exclaimed.[182]

How had Lysenko maintained fairly high standards of milk production on his farm if his methods were so inadequate? He had started with the finest purebred cattle and still enjoyed, several generations later, the effects of that original stock. But a hidden reason for his relative success was—despite his denials—the fact that he was eliminating the poor milk producers through selection. Lysenko had told the Central Committee of the Party he did not eliminate a single cow because of low butterfat content during the decade of his farm's dairy experiments. The investigatory committee concluded, however, that Lysenko was simply incorrect in this assertion.[183] Over the years many members of the herd had departed through sale or slaughter, and those that had remained were "first of all those that gave the most butterfat and also the daughters of those cows with a large butterfat production." [184] Thus, the key to many of Lysenko's claims in dairying remained selection out of heterozygous populations, just as it had been years earlier in his experiments aimed at converting spring into winter wheat.

Lysenko seemed to have learned nothing concerning scientific technique since the early thirties. As one of the investigators described his farm:

There is a complete absence of a methodology of scientific research. There is no selection-pedigree plan. . . . Biometric data are not processed. Reliability is not computed. There is no account of feeding. And not only is there no weighing of food and remainders, which is done even at experiment stations, but even the records of rations that did exist have not been saved. And this, strange as it may seem, was in the laboratory of an Academician, the original leader of the Michurin school, a scientist defining heredity as the property of a living body to demand definite conditions for its development. [185]

Following the report on Lysenko's farm, the science of genetics revived in the Soviet Union. It had, of course, never completely disappeared, but had been forced to hide behind various camouflages, with the result that progress had been very difficult. [186] After 1965 all this changed rapidly. N. P. Dubinin, one of the leaders of the geneticists and a participant in the struggles of the late thirties, became the head of a new Institute of General Genetics. A Soviet journal, *Genetics,* became the theoretical organ of the reborn science. According to Dubinin, in the first two years after the discrediting of Lysenko ten new laboratories were organized in the Institute of Biological Problems. [187] N. V. Timofeev-Ressovskii, the renowned geneticist, became the head of the department of genetics radiation in the new Institute of Radiobiology. American scientists who visited the Soviet Union returned convinced that the Lysenko affair was over and that no longer could one speak of a "Soviet" genetics. Lysenko himself was described as being in semiretirement and refused to grant interviews to visiting delegations and reporters. [188]

Biology and Dialectical Materialism After Lysenko

As far as theoretical biology is concerned, it seems clear that Lysenko's downfall in 1965 was permanent. His demise, however, has not been accompanied by a cessation of Soviet writing on the relation of genetics and dialectical materialism. Indeed, some of the very same scholars who fought against Lysenko have in the last few years been interpreting molecular biology from the standpoint of dialectical materialism. Academician Dubinin, the outstanding Soviet geneticist who as a result of Lysenkoism lost his academic position, had his manuscripts rejected, and saw some of his closest friends imprisoned, wrote an article in 1969 entitled "Modern Genetics in the Light of Marxist-Leninist Philosophy." [189] He defended Marxism strongly and described mutations in terms of dialectical principles.

People with long memories will recall that certain European, Russian, and American geneticists of the 1920's and 1930's saw their science as a brilliant confirmation of the principles of dialectical materialism. Statements by such people as Haldane, Muller, Zhebrak, Agol, Serebrovskii, and Dubinin revealed their basic sympathy, at least in certain respects, with these principles. Indeed, if one reflects on the goals and methods of modern genetics, the feeling grows that it is a major irony that so fundamentally materialist a theory as that of genetics should have been rejected in the name of materialism. The search for the material carriers of heredity, first centered on the gene, now on DNA, is in many ways a lesson in the importance of materialism rather than its irrelevance. As two outspoken Soviet critics of Lysenko commented in 1963, the discovery of the role of DNA "is unquestionably one of the greatest advances in natural science, and one that has explained the material, biochemical nature of the continuity of life on our planet, and perhaps in the whole universe." [190] This view, deeply materialist, is the one that Lysenko rejected in his assertion that it is futile to search for a specific substance of heredity.[191] To refuse to look for the mechanism of heredity is far more akin to religious mysticism or to naïve romantic organicism than it is to materialism.[192]

The philosophers and biologists in the Soviet Union who continued to interpret biology in the light of dialectical materialism after Lysenko's demise were at first divided into "conservative" and "liberal" groups. Both these groups rejected Lysenkoism, but the conservatives continued to look back nostalgically to the days when there was such a thing as a "Michurinist genetics." Some of these authors called for, in effect, "Michurinism without Lysenkoism." An example was the 1965 article of G. V. Platonov in the conservative journal *October*.[193] Platonov was very upset about the "complete" rejection of Michurinism and the "complete" acceptance of formal genetics that he saw coming back to the Soviet Union. A similar view was expressed in 1965 by the author of a candidate's dissertation at Moscow University.[194] Pinter's thesis was an attempt to save Michurinist biology from Lysenko's naïve views, which he saw against the background of the "cult of personality" period. Pinter divided the history of genetic theory of the last half-century into two halves: Michurinist and "so-called formal genetics." Despite the gross errors of Stalinist dogma, he considered the Michurinist to be far more correct. The tragedy of Soviet genetics, according to Pinter, was that after 1948 Michurinist biology in the U.S.S.R. did not have any representatives other than Lysenko.

Underneath the views of people such as Platonov and Pinter lurked the danger of continuing to tie a science to one man—if not Lysenko, then Michurin.[195] Dubinin tried to answer this challenge in a 1966 article in the chief Soviet philosophy journal, in which he showed that Michurin never thought of himself as the founder of a great school in theoretical biology and that significant as Michurin's practical achievements were, genetics had now gone far beyond them.[196] Furthermore, Dubinin noted, in the last part of his life Michurin moved toward Mendelism.

The more liberal camp, to which Dubinin belonged, abandoned the term "Michurinist genetics." To them, there was only one science of genetics, the one known throughout the world. They continued, however, to defend dialectical materialism as a philosophy of science, and believed that it could provide helpful interpretations of biology. They, therefore, were making a careful distinction between "science" and "interpretations of science," as Lenin himself had done.

One of the most influential of these more sophisticated dialectical materialists writing on biology was I. T. Frolov, who in 1968 published a book entitled *Genetics and Dialectics*.[197] Frolov's book deserves consideration for a number of reasons. First of all, he criticized the whole concept of "Party science," firmly stating his opinion that politics concerns only the philosophical interpretation of science, not the evaluation of science itself.[198] He criticized those conservatives such as Platonov who had not, in his opinion, yet seen this distinction.[199] Second, Frolov tried to begin the process of reconstructing an intellectually tenable Marxist philosophy of biology out of the shambles left by Lysenkoism. He drew attention to legitimate philosophical problems of interpretation in genetics: the problem of reductionism, the problem of determinism, and the nature of heredity. He referred to the works of E. S. Bauer and Ludwig von Bertalanffy as examples of interpretations of biology that had similarities to dialectical materialism and that, therefore, should be further explored. And third, Frolov became in the same year that his book appeared the chief editor of the Soviet journal *Problems of Philosophy*. As the editor of the most influential philosophy journal in the Soviet Union, Frolov was able to exert an important influence in the philosophy of science.

Frolov believed that the most important philosophical question in biology was that of reductionism, or the relation of the part to the whole. According to a strict reductionist, the characteristics of an organism can be entirely explained in terms of its parts. Thus, a reductionist would explain life in physicochemical terms. It was around this question that Soviet discussions of dialectical materialism and biology in the late sixties turned.

In Frolov's opinion, dialectical materialism allowed one to have the advantages of studying both the part and the whole, of approaching biology both on the level of physicochemical laws and also on the more general biological or "systems theory" level. Frolov wrote that dialectics "defines a dual responsibility: On the one hand, it opens the way for complete freedom for the intensive use of the methods of physics and chemistry in studying living systems; on the other hand, it recognizes that biological phenomena will never, at any point in time, be fully explained in physicochemical terms." [200] The quantity-quality dialectical relationship had traditionally been interpreted by Soviet Marxists as a warning against reductionism, and Frolov continued to emphasize that warning. However, Frolov gave the caution a somewhat new meaning.

In his discussions of biology Frolov frequently referred to the "organismic" approach of the Austrian (now Canadian) biologist Ludwig von Bertalanffy. Von Bertalanffy, a prolific author, is best known for his two works on philosophical biology, *Kritische Theorie der Formbildung* (Berlin, 1928) and *Das biologische Weltbild* (Bern, 1949). In the period after 1950 von Bertalanffy promoted the interdisciplinary field called general system theory. His approach attracted considerable attention in Europe and America, both complimentary and critical.[201] After the rise to prominence of the fields of cybernetics and biocybernetics in the fifties von Bertalanffy's analyses became particularly well known.

There was much in von Bertalanffy's writings that appealed to dialectical materialists. He believed, like the dialectical materialists, that both mechanism and vitalism are inadequate as approaches to biological phenomena. He looked upon organisms as systems with unique properties not possessed by their components. Von Bertalanffy saw nature as a "tremendous architecture in which subordinate systems are united at successive levels into ever higher and larger systems." On each level, the systems conform to different laws. Thus, both dialectical materialism and von Bertalanffy's "organismic" interpretation of biology contained antireductionist principles.

Frolov highly appreciated von Bertalanffy's biological philosophy, but he also saw some dangers in it. He feared that an exaggeration of the specificity of living nature would result in a return to a form of vitalism. And he even thought that the irreducibility of biological laws to physicochemical laws might be more a result of the limitations of physical and chemical knowledge at any given time than an inherent characteristic of biology itself. As Frolov remarked, the limits of applicability of physics and chemistry to genetics "constantly change in step with the perfection of physicochemical methods and as a result of the development of genetics itself." [202] But since nature is inexhaustible in its complexity and this complexity constantly reveals itself to man at deeper and deeper levels, Frolov thought that this irreducibility would always be present. Thus, the applicability of physics and chemistry to biology would constantly grow, and dialectical materialist scholars should facilitate this growth in every possible way; nonetheless, a moment in time would never arrive when physics and chemistry could entirely explain biology. To Frolov, then, irreducibility in biology seemed more a cognitive or heuristic principle than an absolute characteristic of living matter. In view of the incompleteness of physicochemical explanations of biology, it was necessary to approach the problems from the biological side as well, formulating them in terms of biological laws.

The qualitative specificity of organisms is constant, but man's evidence of this specificity constantly changes in step with the dialectical growth of his knowledge. Frolov had placed von Bertalanffy's scheme in a more dialectical and historical framework. This tendency was continued in many subsequent Soviet analyses. Two Soviet authors writing in late 1969 emphasized that the structural-functional and historical approaches must be combined in order to understand biological phenomena.[203]

The systems approach to biological explanation, developed largely outside the Soviet Union, is at the present time one of the most vital issues in Soviet philosophy of biology.[204] Von Bertalanffy's work falls under the category of the systems approach, as does that of specialists in cybernetics, such as W. R. Ashby and Stafford Beer. Soviet authors have also become interested in earlier work related to the systems approach, such as the concept of "integrative levels" in biology developed by Joseph Needham, R. W. Sellars, A. B. Novikoff, and R. W. Gerard.[205] In a 1970 book on the methodology of systems research a number of Soviet authors tried to steer a middle course between a recognition, on the one hand, of the great value and potentiality of the systems approach, and a warning, on the other hand, that it may be used as an effort to bring vitalistic concepts back into biology.[206]

At the present time not only Soviet genetics itself but also Soviet philosophy of biology are regaining places of respect in the world. There is every sign that this trend will continue and that in those areas of genetics where philosophic problems are of genuine importance, the discussions will be kept within the bounds of philosophy and not allowed uniquely to determine the course of actual experimentation or the elaboration of theory.

VII Origin of Life

With the topic of the origin of life we come to one of the most interesting and least understood of the areas of the interaction between science and Marxist philosophy. Much essential information on this topic is still missing and will be uncovered only by careful monographic studies of the original workers in this field in the twenties and thirties, particularly in Russia and Britain. Recently attention has begun to shift in that direction; in 1968 C. H. Waddington, the distinguished British geneticist, commented:

In the late Twenties and early Thirties the basic thinking was done which led to the view that saw life as a natural and perhaps inevitable development from the non-living physical world. Future students of the history of ideas are likely to take note that this new view, which amounts to nothing less than a great revolution in man's philosophical outlook on his own position in the natural world, was first developed by Communists. Oparin of Moscow, in 1924, and J. B. S. Haldane, of Cambridge, England, in 1929, independently argued that recent advances in geochemistry suggested that the conditions on the surface of the primitive earth were very different from those of today, and were of a kind which made it possible to imagine the origin of systems that might be called "living." . . .[1]

Most scientists and historians of science are very skeptical of easy associations of science and political ideology, and no doubt Waddington did not intend to imply a direct causal link here. In this rather casual comment in a book review he was opening the question of the possible influence of Marxism on the significant theories of life of the first half of the twentieth century, not attempting to answer it. There are many opportunities in the history of science for linking science and politics, but frequently upon close examination the links either dissolve or turn out to be much more complex than thought earlier. As we will see, there are very weighty pieces of evidence against the belief that Oparin and Haldane were applying Marxism in 1924 and 1929. Nonetheless, the question of the interaction of Marxism and biology in the twentieth century is a legitimate and important one.

At the outset it will be useful to compare what non-Marxist and anti-Marxist writers have done with two different developments in the history of science: the Lysenko affair and discussions of the origin of life. In neither case is it self-evident that Marxism as a system of thought was significant in the formation of the interpretations of biological phenomena known, respectively, as the Oparin-Haldane hypothesis and Lysenko's theory of inheritance. In both cases, however, the major actors, at times subsequent to the original development of the hypotheses, explicitly declared that Marxism was an important influence in their biological thought. Lysenko, Oparin, and Haldane all became vocal dialectical materialists. Yet if one mentions "Marxism and biology" to the average educated citizen of western Europe or America, he will think only of Lysenko. This tendency to explain an acknowledged calamity in science as a result of Marxist philosophy while assuming that a brilliant page in the history of biology had nothing to do with Marxism is a reflection, at least in part, of the biases and historical selectivity of anti-Marxist journalists and historians.

The important question still remains: Did Marxism have anything to do with the Oparin-Haldane hypothesis? Although this question cannot be answered at this time, and no doubt will long remain controversial, certain clarifications can be made. Before attempting these clarifications, I would like to make some general comments about A. I. Oparin and the issue of the origin of life. As a Russian, with an active life extending to the present day, Oparin is central to this study, while Haldane falls outside it. Oparin's initial work on this issue was prior to Haldane's entirely independent but similar approach; the British scientist graciously declared in 1963, "I have very little doubt that Professor Oparin has priority over me." [2]

The question of the origin of life is one of the oldest in the history of thought. At almost all periods of time a belief in spontaneous generation was commonly held. This belief was by no means the property of one school of thought; Democritus, Aristotle, St. Augustine, Paracelsus, Francis Bacon, Descartes, Buffon, and Lamarck are only a few of those who expressed support of the concept, but within the frameworks of quite different interpretations of nature. With the development of microscopy the center of attention in discussions of the origin of life shifted to the level of the invisibly small. A famous debate in the 1860's between the French scientists Felix Pouchet and Louis Pasteur over the possibility of the spontaneous origin of microorganisms ended with a negative result that left the subject of spontaneous generation in disrepute for the remainder of the century. To be sure, a few writers such as H. Charlton Bastian continued to consider spontaneous generation possible.[3] Friedrich Engels observed ironically that it would be foolish to expect men with the help of "a little stinking water to force nature to accomplish in twenty-four hours what it has cost her thousands of years to bring about." [4] And most people did not notice that Pasteur himself commented in 1878, "Spontaneous generation? I have been looking for it for 20 years, but I have not yet found it, although I do not think it is an impossibility." [5]

Much of the foregoing would seem to serve as an introduction to A. I. Oparin as a twentieth-century exponent of spontaneous generation. Yet if one understands spontaneous generation to mean the sudden arising of a relatively complicated living entity—whether an organism, a cell, or a molecule of DNA—from nonliving matter, Oparin has actually been an opponent of spontaneous generation. In his opinion, the belief that such an ordered entity as a cell or even a "living" molecule of nucleic acid could arise spontaneously is based upon "metaphysical materialism" and suffers from the same improbabilities (to be discussed below) as Pouchet's arguments.

Aleksandr Ivanovich Oparin (1894–   ) has been a prominent biochemist in the Soviet Union for decades. He graduated from Moscow University in 1917 and subsequently became a professor there. He was closely associated with the Institute of Biochemistry of the U.S.S.R. Academy of Sciences, which he helped to organize in 1935 and of which he became director in 1946. In the same year he became a full member of the Academy of Sciences. In 1950 he received the A. N. Bakh and I. I. Mechnikov prizes. He has worked on several different topics, including such practical ones as the biochemistry of sugar, bread, and tea production, but he is best known both in the Soviet Union and abroad for his theory of the origin of life. Over a period of almost half a century he has published numerous books, revised editions, and journal articles on this topic. As early as 1922 he delivered a talk on the origin of life to the Moscow Botanical Society; he subsequently published these views in a small booklet in 1924.[6] Although frequently cited in the scientific literature, the 1924 publication was an exceedingly rare one and was not translated into English until 1967.[7] Most references by English readers to Oparin's work have been to the 1938 and subsequent editions, which differed from his earliest publications in several ways that are interesting to the historian. These differences will be discussed subsequently. In recent years, Oparin's works have received international attention more rapidly; his 1966 *The Origin and Initial Development of Life* was translated into English by the U.S. National Aeronautics and Space Administration in 1968.[8]

The most prominent contribution by Oparin to the study of the origin of life will probably be seen by historians as his reawakening of interest in the issue. The American biologist John Keosian wrote in his popular text *The Origin of Life,* "Oparin's unique contribution is his revival of the materialistic approach to the question of how life originated, as well as his detailed development of this concept." [9] The British scientist J. D. Bernal commented in 1967 that Oparin's 1924 essay

contains in itself the germs of a new programme in chemical and biological research. It was a programme that he largely carried out himself in the ensuing years, but it also inspired the work of many other people. . . . Oparin's programme does not answer all the questions, in fact, he hardly answers any, but the questions he asks are very effective and pregnant ones and have given rise to an enormous amount of research in the four decades since it was written. The essential thing in the first place is not to solve the problems, but to see them. This is true of the greatest of all scientists. . . . This paper is important because it is a starting point for all the others and, though it is clearly defective and inaccurate, it can be, and has been, corrected in the sequel.[10]

Turning now to the question of intellectual and social influences on Oparin, it can be clearly established that from the early 1930's onward Oparin was influenced by dialectical materialism. The evidence is not only his frequent statements favoring dialectical materialism, but much more importantly, the very method of analysis of his later publications, which are permeated with an assumption of a process philosophy and a concept of differing dialectical levels of regularities in nature. All this is described in his works in the language of dialectical materialism. Oparin has spoken out so frequently on the relevance of dialectical materialism to theories of biological development that almost any of his publications of substantial length contains such statements. To be sure, there is the possibility that these sections of his writings are merely responses to political pressures, but if one reads in chronological order through Oparin's works published over many years at times of greatly varying political atmospheres, one can not avoid the conclusion, it seems to me, that dialectical materialism became an ever-increasing and substantial influence on his work. In 1953 Oparin wrote, "Only dialectical materialism has found the correct routes to an understanding of life. According to dialectical materialism, life is a special form of the movement of matter, which arises as a new quality at a definite stage in the historical development of matter. Therefore, it possesses properties that distinguish it from the inorganic world, and is characterized by special, specific regularities that are not reducible merely to the regularities of physics and chemistry." [11]

And in 1966 in the book translated by NASA Oparin commented:

Regarding life as a qualitatively special form of the motion of matter, dialectical materialism formulates even the very problem of understanding life in a different way than does mechanism. For the mechanist, it consists of the most comprehensive reduction of living phenomena to physical and chemical processes. On the other hand, from the dialectical materialist point of view, the main point in understanding life is to establish its qualitative difference from other forms of motion of matter. Life finds its clearest expression (as a special form of the motion of matter) in the specific interaction of living systems—organisms—with the environment surrounding it, in the dialectical unity of a living body, and in the conditions of its existence.[12]

As I discuss in chronological order a number of Oparin's other publications on the origin of life, more specific effects of his dialectical materialist approach will come to light. This discussion should not be confused with a general history of Oparin's theories, which remains to be written and which would take into account in a much fuller way the development of biochemistry as a whole. In such a history dialectical materialism would play a less prominent role than in this discussion.

Nonetheless, it will always be seen, I think, as one of the important influences upon Oparin.

If we turn to Oparin's 1924 booklet, however, we will find no mention of Marxism. Even more significantly, Oparin's analysis in this small book differed from almost all of his later works in containing no concept of "different levels of regularities," no statement that qualitatively distinct principles govern the movement of matter on different ontological levels. The Oparin of 1924 was a materialist (and here the influence of his political and social milieu no doubt enters in), but he seems to have been an old-fashioned materialist, one who believed that life can be entirely explained in terms of physics and chemistry. Compare, for example, his 1953 statement quoted above, in which he said that life is characterized by special regularities that are not reducible to those of chemistry and physics, to the following statement, which was contained in the 1924 booklet:

The more closely and accurately we get to know the essential features of the processes which are carried out in the living cell, the more strongly we become convinced that there is nothing peculiar or mysterious about them, nothing that cannot be explained in terms of the general laws of physics and chemistry.[13]

And in another place in the same work he commented, "Life is not characterized by any special properties but by a definite specific combination of these properties." [14] Here again, the assumption is a reductionist one, although the "definite specific combination" allowed a bit of room for the later development of the concept of "special biological regularities" distinct from those of physics and chemistry, a concept that became fundamentally important to his work. Ironically, the Oparin who in 1924 campaigned against vitalism in the name of purely physicochemical explanations of life would in the fifties and the sixties defend the uniqueness of biological regularities against those molecular biologists who would try to explain life entirely in terms of the structure of a molecule of nucleic acid—that is, in terms of purely physicochemical explanation.

The fact that Oparin did not possess a knowledge of systematic dialectical materialism in 1924 does not prove that Marxism had nothing to do with the timing or form of his expression. Russia in the early twenties was the home of a victorious revolution carried out in the name of

Marxism. "Materialism" was one of the most popular slogans of the day, and most of it was of the rather elementary, mechanistic sort espoused by Oparin, not the more subtle dialectical materialism developed at a later day. Neither Engels's *Dialectics of Nature* nor Lenin's *Philosophical Notebooks* had yet appeared in print; these are the two books that more than any others have counteracted the severe reductionism of earlier materialism with the concept of qualitatively distinct realms of operation of natural laws. Oparin's materialism developed in parallel with the predominant philosophical views of his society. Communism probably had something to do with Oparin's 1924 statement, but not in the sense of a relationship between dialectics and theories about the origin of life; rather, Communism in Russia in the twenties provided the kind of atmosphere in which the posing of a materialistic answer to the question What is life? seemed natural. The Soviet Union of the twenties was an environment in which speculation about nature on the basis of materialist assumptions was not only welcome but very nearly inevitable. Oparin certainly had nothing to fear from the political and social milieu in expressing such views, and he made no effort to soften their impact. Similar viewpoints were still capable of causing quite an adverse reaction in Britain, as the reception of Haldane's similar views five years later showed; Bernal commented, "Haldane's ideas were dismissed as wild speculation." [15] Haldane assured his readers in 1929 that his opinions (not entirely identical with Oparin's, but similar) were also compatible with "the view that pre-existent mind or spirit can associate itself with certain kinds of matter." [16] Yet it was clear that Haldane did not share this vitalistic view, and his inclusion of it came from his hope that his scientific conception would not be rejected simply because "some people will consider it a sufficient refutation of the above theories to say that they are materialistic." [17]

How close Oparin personally was to Marxist ideas at this early period is still unknown. From 1921 onward he was closely associated with the older Soviet biochemist A. N. Bakh, who was a political revolutionary and former *émigré,* and who published on Marxism as early as the 1880's.[18] But until we know more about Oparin's philosophical orientation in the period 1917–24, not much more can be said about the influence of Marxist materialism on his 1924 publication.

It might be added parenthetically that a similar problem of interpretation exists in the case of Haldane. Like Oparin, he seems to have been

most influenced by Marxist thought *after* he published his first fundamental work on the origin of life. Haldane's first article on the subject
appeared in 1929, as already noted. As late as 1938 Haldane wrote, "I
have only been a Marxist for about a year. I have not yet read all the
relevant literature, although I had of course read much of it before I
became a Marxist." [19] Yet this does not prove, of course, that Haldane
had not learned of Marxist ideas of development by 1929. Since the late
twenties Haldane had been a leader of a group of Cambridge intellectuals who were very interested in Marxism. But there were other influences in the air, too, that were interesting to men like Haldane and relevant to the biological issue, such as the early process philosophy of
A. N. Whitehead. Some writers have argued that Haldane's view of
science grew out of a complicated interaction of reductionist biochemistry, the process philosophy of Whitehead, and Marxism. Whatever the
ingredients and the proportions of the early intellectual influences on
Haldane, a hand-in-hand evolution based on interaction seems to have
occurred.[20] It eventually resulted in his writing a volume entitled *The
Marxist Philosophy and the Sciences*. A similar long-term intellectual
development will be seen in Oparin, with the difference that his interests
in Marxism continued to deepen throughout his career.

The task that faced Oparin in 1924 in biology was as much one of
changing the psychological orientation of scientists as it was altering
their research itself. He had to convince his readers that despite Pasteur's victory over Pouchet years before and the complete inability of
scientists to produce even the most elementary living organisms in a
laboratory, a materialist explanation of the origin of life was still worth
the effort.

In retrospect, Oparin observed, we should not be surprised or fundamentally affected by the outcome of the Pasteur-Pouchet debate.
Pouchet was indeed incorrect, but not because of his materialist assumptions. Even the simplest microorganisms, and certainly those that
Pasteur and Pouchet observed, are extremely complex bits of matter;
they possess an "extraordinarily complicated" protoplasm. How could
Pouchet even suppose that such a highly differentiated form of matter
could "accidentally" have arisen in a few hours or even days from a
relatively formless mixture? To assume such an incredibly improbable
occurrence was unscientific in the deepest sense, a violation of the prin-

ciple of explaining nature in the simplest and most plausible fashion available. As Oparin observed,

Even the simplest creatures, consisting of only one cell, are extremely complicated structures. . . . The idea that such a complicated structure with a completely determinate fine organization could arise spontaneously in the course of a few hours in structureless solutions such as broths or infusions is as wild as the idea that frogs could be formed from the May dew or mice from corn.[21]

How then could one begin to explain an origin of life on the basis of materialist assumptions? Only, said Oparin, by going back to the very simplest forms of matter and by extending the Darwinian principles of evolution to inanimate matter as well as animate matter. The "world of the living" and the "world of the dead" can be tied together by attempting to look at them both in terms of their historical development. Any finely structured entity, alive or dead—whether a one-celled organism, a piece of inorganic crystal,[22] or an eagle's eye—seems inexplicable unless it is examined in historical, evolutionary terms. Pouchet failed because the specimens he thought would arise spontaneously—microorganisms—were already the ultimate products of an extremely long evolutionary history and can be brought into existence only through that chain of material development, not by side-stepping it.

In the section of his 1924 work subtitled "From Uncombined Elements to Organic Compounds," Oparin attempted to reconstruct the historic process that might have led to the origin of life, with the simple always preceding the complex. In order to follow this sequence, it was necessary for him to reject the customary thesis that all organic compounds had been produced by living organisms, a belief still widely held at that time despite the supposed synthesis of urea by Wöhler as long ago as 1828.[23] To postulate that all organic compounds had been produced by living organisms was methodologically faulty, thought Oparin, since organisms themselves were obviously composed of organic compounds, and many of them far more complex than some of the products supposedly produced. Far better, he thought, to assume that at least some organic compounds antedated complete organisms and had been important to the origin of these organisms. An important stimulus to Oparin's thought on this subject was the theory of the carbide origin of petroleum advanced by the great Russian chemist D. I. Mendeleev many years before. Mendeleev had posed the possibility of the origin of the hydrocarbon methane by the action of steam on metallic carbides under conditions of high temperature and pressure, for example:

$$C_3Al_4 + 12\ H_2O \rightarrow 3\ CH_4 + 4\ Al(OH)_3$$

This inorganic source of methane would then have been followed by further transformations leading ultimately to petroleum. Oparin did not accept Mendeleev's hypothesis about the origin of petroleum (and it has not been accepted by geologists generally, although there have been a few attempts to revive it),[24] but the idea provoked him to further thought about the inorganic origin of organic compounds. As late as 1963 Oparin continued to emphasize the importance of Mendeleev's idea to his original conception of the origin of life.[25]

In order to provide the necessary temperatures and pressures and a source of energy, Oparin referred to the theory of origin of the earth that begins with the earth as an envelope of incandescent gas. Oparin maintained, "Only in fire, only in incandescent heat could the substances which later gave rise to life have formed. Whether it was cyan [nitrogen carbide] or whether it was hydrocarbons is not, in the final analysis, very important. What is important is that these substances had a colossal reserve of chemical energy which gave them the possibility of developing further and increasing their complexity." [26] In the 1936 edition of his book, Oparin would tie this view of the origin of the earth to James Jeans's theory of planetary cosmogony, in which a star approaches the sun in such a way as to pull out by gravitational attraction a tidal wave of incandescent solar atmosphere. As we have seen in the chapter on cosmology and cosmogony, this theory came under heavy philosophical criticism in the Soviet Union in later years as "miraculous and improbable." [27] Oparin later abandoned the Jeans theory of planetary cosmogony, finding other available sources of energy for the formation of complex hydrocarbons.

In the last section of his original work Oparin discussed how some of the simple organic compounds could evolve into living organisms. And here he presented one of the paradoxical but now quite believable elements of his theory that has continued to characterize it to the present: The prior nonexistence of life was one of the necessary conditions for the origin of life, and consequently, now that life exists on earth it can not originate again, or at least not in the same way in which it first did. Oparin explained this conclusion quite graphically:

We have already seen in the last section, that at present what are lacking above all are those substances containing much chemical energy which are the only things from which life could develop and which, themselves, could only be formed at extremely high temperatures. However, even if such substances were formed now in some place on the Earth, they could not proceed far in their development. At a certain stage of that development they would be eaten, one after the other. Destroyed by the ubiquitous bacteria which inhabit our soil, water and air.

Matters were different in that distant period of the existence of the Earth when organic substances first arose, when, as we believe, the Earth was barren and sterile. There were no bacteria nor any other micro-organisms on it, and the organic substances were perfectly free to indulge their tendency to undergo transformations for many, many thousands of years.[28]

The posing of the prior nonexistence of life as a necessary condition for the origin of life seemed somewhat more original in 1924 than it does today, since in the interim there has come to light a letter of Charles Darwin's in 1871 mentioning the same hypothesis.[29] Several other scientists also seem to have mentioned this hypothesis in the late nineteenth and early twentieth centuries.[30] In Oparin's later editions this seeming paradox was explained within the framework of dialectical concepts of natural law: On each level of being different principles obtain; therefore, the laws of chemistry and physics that operated on the earth in the absence of life were different from, and superseded by, the biological laws that qualitatively emerged with the appearance of life. In the case of man, the biological laws were transcended, in turn, by social ones.

Oparin continued his hypothetical scenario of the origin of life with a description of the way in which substances with complicated molecules form colloidal solutions in water. Such solutions are inherently unstable, however, and frequently form coagula or gels: "The moment when the gel was precipitated or the first coagulum formed, marked an extremely important stage in the process of the spontaneous generation of life. At this moment material which had formerly been structureless first acquired a structure and the transformation of organic compounds into an organic body took place. Not only this, but at the same time the body became an individual." [31]

This emphasis on the arising of life in a liquid medium by means of the separating out of gels became one of the hallmarks of Oparin's views; once his general materialistic approach to the origin of life had been widely accepted, the gel theory or "coacervate theory" was often considered to be that part of his work that was unique to him. Consequently many current discussions of the validity of Oparin's views revolve entirely around the tenability of the coacervate theory.

The idea of life's arising in a sea jelly of some sort was not new, of course, having been a part of T. H. Huxley's bathybius hypothesis, but Oparin was to present it in a particularly plausible way. In 1924, however, Oparin did not even use the term "coacervate." All that was to come in later editions, after Oparin could make use of the research on coacervation carried out in the early thirties by Bungenburg de Jong. But both in the original 1924 work and in his subsequent publications, Oparin remained steadfast on the principle that life arose on a level of fairly large dimensions: the coagula, the gels, the coacervates, were all definitely multimolecular, and they all possessed a rather complex structure *before* they could be called "alive." After they became alive, a natural selection began that through the costs of survival resulted in increasingly viable and complex organisms.

From the philosophic or methodological point of view the transition from the "nonliving" to the "living" is the crucial moment. Oparin has not been one to attempt rigorous definitions of "life," preferring to speak in metaphors or in terms of varying combinations of characteristics necessary for life, but it is clear that his opinion about the moment when life appeared has changed somewhat with time. In 1924 he described the moment when "the gel was precipitated or the first coagulum formed" and then observed, "With certain reservations we can even consider that first piece of organic slime which came into being on Earth as being the first organism. In fact it must have had many of those features which we now consider characteristic of life." [32] This observation accorded with the reductionist, mechanist approach of the young Oparin, in which a simple physical process—coagulation—could herald a major transition. In later years, he would maintain that the first coacervate droplets were definitely not alive and that a "primitive natural selection" (a concept for which he was much criticized; see pp. 291–2) occurred among these nonliving forms. The transition to life occurred after not only the more commonly named characteristics of life had appeared (metabolism, self-reproduction), but in addition, after a certain "purposiveness" of organization had been achieved.[33] This controversial aspect of his scheme, linked to Aristotelian entelechy by his more aggressive critics, will be discussed in subsequent sections.

One metaphor that Oparin used in his early statement remained constant throughout his works; this was his comparison of life to a flow of liquid. In 1924, he wrote that "an organism may be compared with a waterfall which keeps its general shape constant although its composition is changing all the time and new particles of water are continually passing through it";[34] in 1960, he commented, "Our bodies flow like rivulets, their material is renewed like water in a stream. This is what the ancient Greek dialectician Heraclitus taught. Certainly the flow, or simply the stream of water emerging from a tap, enables us to understand in their simplest form many of the essential features of such flowing, or open systems as are represented by the particular case of the living body." [35] These metaphors, all based on the concept of the constant flow of matter in living organisms, involved Oparin in many discussions about whether relatively static entities sometimes considered alive (crystallized viruses, dried seeds) could be accommodated to his understanding of life.

If one shifts from Oparin's 1924 booklet to his 1936 major work (the first was approximately 35 pages in length, the latter, 270), a number of changes become apparent. The biochemist would notice the much fuller description of the initial colloidal phase and a subsequent section on the development of photosynthesis by the ancestors of vegetative organisms. The historian and philosopher would remark on Oparin's growing philosophic awareness, his refinement of his definitions and his stated shift toward Marxist interpretations.

By 1936 Oparin could take advantage of the recent work by Bungenburg de Jong on "coacervation," a term used by de Jong to distinguish the phenomenon from ordinary coagulation. In solutions of hydrophilic colloids it is known that frequently there occurs a separation into two layers in equilibrium with each other; one layer is a fluid sediment with much colloidal substance, while the other is relatively free of colloids. The fluid sediment containing the colloids, de Jong called the coacervate, while the noncolloidal solution was the equilibrium liquid. Oparin emphasized the interface or surface phenomena that occur in coacervation; various substances dissolved in the equilibrium liquid are absorbed by the coacervate. Consequently, coacervates may grow in size, undergo stress with increasing size, split, and be chemically transformed. In discussing this active role of coacervates, Oparin was attempting to establish them as models for protocells. A "primitive exchange of matter" occurs between the coacervate and the equilibrium liquid, the beginning of that metabolic flow necessary for life, according to Oparin. To initiate life, however, Oparin said that it was necessary for coacervates to acquire "properties of a yet higher order, properties subject to biological laws." [36] He had higher requirements for life in 1936 than in 1924, and his scheme now contained a phase of evolution of the lifeless coacervates.[37]

Oparin's view of coacervate evolution depended heavily on the action of catalysts and promoters, the future enzymes of living cells. These chemicals greatly accelerated the transformations taking place in the coacervates. The "inner chemical organization" of these promoters of chemical reactions "became strengthened in the process of natural selection, insuring a gradual evolution." [38] To be sure, this early "natural selection" was not identical with selection among advanced, truly living organisms, but nonetheless it did occur, and in the following way: Those coacervates containing superior catalysts grew more rapidly and split more often than the others and therefore gradually gained a numerical advantage. But speed was not the only requirement; even more important to Oparin was "a harmonious coordination of the velocities of the different reactions." [39] The bound enzymes in the protoplasm of a living cell are released sequentially as the need for them arises; therefore, in order for cells to develop, a complicated orchestration of temporally separated operations was necessary. It was possible for some necessary reactions to occur prematurely and to decrease survival chances: "If the increase in the rate of a given reaction so affected the coordination between assimilation and degradation as to promote the latter, such an imperfect system would become mechanically unfitted for further evolution and would perish prematurely." [40] Therefore, a very complex interweaving of separate operations gradually developed through natural selection.

In Oparin's 1936 scheme the transition from the nonliving to the living was still not defined clearly. It occurred, he thought, when the "competition in growth velocity" was replaced by a "struggle for existence." This sharpening struggle resulted from the fact that the prebio-logical organic material on which the coacervates were "feeding" was being consumed. Ultimately, this shortage would lead to an important split in the ways in which organisms gained nourishment—resulting in the distinction between heterotrophs and autotrophs—but before that division occurred, the all-important transition to the biological level was reached. As the amount of organic material outside the coacervates lessened, the first true organism appeared. As Oparin described this mo-ment:

The further the growth process of organic matter advances and the less free organic material remains dissolved in the Earth's hydrosphere, the more exacting "natural selection" tends to become. A straight struggle for existence displaces more and more the competition in growth velocity. A strictly bio-logical factor now comes into play.

This new factor naturally raised the colloidal systems to a more advanced stage of evolution. In addition to the already existing compounds, combina-tions and structures, new systems of coordination of chemical processes ap-peared, new inner mechanisms came into existence which made possible such transformations of matter and of energy which hitherto were entirely un-thinkable. Thus systems of a still higher order, the simplest primary organisms, have emerged.[41]

It should be obvious from Oparin's scheme of development that he thought that heterotrophic organisms (organisms that are nourished by organic materials) preceded in time autotrophic organisms (those nourished by inorganic materials). Many scientists had earlier thought that the sequence was the opposite and assumed that carbon dioxide—necessary for photosynthesis by autotrophic green plants—was the pri-mary material used in building up living things. Oparin found this thesis dubious. As evidence against it he cited the fact that heterotrophic or-ganisms are generally capable of using only organic compounds for nourishment, while many autotrophic green plants "have retained to a considerable degree" the ability to use preformed organic substances for their nourishment.[42] The significance here of the word "retain" is obvi-ously one of time sequence; Oparin thought that all organisms had originally been heterotrophic, but that as the supply of organic food diminished, they split along two different paths of development. (This division is not, strictly speaking, the same as that between the plant and

animal worlds, but is close to it, since green plants are largely autotrophic while all the highest and lowest animals and most bacteria and
all fungi are heterotrophs.)

Oparin explained this scheme in a much fuller philosophic framework than previously. By 1936 he had read Engels's *Dialectics of Nature,* and cited it in his footnotes, as well as the earlier-published *Anti-
Dühring.* He commented that Engels had "subjected both the theory of
spontaneous generation and the theory of eternity of life to a withering
criticism." [43] (In his 1924 work, "spontaneous generation" had still
been a positive term, although he had thought the efforts to find it
crude.) Oparin now thought that any effort to explain "the sudden generation of organisms" could rely only on either an act of "divine will" or
"some special vital force." Such a view, Oparin observed, is "entirely
incompatible with the materialistic world conception." [44] On the contrary, "Life has neither arisen spontaneously nor has it existed eternally.
It must have, therefore, resulted from a long evolution of matter, its
origin being merely one step in the course of its historical development." [45]

More indicative of essential changes in Oparin's thought was his shift
away from mechanism. Crude materialism, the belief that all phenomena could be explained in terms of the elemental, was now one of the
objects of his criticism:

Attempts to deduce the specific properties of life from the manner of atomic
configuration in the molecules of organic substance could be regarded as
predestined to failure. The laws of organic chemistry cannot account for
those phenomena of a higher order which are encountered in the study of
living cells.[46]

Although Oparin now frequently cited Engels and thought that he
had been remarkably prescient in his discussions of life, Oparin was also
willing to interpret and modify Engels's formulations. When Engels said
that "Life is a form of the existence of protein bodies," he did not intend
to say, said Oparin, that "protein is living matter." Instead he meant
that protein has hidden in its chemical structure "the capacity for further organic evolution which, under certain conditions, may lead to the
origin of living things." [47] This interpretation by Oparin fits well with his
belief that life is not inherent in a structure, but instead is a "flow of
matter," a process. Structure has a great deal to do with life, he thought,

but to confuse it with life itself would be roughly like confusing a frozen stream of water with a flowing one. This emphasis on process, on "coordinated chemical reactions" rather than on determinate structure, would eventually involve Oparin in controversies with spokesmen from two quite different camps: the ultraorthodox dialectical materialists who wished to stick to Engels's literal word—"protein"—as the essence of life, and the new molecular biologists, who saw the essential features of life in the structure of nucleic acid and whose very terms of description—"template," "code"—carried the sense of the static.

It was the 1936 book, translated into English in 1938, that brought Oparin international stature. At this moment of first impact his primary message was still seen as the legitimacy of a materialistic approach to the study of the origin of life. Consequently, a number of writers who actually differed considerably with Oparin on details considered themselves in agreement. Haldane in his 1929 article, for example, hypothesized—in contrast to Oparin—a primitive earth atmosphere rich in carbon dioxide, described the first "living or half-living things" as "probably large molecules," and did not mention coacervates, coagula, or gels. These are important points of difference. Yet the hypothesis became known as the Haldane-Oparin (or Oparin-Haldane) one and even yet is frequently referred to in that way.

The regarding of all materialist theories of the origin of life as essentially the same was reflected in R. Beutner's comments on Oparin in his *Life's Beginning on the Earth,* published in the United States in 1938.[48] Beutner learned of Oparin's work as his own manuscript was going to press, and was able to add only an addendum taking it into account. Beutner commented that "there is a close resemblance of Oparin's views on the origin of life, with those here evolved." [49] Further, "Oparin draws the same conclusions which this writer tentatively developed in 1933, namely that 'life is just one of the countless properties of the compounds of carbon.' " [50] (It should be clear from the foregoing discussions that Oparin decidedly disagreed with Beutner on this point.) And finally, said Beutner, the small difference between Oparin's and his approaches was "only concerned with the order of the essential events which preceded the appearance of life." [51]

Beutner's reassurances notwithstanding, Oparin's approach was rather different from his, although both were, indeed, materialistic. Beutner was a reductionist, Oparin by 1936 a supporter of the concept of qualitatively distinct levels in nature. Much of Oparin's writing was based on the assumption that life is a multimolecular phenomenon, while Beutner flatly stated that "we know that a 'living molecule' is possible." [52] (He was referring to Stanley's 1935 work with tobacco mosaic virus, a subject of much subsequent discussion in which Oparin would be deeply involved.) And the small "difference of order of the essential events" between Beutner and Oparin was a distinction that Oparin considered fundamental. On this issue Beutner wrote, "The essential feature of Oparin's hypothesis is that structural units with a remote resemblance to living organisms were *first* formed from the organic matter of the early ocean and that subsequently enzymes formed in them so as to enable them to assimilate substances from their environment. . . . Our opinion, on the contrary, was that life-producing enzymes were the first to appear, through the action of electric discharges, without any structure around them—as self-regenerating enzymes." [53] Oparin believed that Beutner's fortuitous and lightninglike appearance of complex and ordered substances capable of coordinating metabolism was merely another variant of spontaneous generation, a concept he could not accept. He felt that it ultimately led to a discrediting of materialism by failing to explain adequately the complexity of life. In contrast to Beutner, Oparin drew a sharp distinction between simple catalysts, perhaps composed of even one chemical element, and complex enzymes:

We must, therefore, conclude that the naturally occurring high-molecular enzymes of living cells are not individual chemical compounds but complexes made up of numerous catalysts and promoters. The tremendous power of enzymatic activity must be attributed exclusively to a favorable arrangement of components in this complex, which, of course, could not have arisen fortuitously but only as a result of a long evolution of living organisms. In this process ever new processes and promoters must have been gradually added on secondarily to the simple and not very active primitive catalyst.[54]

Oparin's 1936 work remained substantially unchanged for twenty years. The 1941 edition contained few modifications; not until 1957 was a third, revised edition published, almost simultaneously appearing in Russian and in English. In the meantime the science of biochemistry was developing extremely rapidly. The new knowledge of molecular biology led to a union of biochemistry and genetics, culminating in the 1953 publication of the Watson-Crick model of the DNA molecule.

The relevance of molecular biology to theories of the origin of life was obvious to most observers in all countries, although just where the developments would lead was a subject of genuine debate.

The topic of viruses was particularly close to the question of the nature of life when viewed from the molecular level; viruses consist of nucleic acid (DNA or RNA) with a protein coat. The relevance of molecular biology to Oparin's work was to center, in part, on discussions of viruses. The most urgent questions could be stated in simple forms: Are viruses alive? If they are alive, in view of the fact that the simplest of them are essentially nucleic-acid molecules, was not Oparin incorrect in stating that life arose on the multimolecular level? Was not the first form of life a molecule of nucleic acid?

In the Soviet Union such issues were discussed in rather difficult circumstances, since the new union of biochemistry and genetics in world science had occurred at approximately the same time that Lysenko and his followers won control over Soviet genetics. Politically, Oparin and Lysenko were linked, however far apart they may have been in intellectual sophistication. Both had won favor from the Stalinist regime, both had built their careers within it, and around both there had arisen schools of biology that were officially described as "Marxist-Leninist" or "Michurinist." They benefited from the government, and they repaid the government in political praise and co-operation. Oparin was active in Soviet political causes in international organizations. As a high administrator in the biological sciences in the Soviet Union during these years he was involved in the perpetuation of the Lysenko school. From 1949 to 1956 he was Academician-secretary of the Department of Biological Sciences of the U.S.S.R. Academy of Sciences, a position that meant he exercised great influence over appointments and promotions at a time when these were keys to Lysenko's continuing power. The Soviet biologist Medvedev wrote in his history of Lysenkoism that in 1955 a petition directed against the administrative abuses of both Lysenko and Oparin was circulated among Soviet scientists.[55] Oparin was for many years a supporter of Lysenko, praising him in print on numerous occasions.[56] Nonetheless, as will be seen, Oparin struggled against several attempts by sympathizers with Lysenko to invade Oparin's field. Medvedev reported that in the final struggle with Lysenko, Oparin took a neutral position.[57]

One of the low points of Oparin's intellectual career came in his praise in 1951 of the new cell theory of Olga Lepeshinskaia. Lepeshinskaia was a mediocre biologist of impressive political stature as a result of her membership in the Communist Party from the very date of its founding and of her personal association with Lenin and many other political leaders. In 1950, a year of great political pressure in the Soviet Union, Lepeshinskaia claimed that she had obtained cells from living noncellular matter. Lepeshinskaia even maintained that she had obtained cells from noncellular nutrient mediums in as short a time as twenty-four hours.[58] Her work won praise from Lysenko himself.[59] It should be obvious from the past discussion of Oparin's views that he was skeptical in the extreme of all hypotheses that supposed the sudden appearance of a finely articulated entity from a less organized medium. Such an error had been made in the past, he thought, by all crude supporters of spontaneous generation. Engels had appropriately ridiculed those people who thought they could compel nature to do in twenty-four hours that for which thousands of years had been necessary. In principle, of course, such an event was possible, but Oparin thought only the best evidence, repeatedly substantiated, would have been convincing. He had in the past allowed room for the creation of life itself by man when his knowledge of biology became much more perfected. But in 1951, far from this moment, Oparin succumbed to the political pressures of Stalinist Russia and praised the "great service" of Professor Lepeshinskaia in "demonstrating" the emergence of cells from living noncellular matter even though Lepeshinskaia's evidence was rejected everywhere outside the Soviet bloc. He even agreed that this emergence of cells from living noncellular matter was occurring "at the present time," although he had opposed such a view many times in the past.[60] As we will see, not until 1953 did Oparin begin to resist these views in print. By 1957, however, he had returned to his flat opposition to spontaneous generation and to the sudden emergence of cellular forms in the manner described by Lepeshinskaia. Between 1953 and 1958 the supporters of Lepeshinskaia, and Lepeshinskaia herself, responded to Oparin's emerging remonstrances by, in turn, increasing their criticism of Oparin.[61]

In the early 1950's some of the most aggressive criticisms of science being made in the name of dialectical materialism were actually crudely materialistic or even idealistic. Certainly Lysenko was incapable of sophisticated philosophic analysis of scientific theory. Oparin was criticized by ideologists who were close to Lysenko in their viewpoints. One of their objects of criticism was his opinion that although life had once arisen on earth, such an event would never again be repeated. Several militant ideologists felt that Oparin's view attributed to life a uniqueness that contradicted uniformitarian and materialist doctrines. These writers were somewhat similar to the materialists of the nineteenth century who felt that the doctrine of spontaneous generation was logically required by materialism. They called themselves dialectical materialists, but they ignored Engels's perceptive criticism of that form of spontaneous generation; the fact that Oparin also opposed spontaneous generation was evidence to them of his philosophical waverings.[62]

In early 1953 Oparin commented on some of these criticisms.[63] "Does life arise now, at the present time?" he asked. No doubt it does, he replied, because matter never stays at rest, but constantly develops ever-new forms of movement. But life is not arising now on the earth—that stage of the development of matter has already been passed through here—but instead on other planets in the universe. He admitted that his critics had made a legitimate point in noting that his books were entitled *The Origin of Life,* as if what happened on earth were the whole story. (The following edition of his book would be entitled *The Origin of Life on the Earth* in recognition of this correction.) But he defended stoutly his belief that the prior nonexistence of life was a necessary condition for its origin.

In 1956 Oparin and the noted Soviet astrophysicist and astronomer V. Fesenkov (whose cosmological views were briefly discussed in Chapter V) co-operated in publishing a small volume entitled *Life in the Universe.*[64] Oparin had been criticized in the Soviet Union on grounds similar to those involved in discussions of James Jeans, upon whose hypothesis of the incandescent origin of the planetary system Oparin had relied in his earlier works. Now, in 1956, Oparin and Fesenkov acknowledged that Jeans's view "inevitably leads to the ideologically erroneous conclusion about the exceptionalism of the solar system in the Universe. Besides Jeans's hypothesis is unable to explain the basic peculiarities of the solar system." [65] Both Oparin and Fesenkov now agreed that O. Iu. Schmidt's idea that the sun had seized part of a dust cloud was a superior approach.[66]

Oparin's critics tried to find other similarities between his views and those of Jeans. Just as Jeans's "near-collision" of stars seemed to them to bestow an exceptional or miraculous character on the origin of the earth, so Oparin's establishing of very special conditions for the origin of life and his insistence that this event could never be repeated on earth seemed to attribute exclusive properties to the origin of life and, ultimately, to man. Oparin attempted to answer this criticism in his joint book with Fesenkov. The origin of life, the two writers said, was a perfectly normal development in the evolution of matter:

In its constant development matter pursues various courses and may acquire different forms of motion. Life, as one of these forms, results each time the requisite conditions for it are on hand anywhere in the Universe.[67]

But just because life is a lawful and normal development does not mean, they observed, that it should be seen everywhere. Those materialists who constantly seek to find evidence of the arising of life around them in order to illustrate the unexceptional character of life are ignoring the very genuine qualitative distinctions that exist in matter, and if carried to extremes, such views will lead to a form of hylozoism. Life must not be seen as an inherent property of matter, they thought, but as a special—yes, exceptional—form of motion of matter:

From the point of view of dialectical materialism life is material in nature, but it is not an inalienable property of all matter in general. On the contrary, it is inherent only in living beings and is not found in objects of the inorganic world. Life is a special, very complex and perfect form of motion of matter. It is not separated from the rest of the world by an unbridgeable gap, but arises in the process of the development of matter, at a definite stage of this development as a new, formerly absent quality.[68]

Just how rare is life in the universe? After a very long and detailed discussion of the physical requirements of life and the known characteristics of the universe, Oparin and Fesenkov came to the tentative conclusion that "only one star out of a million taken at random can possibly have a planet with life on it at some particular stage of development." [69] But such a ratio should not be seen as bestowing anything approaching uniqueness upon life; indeed, the two distinguished scientists continued, in our galaxy there may be thousands of planets on which life is likely and "Our infinite Universe must also contain an infinite number of inhabited planets." [70]

Oparin and Fesenkov's argument here was quite similar to that of astronomers in many other countries. From a methodological standpoint it might be worthwhile to notice that by such calculations, given an infinite amount of time, even so rare an occurrence as Jeans's near-collision of stars would occur with an infinite frequency. (Of course, some astronomers, referring to the red-shift phenomenon, would not grant infinite time to the universe.) But aside from philosophic considerations, there was significant empirical evidence against Jeans's hypothesis by the mid-fifties, and Fesenkov and Oparin, as professional scientists pursuing these problems, knew the evidence well.

In 1957 Oparin published the third revised and enlarged edition of his major work, with a more restricted title, *The Origin of Life on the Earth*.[71] In this volume he attempted to answer a number of recent criticisms advanced against his system, and he incorporated much recent scientific evidence. His original 1924 booklet had now grown, thirty-four years later, to almost five hundred pages.

As in the past, one of Oparin's major points concerned his belief in the erroneousness of the concept of spontaneous generation. The book by Olga Lepeshinskaia, *The Development of Cells from Living Matter*, was in his opinion, "an attempt to rehabilitate Pouchet's experiments and thus to resuscitate the theory of spontaneous generation." [72] Pouchet had expected microorganisms to be spontaneously generated; Lepeshinskaia also looked for spontaneous generation, but of cells from noncellular matter instead of complete organisms. Both had therefore sought the sudden appearance of order out of chaos, and such attempts are "foredoomed to failure." [73] The formation of the "very complicated structure of protoplasm" that is contained in both microorganisms and the single body cell cannot have happened fortuitously, but rather as the result of a long evolutionary process.

Oparin applied a similar sort of criticism to those scientists who would propose the gene, a molecule, or a bit of DNA as the primordial speck of life. Each of these theories is materialistic and therefore commendable, said Oparin, in that it seeks a material basis for life, but it is also mechanistic in the sense of all theories of spontaneous generation: It takes as a starting point a bit of matter that is actually the end point of a long evolution and assumes that the story of life began there. Since no explanation for the origin of this coherent bit of matter is attempted, the whole interpretation, intentionally or unintentionally, acquires a mysterious aura.

By this time J. D. Watson and F. H. C. Crick had suggested the noted double-helix model of the macromolecule of deoxyribonucleic acid (DNA). It was also clear by 1957 that DNA is the inherited material of almost all organisms. The different sequences of the bases (adenine, guanine, thymine, and cytosine) in the nucleotides linking the helixes, and the varying amounts of bases in each organism, led to an almost astronomical number of possibilities of structural combinations. Thus, the DNA macromolecule appeared to be a code of life, varying for each species and indeed for each member of that species. Investigators were now beginning to speak of genes as "sections of DNA," and specialists in the origin of life began to suspect that the first bit of life was a DNA molecule.

Oparin considered the establishing of the structure of DNA to be an event of great importance, and described in detail, with the inclusion of diagrams, the achievement of Watson and Crick. But he was definitely opposed to the talk resulting from the work of molecular biologists about the "first living molecule of DNA." His argument was, at bottom, the same one he had used against spontaneous generation of organisms many years before. Referring to hopes for the appearance of complete microorganisms in infusions, Oparin had written then:

If the reader were asked to consider the probability that in the midst of inorganic matter a large factory with smoke stacks, pipes, boilers, machines, ventilators, etc., suddenly sprang into existence by some natural process, let us say a volcanic eruption, this would be taken at best for a silly joke.[74]

No longer, he acknowledged, did anyone expect the spontaneous generation of complete organisms, or even of complete cells; to stick to the metaphor, no longer did they look for the sudden appearance of the whole factory. But Oparin believed that those people who thought the story of life began with the fortuitous synthesis of DNA were making the same error; they did not pretend that a factory could suddenly spring into existence, but they acted as if they believed it possible for the *blueprint* of that factory to accidentally appear. Yet that blueprint (the molecule of DNA) contained all the information necessary for the construction of the factory; for such a body of coded information to suddenly appear was as wild as to assume the sudden materialization of the factory itself. In their emphasis on a molecule as the starting point of life, many scientists were ignoring the question that to Oparin was the most important of all: "How did the rigidly determinate arrangement of nucleotides in the DNA come into being?" [75]

To Oparin, the synthesis of nucleic acids was a complex process of interaction on the supramolecular level, one requiring the presence of a complicated enzymic apparatus. This supramolecular approach was obviously fitted to his coacervate theory, although he recognized that other supramolecular mechanisms might also account for the origin of life. The important point was that the presence of more than one substance was required and that a hand-in-hand evolution of both nucleic acids and other substances had occurred: "Thus, on the one hand, the synthesis of proteins requires the presence of nucleic acids while, on the other, the synthesis of nucleic acids requires the presence of proteins (enzymes)." [76]

This known interaction had already led biologists to speculate on the relative importance of nucleic acid and protein to the life process. [77] In 1955 J. D. Bernal lectured on the origin of life at Moscow University and at the end of his presentation asked Oparin, "Which came first, nucleic acid or proteins?" Oparin's answer, presented in his 1957 book, provided a clue to much of his thinking:

This question reminds one somewhat of the scholastic problem about the hen and the egg. The problem is insoluble if we approach it metaphysically in isolation from the whole previous history of the development of living matter. Nowadays every hen comes from an egg and every hen's egg from a hen. Similarly, nowadays proteins can only arise on the basis of a system containing nucleic acids while nucleic acids are formed only on the basis of a protein-containing system. The hen and its egg developed from less highly organised living things in the course of their evolution. In the same way, both proteins and nucleic acids appeared as the result of the evolution of whole protoplasmic systems which developed from simpler and less well adapted systems, that is to say, from whole systems and not from isolated molecules. It would be quite wrong to imagine the isolated primary origin either of proteins or of nucleic acids.

Many contemporary authors do not, however, follow this line of thought. They take the view that in the first place nucleic acids arose in some way and that at once, simply by virtue of their intramolecular structure, they were able both to synthesize proteins and to multiply themselves spontaneously. It is, however, clear from all our previous discussion that a hypothesis of this sort is in direct opposition to the facts as they are at present known. [78]

Oparin saw a similarity between the current concentration upon the molecule of DNA as the primordial bit of life and the earlier concentration upon the gene. Furthermore, the view that the gene is a section of DNA molecule allowed room for the coalescence of the two approaches. In each case, however, valid research was being given what Oparin thought was an improper interpretation. "Life" to him was still a process, a flow, an interchange of matter, and could not be identified with any static form. Oparin thought that the new emphasis on DNA was the lineal descendant of earlier erroneous views such as: H. J. Muller's conception of the "random emergence of one successful gene among myriads of types of molecules"; T. H. Morgan's original "gene molecule"; C. B. Lipman's idea of the formation of a "living molecule"; R. Beutner's proposed "self-generating enzymes"; and A. Dauvillier's organic molecules with a "living configuration." [79]

Many molecular biologists were willing to grant that Oparin's evolutionary approach to DNA had its merits, but they felt that his commitment to a definition of life as a multimolecular phenomenon led him to strained, if not absurd, positions on the issue of viruses. Viruses were discussed several times in Oparin's 1957 book and also at an international symposium on the origin of life that he attended in Moscow in August 1957. The topic also appeared in an important book published by Oparin in 1960.[80]

The study of viruses has a great significance for an understanding of nucleic acids and protein synthesis; in viruses we can see with unusual clarity the process of replication that lies at the base of the life process. Furthermore, viruses seem to occupy a curious position, one often described as midway between the living and nonliving. By studying them, scientists have given their definitions of life the most rigorous tests. The most frequently examined virus has been the tobacco mosaic virus (TMV). Ever since W. M. Stanley succeeded in isolating TMV in a crystalline form in 1935, it has been regarded as one of the most suitable types of virus for study. The advantage of the TMV is not only that it can be crystallized and studied in a static form, but the undoubtedly related fact that while many other viruses contain a variety of substances—lipids, carbohydrates, nucleoproteins—the TMV is relatively simple in structure, consisting of nucleic acid and a protein coating. In the mid-fifties several workers succeeded in separating the nucleic acid core from the protein. As two of the leaders in this field of research reported, "The nucleic acid alone can perform all the crucial functions of the virus, namely it can initiate infection and transmit the required genetic information." [81]

Here then was a serious challenge to the earlier concepts of the nature of life, not only to Oparin's, but to those of all biologists. Stanley commented that in their efforts to distinguish the animate from the inanimate, "if viruses had not been discovered, all would have been well." [82] But with viruses, researchers seemed to have come upon a form of "life" that could, at least in some cases, be crystallized and held static indefinitely, that in terms of size was smaller than certain molecules, that could grow and reproduce, and that had the ability to mutate during reproduction. Why not recognize them as fully "living" organisms? Some researchers did. Stanley said of viruses in 1957, "They are all, in short, by definition, alive." [83] Others, including Oparin, believed that there were serious reasons for excluding viruses from the realm of the truly living. And Oparin was particularly firm in his opinion that neither viruses nor any other "living" form on the molecular level should be considered antecedent to all other living organisms. Such molecular forms were to him the products of life, not the producers. He felt that to regard them as the starting point of life would be to begin with the unexplainable and eventually to fall prey to metaphysical, mysterious interpretations of nature. His arguments will be considered in more detail below, after a few additional comments on the nature of viruses.

A clearer view of the action of viruses, and of the central role of molecules of nucleic acid in their function, can be gained by considering bacteriophages, those viruses that prey on bacteria. A particularly suitable example of a bacteriophage is the virus that attacks colon bacilli. These viruses first attach themselves to bacilli and then literally inject their interior nucleic acid molecules inside their hosts, leaving the protein coats outside. Once inside the cell walls of the bacteria, the viruses multiply until the bacteria burst, freeing the viruses for further conquests.

It is important to see that this phenomenon is not identical with the familiar forms of parasitic action in the biological world by which the parasitic organism gains sustenance from a host; there is a much more elemental and striking mechanism at work here. The virus is incapable of metabolic action by itself and, indeed, possesses none of the physiological mechanisms necessary for such action. It uses, instead, the mechanisms of the host to the ultimate degree, bringing to it only information suitable for gaining its goals. One might say that the nucleic acid is no more than a program for using an existing process for a different goal; it is as if an impostor came into an operating chemical plant with only a coded computer card under his arm and, by inserting that card into a central computer, redirected the flow of chemicals to yield a different product, one valued by him. Of course, this analogy, with its inclusion of man's intelligence and its somewhat mechanistic reference to a factory, has its weaknesses, but it may convey some of the strangeness of the situation.

Of Oparin's fairly extensive discussion of viruses, the aspect that most concerns us is his opinion on whether they are alive. He did not flatly declare in his 1957 book that viruses were not alive, but his argument certainly pointed toward that conclusion. It is indeed true, he remarked, that they can replicate themselves quite readily. But replication is not the same as life, he continued, since even inorganic crystals can replicate and grow. Furthermore, viruses cannot even replicate unless they are placed "inside" an existing life process. Oparin commented:

Nobody has succeeded in producing this so-called "multiplication" of virus particles under any other conditions or on any artificial medium. Outside the host organism the virus remains just as inert in this respect as any other nucleoprotein. Not only does it show no sign of metabolism but nobody has yet succeeded in establishing that it has even a simple enzymic effect. It is clear that the biosynthesis of virus nucleoproteins, like that of other proteins, is brought about by a complex of energic, catalytic and structural systems of the living cell of the host plant, and that the virus only alters the course of the process in some way so as to give specific properties to the final product of the synthesis.[84]

Although Oparin doubted that viruses were alive, he seemed to press that point of view less insistently in later publications. There were several potential avenues of compromise between the opposing points of view. Wendell Stanley, the man who crystallized TMV, even suggested at the 1957 conference, "Some may prefer to regard a virus molecule in a crystal in a test tube as a potentially living structure and to restrict the term 'living' to a virus during the time that it is actually reproducing. I would have no serious objection to this. . . ."[85] But Stanley then went on to repeat his belief that viruses are alive, without stipulating the moment in time when they become so. It was unclear what position Oparin would take on the suggestion that viruses are intermittently "alive" and "dead."

As an author primarily committed to the elaboration and defense of a theory on the origin and evolution of life, Oparin's main interest remained not a precise definition of "life," but the description of a development in time during which life appeared at one point or another. The exact moment could be debated. His interest in viruses centered on whether they were in the main path of that development or in a branch. And he believed that the answer to the question was that they were in a branch; whether or not viruses are ever alive, they were not the primordial forms of life from which all the others developed. As he commented at the 1957 symposium:

Today I should like to formulate, in a couple of words, my own viewpoint which I have expounded and substantiated in my book. I assumed that what had arisen primarily, by abiogenetic means, was not the functionally extremely efficiently constructed nucleic acids or proteins which we can now isolate from organisms, but only polynucleotides or polypeptides of a relatively disorderly structure, from which were formed the original systems. It was only on the basis of the evolution of these systems that there developed the functionally efficient forms of structure of molecules, not *vice versa*.

In the opposite case one would have to conceive of evolution as imagined by Empedocles, who held that first there developed arms, legs, eyes and ears and that later, owing to their combination, the organism developed.[86]

And in 1960 he came back to the topic by discussing the tobacco mosaic virus. He emphasized what happens when this virus attacks the cells of a tobacco leaf:

All that takes place is the constant new formation of a specific nucleoprotein with the help of the biological systems of the tobacco leaf. This means that the new formation is only possible in the presence of an organisation which is peculiar to life and consequently the first living thing was not a virus; on the contrary, viruses, like other modern specific proteins and nucleic acids, could only have arisen as products of the biological form of organisation.[87]

In the last phrase of the above quotation Oparin was referring to the well-known fact that parasites frequently become simpler in organization as they become more and more dependent on their hosts and adapted to that ecological niche. All viruses are parasites. Oparin was therefore suggesting that although the coded nucleic acid in viruses is the evolutionary product of more sophisticated organisms, the viruses themselves are the ultimate results of parasitic "devolution." They have lost all but their genetic material itself; they are, so to speak, "escaped" bits of the genetic code, which reproduce themselves by using the metabolic processes of more sophisticated organisms. But, according to Oparin, they could never have arisen without the prior evolution of organisms with a metabolic capability.

The publication by Oparin that most clearly illustrated the refinement of his philosophic views was his *Life, Its Nature, Origin and Development,* published in Russian in 1960 and in English in 1961.[88] In this book the form of dialectical materialism that Oparin had developed over the years permeated his scientific views to a greater degree than in any other of his major publications; dialectical materialism heavily influenced the very structure of his analysis. The careful reader of this volume cannot seriously maintain, it seems to me, that dialectical materialism was merely something to which Oparin paid lip service in prefaces and conclusions as a result of political pressure. Instead, a dialectical and materialist process philosophy, one that he had helped to elaborate, had, in turn, a systemic effect upon his scientific arguments. And all this was written at a time of relatively little political pressure upon scientists in the Soviet Union, during years that were quite different from those of the late forties and early fifties, the only time when most non-Soviet observers have considered dialectical materialism to be significant in Soviet science.

The point to which Oparin returned again and again in the writing of this book is that dialectical materialism is a *via media* between the positions of frank idealists and vitalists on the one hand, and mechanistic materialists, exuberant cyberneticists, and supporters of spontaneous generation on the other. Dialectical materialism was indeed a form of materialism and therefore opposed to the idealistic view that the essence of life was "some sort of supramaterial origin which is inaccessible to experiment." [89] But dialectical materialism was equally opposed to the view that all living phenomena could be explained as physical and chemical processes. To take the latter position would mean, said Oparin, "to deny that there is any qualitative difference between organisms and inorganic objects. We thus reach a position where we must say either that inorganic objects are alive or that life does not really exist." [90] Dialectical materialism provides a means, Oparin continued, of accepting the principle of the material nature of life without regarding "everything which is not included in physics and chemistry as being vitalistic or supernatural." [91] To dialectical materialists, life is a "special form of the motion of matter," one with its own distinct regularities and principles.

Life was to Oparin still a flow, an exchange, a dialectical unity: "An organism can live and maintain itself only so long as it is continually exchanging material and energy with its environment." [92] In a living body, this exchange of material is "strictly co-ordinated in time and space" in an orderly way insuring self-preservation and self-reproduction. But metabolic exchange and reproduction were only necessary, not sufficient, conditions for life. A living organism must have, in addition, the characteristic of "purposiveness." This characteristic figured more prominently in this later work of Oparin's than it had in his earlier writings. He believed that purposiveness pervades the whole living world "from top to bottom, right down to the most elementary form of life." [93] He recognized that his insistence on purposiveness as an essential feature of life had its dangers, since "in one form or another Aristotle's teaching about 'entelechy' has left its mark on all idealistic definitions of life." [94] But Oparin believed that the purposiveness of the organization of life was "an objective and self-evident fact which cannot be ignored by any thoughtful student of nature. The rightness or wrongness of the definition of life advanced by us, and also of many others, depends on what interpretation one gives to the word 'purposiveness' and what one believes to be its essential nature and origin." [95] He thought that dialectical materialists could avoid idealism by always studying this purposiveness in terms of its development, its origins. So long as purposiveness can be understood as a result of a historical interaction between the material organism being studied and its material environment, one need not fear idealism. It is only, Oparin said, when purposiveness is brought in from outside the boundaries of the material world, or is left so unexplained that such an origin seems implied, that biological explanations become idealistic. Hence, Oparin believed that the essential methodological guide through these dangers could be found in the writings of Heraclitus of Ephesus: "One can only understand the essence of things when one knows their origin and development." [96] In this principle Oparin believed that dialectical materialism and Darwinism drew upon a common inspiration, one found in ancient philosophy.

Oparin maintained that his conception of life as a metabolic, purposive material process revealed the faults of a number of twentieth-century views of life. Particularly common at the present time, he thought, were attempts to explain life on mechanistic, crudely materialist bases. These efforts were no doubt the results of the prior successes of science and should be praised for their materialism, but they so oversimplified biology that in the end they might harm its further development. The two best examples of modern mechanism, he said, were those interpretations based on cybernetics and molecular biology. The cyberneticists believed that their new approach was a "fundamentally new and universal road to the understanding of the very essence of life." [97] Helpful as some of their achievements have been, Oparin continued, the cyberneticists were wrong in their belief in the unique superiority of the cybernetic approach. They, like the mechanists of the eighteenth and nineteenth centuries, believed that living things could be identified with machines. They did not take into account the differences between their models and the real thing with the same readiness that they did the similarities. And the differences, Oparin remarked, are very genuine: "However great the complexity or intricacy of its organization, an electronic calculating machine is still further apart in its nature from a human being than is, for example, the simplest bacterium, although this has not got the differentiated nervous system which the machine imitates so successfully." [98]

What, according to Oparin, are these genuine differences between machines and living things? The machine is not metabolic in its composition, it is not constructed of living matter: "The composition of the living body is the very factor which determines its flowing character." [99] And the flow, the metabolism, of each cell in the living body allows it to exist for long lengths of time, to repair itself. In contrast to this, said Oparin, the basic structure of the machine is static.

Another fundamental difference between man and machine, said Oparin, is the origin of its purposiveness. Machines have purposiveness, just as living organisms do, but it is placed in them by man. They will therefore always differ from the "truly living." In order to understand this interesting and debatable insistence by Oparin that life can only be understood in its origin, the following quotation from a science-fantasy that Oparin related is helpful. In this quotation not only will Oparin's emphasis on historical development as a key to understanding emerge more fully, but one will also see his concept of dialectical levels of regularities; for Oparin there exist distinct "physicochemical regularities," "biological regularities," and "social regularities." Only human beings display all three:

Let us imagine that people have succeeded in making automatic machines or robots which can not only carry out a lot of work for mankind but can even independently create the energetic conditions necessary for their work, obtain metals and use them to construct components, and from these build new robots like themselves. Then some terrible disaster happened on the Earth, and it destroyed not only all the people but all living things on our planet. The metallic robots, however, remained. They continued to build others like themselves and so, although the old mechanisms gradually wore out, new ones arose and the "race" of robots continued and even, perhaps, increased within limits.

Let us further imagine that all this has already happened on one of the planets of our solar system, on Mars, for example, and that we have landed on that planet. On its waterless and lifeless expanses we suddenly meet with the robots. Do we have to regard them as living inhabitants of the planet? Of course not. The robots will not represent life but something else. Maybe a very complicated and efficient form of the organization and movement of matter, but still different from life. . . . It is impossible to grasp the nature of the "Martian robot" without a sufficient acquaintance with the social form of the motion of matter which gave rise to it. This would be true even if one were able to take down the robot into its individual components and reassemble it correctly. Even then there would remain hidden from our understanding those features of the organization of the robot which were purposefully constructed for the solution of problems which those who built them envisioned at some time, but which are completely unknown to us.[100]

In this passage, Oparin's view of life emerges in a particularly colorful way. It is obvious that he would not accept a purely functional definition of "life." Less obvious is how he would meet the arguments of a functionalist. How, for example, would a man who meets such robots on Mars know that they are, indeed, robots? How would he know that he does not "have to regard them as living inhabitants of the planet," as Oparin put it? Surely such an explorer would expect extraterrestrial life, existing in conditions quite different from those of the earth, to have a quite different appearance from life he had already witnessed. How would he know to be suspicious even if what he saw displayed baffling characteristics? Evidently Oparin would answer that man might indeed make such a mistake, but upon further study he would probably begin to realize that the robots had a social origin, even if he never learned very much about that disappeared society.

The other form of mechanistic interpretation of biology that Oparin resisted in his 1960 book was that coming out of molecular biology. He believed that Erwin Schrödinger in his important and stimulating little book *What Is Life?* had both helped biologists and hindered them; by calling attention to the importance of molecular structure to genetics, he had helped spur one of the great scientific searches of the twentieth century, leading to the description of the DNA molecule. But Oparin thought that at the same time he had thrown at least some biologists back upon static approaches to life.[101] It was Schrödinger who described life as "clockwork," reminding Oparin of eighteenth-century mechanism. And Schrödinger had written, "Now, I think, few words more are needed to disclose the point of resemblance between a clockwork and an organism. It is simply and solely that the latter also hinges upon a solid—the aperiodic crystal forming the hereditary substance, largely withdrawn from the disorder of heat motion." [102]

Oparin believed that here Schrödinger was making a methodological error. He ignored the room for compromise contained in Schrödinger's phrase *"largely* withdrawn from the disorder . . ." (italics added), which indicated that Schrödinger saw the hereditary substance as only relatively, not absolutely, static. But even if one admitted that the difference between the two men was a relative one, a difference remained. Oparin pointed out that he was not alone in believing that the mechanistic assumptions of molecular biologists were leading them astray. Erwin Chargaff, a biochemist from Columbia University, had warned against mechanism in a paper at the 1957 congress, when he asked,

Is the cell really nothing but a system of ingenious stamping presses, stencilling its way from life to death? Is life itself only an intricate chain of templates and catalysts and products? My answer to these and many similar questions would be No; for I believe that our science has become too mechanomorphic, that we talk in metaphors to conceal our ignorance, and that there are categories in biochemistry for which we lack even a proper notation, let alone an idea of their outlines and dimensions.[103]

And Chargaff had also seen, along with Oparin, that the mechanistic interpretations were leading to an improper emphasis on the nucleic acid molecule as the "patriarch" of life:

Thus, what started cosmically with beautiful and profound legends has come down to a so-called "macromolecule." If poetry has suffered, precision has not gained. For we may ask whether a molecule that merely provides for one cell constituent continually to make itself can teach us much about life and its origins. We can also ask whether the postulation of a hierarchy of cellular constituents, in which the nucleic acids are elevated to a patriarchal role in the creation, is justified. I believe there is not sufficient evidence for so singling out this particular class of substances.[104]

Thus, Oparin's denial of the primordial role of nucleic acid found sympathy in other ears and in other countries. An old-fashioned theist would have agreed with Oparin's and Chargaff's criticism of mechanism in biology, supplementing it with a story of divine creation. Chargaff's last statement above would fit very well with such an interpretation. But Oparin apparently would not have been troubled by this fact; since his dialectical materialism occupied a middle position between vulgar materialism and idealism, it followed that both the idealists and the dialectical materialists opposed mechanistic interpretations of life.

In October 1963 Oparin attended a conference on "The Origins of Prebiological Systems" at Wakulla Springs, Florida, sponsored jointly by Florida State University, the University of Miami, and NASA.[105] At this meeting P. T. Mora, of the National Institutes of Health, submitted current theories of the origin of life, including Oparin's, to a methodological critique.[106] He showed what has frequently been noted by philosophers of science, namely, that questions of singularity, or origin, are not in principle resolvable by experimental science. Thus, from the standpoint of strict logic and the methods of empirical science the question to which Oparin had devoted his life was not answerable. Mora was particularly critical of the application of the term "natural selection" (in the manner of Oparin) to nonliving systems.[107]

Indeed, Mora said that the gap between physical science and biology is "too big to bridge." [108] Consequently, Mora was extremely skeptical of

attempts such as Oparin's to throw a bridge across the gap, and believed
that it could be achieved only by committing methodological errors:

I believe that this accounting for the appearance of the first persistently
self-reproducing unit in a prebiotic system is an unwarranted extension of
the meaning of the word selection, used by Darwin in a valid, but different
operational sense. Remember, the Darwinian selection and evolution concept
was arrived at empirically, by observing the spectrum of living species. . . .
Of course, this Darwinian concept of selection is a necessary attribute of the
living, which allows it to adjust to the changing milieu. But it is not a sufficient
concept. Mutability can occur only when first there is present a persistent
reproducibility. Thus, mutability should be considered as a consequence of
the persistent reproducibility and not vice versa.

Thus, two operationally different meanings of the word, selectivity—one
in the physicochemical sense and the other in the Darwinian sense, each of
them valid when used in its proper context—can give an illusion that the
word may operate in a third sense. I believe this is the origin of the unwarranted speculation that some kind of molecular selection may account for
the appearance of the first persistently self-reproducing, internally controlled, living unit.[109]

What alternative explanation did Mora offer? He remarked that he
believed that "we have to have a teleological orientation in our thinking." [110] But when asked if he could describe this teleological orientation
more fully, he replied, "No. I am very destructive in my attitude. I can't
say anything conclusive and constructive and useful. I can only tell you
that I consider that the impression of physics on scientists is so strong
nowadays that it is difficult to think, without using the same concepts. In
my mind, this almost parallels the situation which existed around the
end of the scholastic period, when teleology was so overwhelming that
considerations and approaches that physics now uses were very difficult
even to think about." [111]

Mora's presentation aroused considerable controversy at the Florida
conference.[112] He posed a very old and very important problem in the
history of science. This issue is one of the fundamental problems of explanations of development, which can be simply stated in terms of
Mora's thesis: We cannot obtain a higher degree of order or organization than that present in the interacting elements and the environment.

It was not Oparin but Bernal who took the responsibility for answering Mora at the Florida conference. (Oparin would reply in a later pub-

lication.) Like Oparin, Bernal favored a materialistic, developmental explanation of the origin of life. He differed with Oparin by doubting the fundamental role of coacervation, favoring instead a process of clay mineral adsorption, but their two approaches both assumed the fruitfulness of trying to bridge the gap between the nonliving and the living. Bernal agreed with the validity of much of Mora's argument, and specifically, that questions of origin can not be explained on the basis of logic. They have instead, Bernal said, "a logic of their own." But, said Bernal, Mora

draws a conclusion which is the opposite to the one which I would draw. The present laws of physics, I would agree with him, are insufficient to describe the origin of life. To him this opens the way to teleology, even, by implication to creation by an intelligent agent. Now both of these hypotheses were eminently reasonable before the fifteenth or possibly even before the nineteenth century. Nowadays they carry a higher degree of improbability than any of the hypotheses questioned by Dr. Mora. . . .

I do not agree with the criticisms of the limitations of scientific method which Dr. Mora puts forward, but I think he has done a very valuable service in stating them. The contrast between a Cartesian physics with material causes and a teleological biology with final causes which he poses, I think is false. Nevertheless, it contains the truth of the different laws for different levels, an essentially Marxist idea.[113]

But the real difference between Mora and Oparin-Bernal was not whether different laws exist on different levels. In fact, Mora believed in the existence of such different levels even more strongly than Oparin and Bernal, for he thought the gap between physics and biology was unbridgeable, and, therefore, the distinction between the two levels was absolute. Oparin and Bernal, on the other hand, saw the distinction as a relative one.

Oparin took up the question of the way in which the transition from one level to the next higher one occurs in his 1966 book *The Origin and Initial Development of Life,* translated by NASA in 1968.[114]

In this work Oparin sketched out the "prebiological" state in greater detail, incorporating more of the recent evidence. He allowed more room for noncoacervate prebiological systems, pointing toward compromise with views such as Bernal's. The "coacervatelike" droplets, which had a complex and advanced organization but which are still simpler than "the most primitive living beings," Oparin now called protobions. The protobions went through a further evolution promoted by a process to which Oparin still insisted on giving the name "primitive natural selection." And in this section of his book Oparin cited Mora's criticism of his scheme at the 1963 Florida conference and attempted to answer it. He maintained that the "logic of its own" by which Bernal said the origin of life must be explained was, in fact, the logic of dialectics. As Oparin observed:

At present, a number of opinions have been expressed in the scientific literature on the competency of the use of the term "natural selection" only in respect to living beings. It is an opinion widely held among biologists that natural selection cannot be extended to objects which are not yet alive, and particularly not to our protobions.

It is, however, erroneous to think that living bodies first originated and then biological laws or vice versa, that in the beginning biological laws were formulated and the living bodies arose. . . .

Dialectics obliges us to consider the formation of living bodies and the formulation of biological laws as proceeding in indissoluble unity. It is therefore quite permissible to assume that protobions—those initial systems for the formation of life—evolved by submitting to the action not only of intrinsically physical and chemical laws, but also of incipient biological laws including also prebiological natural selection. Here we may cite an analogy with the formation of man, i.e., with the rise of a social form of the motion of matter which is even higher than life. As is known, this form took shape under the influence not so much of biological as of social factors, chiefly the labor of our ancestors, coming into being at a very early state of homogenesis, and improving more and more. Therefore, just as the rise of man is not the result of the operation of biological laws alone, so the rise of living bodies cannot be reduced to the action of only a few laws of inorganic nature.[115]

In the above quotation Oparin's belief in a hierarchy of laws in nature is revealed particularly clearly; social, biological, and physico-chemical laws all operate on different levels. The most difficult conceptual problem within the framework of Oparin's scheme is the transition from one realm of law to another. If one assumes, as Oparin did, that living matter evolved from nonliving matter and human beings with their social life evolved from lower orders of animals, some method of explaining these transitions must be found. Oparin relied on a dialectical concept of the emergence of qualitative distinctions; he believed that "incipient forms" of laws of a higher realm could be found in the realm immediately below it. It is the sort of concept that has appealed to many thinkers in the past—C. Lloyd Morgan's concept of "emergent evolution" was somewhat similar, but supplemented with the presupposition of God—and it has a certain persuasiveness. Nonetheless, it should be admitted that Oparin's philosophy of biology suffered from a lack of precision in definition, on which critics such as Mora correctly centered their attention; furthermore, Oparin's emphasis on the irreducibility of biology to physics and chemistry and his increasing attention to "purposiveness" skirted ever more closely to the very real dangers of vitalism (although Mora did not criticize these aspects of Oparin's thought, no doubt because of his own emphasis on teleology).

The Soviet philosopher I. T. Frolov recognized these pitfalls in his 1968 book on genetics and dialectics (discussed at the end of the previous chapter of this study) when he described the irreducibility of biology as more a result of man's incomplete knowledge than a characteristic of living matter itself. In Oparin's approach, living matter is inherently distinct from nonliving matter and cannot, in principle, be reduced to physics and chemistry. Frolov was less adamant.

It should be noticed that nothing in the philosophic system of a materialist absolutely requires him to believe that living matter on earth evolved from nonliving matter. Materialists have usually supported this view since it seemed the best explanation for the origin of life on earth without reliance on a divine agent. But strictly speaking, there is another alternative available to the materialist; he can maintain that matter has existed eternally in the universe in *both* its living and nonliving forms. Whether nonliving matter actually evolves into living matter can then be left an open question without violating any assumptions of philosophic materialism. The life that exists on earth can be explained by saying that it resulted from the depositing of primitive organisms on the surface of the globe from elsewhere at some past moment in its history. Such a hypothesis is frequently called panspermia and has been posed in the past in various forms by such well-known scientists as Liebig, Helmholtz, and Kelvin.

A few Soviet scientists began to re-examine panspermia in the late sixties. A geologist, B. I. Chuvashov, wrote in 1966 in *Problems of Philosophy* that in his opinion life had existed in the universe eternally.[116] He cited the recent criticisms of Oparin's application of the term "natural selection" to prebiological systems as one of the reasons for his dissatisfaction with Oparin's theory and his consequent interest in panspermia. Nonetheless, Chuvashov thought that nonliving matter may also occasionally develop into living matter, but perhaps only once in each planetary system or galaxy. Life was then spread throughout the neighboring area in the form of spores by meteorites and dust.

This view has been held by no more than a small minority of scientists in the Soviet Union; it enjoys similar modest support elsewhere. Some of the scientists now interested in panspermia have cited the presence of carbonaceous chondrites in the lunar soil samples brought back to the earth by the Apollo 11 expedition as support for their hypothesis.[117] This evidence, however, is subject to differing interpretations and does not yet warrant conclusions.[118]

The emergence of life from nonliving matter remains the favored view of Marxist philosophers and biologists. Dialectical materialism has been deeply penetrated by the concept of an over-all development of matter, with no impassable barriers.

Discussions of the origin of life are currently developing at a rapid rate throughout the world. We can expect a continuing debate of the topic with full Soviet participation. Soviet scientists have already made some of the most important contributions to past discussions of the issue.

VIII Structural Chemistry

The nature of bonds between atoms is of fundamental importance to chemistry, since that science is to a large degree a study of the alteration of such bonds. Yet the inadequacy of bond diagrams in depicting important chemical compounds has been known from the very beginning of structural chemistry. Succeeding diagrammatic systems were discarded because of the failure of each system to account for certain phenomena. The ancient contest between idealism and materialism entered the discussion when chemists began using models that seemed physically inconceivable to some other chemists.

The formulas and models constructed by chemists must explain not only the composition of chemical compounds, but also their properties. In the first half of the nineteenth century no single convention or method of representing compounds was accepted. J. R. Partington remarked, "It was apparently considered a sign of independence of thought for every chemist to have his own set of formulae." [1] As late as 1861 Friedrich August Kekulé gave nineteen different formulas for acetic acid. [2]

The reason for the fragmentation of theories and the use of multiple formulas lay in the inability of chemists to observe or measure molecules directly. Chemistry in general, and organic chemistry in particular, were frighteningly unknown. In 1835 Wöhler wrote to Berzelius, "organic chemistry appears to me like a primeval forest of the tropics, full of the most remarkable things." [3] During the next thirty years chemists collected an astonishing amount of data and isolated many compounds, but the formulas for compounds were still conjectures based on very incomplete experimental data.[4]

The nineteenth-century chemists soon discovered that many compounds could not be represented by a single formula that would explain all their known reactions. One formula accounted for one particular reaction, and another explained a different reaction. Perhaps by the use of four or five dissimilar models of the molecule of one compound a chemist could account for all of the known reactions of that compound, but this method merely pointed up the dilemma: Common sense indicated to the chemists that any substance should have molecules of a particular shape, which could be reproduced by a model (leaving aside for the moment isomers and tautomers, which are a separate topic: see note 7). But there was only a certain number of ways, geometrically speaking, that a particular model of a molecule could be constructed, and no one of these ways explained all the reactions of that particular substance. This is the case for many compounds at the present time, the most familiar being benzene.

When Kekulé sketched out the simple hexagon still used to represent the starting point for aromatic compounds, he immediately ran into the problem of the location of the bonds.[5] Kekulé believed that carbon atoms were quadrivalent, so each carbon atom had one bond unaccounted for. Kekulé adopted the idea of alternate double and single bonds:

This formula, although still utilized almost universally, is not satisfactory. If benzene actually were so constituted, it would mean the following two isomers would be possible:

Upon examination of the above diagrams one will see the difference: In the first case, there is a double bond between the two added chlorine atoms, while in the second, there is a single bond. But two such isomers do not exist, not with chlorine or any other added groups; we know that isomers of orthodisubstituted compounds of benzene can not be created.

In 1872 Kekulé introduced the concept that the bonds are constantly "flapping between alternate sectors like a pair of swinging barn doors." [6]

Rather than draw such a complicated model for benzene every time, chemists usually draw two formulas, showing the two positions. These two diagrams are usually called the "ideal Kekulé structures." [7]

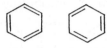

The explanation that the bonds of benzene are shifting back and forth satisfied the needs of chemists for many years. Abbreviated or out-of-date histories of chemistry sometimes stop with the story of benzene at this point. However, chemists found that a connection between opposite carbon atoms, or a link between the para positions, also must exist:

This formula was originally suggested by Sir James Dewar, who intended it to supplement the two original Kekulé structures.[8] Now there were three formulas for benzene, and a mental picture of "flapping barn doors" was becoming increasingly difficult. Furthermore, other variants were added. And most perplexing, it became apparent that no actual movement between simple bond configurations was occurring in the benzene molecule.

The resonance theory of valency, developed around 1930 by Linus Pauling and further expanded by G. W. Wheland, is an attempt to explain the structure of molecules such as benzene.[9] The significance of the resonance theory according to Wheland is that "it is considered possible for the true state of a molecule to be not identical with that represented by any single classical valence-bond structure, but to be intermediate between those represented by two or more different valence-bond structures." [10] Such an intermediate structure is known as a "resonance hybrid." Structural chemists have described many such hybrids. The two valence-bond structures "contributing" to the carboxylate ion are:[11]

$$R - C \underset{\ddot{\ddot{O}}:^-}{\overset{\ddot{O}:}{=\!\!\!=}} \qquad\qquad R - C \underset{\ddot{\ddot{O}}:}{\overset{:\ddot{O}:^-}{\ldots}}$$

The resonance hybrid for the carboxylate ion is usually drawn:

$$R - C \underset{O\ -\frac{1}{2}}{\overset{O\ -\frac{1}{2}}{=\!\!\!=}}$$

In the case of benzene, five different structures, the Kekulé-Dewar ideal forms, are considered to be contributing to the hybrid:[12]

For other compounds, many more models are used. To explain the reactions of anthracene, over four hundred diagrams are utilized.

Wheland repeatedly reminded his readers that resonance should be regarded not as any sort of oscillation between the various structures, but as a word referring to a molecule in a permanent hybrid state.[13] The five structures are merely aids for descriptive purposes and never, in fact, exist.[14] On this point Pauling has commented:

> We might say . . . that the molecule cannot be satisfactorily represented by any single valence-bond structure and abandon the effort to correlate its structure and properties with those of other molecules. By using valence-bond structures as the basis for discussion, however, with the aid of the concept of resonance, we are able to account for the properties of the molecule in terms of those of other molecules in a straightforward and simple way. It is for this practical reason that we find it convenient to speak of the resonance of molecules among several electronic structures.[15]

According to the theory of resonance, the bonds between the carbon atoms in benzene would be neither single nor double bonds, but a type of bond between the two, roughly described, perhaps, as a $1\frac{1}{2}$ or $1\frac{1}{3}$ bond. Such a description is supported by electron diffraction and infrared spectroscopic examinations, which show that while the distance between carbon atoms connected by a single bond is about 1.54 Angstrom units and by double bonds is 1.33 units, the measurement for benzene bonds is 1.40, between that of a double and that of a single bond.[16]

Although, as Pauling emphasized, the theory of resonance does not rest upon quantum mechanics in its conception, a quantum-mechanical method of calculation is utilized in computing certain properties of the molecules, such as stability during reactions. A wave function, or Schrödinger equation, is written for each of the idealized, or resonance, structures, and then the wave functions are combined in a purely linear fashion, that is, by simple addition, with a weighting factor applied to each equation depending on the amount of "influence" each ideal structure exercises.

The theory of resonance and Pauling's elaboration of it were known in the Soviet Union long before World War II; many years passed before the theory of chemical bonds attracted any particular attention. The theory of resonance became popular among chemists in the Soviet Union. Prominent chemists such as A. N. Nesmeianov,[17] R. Kh. Freidlina, D. N. Kursanov,[18] E. N. Prilezhaeva,[19] M. I. Kabachnik,[20] and many others utilized the theory of resonance in their research and in their published works. In 1946 two Soviet chemists about whom we will hear considerably more, Ia. K. Syrkin and M. E. Diatkina, appeared with their own treatment of resonance in the book *The Chemical Bond and the Structure of Molecules,* which Pauling described as an "excellent work." [21] He added that in his opinion Syrkin and Diatkina were "among the most able [chemists] in Russia today." [22] The two authors' book was adopted by the Ministry of Higher Education of the U.S.S.R. as a study aid for the chemistry faculties of the universities and received widespread distribution. It subsequently was translated into English for use in the United States.[23] The year after they published their own book, Syrkin and Diatkina translated Pauling's *The Nature of the Chemical Bond* into Russian; again, in the following year, they worked together in translating Wheland's *The Theory of Resonance and Its Applications to Organic Chemistry,* this time with Syrkin as editor and Diatkina as translator.

The resonance controversy was initiated by a zealous, ambitious, but undistinguished chemist, Gennadi V. Chelintsev, who was later accused of trying to gain the supreme position in chemistry that Trofim D. Lysenko had won in biology. Although the eventual outcome of the controversy would be considerably different from what Chelintsev called for, he was a central figure throughout the discussions in the following years.

In 1949 Chelintsev, a professor of chemical warfare at the Voroshilov Military Academy, published a book entitled *Essays on the Theory of Organic Chemistry;* the controversial nature of the small book was foreseen in the publisher's prefatory remarks that it would

open up a discussion on the crucial and genuine problem of organic chemical theory. Therefore, the publisher . . . gives Soviet chemists the possibility of becoming acquainted with the author's views and arguments; the publisher expresses the hope that the book will be met with interest and that it will serve a useful purpose in the further creative development of Soviet chemical science.[24]

Chelintsev proposed to explain molecular structure in a way that would not include the approximate methods of quantum mechanics and would not require the use of more than one formula for one compound. In particular, he said the formula for benzene should be drawn, not on the basis of covalent bonds, but on the basis of electrovalent, or ionic, bonds.[25] Chelintsev would represent benzene in the following way:

$$
\begin{array}{c}
+ \\
-\,\bigcirc\,- \\
+\quad\quad+ \\
-
\end{array}
$$

The dotted line signified, to Chelintsev, "the levelling out" of the electronic charge. According to him, there were no double bonds in benzene at all. He maintained that the theory of resonance not only was sterile methodologically, but also introduced a mechanistic concept into chemistry, filling a gap in man's knowledge with an unrealistic but comforting mechanical description.

The appearance of Chelintsev's book elevated the theory of resonance to the platform of philosophical discussion at a moment of ideological militance in the Soviet Union. Resonance theory's use of multiple ideal structures, which Chelintsev called mechanistic, made the theory appear susceptible to criticism on philosophic grounds.

That the theory of resonance would be considered philosophically untenable to authors other than Chelintsev was made clear by V. M. Tatevskii and M. I. Shakhparanov in an article in the fall of 1949 entitled "On a Machist Theory in Chemistry and Its Propagandists." [26] The two writers selected for particular criticism Wheland's description of resonance as a man-made concept that "does not correspond to any intrinsic property of the molecule itself, but instead is only a mathematical device, deliberately invented by the physicist or chemist for his own convenience." [27] It would have been possible to center the criticism upon the philosophic implications of this statement rather than upon the resonance theory itself; one could easily maintain, according to a realist philosophy, that the resonance structures had some, perhaps quite indirect, relationship to the actual structure of the molecule, but that this relationship remained obscure. Until more information on the actual structure was obtained, the resonance theory could be used without necessarily subscribing to Wheland's philosophic remarks. Instead, Tatevskii and Shakhparanov affirmed that it would be philosophically incorrect to describe molecules in terms of ideal structures that were physically inconceivable. The primary fault of the resonance theory, according to the authors, was that it utilized more than one structure while insisting that no transformation back and forth between the forms occurred. Thus, the theory of resonance had become "divorced from reality." Tatevskii and Shakhparanov maintained that Wheland and Pauling had tried to cover up their ignorance of the true nature of molecules with a clever creation containing false philosophic assumptions:

The theory of resonance may serve as one example of the Machist theoretic-perceptional tendencies of bourgeois scientists, which are hostile to the Marxist world view and which lead them to pseudoscientific conclusions concerning the solution of concrete physical and chemical problems.[28]

The position taken by Tatevskii and Shakhparanov was paralleled closely by unsigned articles that appeared in the *Journal of Physical Chemistry* and in *Pravda* commemorating Stalin's seventieth birthday.[29] The authors of the articles asked that defects in Soviet science, especially in chemistry, be eliminated.

The discussion of resonance theory contained many references to a prominent and able nineteenth-century Russian chemist, Aleksandr M. Butlerov. Butlerov, a professor of chemistry at the universities of Kazan and St. Petersburg, and a member of the Imperial Academy of Sciences from 1874 to 1880, is only very rarely mentioned in non-Russian textbooks or histories of chemistry. Little doubt exists that he deserves much more attention.[30] In 1940, before the resonance controversy began in the Soviet Union, a noted American historian of chemistry, Henry W. Leicester, wrote a biographical article lauding Butlerov for his advanced studies in the field of organic chemistry.[31] In 1953 a French chemist, J. Jacques, asserted that the name of Butlerov should have equal prominence with that of Friedrich Kekulé in the development of theories of molecular structure.[32] In the 1960 edition of his *The Nature of the Chemical Bond* Linus Pauling, who had recently been criticized severely by Soviet philosophers of science, gave credit to Butlerov for his work on valency.[33] In earlier editions of the same work, Pauling had not mentioned Butlerov, evidently because he had at that time not known of his work.[34] Although Butlerov's conception of molecular structure has still not been thoroughly evaluated outside the Soviet Union, Professor I. M. Hunsberger gave what may serve as an interim judgment: "There is no doubt whatsoever that Butlerov has not received the credit he richly deserves and that his monumental contributions to organic structural theory have been for the most part virtually overlooked. . . . Butlerov's contributions certainly equal those of Kekulé and Couper, but it is ridiculous to maintain that Butlerov is the *sole* author of structural theory." [35] In this light, Butlerov began to receive slightly more attention outside the Soviet Union in the 1950's. As late as the 1955 edition, however, the *Encyclopedia Britannica* did not even list Butlerov, although it gave a whole column to Kekulé. The 1963 edition contained a paragraph on Butlerov.

Butlerov's philosophic views differed from those of such chemists as Charles Gerhardt, who did not believe that chemical formulas represented any sort of reality. Kekulé himself never attributed much physical significance to his formulas; he tended to regard them only as symbols for explaining reactions.[36] Butlerov, on the contrary, believed that one substance should have one structural formula with a real, even if indefinite, relationship to the actual structure of the substance. He remarked,

If we attempt to define the chemical structure of substances and if we have success in expressing it by our formulas, then those formulas will be, although not completely, real rational formulas. . . . For each substance there will be, therefore, one rational formula, and when the general laws of the properties of substances are well known, such a formula will express all of these properties.[37]

Quotations such as this were convenient for authors who wished to use Butlerov to criticize the multiple forms of resonance theory. Many of these authors ignored Butlerov's further statement on the meaning of chemical formulas that "what matters is not the form, but the essence, the conception, the idea. . . . [I]t is not difficult to realize that any method of notation is good as long as it expresses these relationships conveniently." [38]

From February 2 to 7, 1950, the Institute of Organic Chemistry of the Academy of Sciences conducted a discussion on modern theories of organic chemistry.[39] Out of this discussion came a report entitled "The Present State of Chemical Structural Theory," written by D. N. Kursanov, chairman, and a committee of seven other chemists.[40] Later, Chelintsev criticized most of these men.

The committee report referred to the Communist Party's interest in the present discussion and the direct connection to the controversy in biology:

The decisions of the Central Committee of the VKP(b) [all-Union Communist Party (Bolsheviks)] in regard to ideological problems and the sessions of the VASKhNIL [all-Union Lenin Academy of Agricultural Sciences] have mobilized Soviet scientists for the solution of the problem of a critical analysis of the present state of theoretical concepts in all fields of knowledge and the struggle against the alien reactionary ideas of bourgeois science.

The crisis of bourgeois science, connected with the general crisis of the capitalistic system, has been illustrated by the theoretical concepts of organic chemistry now being developed by bourgeois scientists and has led to the appearance of methodologically faulty concepts, which are slowing down the further development of science.[41]

But while the committee criticized resonance theory and even its own members for using the theory, it forthrightly asserted that Chelintsev's views were based on false scientific grounds. Chelintsev was helpful in the sense that he drew "the general attention of Soviet chemists to the necessity for a critical analysis of resonance theory," but he then "incorrectly identified the theory of resonance with the entire contemporary theory of chemical structure. . . ." [42] By so doing, Chelintsev tried to halt the application of quantum mechanics to chemistry, which was, in the view of the committee, actually "a further development of Butlerov's theory which made it concrete. . . ."

As for Chelintsev's "New Structural Theory," Kursanov and his colleagues had no kind words:

An understanding of the nature of the chemical bond requires the application and consideration of all the data derived by modern chemistry and physics. G. V. Chelintsev's "New Structural Theory" is an attempt to construct a new theory of the chemical bond without considering these facts. . . . Significantly, even the author himself does not apply his "New Structural Theory" in his works. This theory should be rejected.[43]

In this report and in several other articles that appeared at this time, four main points seem paramount: (1) Butlerov was the true founder of the theory of chemical structure; (2) the theory of resonance is idealistic and, therefore, unacceptable; (3) although resonance must be rejected, quantum mechanics is essential for scientific research, and a clear line can be drawn between the theory of resonance and quantum mechanics; (4) G. V. Chelintsev is not a competent scientist. As time progressed, points three and four became increasingly important. An effort was clearly being made during 1950 to discredit Chelintsev and simultaneously rally as many scientists as possible to the defense of quantum mechanical methods of calculation in chemistry.

In an article that appeared in January 1951, O. A. Reutov recognized that too strict an adherence to Chelintsev's views would result in a simple demand for static mechanical models for molecules. He hinted that past discussions of the theory of resonance had emphasized materialism while ignoring the dialectic. Reutov affirmed that "there are two sides to Butlerov's theory. One is related to the unconditional recognition of a definite structure of molecules. The other aspect of this doctrine asserts the presence of interactions between atoms. . . ."[44] Reutov indicated that any description of molecules must not be of a static model, but of constantly changing ones, the result of the interaction of opposing forces, a truly dialectical process.

On the page following Reutov's article in the January 1951 issue of the *Journal of General Chemistry* was printed a notice announcing a forthcoming all-Union conference on the theory of chemical structure. The topic would now be discussed, not only by chemists, but also by hundreds of physicists, philosophers, and educators. The Bureau of the Division of Chemical Sciences of the Academy of Sciences organized a commission headed by the president of the Academy, A. N. Nesmeianov, to prepare the major report, which was to concern Butlerov's views of structural chemistry, a critique of resonance, and the future development of the theory of chemical structure. Readers were invited to write in comments or suggestions.

The conference was held in Moscow on June 11–14, 1951, under the chairmanship of M. M. Dubinin.[45] The main report was given by A. N. Terenin, rather than by Nesmeianov, who was ill and did not attend. A total of forty-four delegates gave speeches, many of them similar in content, but there were several heated debates despite the fact that not one person defended the theory of resonance. The previous articles and discussions had already so set the stage for the conference that it seems to have been a foregone conclusion that resonance would be rejected. The real issue was finding an alternative. Chelintsev, who evidently would have jettisoned the methods of quantum chemistry as a whole, became increasingly pathetic as he continued to make ineffective attacks against his fellow chemists.

Syrkin, Diatkina, Volkenshtein, and Kiprianov retracted their earlier defenses of the theory of resonance and said that they had been mistaken. Syrkin said that when he had written his book, he had not known about Butlerov's views and therefore had not been aware of the correct trend of development in chemistry. Diatkina revealed that at an earlier point she had tried to defend resonance chemistry in terms of dialectical materialism, referring to the "qualitative and quantitative aspects of the theory of resonance." [46] Her effort had failed, and she now called it "a confusion of irrelevant matters." Diatkina's attempt to illustrate the philosophical acceptability of the resonance theory by referring to the dialectic was a parallel, perhaps unknown to her, of the British scientist J. B. S. Haldane's opinions in 1939, when he wrote that the theory of resonance was "a brilliant example of dialectical thinking, of the refusal to admit that two alternates (two contributing structures) which are put before you are necessarily quite exclusive." [47]

Terenin's report, to which Diatkina referred and which was the basis of discussion for the conference, was very similar to the report that the Institute of Organic Chemistry had given in February 1950. This similarity is not surprising, since several of the same scientists were in charge of both reports. One difference was obvious, however; Terenin's commission was assigned the task not only of criticizing the resonance theory on the basis of Butlerov's writings, but also of planning the future work of Soviet chemists—that is, suggesting a substitute for resonance theory.

Terenin and his co-workers isolated the error of resonance as being specifically the use of ideal, fictitious resonance structures.[48] These fictitious structures, such as the Kekulé forms of benzene, had been suggested by chemists in an effort to explain the various reactions of compounds. Terenin and the other authors of the report pointed out that this empirical data on the properties of molecules was, of course, perfectly acceptable and useful. Therefore, as long as the Soviet chemists did not resort to computations derived from fictitious structures, they could use all of the data that they could collect concerning molecules, and also the mathematical expressions that, as represented in physical terms by supporters of the theory of resonance, led to the contradictions of incompatible physical forms. This alternative approach, which avoided the ideal structures, was named the "theory of mutual influences," borrowing a phrase from Butlerov. The explanation that Terenin and his colleagues gave for this apparent contradiction was the inadequacy of man's knowledge of the structure of matter.[49] Eventually structural chemistry will advance to the point where this contradiction will be resolved. Whatever that more complete answer may be, it cannot possibly be the theory of resonance, which is a blind alley postulating that the form of a molecule is physically inconceivable. The theory of resonance leads to agnosticism, said the authors of the report, which they defined as the Kantian belief that man cannot know his surroundings.

The approach that the report writers suggested to molecular structure would permit the chemist to use all of the data leading up to the theory of resonance as long as he stopped short of representing molecules as hybrids of ideal graphic forms. But he could use the equations themselves, which are the essentials for utilizing the resonance theory.

The difference between the forbidden theory of resonance and the permitted theory of mutual influences was exceedingly subtle. Epistemologically, there was a difference; chemists following the theory of mutual influences as described by the authors of the report could not conclude that molecules are merely intellectual forms, nor that molecules can be explained only in terms of intellectual forms, no matter how persuasive the evidence for one of these alternatives might be. The primary practical distinction between the method suggested by the Soviet chemists and the theory of resonance was that scientists following the first approach would be deprived of the use of resonance forms as visual aids in the classroom and the laboratory.

The theory of mutual influences had a parallel in other countries in an approach known as the "molecular orbit" method, which does not postulate the exact location of certain molecular bonds. Consequently, many chemists believe that it does not explain certain reactions as satisfactorily as the theory of resonance. Wheland commented, however, that in mathematical form the molecular orbit method becomes virtually identical with the resonance theory.[50]

Professor Chelintsev fiercely tore into the report. The whole conference, he maintained, was failing its task; it was supposed to reject the theories of Pauling and C. K. Ingold, but on the contrary, it had been taken over by the partisans of these two foreign scientists. He added that the theory of mutual influences was only a modification in nomenclature of the mesomerism-resonance theory.[51] "The authors of the report," he accused, "were given the assignment of saving the beloved heart of Ingold's and Pauling's mesomerism-resonance theory." [52] Chelintsev said that the leaders of the conference had suppressed his articles and had camouflaged the resonance theory.[53] No doubt feeling frustrated, Chelintsev commented:

This is the first case in the course of the recent scientific-methodological discussions when the approved report advances not a criticism of the mistakes but a plea of guilty in their behalf, representing, moreover, not an isolated person, but the commission that was named by the Department of Chemical Sciences of the Academy of Sciences of the U.S.S.R., and approved, surely, by the Presidium of the Academy.[54]

Chelintsev noted that he considered it his duty to name the most active propagandists of the resonance theory; he began with the President of the Academy of Sciences, A. N. Nesmeianov, and named twenty-six chemists,[55] a good number of them the leaders of their fields. Almost all the authors of articles denouncing resonance that had appeared in the previous two years were on the list, including Tatevskii and Shakhparanov, whose article in late 1949 had indicated that the criticism of resonance would go beyond that of Chelintsev. Five of the nine members of the commission of the department of chemical sciences, which had investigated the resonance theory in February 1950, were named, as were six of eleven scientists designated to compile the report for the June 1951 all-Union conference.

After his speech Chelintsev faced a series of questions from the floor. One questioner sarcastically inquired, "You have read off the defenders of idealism in Soviet chemistry. Who, in your opinion, in all of Soviet chemistry, are representatives of dialectical materialism? (laughter in the hall)" [56] Chelintsev replied that it would be impossible to name all of the defenders of dialectical materialism because there were only twenty or thirty men—the ones he named—who ignored dialectical materialism. Applause from the floor met this reply.

A few supporters of Chelintsev spoke at the conference. S. N. Khitrik defended Chelintsev's views and also pointed out the "irrefutable fact" that Chelintsev was the first person to unmask "the idealistic essence" of the theory of mesomerism-resonance.[57]

One of the speakers who followed Chelintsev was A. A. Maksimov, a member of the editorial committee of *Problems of Philosophy*. Maksimov had been involved in a long series of squabbles concerning dialectical materialism and science; he has already appeared in this study in the discussion of relativity physics, in which he attacked not only general but special relativity. Maksimov's presence, along with that of the journalist V. E. Lvov, was resented by some of the other participants at the conference. The chemist Volkenshtein asked, "Why did this journalist drop in here?" [58] Volkenshtein observed that the year before, Lvov had been expelled from the midst of the Leningrad physicists for stirring up trouble. This is an indication that some additional scientists may have united against ideological demagogues as early as 1950.

Maksimov, however, had shifted from the offensive to the defensive in the debate over idealism in science. Rather than supporting Chelintsev, he criticized him. He plumbed Chelintsev's motives in his remark: "According to Chelintsev, he is assuming the role in chemistry of T. D. Lysenko, and the 'Pauling-Ingoldites' named by him are playing the role of Weissmann-Morganists." Maksimov also affirmed, "I know the members of the commission and the 'Pauling-Ingoldites.' . . . I consider them honest, devoted Soviet chemists, sparked by desires for Soviet chemistry to flourish." [59]

When the conference prepared to vote on the final resolution, Chelintsev rose to say that although he had been a member of the commission in charge of drafting the resolution, he had been completely outnumbered by the Paulingites and Ingoldites (he did not cease using these terms although members of the conference had asked him to), and consequently his voice had not been influential.[60] In the final vote the three or four delegates who had supported Chelintsev in earlier procedural matters deserted him and Chelintsev alone opposed the resolution.[61]

The resolution approved the essence of the report of the chemical section of the Academy of Sciences, which Professor Terenin had read at the beginning of the conference, but noted several "serious defects" in that report. First of all, the report had not illustrated that the perversions in chemistry were closely connected with those in biology and physiology and that all these hostile theories "present a united front in the fight of reactionary bourgeois ideology against materialism." [62] Another defect of the report of Terenin and his partners was their failure to describe adequately the great progress of Soviet chemistry.[63]

The conference resolution also reprimanded Syrkin, Diatkina, Volkenshtein, and Kiprianov for not giving complete criticisms of the theory of resonance and for not detailing their errors. Volkenshtein's and Kiprianov's pleas of ignorance were rejected. The resolution noted, however, that all four of the chemists had admitted their mistakes.

Soviet philosophers, chemists, and physicists were all criticized, each group for a slightly different reason. The philosophers had not been active on the chemical front. The resolution pointed out that chemists, not philosophers, had discovered the ideological weaknesses. The chemists were criticized, nonetheless, for not giving adequate attention to the methodology of science. The resonance theory had long been tolerated when it should have been expelled. The criticism of the physicists indicated an undercurrent of unpublished opposition or neutralism to the conference. The resolution stated that "the conference notes the withdrawal of the majority of physicists from participation in the struggle for the creation of the progressive theory of chemical science. It is also unsatisfactory that practically none of the leading theoretical physicists has taken part in the work of the present conference." [64]

In addition to calling on all Soviet scientists to pay more attention to methodological matters, the resolution recommended the prompt publication of the stenographic report of the conference and the publicizing of the resolution, future discussions in scientific circles of the theory of chemical structure, and publication of new chemical textbooks and monographs incorporating the correct viewpoint on chemical structure.[65] The resolution ended with an ebullient tribute to Stalin.

Although Chelintsev had not been mentioned in the resolution of the all-Union conference, his theory did not escape what amounted to official condemnation, at the hands of Nesmeianov himself. In January 1952 Nesmeianov carefully analyzed Chelintsev's position in a technical article[66] and concluded that although Chelintsev had performed a useful service in drawing attention to the theory of resonance, the delegates at the conference had been correct in describing his scientific theories as worthless. Nesmeianov, an experienced chemist as well as president of the Academy of Sciences, considered Chelintsev an enthusiastic materialist who was not sufficiently grounded in either chemistry or dialectics. He said that Chelintsev's theory concerning the possibility, at that time, of reproducing the structure of every molecule was absolutely incorrect. Nesmeianov asked that Chelintsev produce the fruit of his labors. If he had a correct structural theory, he should be able to predict reactions. Where are these predictions? Nesmeianov hinted that not even Chelintsev could sincerely believe in his naïve pronouncements. He concluded with a barb aimed both at partisans of resonance theory and supporters of Chelintsev: "Our chemistry must be thoroughly cleared of all the unhealthy influences of corrupt bourgeois philosophy and science. It must be cleared also of vulgarizations of home origin." [67]

Chelintsev's audacity in the face of such criticism from the scientist with the most institutional authority in the Soviet Union seems rather surprising. In the same issue as Nesmeianov's article, Chelintsev repeated his accusations about the monopoly of Paulingites in chemistry, including Nesmeianov. He rejected Nesmeianov's indication that he did not believe in his own theory:

As far as my conviction of the worthlessness of the New Structural Theory is concerned, is it really possible to suppose that for a number of years I could bear the whole weight of the struggle with the Ingold-Paulingites who were monopolizing Soviet chemical science if I were not deeply convinced of the correctness and usefulness of my idea? [68]

The fact that the irrepressible Chelintsev's article appeared and that he was so bold may indicate that he had support somewhere in the Soviet Union. Furthermore, Nesmeianov would not have bothered to criticize Chelintsev at length if the rebellious Soviet chemist had been as isolated as Nesmeianov indicated. Chelintsev still hoped, apparently, that his views would win the favor of the Party officials. In his continuing battle Chelintsev finally exhausted the patience of his fellow chemists. In January 1953 two Soviet chemists, B. A. Kazanskii and G. V. Bykov, lashed out at Chelintsev: "We have no thought at all of denying the right of G. V. Chelintsev, as well as that of any Soviet scientist, to put forward hypotheses relating to any field of chemistry. We think that some of our chemists avoid expressing their views on theoretical matters due to fear of criticism, especially when it is so crude and tendentious as that of G. V. Chelintsev, or due to diffidence about the results of discussions, or to disinclination, in general, to enter into polemics." [69] Nevertheless, the authors continued, the scientific worthlessness of Chelintsev's theories was apparent. They accused Chelintsev of masking his scientific inadequacies with demagoguery. Describing Chelintsev's tactics, Kazanskii and Bykov asked:

Who dares to criticize the New Structural Theory? How can there be a battle of views between other chemists and this author? Any opposition to his "theory" is made to appear a reprehensible heresy. Anybody opposing Chelintsev's theory or ignoring it is readily accused by him of "mechanism," "agnosticism," "Machism," and similar "isms." Chelintsev prosecutes the whole campaign for this theory with a vociferous and defamatory slogan: He who is against me is against dialectical materialism. Under the guise of the fight for the dialectical materialist development of the Butlerov doctrine of chemical structure, G. V. Chelintsev is at present engaged in defaming existing and potential opponents of his "theory," resorting to methods that cannot fail to shame Soviet chemistry.[70]

This admonition seems to have had the desired effect; Chelintsev's articles disappeared from the pages of the journals. Whether Chelintsev voluntarily withdrew or the editors refused to print any more of his polemics is not known, but Kazanskii and Bykov's article marked the end of the running battle. Chelintsev certainly did not change his mind, because four years later he briefly reappeared, sawing away at his old thesis.

But after January 1953 a relative silence fell upon the controversy, punctuated by an occasional article reaffirming the now-official position on the theory of resonance, illustrating that the issue was not dead. An article in the October 1953 issue of the *Journal of General Chemistry* honored the 125th anniversary of the birth of Butlerov, but only briefly mentioned the familiar alleged contradiction between Butlerov's views and the theory of resonance, and did not mention Chelintsev in the twenty-seven footnotes.[71] It is quite likely that an attempt was being made to smother Chelintsev with silence.

By 1954 the tone of the few articles that touched upon the theory of resonance had changed discernibly, perhaps because of a freer atmosphere of discussion following the death of Stalin. But this change was a subtle one; no one even hinted that the validity of the theory of resonance should be reconsidered. A few of the scientists who had been accused of ideological errors in the period 1949–51 started striking back at their critics. Thus, in the January 1954 issue of the *Journal of General Chemistry*[72] A. A. Kalandiia pointed out that his earlier critic G. B. Tsitsishvili[73] had concealed the fact that

in all his works . . . on the so-important topic of chemical structure of compounds not only was the Butlerov theory of chemical structure never considered, but there was an over-all silence concerning the very existence of A. M. Butlerov's structural theory. On the basis of the substance of the works of Tsitsishvili and also of his recent communications concerning my work one is forced to conclude that Tsitsishvili was trying to create the impression that he is an opponent to the "theory of resonance," thereby distracting attention from his own methodologically defective Machist-idealist concepts.[74]

Kalandiia admitted that he had made a mistake when, in his 1949 article, he had spoken of the state of "mutual superposition" of molecules, but he added, "I recognize it to be erroneous and not proceeding from the results of my work and have so declared at the opportune time."[75]

After the death of Stalin in 1953 the chauvinistic praise of Soviet chemistry greatly decreased. In Soviet histories of chemistry and in chemistry textbooks Butlerov remained the founder of the theory of chemical structure, but the more strident criticism of west European chemists frequently credited with establishing the nomenclature and formulas of modern chemistry began to disappear. This decline in national exaggeration was in part a result of a general diminishing of ideological fervor. Some chemists who had opposed the criticism of resonance theory back in the period 1949–51 or felt neutral about the issue had at that time used praise of Butlerov as a means of avoiding the necessity of direct criticism of resonance. A. E. Arbuzov, a member of the Academy of Sciences and a dean of Soviet chemistry, who was forty years old at the time of the Revolution, spoke at length at the June 1951 conference without even discussing the theory of resonance. He confined himself to an exposition of Butlerov's significance in the history of chemistry. Since Butlerov was truly a great chemist, according to all Soviet and non-Soviet scholars who have studied his work, Arbuzov preserved his sense of academic integrity and yet loyally supported the dialectical materialists.[76] Chelintsev described Arbuzov as a chemist who had been essentially neutral toward the theory of resonance.[77] In the new atmosphere that existed after 1953, the absence of compulsion to speak out on the issue resulted in fewer praises of Butlerov by people of Arbuzov's position.

An article illustrating the new mood was Nesmeianov and Kabachnik's "Dual Reactivity and Tautomerism," published in January 1955.[78] The problems discussed in this article were closely connected with the theory of resonance, but the old issues were not revived. After describing a series of chemical reactions, the two authors admitted their inability to postulate the structure of the molecules of the compounds involved. They remarked, "Many of these problems, which of late seemed to be solved, were solved incorrectly, and new investigations are required. . . ."[79]

The theory of resonance itself remained prohibited, at least in name. In August 1957 Chelintsev briefly reappeared, repeating all his old charges.[80] He directed his criticisms at the second edition of the report on chemical structure by the chemical section of the Academy of Sciences, which had been published in 1954 and which differed very little from the original report read by Professor Terenin at the June 1951 conference.[81]

In 1958 a rather complete criticism of the resonance theory appeared in M. I. Shakhparanov's booklet *Dialectical Materialism and Several Problems of Physics and Chemistry*.[82] Shakhparanov was the same chemist who assisted Tatevskii in writing the signal 1949 article in *Problems of Philosophy* near the beginning of the controversy, and who at one time utilized the theory of resonance himself. Shakhparanov's publication was notable for two reasons: He de-emphasized philosophical criticism of the theory of resonance, maintaining that it was incorrect also for obvious scientific reasons; and he presented slightly different philosophic objections to the theory.

Shakhparanov noted that the discussion of the theory of resonance in the Soviet Union had been debated abroad, and that some foreign chemists, particularly in England and Japan, had also criticized the theory. An American commentator had noted, "The resonance theory stands in danger of being largely discredited, at least in so far as it has been applied hitherto. . . . It must never be forgotten that the theory ultimately depends upon the use of limiting structures which, by admission, have no existence in reality." [83] Although these non-Soviet critics did not propose superior alternatives to the theory, they believed that the use of multiple fictitious structures for explaining the properties of compounds was only a temporary method.

In August 1959 Ia. K. Syrkin, one of the two chemists who were criticized most sharply in the earlier debate, published an article in *Progress of Chemistry* entitled "The Current Situation in the Problem of Valency." [84] Syrkin had ceased working in the area of structural chemistry after the controversy, but in the late fifties returned to the field. Syrkin did not present any new views in the article, but confined himself to a moderately enthusiastic description of the molecular orbit method of describing compounds and did not attempt to define the location of specific bonds.

Speaking of the molecular orbit method, Syrkin remarked, "To our regret it does not possess sufficient descriptiveness and does not permit us to portray the formulas of chemical bonds. . . . But quantum mechanics has already accustomed us to the fact that descriptive pictures are often not possible, and although we do not make representations of the wave properties of electrons, we know nonetheless that they are real." [85]

The specific problem of the nature of atomic bonds in benzene was taken up in November 1958 by M. I. Batuev in a technical article in the *Journal of General Chemistry*. Batuev attempted to disprove the theory of resonance on the basis of physical measurements by electron diffraction and X rays. According to resonance, all six bonds of benzene are equivalent, and therefore should be of equal length. Batuev, however, maintained that the benzene molecule consisted of not six equivalent bonds, but alternating "three slightly elongated double bonds and three slightly shortened single bonds." The dimensions he gave (1.382 Angstrom units for single bonds, 1.375 for double bonds) were extremely close (.007 units), especially since $\pm$ .005 Angstrom units was often given as the experimental error inherent in the measurement process.[86]

Batuev's article was particularly interesting in that the criticism of resonance in benzene was based entirely on empirical data. This data concerned only benzene among the various molecules to which resonance theory has been applied, and it was not unambiguous. Nonetheless, it helped place the issue on a more normal scientific plane.

When I studied at Moscow University during the 1960–61 academic year, I found that Soviet chemists at that time spoke of the controversy as having "blown over," but observed that the term "resonance" was not used in chemistry lectures on valency and that standard textbooks in structural chemistry continued to avoid the resonance theory. In many cases these precautions were mere terminological modifications.

In the early sixties attitudes toward resonance theory in the Soviet Union continued to evolve.[87] In November 1961 Linus Pauling lectured on resonance at the Institute for Organic Chemistry in Moscow to an audience of about twelve hundred people.[88] The large audience was due no doubt to attractions of a poignantly opposite nature: Pauling was simultaneously respected in the Soviet Union by many people of an internationalist frame of mind for his opinions on peace and atomic weapons, and, of course, for his stature as a scientist, while he had also been the object of severe criticism in the Soviet Union because of his authorship of the resonance theory. Pauling later observed that his lecture was favorably received.

Of all Soviet discussions of science and philosophy described in this book, the one over the theory of resonance has become the most quiescent. Indeed, it has almost disappeared entirely. Nonetheless, as late as 1969 a text for Soviet teachers of chemistry in the secondary schools pointedly avoided mention of the theory of resonance, even though the inadequacy of classical structural diagrams was carefully described.[89] The text, a quite sophisticated one for teachers at the secondary level, included a discussion of the molecular orbital method and the "method of superposition of valence schemes," a phrase used to describe the theory of resonance method without using the actual term. The Soviet author, G. I. Shelinskii, criticized the "superposition method," observing quite sensibly that it approached the problem of the delocalization of the electron charge too indirectly. He emphasized that chemists now speak more and more of "electron clouds" rather than "bonds"; he further commented that retention of the old bond diagrams for such compounds as anthracene became almost impossibly complex, since hundreds of diagrams were needed to describe one compound. Thus, the superposition method loses even its advantages as a visual aid, the argument usually given in its favor. Shelinskii preferred abandoning graphic models entirely when working with the more complex molecules, relying entirely on the mathematical descriptions of the molecular orbital method. He made no comment on the philosophic implications of such a position, apparently believing, again quite correctly, that it could easily be reconciled with a realist or materialist point of view. The fact that the microworld can not be adequately described by models in the macroworld is certainly no proof that the microworld does not exist. Shelinskii's position, therefore, displayed a few almost indistinguishable traits of past discussions in the Soviet Union—specifically avoidance of the word "resonance"—but it was an intellectually defensible argument and one quite understandable to chemists everywhere.

In recent years the attention of Soviet scholars interested in the philosophic problems of chemistry has begun to shift from the theory of resonance to more general questions. Although the philosophy of chemistry is a quite underdeveloped area in the Soviet Union, a few new works of some interest have appeared in the last few years.[90] Iurii Zhdanov in his 1960 book entitled *Essays on the Methodology of Organic Chemistry* was still quite critical of the epistemological basis of resonance theory (he accepted, of course, the quantum mechanical calculations of the molecular orbital method), but he began directing his major attention to broader problems of the meaning of chemical formulas, the meaning of homology in chemistry, and the validity of modeling.[91]

This trend has continued in the most recent works. N. A. Budreiko, in a 1970 book entitled *Philosophic Problems of Chemistry*, concentrated primarily on issues such as the definition of the terms "chemistry" and "chemical element," the philosophic significance of Mendeleev's periodic table, and the presence in chemistry of dialectical laws of nature.[92] Unfortunately, Budreiko's book was somewhat elementary and mechanistic; his easy perception of the dialectical laws in chemistry reflects some of the more superficial aspects of nature philosophy. Other recent books of more value have concerned the philosophic significance of certain issues of importance in the history of chemistry, such as the atomistic views of Dalton, Gibbs, and Mendeleev.[93]

By far the most interesting recent philosophic work by a prominent Soviet chemist was N. N. Semenov's 1968 article entitled "Marxist-Leninist Philosophy and Problems of Natural Science." [94] Semenov, a recipient in 1956 of the Nobel Prize, is probably the best known of all Soviet chemists currently at work. His recent articles have revealed a deep interest in the philosophy of science; furthermore, he stoutly defends the materialist dialectic. He remarked in his 1968 article that since the materialist dialectic is the method of man's cognition, of man's thought, it is "applicable to the development of all the sciences, both the social sciences and the natural sciences. Dialectical materialism lies at the base of the conscious transformations of society, its production and its culture." [95] Nonetheless, the particular interpretation that Semenov has placed upon the relation of dialectical materialism to science is highly controversial in the Soviet Union. While many of the authors previously discussed have believed that dialectics are inherent *in* nature, Semenov apparently believes that dialectics are characteristic primarily of man's thought, not of nature existing outside of his thoughts.

Semenov wrote that philosophy can play its proper role only if it occupies an independent and definable position among the other sciences, with its own carefully delineated subject matter. And this subject matter of philosophy can not be, he continued, "the world as a whole," as many philosophers have maintained, for the world as a whole is studied by the entire system of natural and social sciences, rather than by philosophy. If philosophy attempts to study the world as a whole, it will, in practice, study nothing at all. Furthermore, Semenov continued, such an attempt may lead to restrictions on science itself, such as the condemnation of the theory of resonance. Instead it should concentrate on problems of Logic (*sic*) and the theory of knowledge.

Semenov attempted to answer the criticisms of his view by the dialecticians of nature:

Some philosophers sometimes express the following fear: How is it possible for us to consider Marxist-Leninist philosophy as Logic, as the theory of knowledge? Won't such a view lead to a loss of the meaning of Marxist philosophy as a world view, to a depreciation of its role, and even to a "breakaway of philosophy from natural science"?

If one understands Logic in a Leninist fashion, one need not fear. Indeed, the reverse is true: Actually, all our sciences, all our culture, develop with the aid of thought, based on human practice, and therefore the science of thought preserves its universal meaning, its primary role in the development of a scientific understanding of the world.[96]

Semenov saw an inherently dialectical process in the development of man's ideas about nature. He described the evolution of scientific ideas in the following fashion: A certain theory about nature develops steadily until new evidence confronts it with contradictions. This new evidence cannot be accommodated within the logical structure of the old theory. Nevertheless, the evidence steadily accumulates, and the contradictions become sharper. These contradictions become the "vectors of the movement of thought in the process of the formulation of hypotheses," and in this fashion, contradictions serve as the "motors" of development, as Hegel himself had emphasized.[97] Eventually, a crisis, a turning point, is reached. At this moment, said Semenov, formal logic is of no avail to the scientist. Instead, he needs to have a knowledge of dialectical logic so that he can make a creative leap to a new formulation, a new means of understanding. Semenov maintained that the dialectical description of the development of thought could be illustrated by his own work in the late twenties, when he finally overcame the contradictions in his area of chemistry by developing a new theory of chain reactions. Thus, dialectical logic, Semenov continued, "helps a scientist to understand and 'make more precise' the real movement of creative, fruitful scientific thought." [98]

Semenov recommended, therefore, that Soviet graduate students in the sciences place less attention on the formalistic identification of dialectical contradictions in nature, and more attention on dialectics as a method of cognition:

A scientist or graduate student who talks about dialectical materialism in a public lecture or during an oral examination usually can expound beautifully on the whole question of the role of contradictions in the development of nature and thought. However, in his work he usually doesn't even think about such things, apparently believing sincerely that for actual scientific work philosophy is not necessary. Of course, we must agree with such a scientist that the Marxist dialectic is not a collection of rules: You don't just directly apply it to a specific problem and receive a direct answer. No, the Marxist dialectic is something else; it is a general orientation and culture of thought that helps each person to pose a problem with greater clarity and purpose and thereby helps him to solve the riddles of nature.[99]

Traditional dialectical materialists would consider Semenov's description above an exaggeration of the role of ideas in the development of science. They would maintain that man's thoughts are also a part of nature and are ultimately subject to the same regularities as the rest of nature. They would then continue to search for these regularities both in thought and external reality, terming them, as before, "the dialectics of nature."

Semenov's article appeared in both the *Herald of the Academy of Sciences* and the Party journal *Communist*. Soon thereafter there appeared in the latter journal an article favoring the traditional approach, in which the dialectic was seen as a generalization of specialized scientific knowledge and therefore inherent in nature.[100]

If the interpretation of the resonance controversy presented here is correct, several older interpretations should be modified. For example, one author described the resonance dispute as being ideologically and politically the same phenomenon as the biology discussion, with the same result. He called the outcome "Lysenkoism in chemistry." [101] One can understand this description, but the result of the discussion was a defeat for Lysenkoism, in contrast to its victory in biology. The initial attack on the resonance theory was a manifestation of Lysenkoism, but the significant feature of the resonance controversy was that the chemists successfully defended themselves against the most serious attack. The modifications of theory were primarily terminological.

Gustav Wetter's short discussion of the resonance controversy also did not mention the central position of Chelintsev.[102] Wetter indicated that the real source of the resonance controversy was central direction from above: "One gets the impression that at this period virtually everything was seized upon and correspondingly inflated, which could in any way offer a handle for convicting Western theories of 'idealism,' 'Machism,' etc.; just as the possibility of embracing Butlerov's theory offered a welcome opportunity for display of 'Soviet patriotism.' " [103]

Rather than being a controversy initiated from above, it appears that the resonance issue sprang up from below, nurtured and prompted by zealous, ambitious chemists who hoped to win Party support in discrediting the scientists who utilized the resonance theory. This initiative probably embarrassed rather than pleased Soviet scientific administrators, who did not want to harm the productivity of the chemists.

The majority of these chemists were willing to defend the tools necessary for the practice of their science, and the quantum mechanical approach, condemned by Chelintsev, was one of these essential tools. Therefore, they decided to accept Chelintsev's diagnosis of the methodological disease in chemistry, but to reject his recommended cure. Without using the fictitious resonance structures, they would preserve the mathematical core of the theory, maintaining that such a solution was compatible with the dialectical materialist approach to science and Butlerov's approach to chemistry.

Very little has been said here in defense of the theory of resonance, although the criticisms of it have been described in considerable detail. Actually, the theory of resonance has already proved its usefulness in science. If the theory of resonance were replaced by a new theory tomorrow, the concept of resonance would have served an important and useful purpose. The originators of the theory have warned repeatedly that no physical significance should be assigned to the resonance structures, which are primarily helpful descriptions. Nevertheless, it is true that some chemists mistakenly think of resonance as a mechanical phenomenon.[104] The theory of resonance is a man-made system for organizing and understanding the complex data collected from chemical reactions—a system that can be thought of, if one prefers a realist epistemology, as bearing a certain resemblance to the structure of a molecule, but that is far from being identical with that molecule.

Yet one should add that underneath the debate over resonance theory there is a philosophic issue of considerable interest. As in the case of quantum theory, the interpretation in terms of a model given to a mathematical formalism is sufficiently far from customary descriptions of physical nature to cause uneasiness to some scientists and philosophers. The essential philosophical problem here, then, is that of the use of models in scientific explanation. This is a serious topic on which philosophers of science have written a great deal; the fact that the participants in the Soviet discussion of resonance never brought out fully the underlying issue does not contradict the fact that it was there. One hopes that in the future Soviet authors will discuss resonance from the standpoint of philosophical analysis, with no question of interfering with the work of scientists arising.

Mary Hesse described the intellectual issue involved in the use of models in the following way:

The main philosophical debate about models concerns the question of whether there is any essential and objective dependence between an explanatory theory and its model that goes beyond a dispensable and possibly subjective method of discovery. The debate is an aspect of an old controversy between the positivist and realist interpretations of scientific theory. Many episodes in the history of science may be regarded as chapters in this controversy, including application of Ockham's razor to scientific theories, the Newtonian-Cartesian controversy over the mechanical character of gravitation, nineteenth-century debates about the mechanical ether and the existence of atoms, and Machian positivism.[105]

Among the scholars who have described models as merely dispensable aids for the construction of theories are Ernst Mach, Heinrich Hertz, and Pierre Duhem. Among those who have argued that without some form of material analogy there is no valid grounds for prediction are N. R. Campbell, E. H. Hutten, and Hesse herself.

In the case of the theory of resonance, a realist or materialist need not be disturbed by his inability to construct a model that adequately explains all the reactions of certain chemical compounds. Indeed, the scientific theory that stands behind such classical models—the theory of valency in which chemical bonds are highly localized—has long since been abandoned by chemists. The current theories of valency—in which electrons are recognized as micro-objects in terms of quantum theory and hence have both a wavelike and corpusclelike character—do not permit such structural diagrams. But quantum theory has already acquainted us with this problem of visualization. Thus, although there is much of intellectual interest in the interpretation of chemical valency, there is little reason to believe that it presents obstacles of a uniquely difficult character to supporters of philosophic realism or materialism. As Hesse commented, "We ought to be prepared for, rather than surprised at, the inadequacy of familiar models in much of modern physics." [106]

IX   Cybernetics[1]

Cybernetics enjoyed more prestige in the Soviet Union in the 1960's than in any other country in the world. This statement may seem surprising to non-Soviet observers, particularly to those who know that the field was at first criticized in the Soviet Union. Furthermore, the Soviet Union lagged behind the United States in computer production, both quantitatively and qualitatively. How, then, could Soviet writers constantly speak of the unique roles that they believed cybernetics would play in their society? To try to answer this question, we must begin by analyzing the essential concepts of cybernetics against the background of traditional Soviet social aspirations and the philosophic framework of dialectical materialism. To its Soviet supporters cybernetics was a new chapter in the history of materialistic approaches to nature that promised both better ways to conceptualize the world and also to achieve social goals.

The cybernetics issue bore some resemblances to past Soviet debates in science, such as those over biology and relativity physics, but it was distinguished by several new characteristics.[2] The most prominent of these was that the cybernetics controversy resulted in an overt effort to develop the field more rapidly, in contrast to the past retardation of genetics by reason of the Lysenko affair and the essentially neutral effect of discussions in such fields as physics on the actual rate of development of Soviet science.[3]

The Soviet Striving for Rationality

An original promise of the Russian Revolution, for those who supported it, was the rational direction of society. Marxism as an intellectual scheme was heir to the optimism of the French Enlightenment and the scientism of the nineteenth century; one of its primary characteristics was the belief that the problems of society could be solved by man. Nature was not so complicated but that it could be controlled if only the artificial economic barriers to that control erected by capitalism were removed.

The key to progress, then, according to the Marxists, was social reorganization. The Bolsheviks considered the Revolution of 1917 to be the decisive breakthrough toward that reorganization. They admitted, of course, that progress toward efficient administration would be very difficult to achieve in Russia as a result of its primitive state. Even in the early years of Soviet Russia, however, there were at least a few theorists who hoped to achieve centralized, rational direction. The first attempt toward this goal was made during the period of War Communism (1918–21). However important the civil war may have been in forcing a command economy, it is quite clear that the ideological urge to create a planned communist society also played an important role. From this standpoint the New Economic Policy (1921–27), with its relaxation of economic controls, was a definite retreat. The rapid industrialization that succeeded the New Economic Policy might have been carried out in accordance with any one of several different variants, but all assumed greater planning and centralization.

After the 1930's, however, the goal of a rationally directed society became more remote. The fact most disheartening to the Soviet planners was that the more the early difficulties of industrial underdevelopment were overcome, the more distant seemed the goal of rational, centralized control. By the time of Stalin's death in 1953 the economy had become so complex that it seemed to defy man's ability to master and plan it. It would have been convenient to attribute these troubles to the irrationalities of Stalin himself rather than to the inability of Soviet man to control his affairs. Yet by 1957, four years after Stalin's death, it was clear that the trouble lay not in the aberrations of one man but in the entire concept of centralized planning.

Notwithstanding the rhetoric surrounding the reform, the decentralization of industry that occurred in 1957 was again a defeat for rationality—at least as rationality had originally been conceived in the Soviet Union.[4] There were very serious grounds for the belief that a complex modern industrial economy simply could not be centrally directed. Every modification of the quantity of one commodity to be produced called for unending modifications in the quantities of others. Even a relatively decentralized economy seemed to have an insatiable demand for bookkeepers and administrators. Academician Glushkov said that if things continued as they were going, by 1980 the entire Soviet working population would be engaged in the planning and administrative process. To use a cybernetic term, the entropy of the system was multiplying at a horrifying rate.

It was at this time in the history of the Soviet Union that cybernetics appeared. Leaving aside temporarily the initial Soviet hostility toward cybernetics (which has been exaggerated outside the Soviet Union), the promise of cybernetics, as it appeared to Soviet administrators and economists, was twofold: First, it held out the hope of rational control of processes that previously had been reluctantly judged uncontrollable because of their complexity; second, it offered a redefinition of what rationality is, at least as far as the direction of complex processes is concerned.

The new hope for rationality in cybernetics seems obvious enough. The subject matter of cybernetics—the control of dynamic processes and the prevention of increasing disorder within them—was exactly the concern of Soviet administrators. Perhaps through the new science of cybernetics, they thought, genuine control of the immensely complex Soviet economy and government could be achieved. Whether that hope was justified and whether cybernetics is a true science are, of course, still unanswered questions.

The second result of cybernetics—the redefinition of rationality in controlling complex mechanisms—arose from the very nature of cybernetics. It is necessary, therefore, to spend a little time in defining the subject.

The Science of Rational Control

The term "cybernetics" is often improperly understood as being synonymous with "automation." It brings to mind discussions of unemployment and impressive statistics about the number of operations a computer can perform in one hundredth of a second. In its original sense, however, cybernetics meant something quite different. The founders of cybernetics—Norbert Wiener, Arturo Rosenblueth, Julian Bigelow, Walter B. Cannon, Warren S. McCulloch, Walter Pitts, W. Ross Ashby, Claude Shannon, and John von Neumann—believed they were advancing a generalized theory of control processes.[5] To them, a control process was the means by which order is maintained in any environment—organic or inorganic. In terms of this view of cybernetics, a computer by itself is not a cybernetic device. It can become a part of a cybernetic system when it is integrated with the other components of that system in accordance with a control theory.

The aspiring scientific discipline of cybernetics did not base itself upon the technological innovations that permit the construction of modern computers. Instead, it rested on the concept of entropy, taken from thermodynamics and broadened to mean the amount of disorder in any dynamic system. According to this approach, all complex organisms are constantly threatened by an increase in disorder, with the end point complete chaos. However, certain organisms are arranged in such a sophisticated and efficacious manner that through a dynamic process they can resist, at least temporarily, the tendency toward disorder. Cybernetics studies the common features of these organisms, particularly their use of information to counter disorder. The more enthusiastic supporters of cybernetics view human society, which also obviously places a premium on order, as a particular type of cybernetic organism. In sum, cybernetics is the science of control and communication directed toward fending off increasing entropy, or disorder.

Cybernetics fits well with materialistic assumptions. It postulates that the control features of all complex processes can be reduced to certain general principles. Yet its mode of operation differs distinctly from the science of the eighteenth and nineteenth centuries out of which the scientific optimism of Marxism arose. In the terms of the Enlightenment, rationality came through the knowledge of quantitative laws that would permit the prediction of the future. Such rationality was perhaps best symbolized by the celestial mechanics of Laplace. Control of a process, according to this early view, was based on knowledge of all physical laws and variables needed to predict future states of the process, and on the ability to change the magnitude of the variables. Even the indeterministic nature of modern physical theory, troublesome as it was, did not destroy the belief that rationality is essentially a theoretical rather than an empirical approach. In economics this concept of rationality led to the belief that if a centralized economy were not running smoothly, the difficulty must be either inadequate authority or insufficient knowledge at the center of local conditions and of the necessary economic laws for the changing of these conditions.

Cybernetics—which is based on analogies among all complex self-perpetuating processes, with living organisms the ultimate examples of success in self-perpetuation—does not emphasize exact prediction of future states or conditions. Nor does it call for strict centralized control. The executive or command organs in all truly sophisticated cybernetic mechanisms are arranged in hierarchies of authority, with semiautonomous areas. Furthermore, rather than trying to predict indefinitely the results of its executive actions, a cybernetic system makes constant empirical checks of these results through feedback, and it adjusts its commands on this basis. As Norbert Wiener said, cybernetics derives from control on the basis of actual performance rather than expected performance. Cybernetics thus places a premium on combining two seemingly contradictory principles: local control based upon empirical evidence, and overriding centralized purposes.

It is a mistake to believe that cybernetics makes it possible to control the most complex processes by collecting in a central location enormous amounts of information. Indeed, cybernetics holds that barriers to information matter as much for control of processes as do free-flowing avenues of information. The best example of this paradox can be found in the human body, in many ways the paragon of a cybernetic mechanism. If we were conscious of everything that goes on in our stomachs, or even just the information that some part of our bodies must be aware of in order for proper digestion to take place, we would be very neurotic indeed. Yet the human body represents the greatest victory of control over a complex process to which cybernetics can point; the features of its organization are basic to an understanding of cybernetic systems.

The Rebirth of Hope

The lesson of cybernetics for the Soviet Union, and especially for its economy, seemed clear. If Moscow knew everything occurring in its factories in Omsk, it would also be "neurotic," as indeed it was when it attempted to do so. Cybernetics taught the lessons of *selectivity* of information and *decentralization* of control. The point is not, however, that cybernetics was the cause of the decentralization of the Soviet economy in the late 1950's. That reform was implemented prior to the recognition of cybernetics, and flowed from doubts about past Soviet planning. Cybernetics came later, and restored faith in over-all planning by presenting clear analogies for the combination of decentralization with overriding central purposes.

Cybernetics revitalized, at least temporarily, the Soviet leaders' confidence that the Soviet system could control the economy rationally. This renewal came exactly at the moment when the possibility seemed to be irretrievably vanishing.[6] This rebirth of hope was the explanation of the intoxication with cybernetics in the Soviet Union in the late fifties; in the period after 1958 thousands of articles, pamphlets, and books on cybernetics appeared in the Soviet press.[7] In the more popular articles the full utilization of cybernetics was equated with the advent of communism and the fulfillment of the Revolution.[8] If the curious mixture of ideology and politics in the Soviet Union can upon occasion affect certain sciences adversely—as it did at one time with genetics—it can also catapult others to unusual prominence.

One can find no other moment in Soviet history when a particular development in science caught the imagination of Soviet writers to the degree to which cybernetics did. Perhaps the closest parallel occurred in the 1920's, when GOELRO, the State Commission for Electrification, was made the subject of poetry.[9] At that time, too, the industrial time-and-motion studies of Frederick Winslow Taylor were applied widely and somewhat indiscriminately, and the general enthusiasm for industrialization expressed itself on occasion in such unusual forms as concerts for the workers in which the instruments were factory whistles.[10] But even the twenties will not serve as a parallel. For cybernetics was held out by its most ardent advocates as a far more universal approach than any of the diverse theories of the twenties.

It was quite common in the Soviet Union in the early sixties to find articles on the application of cybernetics in such surprising fields as musicology and the fisheries industry, although frequently such expositions involved distortions of the meaning of the term "cybernetics." A number of the normally stolid and reserved Academicians of the Academy of Sciences were the most exuberant disciples of the new field. The Communist Party itself in 1961 endorsed cybernetics as one of the major tools for the creation of a communist society.[11]

Even before formal endorsement the movement toward cybernetics began to take on the dimensions of a landslide. In April 1958 the Academy of Sciences created the Scientific Council on Cybernetics, headed by Academician A. I. Berg, which included mathematicians, physicists, chemists, biologists, physiologists, economists, psychologists, linguists, and jurists. The Academy's Institute of Automation and Telemechanics began directing most of its research toward cybernetic applications. The Moscow Power Institute, one of the largest and oldest engineering institutions in the country, with an enrollment of seventeen thousand students, devoted approximately one third of its instruction and research to cybernetics.[12] Soviet students were urged to major in cybernetics; science fiction was filled with descriptions of "cybernetic brain-modeling" and the "cybernetic boarding schools" of the future. The Academy of Pedagogical Sciences of the Russian Republic established such an experimental boarding school in Moscow to prepare children from an early age for careers in cybernetic programming.[13]

In 1961 Academician Berg edited a book entitled *Cybernetics in the Service of Communism,* in which Soviet scientists outlined the potential applications of cybernetics in the national economy.[14] In his introduction he argued that no country would be able to utilize cybernetics so effectively as the Soviet Union; since cybernetics consists largely of the selection of optimum methods of performing operations, only a socialist economy could incorporate these methods universally. "In a socialist planned economy," said Berg, "all conditions are present for the best utilization of the achievements of science and technology on behalf of all members of society rather than for various competing groups and the privileged minority." [15] He then discussed the most promising uses of cybernetics in the Soviet Union.

One important application of cybernetics envisaged by Academician Berg was in the field of economic planning; here cybernetics was said to be of much greater utility in the Soviet Union than in any capitalist country. Soviet analysts pointed out that Western cyberneticists themselves noted the "absence of purposefulness" in the development of Western economies and the consequent inapplicability of the cybernetic approach. The Soviet Union, however, had a national economic plan that provided national purpose. The most important economic application of cybernetics, according to the Soviet economist V. D. Belkin, was the control of the supply of products and raw materials to industrial enterprises.[16] Belkin noted, however, that as late as 1961 some Soviet economists opposed the use of cybernetics in economic planning; he considered such opposition "intolerable." [17] Another economist and partisan of cybernetics, A. I. Kitov, called for the creation of a unified national system of computing centers for control of the economy.[18]

The authors associated with Berg also noted that the theory of information could be applied in a multitude of ways to industry. For example, the amount of electrical power needed to supply machinery is usually calculated as the sum of the maximum power needed for each establishment. However, as is well known, electrical machinery goes through cycles; only at certain moments is maximum power needed. Providing maximum power constantly throughout the entire network results in wasted energy and requires excessively strong cables and transformers. On a statistical basis, computers could calculate how many consumers were likely to require maximum power at any one moment and so permit economy in power network construction and utilization. The same statistical approach was used for analyzing the service that a repairman gave to automated machinery. Reliability statistics were used to calculate not only how many machines one man could efficiently service on an automatic line, but also how many spare parts needed to be brought to a warehouse at the beginning of each year in order to service the line, and even where the spare-parts bins should be located to provide the most efficient service when breakdowns occurred. This "service time" approach (the Danish scientist A. K. Erlang and the American engineer Shannon elaborated the fundamentals of the concept) could be used to analyze the service of passengers in transportation terminals and stores.[19] It also had obvious applications in military operations.

Paradoxically, the problem of reliability in machinery for which a cybernetic approach seemed so fruitful was also one of the major obstacles to the use of computers. The Soviet authors noted that the cybernetic systems with their controlling computers were so complex that even if each component had a very long average life, the reliability of the entire mechanism could be quite poor. The chance of one particular component failing was extremely low, but the probability of any one of thousands of components failing was always high. Several Soviet authors believed that the problem of reliability should be regarded as one that would be resolved only by evolution and natural selection. Over millions of years only those living organisms survived that possessed a high reliability in spite of their complexity. Since the construction of cybernetic machines had gone on only a few years, the selection of reliable mechanisms had hardly begun; the pace of the cybernetic revolution would be governed, they thought, by the rate of this evolution. The disillusionment that many cyberneticists had experienced had often come, they thought, not so much from their inability to construct machines with almost infinite possibilities of information storage as from their inability to operate them for an appreciable length of time.

Another important subject of study for Berg and his assistants was the problem of communication between man and his machines. They noted that a man in charge of an automated process found himself a link in a new sort of chain operation consisting of machine–computer controlling it–man. The computer was capable of thousands of operations per second, while the man's speed was much less. This presented him with an unusual problem: All the control operations he performed must be within his physical and psychological capabilities, even though many of the operations performed by the machine under his command were beyond these capabilities.

Still, the Soviet authors pointed to the opportunities to benefit man presented by cybernetics that made all the more important efforts to master them. One such opportunity envisaged by Academician Berg was medical diagnosis. As early as 1956 Soviet researchers were feeding clinical information to electronic computers that were then asked to provide diagnoses based on various complexes of symptoms noted in the patients. The advantages of computers in medical diagnosis were that they had a much greater memory capacity than an individual doctor and could approach diagnosis with a statistical exactness that a doctor could only approximate. Thus, on the basis of clinical histories, it was shown that a certain symptom appeared in ninety-eight per cent of the cases of one disease, in thirty-seven per cent of another, and in only two per cent of another. If a patient displayed a series of symptoms, each with a different disease-probability ratio, a computer could arrive at a probable diagnosis that was a valuable supplement to the diagnostician's own judgment. The machines were capable of producing several possible diagnoses for one patient, and on a few occasions indicated that one patient was suffering from more than one disease, contrary to earlier opinion.[20]

An example of the application of cybernetics in a primarily nontechnological field, said the Soviet authors, was in education. Programmed learning and teaching machines were viewed in the Soviet Union as "cybernetics in education," a topic of intense interest when I visited the Soviet Union in the spring of 1963 as a part of a delegation of the U.S. Office of Education. Teaching machines permitted the student to check on his progress (to have feedback) at each stage of his development; some of the more sophisticated ones varied the questions given to students on the basis of their previous records. An outstanding student would be rapidly jumped ahead in the lesson sequence, skipping certain sections, while a poor student might be required to repeat lessons many times. Soviet educators were particularly interested in foreign-language teaching machines, some of which were described as being capable of speaking to the student, correcting his pronunciation, and regulating his progress.[21]

Many of these applications of cybernetics in which the Soviet Union was interested were quite well known in non-Soviet areas. The mechanical translation of languages was a subject of study in many countries— and a very frustrating one. The use of computers for surveying immense bodies of legislation or national statistics so that lawyers or statisticians could find vital facts, legal precedents, or discern the need for new laws, as well as find conflicts with old ones, was also a familiar problem of information retrieval.[22]

Soviet cyberneticists also applied cybernetics in the sciences themselves, as was being done in other countries in these years. Weather prediction, for example, was facilitated by the use of computers. Somewhat more novel at this time was the use of computers in organic chemistry. Soviet engineers worked on computers that would translate the familiar structural formulas of chemical compounds into a machine language. By using the new language, machines would hopefully suggest optimum means of synthesizing new organic compounds. Their proponents maintained that the machines would improve their knowledge of structural chemistry on the basis of past experience. The need for computers in chemistry was adequately illustrated by the fact that there were over a million chemical compounds in the chemical literature and many millions of different chemical reactions.

These examples of the application of cybernetics give a sketchy outline of Soviet interests in the field in the early sixties. Another interest not discussed here was the utilization of computers in planning military operations, a subject not covered by Academician Berg in his surveys. The statistical approach to service time and priority, the computation of optimum transportation methods, and the assignment of effectiveness and reliability ratios could all be applied to military planning. Several Soviet authors criticized the utilization of computers in military planning on the ground that these calculations did not take sufficient account of morale and national pride, but there was no doubt that the Soviet Union was pursuing research in this area.[23]

In reviewing Soviet interests in cybernetics, it becomes clear that engineers and scientists in that country had, as in the United States, blurred the original meaning of the term. While to the founders of the field "cybernetics" had meant a control theory for self-regulating dynamic processes, to many less theoretically inclined engineers and administrators it had come to signify merely the application of computers to any problem where they seemed appropriate and profitable. Norbert Wiener would never have regarded the application of computers to simple information-retrieval problems such as the search of legal literature as a proper use of his term. Homeostasis, or the maintenance of equilibrium through feedback, was important to the earlier understanding, not the mere utilization of computers to solve difficult problems.[24]

A number of Soviet theoreticians remained loyal to the original meaning of the term. To many of the economists among them the possibility of creating a self-controlling economy that would contain largely autonomous areas and yet be responsive to national goals remained very attractive.

The combination of centralized purposes with decentralized organization in a national economy obviously contained contradictory strains. A number of non-Soviet commentators observed that the degree of success that the Soviet Union obtained in the one direction would be accompanied by the corresponding degree of failure in the other. There is no doubt some truth in this observation. Soviet administrators were not alone, however, in being attracted by the new possibilities of combining what appeared to be opposing principles. The United States was the country that supplied the Soviet administrators with models for the organizational forms of industry, just as it had before World War II. The rhetoric in the following description in an American text published in 1963 of the organization of the Sylvania corporation could be transferred to a Soviet publication of the same period with no difficulty:

In a revolutionary hookup Sylvania has 12,000 miles of communication network connecting 51 cities, to produce what spokesmen for that company call a step in "administrative automation." Their network of operation aims ultimately to include payroll, supply maintenance, auditing, statistical services and the byproducts thereof, in their various plants, sales offices, and warehouses. This form of integration secures many of the advantages of centralized control in decentralized locations, a feat which previously seemed tantamount to having one's cake and eating it too.[25]

If cybernetics provided a stimulus for the hopes of the Soviet planners, it did so in a rather ironic fashion. Traditionally, Soviet leaders maintained that the key to their progress was the socialist organization of society; with the advent of cybernetics they came close to saying that science would *permit* the organization of such a society. Soviet economists during the 1930's spoke of how the communist state would provide for the welfare of science, a part of the cultural superstructure; in the 1960's, they spoke of how science, which according to the Party was becoming a direct productive force and therefore a part of the economic substructure, would provide for the welfare of the communist state.[26]

In other words, while in the 1930's it was possible to speak of the Bolshevization of science, in the 1960's it was possible to speak of the scientization of Bolshevism. To the multitude of problems faced by the leaders of the Soviet Union—the mechanics of planning an intricate economy, the need for producing ever greater quantities of agricultural and industrial goods, the subordination of a land with climatic and other natural disabilities—science seemed to provide the only possible solutions. The products of physical and social science research presented the latest, and probably the greatest, opportunity for the Soviet state to be as rational, and its fruits as bountiful, as the early partisans of the centrally planned state had hoped.

Philosophical Discussions

Cybernetics coincided with the materialism and optimism of Marxism, but it also raised a number of serious philosophical and sociological problems. Cybernetics had obvious applications in a number of fields—psychology, econometrics, pedagogical theory, logic, physiology, and biology—disciplines subjected to ideological restrictions under Stalin. These connections a priori confirmed the sensitivity of the subject. In the early 1950's Soviet ideologists were definitely hostile to cybernetics, although the total number of articles opposing the field unequivocally seems to have been no more than three or four.[27] This number is far fewer than the number of ideologically militant publications that appeared in the other controversies discussed in this volume, a fact that is largely explained, no doubt, by chronology: By the time cybernetics became widely known, the period of severe ideological interference in Soviet science had passed. On the other hand, Soviet scientists and engineers had for many years worked on the mathematical and physiological foundations of cybernetics. Such Soviet scientists as I. P. Pavlov, A. N. Kolmogorov, N. M. Krylov, and N. N. Bogoliubov must be counted among the men who prepared the way for the development of cybernetics, but they did not advance a generalized theory of control processes in a great variety of organic and inorganic environments. The construction of such a theory, which is the heart of cybernetics, instead fell largely to people in North America.

Until the early 1950's the reception of cybernetics in the Soviet Union was silence; not until 1952, a year before Stalin's death, was cybernetics openly attacked, although a few earlier articles questioning mathematical logic could be seen as implied criticism of cybernetics.[28] A 1953 article, which appeared in *Literary Gazette,* labeled cybernetics a "science of obscurantists" and ridiculed the view that a machine can think or duplicate organic life. The author particularly criticized the efforts of cyberneticists to extend their generalizations to explicate the collective activities of man. In addition, the critic attributed to cyberneticists in capitalist society the hope that their new machines would perform their society's unpleasant tasks for them: The striking and troublesome proletariat would be replaced by robots, bomber pilots who object to bombing helpless civilians would be replaced by "unthinking metallic monsters." [29]

In October 1953 a very critical article entitled "Whom Does Cybernetics Serve?" appeared in the leading Soviet philosophy journal; in later years this article was often referred to by Soviet defenders of cybernetics as the most typical statement of the opposition to cybernetics in the early 1950's. The author of the article, who identified himself only as "Materialist," advanced a criticism of cybernetics based on the dialectical materialist belief in the qualitative difference in matter at different levels of development; thus, a difference in principle existed between the human brain and even the most sophisticated computer. Such authors as Claude Shannon and Grey Walter, who attempted to construct mechanical devices that would display "social behavior," were falling into the same error as the materialists of the eighteenth century, such as La Mettrie and Holbach.[30] But while the views of the latter men were "progressive" in the eighteenth century, since they were directed primarily against religious beliefs, continued the Soviet critic, in the twentieth century such views were clearly reactionary. And, finally, "Materialist" returned to the earlier-expressed view that cybernetics represented a particularly pernicious effort by Western capitalists to extract more profits from industry by eliminating the necessity to pay wages to the proletariat.

Just as the initial hostility of the Soviet writers toward cybernetics can be related to the intellectual scene characteristic of Stalinism, so can the beginnings of a discussion of its merits be best explained by noticing the changes in position of the Communist Party toward the natural sciences after Stalin's death. The influential position of the Party should not obscure the fact, however, that many scientists and engineers in the Soviet Union were skeptical of the claims of cyberneticists in the United States for reasons that were not, in many cases, uniquely Marxist in viewpoint.

In the spring of 1954 the Central Committee promoted a policy of much greater leniency on ideological issues in the sciences; a primary criterion for judgment was to be the empirical results of the utilization of scientific theories.[31] This position, while not totally new, was probably connected with the criticism of Lysenko's theories on genetics; it also allowed a more liberal discussion of cybernetics.

The first person to espouse a positive view toward cybernetics seems to have been the Czech philosopher and mathematician Ernst Kol'man, who lived in Moscow for long periods of time and often wrote on questions of the philosophy of science. Kol'man should by this time be a familiar name; involved in the debates over science for over three decades, he rather frequently took the more liberal side in various controversies, despite something of a reputation to the contrary. On November 19, 1954, Kol'man gave a very important lecture to the Academy of Social Sciences of the Central Committee of the Communist Party, in which he specifically attacked "Materialist" 's 1953 article.[32] Only later would the full irony of this occasion become clear; Kol'man, who assumed the role of champion of cybernetics, was later outpaced in his enthusiasm for the new field by many Soviet scholars, and in subsequent publications appealed for restraint in evaluations of the potentialities of cybernetics.[33] The major point of Kol'man's talk to the Academy of Social Sciences was his belief that the Soviet Union stood in danger of overlooking a technological revolution by discounting cybernetics. The new computing machines could be compared in significance, he said, to the implementation of the decimal numeral system or to the invention of printing. The Soviet Union must master new processes and use them for its own goals, he continued.

Kol'man's speech, later published in *Problems of Philosophy,* was the beginning of a debate in the Soviet Union over the legitimacy of cybernetics, which lasted from 1954 to 1958. The first stage in this discussion was an exploration of the reasons for the initial coolness of Soviet Marxists to cybernetics. A group of authors strove for an explanation in mid-1955:

Several of our philosophers made a serious mistake: Not understanding the essence of the problem, they began to deny the significance of this new development in science basically because of the fact that around cybernetics abroad there was raised a sensational clamor, and because several ignorant bourgeois journalists promoted publicity and cheap speculations about cybernetics.[34]

The discussion of cybernetics soon turned to attempts to define the field and the technical terms used in it, such as "information," "quantity of information," "noise," "control," "feedback," "neg-entropy," "homeostasis," "memory," "consciousness," and even "life." Many articles seeking such definitions appeared. The adoption of cybernetic methods could not await the formulation of ideologically correct definitions, however; with the peculiar insistence of modern technology, computers found their way into many areas of the Soviet economy, including the defense and space efforts. Thus the Soviet Union turned toward cybernetics rather rapidly even though the new field contained many new concepts that had not yet received philosophic interpretation. This movement was led by scientists and engineers, accompanied by those philosophers, such as Kol'man, who shared their enthusiasm for the adoption of the most modern methods in the Soviet economy and who also saw, quite correctly, no inherent contradiction between Marxism and cybernetics.

Gradually the support for cybernetics became more impressive. Well-known scientists, such as Academician S. L. Sobolev, presented elementary and positive explanations of cybernetics to the philosophers and social scientists. Other scientists publicly underwent obviously sincere changes in their attitude toward cybernetics. As late as October 1956, Academician A. N. Kolmogorov, whose work on the theory of automatic control was a genuine contribution to cybernetics, refused to accept its validity as a separate field; by April 1957, however, he declared at a meeting of the Moscow Mathematical Society that his earlier skepticism toward cybernetics had been mistaken, and in 1963 he wrote that it is theoretically possible for a cybernetic automaton to experience all activities of man, including emotion.[35]

The three main questions of philosophical concern that cybernetics raised were: (1) What is cybernetics, and how general is its application? (2) Can life processes be duplicated? (3) What is "information," and what is its connection with thermodynamics? In the early stages of the discussion of cybernetics in the Soviet Union questions one and two, which are related, received the most attention. After a certain degree of sophistication was attained, however, the question concerning information seemed the most pressing. Indeed, the problem of information, which may seem quite narrow at first glance, was basic to the whole

debate; the answers given to this question affected the other two in unexpected ways.

What Is Cybernetics?

The initial question, which concerns the universality of application of
cybernetics, was one of the first aspects of the new field to concern the
Soviet Marxists. The spectrum of the debate ranged from those who
believed cybernetics to be no more than a loose word for process engineering to those who saw it as a new science providing the key to literally every form of the existence of matter. In the hands of its most
enthusiastic proponents, cybernetics became an all-embracing system
including even human society. Non-Soviet cyberneticists whose research originated in mathematics and engineering, such as Norbert
Wiener and W. Ross Ashby, often spoke of the homeostatic properties
of society. A homeostat is a random mechanism capable of adapting itself in such a way that it arrives at equilibrium and appears to be "purposeful." [36] Wiener believed that the controlling mechanism of society is
its legal system, and that society constantly adjusts its laws on the basis
of feedback information concerning the degree of disorder in society.[37]
American political scientists, such as Karl Deutsch of Yale University,
quickly followed with models of political behavior taken from cybernetics.[38] Others began to use the cybernetic approach to sociology, history, and public administration.[39] The range of cybernetics loomed so
great that the discipline seemed to some Soviet scholars to be a possible
rival to Marxism, which advances a philosophy of both the natural and
social sciences; the advent of this new field alarmed the more conservative Soviet philosophers. As one author commented:

The subject of cybernetics is organic and inorganic nature (and technology),
social processes, and phenomena concerned with consciousness. . . . But
doesn't this mean that cybernetics opposes [Marxist] dialectics, that it is
attempting to take its place as a new ideology? If such is the case, . . . then
the question would be: either dialectics or cybernetics. . . . The attempts
to convert cybernetics into some universal philosophical science are completely baseless. Marxists must reject them out of hand.[40]

But it quickly became clear that rather than choose between cybernetics and Marxism, certain Soviet writers wished to unite them. L. A.

Petrushenko, for example, discussed productive labor as a series of processes performed on the basis of feedback.[41] Two other authors discussed the succeeding stages of history according to Marxism as epochs containing progressively smaller amounts of entropy.[42] V. N. Kolbanovskii criticized such extensions of cybernetics, but even he referred to the Marxist "withering away of the state" as a cybernetic phenomenon, a moment when society becomes self-regulating.[43]

Many Soviet scholars applied themselves, however, to the task of defining cybernetics in such a way that it did not even impinge upon Marxism as a general approach to phenomena. Of those scholars who attempted to define cybernetics (Berg, Kol'man, Novik, Shaliutin, Kolmogorov), most emphasized that cybernetics is the science of control and communication in complex systems, while Marxism is the science of the broadest laws of nature, society, and thought. According to this approach, Marxism is so much more general an intellectual system than cybernetics that the two do not conflict. This solution of the relationship of Marxism and cybernetics by placing them on entirely different levels was achieved in 1961 and 1962, when several important studies of cybernetics appeared.[44] However, the "two-plane" solution was not subscribed to by all authors. A number of contributors to Soviet philosophy journals continued to postulate that cybernetic analysis could be applied literally to all phenomena and that "information" was a property inherent in matter. This attempted expansion of cybernetics included occasionally a criticism of Friedrich Engels's writings on science, sometimes openly stated; others wrote that the works of Karl Marx reveal an understanding of the "cybernetic organization of matter," though the term itself was of course unknown to Marx.

Can Life Processes Be Duplicated?

Many persons approaching cybernetics for the first time, both in the Soviet Union and abroad, saw the entire controversy in the questions Can a machine think? and Can cybernetic mechanisms be considered alive? To narrow all the controversies of cybernetics to these queries is to impoverish the intellectual content of the recent discussions; nevertheless, the questions were seriously considered by most cyberneticists, and they played important roles in the debates in the Soviet Union.

Wiener himself was properly cautious on these issues, but even he felt that in the light of cybernetics research some new definitions of "life" may be necessary. He noted that living organisms and inorganic cybernetic systems are similar in that they are both islands of decreasing entropy in a world in which disorder always tends to increase. He observed that "the problem as to whether the machine is alive or not is, for our purposes, semantic and we are at liberty to answer it one way or the other as best suits our convenience." [45] The question of whether machines are "alive" is clearly not identical with that of whether they "think," but Wiener's answers were similar; he remarked that whether a machine can think is merely a "question of definition." More exactly phrased, the problem of thinking machines was usually posed as: Do computers perform functions that are merely analogous to thinking, or are these functions structurally identical with thinking? With the question phrased in this way, the English logician A. M. Turing was quite willing to face the possibility of thinking machines. He believed that a machine could be considered to think if a man separated from a machine by an opaque partition could not determine whether he was facing a machine or another man, basing his conclusions on the answers that he received to questions addressed to the machine.[46]

Several of the Soviet scholars who first supported cybernetics, such as Kol'man and Berg, later warned against the concept of thinking machines; they admitted only an analogous relationship between the machines' functions and thought. Thus, Kol'man commented, "Cybernetic machines, even the most perfected, handling complex logic processes, do not think and do not form concepts." [47] Berg was even more unequivocal: "Do electronic machines 'think'? I am sure that they do not. Machines do not think, and they will not think." [48] These views were supported by Todor Pavlov, honorary president of the Bulgarian Academy of Sciences, who wrote, "Even the most complex robot does not assimilate, does not sense, does not remember, does not think, does not dream, does not fictionalize, does not seek." [49]

The theoretical explanation of the inability of computers to have consciousness was the same as that given by "Materialist" in 1953: Matter at different levels of development possesses qualitative differences; to attribute mental powers to a mechanical agglomeration of transistors and circuits would be, the critics said, to make the mechanistic mistake of believing that all complex operations can be reduced to combinations of simple ones, a belief specifically denied by their interpretation of dialectical materialism. They maintained that sophisticated organisms differ qualitatively from less complex ones, and cannot be reduced to the same components.

As in the case of the problem of defining cybernetics,[50] it appeared by 1961 that considerable agreement had been reached, in this instance specifically denying the possibility of thinking machines; after that time, however, cybernetics enthusiasts returned with their most effective statements to date. In a 1964 article prefaced with the slogan "Only an Automaton? No, a Thinking Creature!" Academician Kolmogorov commented, "The exact definition of such concepts as will, thinking, emotion, still have not been formulated. But on a natural-scientific level of strictness such a definition is possible. . . . The fundamental possibility of creating living creatures in the full sense of the term, built completely on discrete (digital) mechanisms for processing information and for control, does not contradict the principles of dialectical materialism." [51] Such articles alarmed a number of nonscientists, one of whom, B. Bialik, wrote an article entitled "Comrades, Are You Serious?" in which he refused to believe that a machine can experience emotion, appreciate art, or possess genuine consciousness. Academician S. L. Sobolev, director of the Institute of Mathematics and Computation Center of the Siberian Branch of the Academy of Sciences, answered Bialik in an article entitled "Yes, This Is Completely Serious!" This was probably the most outspoken favorable article on cybernetics by a responsible author to appear in the Soviet press. Sobolev straightforwardly called man a cybernetic machine and posed the possibility of man's creating other machines that would be alive, capable of emotion, and probably superior to men.[52]

Although the question of the ability of cybernetic devices to duplicate living organisms remained controversial in the Soviet Union, an affirmative answer to this question did not receive much support from philosophers. Dialectical materialism may not specifically deny the possibility of thinking machines, but the anthropocentric or humanistic nature of historical materialism was a genuine obstacle to such an opinion. According to a number of Soviet authors, the main difference between man and machine was not technical, but social. As Kol'man remarked, "Those who maintain that man is a machine and that cybernetic devices think, feel, have a will, etc., forget one 'trifle'—the historical approach. Machines are a product of the social-labor activity of man." [53] This view was expressed even more forcefully by N. P. Antonov and A. N. Kochergin:

It is necessary to emphasize that man works, and not the machine. One can say that the machine functions, but not that it labors. The machine cannot become the subject of laboring activity because it does not and cannot be possessed of the necessity to work, and it has no social requirements that it must labor to satisfy. This is the main and principal difference between machine and man.[54]

A question even more important than the ability of machines to duplicate man's functions was that of the moral responsibility of man for the actions of his machines. On the whole, non-Soviet cyberneticists were, at least publicly, more fearful than their Soviet counterparts of the possible results of their employment of computers. In 1960, when Norbert Wiener visited the editorial offices of the leading Soviet journal in philosophy, *Problems of Philosophy* (where he received a very warm reception), he commented:

If we create a machine . . . that is so "intelligent" that in some degree it surpasses man, we cannot make it altogether "obedient." Control over such machines may be very incomplete. . . . They might even become dangerous, for it would be an illusion to assume that the danger is eliminated simply because we press the button. Human beings, of course, can press the button and stop the machines. But to the extent that we do not control all the processes that occur in the machine, it is quite possible that we will not know when the button should be pressed. Thus, the programming of "thinking" machines presents us with a moral problem.[55]

Wiener's uneasiness was expressed in different terms by other cyberneticists who spoke of the possibility of a dictator controlling society through the use of cybernetic machines, while still others referred to the computer as the demon that turns on its master.[56]

These pessimistic views of authors in western Europe and North America were by and large rejected by Soviet writers. Like the philosophers, Soviet scientists were, with very few exceptions, optimistic in their statements about science. If any of them, in Oppenheimer's phrase, "came to know sin" as a result of their research, they kept this encounter to themselves. Indeed, several Soviet scholars said that the essential difference between man and machine was the fact that man sets his own goals while the machine strives only toward those for which it has been programmed. If society places a premium on worthwhile goals, said the Soviet authors, the machines of that society will be assigned similarly meritorious functions. These writers suggested that cyberneticists in capitalist societies betrayed a lack of confidence in those societies when they were unsure of the roles their computers would be asked to play.

What Is "Information"?

Cybernetic systems operate on the basis of the collection, processing, and transmission of information. The development of increasingly sophisticated means of evaluating and measuring information was one of the important factors determining the progress of cybernetics. Yet, interestingly enough, no one devised a thoroughly satisfactory definition of information. Norbert Wiener once observed, perhaps with no great intent, that "information is neither matter nor energy," it is just "information." [57] W. Ross Ashby also warned of the dangers of trying to treat information as a material or individual "thing": "Any attempt to treat variety or information as a thing that can exist in another thing is likely to lead to difficult 'problems' that should never have arisen." [58]

Dialectical materialism asserts that objective reality consists of matter and energy in various forms. If information is neither matter nor energy, then what is it? In the years since the early sixties the attention of Soviet philosophers has shifted, at least relatively, from the broader questions of the nature of cybernetics and life to the more narrow problem of the nature of information. They advanced several reasons for this shift of emphasis. In the first place, the more restricted question of the nature of information can be treated more rigorously than such a question as Can machines think? Second, upon investigation the problem of information proves to be the key to many of the broader questions raised earlier.

The problem of the philosophic interpretation of the concept of information was a genuine and troublesome one. If information can be measured, some Soviet scholars reasoned, then it must possess objective reality. As early as 1927 R. V. L. Hartley observed that the amount of information conveyed in any message is related to the number of possibilities that are excluded by the message. Thus, the phrase "Apples are red" carries much more information than the phrases "Fruits are red" or "Apples are colored" because the first phrase eliminates all kinds of fruits other than apples and all kinds of colors other than red. This exclusion of other possibilities increases informational content.[59] In later years the basic principle enunciated by Hartley was refined and elaborated on a mathematical basis. In 1949 in the fundamentally important publication *The Mathematical Theory of Communication* Claude Shannon and Warren Weaver presented a formula for the calculation of quantity of information in which information increases as probability of the particular message decreases. In this method, information is defined as a measure of one's (or a system's) freedom of choice in selecting a message. Thus, in a situation where the number of likely messages from which to choose is large, the amount of information produced by that system is also large. To be more precise, the amount of information is defined (in simple situations) as the logarithm of the available choices. Shannon's and Weaver's 1949 formula was

$$H = - K \sum_{i=1}^{n} p_i \log p_i$$

where H is the amount of information in a system with a choice of messages with probabilities ($p_1, p_2 \ldots p_n$), and K is a constant depending on the unit of measure.[60] This formula is functionally the same as that for thermodynamic entropy devised by Max Planck at the beginning of the century: $S = K \log W$, where S equals entropy of the system, W equals the thermodynamic probability of the state of the system, and K equals Boltzmann's constant.[61]

Some scientists considered the potential implications of this coincidence to be immense. The possibility of an analogy or even a structural identity between entropy and information generated a heated debate among physicists, philosophers, and engineers in many countries. Weaver commented, "When one meets the concept of entropy in communication theory, he has a right to be rather excited—a right to suspect that one has hold of something that may turn out to be basic and important." Louis de Broglie called the link between entropy and information "the most important and attractive of the ideas advanced by cybernetics." [62] If one could demonstrate that the relation between neg-entropy and information were more than functional similarity, and was instead an identity, the construction of a general theory of matter by which all complex systems—inorganic and organic, including human— could be mathematically described seems at least conceivable. The more venturesome dialectical materialists tended to welcome such a possibility, since it seemed to them a vindication of materialistic monism. A literal rephrasing of the three basic laws of the dialectic in cybernetic terms was attempted by the author of an unpublished doctoral dissertation at Moscow University.[63] The more orthodox majority, however, was unsettled by the difficulties of fitting such an ambitious theory into the principles of dialectical materialism.

An additional problem in interpreting information theory in terms of dialectical materialism concerned the supposed "subjective" nature of quantity of information. The proponents of the subjective approach (Ashby and L. Brillouin among others) pointed out that one can hardly speak of the quantity of information in any message in absolute terms, since a certain message will carry much more information to one observer than to another, depending on the prior knowledge of the observer. Following this approach, several non-Soviet authors called for the attaching of qualitative coefficients to calculations of quantity of information, based on the value of information, the degree of certitude, and meaningfulness. But if information (variety) is to be quantitatively measured, the Soviet philosophers insisted, it must be part of objective reality, and not conditioned by subjective considerations. A. D. Ursul's comment on this topic was appropriate: "First of all we must notice that in a finite system the quantity of variety inherent to it does not depend on the observer and is always finite. . . ."[64]

Not only in the Soviet Union were scholars cautious about information theory; for every enthusiast who might attempt to identify information with neg-entropy, another sober-minded individual added a cautionary note. Ashby, for example, commented: "Moving in these regions is like moving in a jungle full of pitfalls. Those who know most about the subject are usually the most cautious in speaking about it."[65] And yet, despite the warnings, the general trend among cyberneticists in the early sixties was a greater acceptance of the conception that there exists some essential link between entropy and information.

I. B. Novik was one of the more energetic Soviet philosophers who attempted to define information in terms of dialectical materialism. In his book *Cybernetics: Philosophical and Sociological Problems* Novik tried to present a systematic treatment of cybernetics from the standpoint of enlightened Marxism.[66] From the outset he aligned himself with the partisans of cybernetics; he insisted that there was no conflict between this new field and dialectical materialism. Wiener to him was an unconscious dialectician. Novik explained that cybernetic information is a property of matter, a property directly connected with Lenin's copy-theory of epistemology. Lenin wrote in *Materialism and Empirio-Criticism* that materialism is based on a recognition of "objects in themselves," and that objects exist "without the mind." According to Lenin, ideas and sensations are reflections of these objects; all matter has this property of "reflection." Novik then postulated that "quantity of information" is a measure of the order of the reflection of matter. "In my opinion," he wrote, "in information, and especially in the structure of its symbols, is expressed the order of reflection. . . . In this sense, noise is connected with disordered, chaotic reflection." [67] Novik then called for the creation of a science of the "physics of reflection," and in order to hasten the development of this new field, he proposed a rudimentary law of the conservation of information patterned on the law of conservation of matter, since information is "inseparably linked" to matter.[68]

Other authors, following the lead of the philosophers who in 1961 and 1962 attempted to separate the realms of applicability of cybernetics and dialectical materialism, denied that the concept of information can be related to entropy or to states of matter outside narrowly defined control systems. Thus, N. I. Zhukov commented, "Certain authors consider that information processes are characteristic of all processes in inorganic nature. Such a universal understanding of this concept creates difficulties in the development of the theory of information and cybernetics. . . . Information, in our opinion, may be precisely defined as an adjusted change used for the purposes of control." [69]

Nonetheless, the enthusiastic proponents of cybernetics continued to maintain that information is a property of all matter and that the evolution of matter, from the simplest atom to the most complex of all material forms, man, may be seen as a process of the accumulation of information. Thus, these authors tied cosmogonical, geological, and organic evolution together in one process of the tendency of matter, at least in certain *loci,* to increase its informational content. The result was a sort of great chain of being, a ladder in nature of ascending complexity, although evolutionary instead of static.

In an article that appeared in *Problems of Philosophy* in 1965 E. A. Sedov, an engineer, maintained that insufficient attention had been paid to the application of information theory to the physical processes of inert matter. The main cause for this neglect, said Sedov, was the widespread belief among many scientists that the main similarity between neg-entropy and information was merely mathematical form. Sedov, however, believed that information is literally a form of neg-entropy and that neg-entropy is a measure of the order of material systems in both communications and ordinary physical systems. Thus, the competence of thermodynamics would be extended over all matter, including complex life. Furthermore, Sedov believed that an increase in orderliness was a natural evolutionary tendency in the material world, "beginning with the origin of the galaxies and planetary systems" and ending with organic structures. Sedov said that this natural tendency toward increased orderliness of movement was a compensating factor for the irreversibility of energy transformations according to the Second Law of Thermodynamics and therefore a refutation of the "heat-death" theory of the universe.[70]

A. D. Ursul, a mathematician, continued Sedov's viewpoint in a later issue of *Problems of Philosophy;* Ursul believed that increase of entropy (described by the Second Law of Thermodynamics) and increase of information (described, Ursul believed, by a cybernetic interpretation of material evolution) are countering forces, one "progressive," the other "regressive," which together eliminate the possibility of an end point in energy transformations. Ursul found what he considered evidence of the progressive evolution of matter on the most fundamental levels:

Atoms contain less information than molecules, molecules contain (including the most complex organic ones) less than one-cell and other simple organisms, and the latter contain less than developed multicell organisms. It stands to reason, then, that the quantity of information objectively contained in qualitatively different material systems may be connected with the degree of development of natural material systems.[71]

Yet the debate between the enthusiasts and the skeptics continued. The Soviet Union was witnessing one of the healthiest debates in the philosophy of science in many years. Some writers obviously agreed with Kol'man in his criticism of "universal cybernetics" and with Todor Pavlov's comment that some authors had "absolutely and metaphysically exaggerated" the significance of cybernetics.[72] In the same issue of *Problems of Philosophy* in which Ursul applied information to all levels of matter, another writer, B. S. Griaznov, commented that cybernetics "does not study phenomena in which there are no processes of control or transfer of information—that is, processes of inorganic nature." [73] A view similar to that of Griaznov was taken by N. Musabaeva, who wrote in a 1965 book that while all matter possesses the Leninist property of "reflection," nonliving matter does not contain "information," with the exception of cybernetic devices constructed by man.[74] "Information," Musabaeva continued, is a special form of "reflection" possessed by complex, highly organized material systems that are capable of adapting to their external surroundings.

In 1968 the same Ursul mentioned above published an interesting book entitled *The Nature of Information,* in which he defended very strongly his belief that information is a characteristic of all matter, from the simplest inorganic forms to human society.[75] Ursul tied this conception of the unity of nature closely to dialectical materialism, arguing that the dialectical laws help one to understand information processes.[76] But he also thought that information theory had added new content to dialectical materialism; he posed the possibility of making a few changes in Marxist philosophy as a result of the contribution to man's knowledge of information theory. In particular, he believed that there was a good argument for converting the concept of information from a scientific-technical one to a general philosophic category, adding it to the existing list of categories in the Marxist dialectic.[77] But he also recognized that in view of the short length of time that Marxist philosophers had studied the concept of information, it might still be premature to call for universal acceptance of it as a Marxist category.

Ursul believed that those writers who refused to accept the application of information theory to inorganic systems, on the grounds that these systems do not "use" information, were overlooking an extremely fruitful approach to nature. Information may be either "used" or not "used" in a functional sense, but it still exists. Furthermore, an information approach to molecules can even help us to understand the difference between the inorganic and organic worlds: "If the information content of an object is several dozens of bits on the molecular level, then it probably is an object of inorganic nature. If the object contains 10^{15} bits on this level, then we are dealing with a living object." [78]

Ursul recognized, however, that analysis of this sort carried with it the dangers of reductionism—the elimination of qualitative characteristics on different levels of matter. He believed, however, that information cannot be entirely described by one method, such as mathematical probability, but instead must be approached from standpoints that include qualitative characteristics, such as topology.[79] He also urged that information theory be supplemented with an understanding of dialectical levels of nature. Not all information is the same; it possesses qualitative characteristics, and two different types of information cannot be compared. According to Ursul, each level of nature possesses "its own" information.[80] Ignoring this specificity accounts for the incautious way in which many philosophers and scientists have extended concepts derived from physics, such as entropy, to other realms. Instead, Ursul favored "classifying" information (neg-entropy) into different types, each with its own realm of applicability. The Soviet author V. A. Polushkin had already made such an attempt when he divided information into "elementary," "biological," and "logical" types. In this scheme, elementary information was understood as information in nonliving nature. Ursul thought that Polushkin's effort was in the right direction, but suggested that much further work would have to be done; to him, "human" or "social" information was also a type, and within human information he further distinguished at least two aspects: semantic (content) and pragmatic (value).[81]

Ursul thought that the transition from one realm of information to another was, in evolutionary terms, a qualitative jump. He pointed to the possibility of combining this approach with elements of the biological philosophy of Ludwig von Bertalanffy, the Canadian scholar.[82]

In recent years a certain decline in enthusiasm for cybernetics seems discernible in the Soviet Union, although interest remains high. Many people, outside the Soviet Union as well as inside that country, who have encountered cybernetics as a conceptual scheme have gone through a cycle of intoxication, or at least elation, followed by a period of disillusionment, or at least retracted ambitions. Initially enthusiasts, they become more aware that cybernetics will not solve all the complex problems that they had hoped to attack with its assistance.

For all the persuasiveness of cybernetics upon first contact, it is a very incomplete science.[83] Cybernetics may dissolve into less dramatic subareas of information theory and computer technology. As a French specialist in cybernetics observed, "As an adjective, 'cybernetic' threatens to go the way of 'atomic' and 'electronic' in becoming just another label for the spectacular." [84] Certain scientists find the use of the term embarrassing. Furthermore, it is now clear that there were genuine defects in the writings of several of the founders of cybernetics who, in their enthusiasm, often confused certain technical terms, such as "quantity of information" and "value of information." [85] And finally, cybernetics proceeds on the basis of analogical reasoning, which by itself leads not to logical or scientific proofs, but instead to inferences that may or may not be significant and fruitful.

The strength of such reasoning depends upon the similarities that can be identified between the two entities being compared. Soon after the development of the methodology of cybernetics, the comparison of the human body as a control system to an economic system, a city government, or an automatic pilot seemed to result in the identification of truly striking similarities. The longer one dwells upon such analogies, however, the more clearly emerge the very genuine differences that exist between the entities being compared.

There are cyberneticists, such as Stafford Beer, who maintain that cybernetics goes far beyond analogy. They argue that if one abstracts the control structures of two dissimilar organisms, the relationship between these structures may be one of identity rather than of analogy.[86] The control structure of a complex industry and that of a living organism may be identical, according to this view, in the same way that the geometrical form of an apple and that of an orange may be identical circles. This approach may be correct on an abstract level, but it has not resulted in as many discoveries of fruitful similarities and avenues of research, beyond those originally identified, as early proponents of cybernetics hoped.

In the history of science we are familiar with an exuberant seizing upon the latest conceptions of science as models for explanations in areas rather distant from their points of origins. It is not accidental that the popularization of physics in the seventeenth and eighteenth centuries was followed by the application of physical concepts in a universalist fashion, with the human body described as a mechanical assemblage and the relations of European states analyzed by the use of such terms as "balance of power" and "fulcrums" of diplomatic pressure. Such examples could be multiplied almost endlessly. The Copernican heliocentric (actually heliostatic) theory influenced thinkers in such distant fields as physiology (Harvey). Darwinian theory was misappropriated by capitalist businessmen wishing to justify economic competition. In the 1930's the concepts of complementarity and indeterminism arising from the study of subatomic particles were applied by certain physicists and philosophers in attempts to explain cultural relativism and freedom. With the development of cybernetics, our vocabulary is enriched with models of political behavior containing "feedback" and models in which the achievement of "homeostasis" is the desired goal.

In the history of science many of these various models have been of scientific importance. Some have been of heuristic value in fields other than the science in which they were originally developed. Others, particularly when transferred far afield, have had negative effects. But all have been followed by periods of greater appreciation of the failures and inadequacies of such models as well as of their successes. Every model of behavior emphasizes certain attributes of an organism at the expense of others. The greatest heuristic value of such models is likely to occur at the first moment of genuine understanding.

In cybernetics, one might suspect, the absence of dramatic theoretical breakthroughs is apt to lessen the persuasiveness of its conceptual scheme as an explanation of all dynamic processes. In the United States, where computers are applied very widely and where their sociological and economic consequences are still topics of vigorous debate, the decline in interest in cybernetics as a conceptual scheme is clearly evident. The postcybernetic epoch involves not a renunciation of cybernetics, but only a more sober appraisal of its potentialities. The original zeal might be renewed by future developments in theory, but one obviously cannot foretell such events.

Whatever the destiny of cybernetics as a field of investigation in the Soviet Union, there will no doubt be a lasting residue of renewed belief in man's ability to control intricate processes based on the new wave of genuine achievements in this area. Thus, the scientific optimism to which Marxism is an heir received a new impetus.[87] True, a new order of complexity of problems—and a new need for ways of overcoming complexity—now seems to face all nations in such areas as environmental pollution and overpopulation. In their devotion to scientific optimism and their emphasis on man's ability to transform nature, Soviet writers have not yet demonstrated a sufficient awareness of the fact that man's very effort to transform nature may lead to increasing disorders. In their lack of attention to the value of undisturbed nature, they seem rather similar to the capitalist countries of the West, although the different economic system results in somewhat different patterns of disruption.[88] But Soviet writers, like their colleagues in other countries, are now beginning to awaken to such problems, and new discussions of ecology—which with its emphasis on the interconnectedness of the world has close ties to both cybernetics and dialectical materialism—are beginning to appear.[89]

X Physiology and Psychology

In no other scientific field discussed in this volume does there exist an identifiably Russian tradition of interpretation to the degree that there does in physiology and psychology. Long before the Revolution the study of physiology and psychology in Russia was known for its materialism. To be sure, there were many supporters of idealistic psychology in pre-Revolutionary Russia, but materialism in psychology received unusual support there at a fairly early date. In 1863 Ivan Sechenov (1829–1905) published his *Reflexes of the Brain,* a book the true purpose of which is better revealed by the title that Sechenov originally gave it, but that was disapproved by the tsarist censor: *An Attempt to Establish the Physiological Basis of Psychological Processes.*[1] Sechenov wrote in this work that "all acts of conscious or unconscious life are reflexes."

Surrounding Sechenov's views there soon grew up a controversy among the St. Petersburg educated public. The particular political and ideological scene of late-nineteenth-century Russia influenced the course of the debate, with the radical intelligentsia usually, but not always, responding favorably to Sechenov's opinions and the government bureaucracy usually disapproving. In 1866 the book was prohibited for sale by the St. Petersburg censors, and Sechenov himself was threatened with court action for allegedly undermining public morals. Eventually Sechenov escaped trial, but the already existing link between materialism in science and radical politics was strengthened and made more apparent.

Although materialism was strong in Russian psychology before the Revolution, it by no means monopolized the field. Sechenov was thought of primarily as a physiologist, not a psychologist. Opposing the views of Sechenov and some of his pupils were not only the censors of St. Petersburg and representatives of the Church, but also many university professors of philosophy and psychology. Indeed, Sechenov was outside the mainstream of academic psychology in Russia. Nonetheless, the essential issues that he raised concerning the nature of the psyche and the relationship of the physiological to the psychological were hotly debated among Russian psychologists, physiologists, philosophers, and political activists in the last decades of the nineteenth century.[2] The history of these debates is still insufficiently explored, but even a cursory examination reveals that some of the features of these polemics—not only between materialists and idealists, but among members of each camp and of other groupings as well—resemble discussions that have continued throughout the Soviet period.

The materialist tradition in physiological psychology was also strong in France; Sechenov's views have consequently been traced back to the French tradition through such men as the great French physiologist Claude Bernard, in whose laboratory the young Sechenov once worked. No doubt such an influence was genuine in Sechenov's development. One awaits further scholarship on the connections between Russian and west European schools of thought in this interesting area. Nevertheless, the originality of Sechenov seems clear; as Edwin G. Boring wrote in his *A History of Experimental Psychology,*

Sechenov became the Russian pioneer in reflexology. We must, moreover, remember that he was far ahead of western European thought on this matter. His papers of 1863 coincided with Wundt's *Vorlesungen über die Menschen- und Thierseele,* and 1870 was four years before Wundt's *Physiologische Psychologie.* In no other country did reflexes yet seem to provide a means for studying cognition. Later, Pavlov read Sechenov's work and was, as a young man, greatly influenced by his argument.[3]

Much later, Sechenov's significance was further emphasized by one of the developers of cybernetics, Walter A. Rosenblith, who in 1964 called the Russian "a too little appreciated forebear of Norbert Wiener." [4]

The most important influence on Russian physiology and psychology was Ivan Pavlov (1849–1936), a great figure in world science. Although it is impossible and inappropriate to summarize Pavlov's views here, some aspects of his work must briefly be discussed, particularly those that would later become the subject of philosophical and method- ological discussion in the Soviet Union.

From the standpoint of the history and philosophy of science the greatest significance of Pavlov derives from his success in bringing psy- chic activity within the realm of phenomena to be studied and explained by the normal objective methods of natural science. In contrast to the introspective approach of many investigators of mental activity at the turn of the century, Pavlov's method was based on the assumption that psychic phenomena can be understood on the basis of evidence gath- ered entirely externally to the subject. He was not entirely original in his intention to proceed in this manner, but as a great experimentalist he was able brilliantly to combine this methodological assumption with un- usual skill in devising and conducting experiments with animals. On the basis of these experiments he erected a theory of nervous activity that presented general principles aimed toward the eventual explanation of man's psychic activity on a physiological foundation.

Pavlov is, of course, best known for his theory of conditioned and unconditioned reflexes. Unconditioned reflexes, he said, are inborn forms of nervous activity and are transmitted by inheritance. Condi- tioned reflexes are acquired during the life of an organism and are based on a specific unconditioned reflex; conditioned reflexes are not normally inherited, although Pavlov believed that in some cases they could be- come hereditable.

In the classic case of the dog and the bell, the unconditioned reflex is the natural, inborn salivation of a dog in response to the stimulus of food. The conditioned reflex, salivation in response to a bell alone, is created by the prior repeated juxtaposition of the bell and the food. Pavlov further illustrated that "conditioned reflexes of the second order" could be created by using the conditioned response to the bell as a basis for the formation of yet another conditioned reflex to a third stimulus, such as a light. In the latter case, it must be emphasized that at no time was the original stimulus (food) combined with the stimulus triggering the second-order reflex (the light). In this fashion Pavlov was able to point to the quite indirect ways through association by which reflexes could be created. He believed that the psychic activity of man could be interpreted in this way, or at least on this foundation. This theory of the broad significance of conditioned reflexes Pavlov called the Theory of Higher Nervous Activity, and this phrase is a part of the standard terminology of Soviet physiologists and psychologists. Pavlov's view is supported by numerous contemporary non-Soviet scholars as well; their opinion is summed up in the words of a recent French biographer of Pavlov: "It is beyond doubt that many of our thoughts—if not all—are, or at least originate in, reflexes acquired from upbringing, education and culture, and strengthened by experience." [5]

The inner structure of reflex action was described by Pavlov in terms of a "reflex arc," a term that would be the subject of much later discussion. The reflex arc had three links: the *affector* neurons, the *nervous centers,* and the *effector* neurons. The original excitation caused in the sense organs by an external stimulus travels inward along a chain of affector neurons to the nervous centers; then another stimulus travels outward along the effector neurons to specific muscles or glands, causing a response to the original stimulus. The three links in this arc are sometimes described as sensor-connector-motor.

In the case of the formation of conditioned reflexes in man Pavlov believed that the nervous centers are located on the cortex of the cerebral hemispheres. "Temporary connections," an inclusive term embracing conditioned reflexes and other more rudimentary or fleeting linkages, are formed as a result of "irradiation" of stimuli reaching the hemispheres. In other words, stimulation is "generalized" in the hemispheres in such a way that other areas of the cortical region now react in the same way as that concerned in the original stimulus. Thus, the area of the cortex receiving nervous responses to light signals may be incorporated into reflex action originally based only on sound signals. As Pavlov wrote, "The fundamental mechanism for the formation of a conditioned reflex is the meeting, the coincidence of the stimulation of a definite center in the *cerebral cortex* with the stronger stimulation of another center, probably also in the cortex, as a result of which, sooner or later, an easier path is formed between the two paths, i.e., a connection is made." [6]

By a process of training, inhibition, the reverse of irradiation, can also be illustrated. Physiologically, the area of the cortical region that has been irradiated is reduced by teaching the subject to discriminate not only between very different signals, such as sound and light, but between sounds of different vibrations. Thus, Pavlov was able to teach a dog to respond to a tempo of one hundred beats a minute but not to ninety-six, as a result of producing food only after the more rapid signal. After this process of inducing inhibition, Pavlov concluded that "the nervous influx produced by the stimulus is now communicated to only a very limited area of the cortical zone under consideration."

One of the most flexible concepts that Pavlov advanced, and one still exploited only to a rather small degree, was that of the "second-signal system," a feature unique to the psychic activity of man. Most of Pavlov's research was based on experiments with dogs, but in the latter part of his life he worked with monkeys and gorillas, and his interests were shifting more and more to what he considered the ultimate goal of neurophysiology—the study of man. Man has fewer instincts than animals; Pavlov believed, therefore, that his behavior would be governed by conditioned reflexes to a much higher degree. Both animals and man can be conditioned in similar ways, but man, in addition, possesses the almost infinitely rich instrument of language. While animals responded to simple ("primary") signals or symbols (even a dog responding to a word command reacts to it in a fashion not dissimilar to the response to a bell or light), man responds to the meanings and incredibly rich associations conveyed by speech and writing ("secondary signals"). The language message that any one human subject receives will contain meanings and associations unique for him, given a message of even minimal complexity. And Pavlov saw the second-signal system as infinitely more complex than the primary one: "There is no comparison, qualitative or quantitative, between speech and the conditional stimuli of animals." Thus, Pavlov cannot be described fairly as a person who believed that human behavior can be reduced to the simple stimulus-response action of the noted experiments with dogs. He fully recognized that human beings were qualitatively quite distinct from other animals. But he believed, nevertheless, that human behavior is amenable to investigation on the basis of physiology, an assumption sensible and necessary in order for physiologists to investigate the human nervous system.

Pavlov's attitude toward psychology has been the subject of numerous inaccurate statements, many of which imply that Pavlov was opposed to the very existence of psychology. Pavlov did object to the concept of animal psychology, since he felt that there was no way for man to gain access to the inner world of animals. He was, further, deeply critical of what he considered metaphysical concepts presented in psychological terms. In his early years he was doubtful of the scientific validity of much that was presented as psychological research. As he grew older, and as experimental psychology steadily developed as a discipline, Pavlov became more and more kindly disposed toward psychology. In a speech given in 1909 Pavlov said:

I should like to elucidate that which might be misunderstood in these statements concerning my views. I do not deny psychology to be a body of knowledge concerning the internal world of man. Even less am I inclined to negate anything which relates to the innermost and deepest strivings of the human spirit. Here and now I only defend and affirm the absolute and unquestionable rights of natural scientific thought everywhere and until the time when and where it is able to manifest its own strength, and who knows where its possibilities will end! [7]

But even in this statement affirming the right of psychology to exist, one can detect Pavlov's skeptical view of psychology. The last sentence implies a distinction between psychology and "natural scientific thought," which most psychologists would reject. And when Pavlov spoke of a fusing in the future of physiology and psychology, many psychologists thought that he was actually referring to an absorbing of psychology by physiology after the necessary progress in physiology had occurred. One must admit that Pavlov remained somewhat dubious about psychology as a science, although he was by no means so hostile as many later commentators have implied. Despite his frequent warnings against reductionism, his call for the study of the "whole organism," and his belief in the "qualitative and quantitative uniqueness" of man, Pavlov tended to see psychic phenomena, and especially the reflex arc, in somewhat mechanistic and elementary terms. This tendency was probably inevitable in the period when psychology was, indeed, heavily influenced by idealistic concepts and Pavlov had to struggle to establish his teaching on conditioned reflexes, now recognized as one of the great achievements of both physiology and psychology.

Pavlov was not a Marxist and did not defend his system in terms of dialectical materialism. For many years after the Revolution he stoutly resisted Marxist influences in educational and scientific institutions and even criticized Marxist philosophy.[8] In the last years of his life, however, his views changed; he praised the Soviet government for its support of science, and he was impressed by the intelligence of individual

Bolshevik leaders, such as Nikolai Bukharin. One of his pupils, P. K. Anokhin, a man whose views will be separately discussed, maintained that once in a conversation he tried to show Pavlov that his teaching about the contradictory but necessary effects of irradiation and inhibition was deeply dialectical and revealed the struggle and unity of opposites. To this observation Anokhin said that Pavlov responded, "There you are, it turns out that I am a dialectician!" [9]

There are many aspects of Pavlov's thought that appeal to dialectical materialists. First of all, his primary goal, the explanation of psychic phenomena on the basis of physiological processes, is one that materialists have traditionally and understandably supported. His emphasis on the necessity to study organisms as a whole, "in all their interactions," rather than by isolating out one portion or one phenomenon has been praised by Soviet writers for being in agreement with the dialectical principle of the interconnectedness of the material world. His emphasis on the unique qualities of man, with his second-signal system, has been termed an understanding of the qualitative differences of organisms at different levels of complexity, based on the principle of the transformation of quantity into quality. His description of the human body as a system "unique in the degree of its self-regulation" has been seen both as a prefiguring of cybernetic concepts of feedback and as an understanding of the dialectical process of development.

Scholars in the Soviet Union, on the one hand, and those outside that country, on the other, frequently look upon Pavlovianism in different ways and almost as different things. Non-Soviet scientists often consider it a rather restricted body of experimental data and hypotheses concerning conditioned and unconditioned reflexes. To some of them his name is nearly synonymous with the mental picture of salivating dogs. Soviet scholars, on the other hand, see Pavlovian theory not only as this body of facts and conclusions, but also as an approach to nature in general and to biology in particular. Pavlov himself contributed to this latter understanding in a conversation with the American psychologist K. S. Lashley; when Lashley asked Pavlov to define the concept of "reflex," Pavlov replied:

The theory of reflex activity operates on three basic principles of exact scientific research: first, the principle of determinism, i.e., of a stimulus, a cause, a reason for every given action or effect; second, the principle of analysis and synthesis, i.e., of an initial decomposition of the whole into parts or units and then the gradual building up anew of the whole from its units or elements; finally, the principle of structure, i.e. the distribution of the actions of force in space, the timing of dynamics to structure.[10]

In reply to this statement by Pavlov, Lashley observed that this definition of reflex was so general that it could be taken as the general principles of all science. But Pavlov stuck to his formulation, which is often quoted in Soviet discussions of the significance of reflex theory.[11]

Some Soviet authors distinguish between the reflex principle in a philosophic sense and the reflex principle in a concrete, physiological sense, thus opening up considerable possibilities for recognizing certain limitations in Pavlov's teaching while retaining its methodological content. In 1963 F. V. Bassin, a Soviet scholar who called for much greater attention to the subconscious realm and pointed to certain elements of value in Freud's work at a time when this was rare among Soviet scholars, made this distinction; in his opinion, the most valuable aspect of Pavlov's work was the underlying idea of the essential dependence of biological factors on the environment. Bassin wrote:

That person who abandons the reflex theory in its philosophic sense abandons more than Pavlov's teaching: He abandons the dialectical materialist interpretation of biological processes in general. This is undoubtedly so, since the primacy of the reflex principle in its philosophic sense (i.e., the idea of the dependency in principle of biological processes on factors of the environment) is that basic, that most profound element that distinguishes us from the supporters of idealistic biology, with its emphasis on immanence, spontaneity, and consequently, the absence of the reflex principle in life processes. . . . I mention this because it is necessary to see the difference between the reflex principle in its general philosophic meaning and as a concrete understanding of physiological structure. . . .[12]

The history of psychology in Russia in the years after the Revolution is a very rich and contradictory story; since our center of attention in this volume falls on the years after World War II, it will be impossible to discuss the earlier period in detail. However, some of the features of the work of L. S. Vygotsky, who continues to be influential, will be considered below. More detailed discussions can be found in A. V. Petrovskii's *History of Soviet Psychology* (in Russian) or in Raymond Bauer's *The New Man in Soviet Psychology*.[13]

Immediately after the Revolution members of several different schools of psychology could still be found in Russia. Those with the closest links to introspection and idealistic psychology were N. Lossky and S. Frank, both of whom lost their positions shortly after the Revolution. Another group was made up of experimental psychologists who had been heavily influenced by subjective psychology, but who moved after the Revolution to a position of neutral empirical psychology, hoping in that way to remain clear of the controversies. They included G. I. Chelpanov and A. P. Nechaev. A third group was made up predominantly of physiologists, such as V. M. Bekhterev, and hoped to reconstruct psychology on an objective, scientific basis. They usually doubted the validity of the term "psychology."

The first psychologist to call for an application of Marxism to psychology was K. N. Kornilov, a scholar with an interesting history in the discussions of the twenties and the thirties. At congresses of psychoneurologists in 1923 and 1924 Kornilov attempted to discern the operation of the materialist dialectic in his psychological research. He maintained that the dialectical principle of universal change could be seen in psychology "where there are no objects, but only processes, where everything is dynamic and timely, where there is nothing that is static." [14] The dialectical principle of interconnectedness is illustrated, he continued, by the tendency toward "extreme determinism" in psychology, including the determinism of the Freudian school. This principle accorded well, further, with the views of Gestalt psychologists and with the emphasis on the importance of total patterns rather than discrete bits of experience. And a third principle—the transition from quantity to quality by leaps—is illustrated in many ways: color discernment, in which quantitative differences in frequency of light waves result in qualitatively distinct perception of colors; the concept of thresholds of perception, in which one senses change only after a considerable amount of quantitative stimulation of sense organs; and the Weber-Fechner law of weight and auditory discrimination.

Like Engels in his more enthusiastic moments, Kornilov seemed to see the operation of the principles of the dialectic on every hand. Not surprisingly, Kornilov was soon criticized for applying the dialectic in a "purely formal" fashion, for using it simply as a means of justifying research on which he was already embarked rather than as a methodology basically affecting the course of his work. He was particularly criticized for maintaining that "reactology"—his term for his approach to psychology—was a dialectical synthesis of the subjective and objective trends in Soviet psychology, one that would preserve a concept of consciousness, of the psyche, while at the same time utilizing the rich findings of the physiologists in the study of reflexes. One of his critics, V. Struminskii, observed that Kornilov was applying the materialistic dialectic to the world of ideas rather than to matter, combining intellectual currents in a Hegelian fashion: "[T]his synthesis is not materialist, but a completely idealistic construction. . . . The materialist dialectic does not concern itself with psychological currents as independent factors and magnitudes." [15]

Despite Kornilov's attempt to identify the dialectic in his research, Marxism was not a major influence on his work. His effort to combine elements of subjective psychology and the newer physiological study of reflexes stemmed from his opinion that both possessed advantages. He thought that the physiologists and behaviorists were abdicating the responsibility of psychologists by occupying themselves exclusively with muscular responses. The traditional psychologists, on the other hand, were just as blindly ignoring the significant work of Pavlov, Bekhterev, and their followers. After 1923 Kornilov headed the Moscow Psychological Institute, where he worked with other scholars of later prominence, such as N. F. Dobrynin, A. N. Leont'ev and A. R. Luria. They were also in close communication with groups led by P. P. Blonskii and M. A. Reisner. All of these men at this time were experimenting eclectically with various currents in psychology in a manner that later became impossible because of ideological pressures.

In addition to reactology, the other major tendency in Soviet psychology and physiology at this time was the "reflexology" of M. Bekhterev. It contrasted sharply with reactology in its refusal to use subjective reports and such traditional terms as "psyche," "attention," and "memory." This school drew heavily on two different sources: the materialist tradition in Russian physiology stemming from Sechenov through Pavlov and Bekhterev himself, and American behaviorism. Bekhterev (1857–1927) had long before the Revolution maintained that every thought process, conscious or unconscious, expresses itself sooner or later in objectively observable behavior. On this basis he and his followers hoped to create a science of behavior. In the twenties their approach was so popular that the existence of psychology as a discipline was threatened. In the Ukraine in 1927 higher educational institutions replaced the term "psychology" with "reflexology" as a description of courses of study.

There was also in the twenties a genuine interest in Freudian psychology and much controversy over how well it fit with Marxist interpretations. It was by no means clear in this early period that Freudianism would become a pejorative term to Soviet Marxists. Part of the interest in Freud was simple curiosity; many of the articles in political and literary journals contained elementary descriptions of his work. Freud had not yet published his later, more speculative, works such as *Civilization and Its Discontents,* in which, in addition to some dubious psychological theorizing, there appeared a criticism of communism.[16] To some Soviet writers Freud's teachings appeared as a victory of determinism, an end to free will. Writing in the major Marxist theoretical journal in 1923, the Soviet author B. Bykhovskii commented, "We conclude that despite the subjective casing in which it appears, psychoanalysis is at its foundation imbued with monism, with materialism . . . and with the dialectic, i.e., with the methodological principles of dialectical materialism." And further, "I am not inclined to exaggerate the value of psychoanalysis. I know that it contains much that is controversial and that several of Freud's conclusions are arbitrary and one-sided. . . . But I have tried to bring out the healthy kernel of psychoanalysis, and I think it is rather valuable and significant." [17] Similar comments were made by such intellectual and political leaders as M. A. Reisner, A. P. Pinkevich, and Trotsky himself. But by the latter years of the twenties, discussion of Freud had shifted to open criticism.

Aside from the question of Freudianism, a new trend in Soviet psychology became discernible by the end of the 1920's.[18] This trend stemmed from the rather widely held realization that with the defeat of the supporters of subjectivism and introspection in Soviet psychology, the greatest danger was now from the left—from those militant materialists who hoped to swallow up psychology in a purely physiological understanding of mental activity. The defenders of psychology rallied around the concept of *psikhika* (psyche) and *soznanie* (consciousness) in what has been called a "great struggle for consciousness." This controversy, which ended in victory for the defenders of psychology and consciousness, bore many characteristics peculiar to the Soviet environment. It is well, however, to guard against the tendency of non-Soviet historians to look upon all events in the Soviet Union as *sui generis,* as irrelevant to intellectual history as a whole. These were years in which the validity of the concept of consciousness was being discussed in many countries. According to Boring:

The attack on old-fashioned analytical introspectionism was successful, and in the late 1920's Gestalt psychology and behaviorism found themselves practically in possession of the field. With their missions thus more or less accomplished, both these schools tended to die out or at least to lose their aggressiveness during the 1930's. Psychological operationism came in at this time to supplant behaviorism as a more sophisticated view of psychology, and the outstanding systematic issue in the early 1940's seemed to be whether the Gestalt psychologist could save consciousness, as observed in direct experience, for psychology, or whether the operationists would succeed in having it reduced to the behavioral terms which define the manner of its observation.[19]

Echoes of these changes in psychology internationally were reverberating within the Soviet Union. There, too, the criticism of introspectionism had been successful—indeed, to the point of overkill resulting from the peculiar political instruments at the disposal of the Communist Party, such as increasing control of faculties and editorial boards. In the Soviet Union, as abroad, the crude mechanism of early behaviorism was being succeeded by a more sophisticated approach that, nonetheless, did not deny the achievements of the behaviorists.

In the Soviet Union there were other unique elements as well. The debates were increasingly cast in the terms of theoretical Marxism. Furthermore, the policy decisions of the Communist Party were beginning to have a direct influence on the course of the psychological discussions. The decision to embark upon a rapid industrialization program required great effort on the part of Soviet citizens and enormous will power. A psychology that left more room for voluntarism, for personal resolve and dedication, was welcome on this scene. This shift in Soviet psychology has been frequently discussed by previous authors, such as Raymond Bauer, who entitled his chapter describing these events "Consciousness Comes to Man." [20] In terms of Marxist theory the shift was explained on the basis of the Leninist "theory of reflection," which maintains that the mind, or consciousness, plays an active role in the process of cognition.

In the early 1930's the place of psychology in the Soviet Union became more secure, while Bekhterev's reflexology gradually lost its popularity. As we shall see in the cases of Vygotsky and Rubinshtein, Marxism was incorporated into psychological theory in a more sophisticated way. The increase in the subtlety of psychological theory was accompanied, perhaps surprisingly, by an increasing concern with such practical activities as industry and education. Industrial psychology, the scientific organization of labor movement (NOT), and psychotechnics all grew impressively. Educational psychology was also very important in the early thirties.

The issue lying immediately behind the famous decree of the Central Committee of the Communist Party of July 4, 1936, "On Pedological Perversions in the System of the People's Commissariat of Education," seems to have been one of social class. The decree accused pedologists of attempting "to prove from the would-be 'scientific,' 'biosocial' point of view of modern pedology that the pupil's deficiency or the individual defects of his behavior are due to hereditary and social conditioning." [21] It was the perennial issue of environment versus heredity and the practical question of how an educational system can overcome the deleterious effects of both. In the thirties in the Soviet Union a great effort was being made to achieve literacy among a backward population. From the standpoint of performing this monumental educational task, what was most needed were concrete suggestions in the field of elementary pedagogy, not theoretical and inconclusive discussions of the determining elements of intelligence. The American scholar Bauer seems quite correct in his observation that much of the criticism of the educational psychologists stemmed from the fact that they appeared, at least to their critics, to be "professionally more oriented toward finding an excuse than toward the development of a cure." [22] From the standpoint of social reform, this was no inconsequential issue; similar controversies of great implication over the need to link theoretical analysis with practical reform could be found in the United States in the sixties. Academic social science frequently does become aloof to social needs, occasionally to an immoral degree. In the Soviet Union in 1936 the issue was resolved not so much by discussion from below as by political order from above.

The political atmosphere of the Soviet Union in the late thirties was grim, and the situation would be even worse immediately after World War II. The Stalinist system of control became firmly established. The great purges within the Communist Party eliminated several early defenders of innovative psychology. Soviet historians now openly admit that political controls did serious damage to many fields, including psychology. As M. G. Iaroshevskii, a Soviet historian of psychology, wrote in 1966: "The violation of Leninist principles in social life and in science hindered the development of the psychology of labor in the U.S.S.R. (the Psychotechnical Society and the journal *Soviet Psychotechnics* were abolished, and the teaching of psychotechnics in the universities ceased. . . . In 1936 almost all laboratories concerned with psychotechnics and the psychology of labor were closed.)" [23] Several pages later he continued in the same vein, "The criticism of pedology occurred in the complicated environment of the second half of the thirties and frequently was accompanied by a denial of all that was good in the work of Soviet scholars in pedology, and even in pedagogy and psychology, which had been developing in a very creative fashion." [24]

Fortunately, important work had been done in the Soviet Union before these controls were imposed. A case is the achievement of L. S. Vygotsky, who did his research in the late twenties and early thirties. To the work of this significant Soviet psychologist we must now turn.

Lev Semenovich Vygotsky (1896–1934)

L. S. Vygotsky is one of the most important influences in Soviet psychology; in recent years his ideas have begun to spread outside the Soviet Union, particularly with the publication in English in 1962 of his *Thought and Language*. His influence is particularly remarkable in view of the fact that Vygotsky died of tuberculosis in 1934 at the age of thirty-eight; he rushed to completion some of his most important writings in his final illness. One of Vygotsky's best-known pupils, A. R. Luria, is supposed to have remarked many years later, "All that is good in Russian psychology today comes from Vygotsky." [25] Luria dedicated his important monograph *Higher Cortical Functions in Man,* published in Moscow in 1962, to Vygotsky's memory, and remarked that his own work could in many ways "be looked upon as a continuation of Vygotsky's ideas." [26]

Vygotsky has not always enjoyed the esteem of official circles in the Soviet Union, however. From 1936 to 1956 his writings were in disfavor. In 1950 Vygotsky's theories on the relationship of language and thought were contradicted by Stalin himself, as will be related below. Even in recent years, when Vygotsky regained his earlier popularity, he has frequently received a mixture of praise and criticism from Soviet historians of psychology.[27] In 1966 A. V. Brushlinskii commented that Vygotsky underestimated the epistemological aspect of mental activity, but that nonetheless Soviet psychology was heavily in debt to him for his being the first to discuss in a detailed fashion the influence of sociohistorical factors on the human psyche. Vygotsky is now widely praised in the Soviet Union for this service, described as an important introduction of the Marxist approach to psychology. His works have been published and circulated widely.

In his introduction to Vygotsky's *Thought and Language* Jerome Bruner sensibly warned non-Soviet scholars against interpreting Vygotsky solely in terms of Soviet conceptions of man; Vygotsky, he said, is an original who transcends ideological divisions.[28] To identify some of the intellectual currents to which Vygotsky responded is not, however, the same as reducing him solely to those influences. In any case, there seems little question that Vygotsky was influenced by Marxist philosophy, as he interpreted it. Non-Russian readers of his works may not believe that the influence of Marxism on Vygotsky was genuine, and for a very understandable reason; when Vygotsky's works were translated from Russian to English for publication in an abridged version in the United States, most of the references to Marx, Engels, and Lenin were omitted. Lenin disappeared completely. The translators believed, evidently, that the references to Marxism were extraneous to the scientific content of Vygotsky's writings and could be dropped without damage.[29] As a result it is almost impossible for the historian of psychology without knowledge of the Russian language to understand the initial assumptions of Vygotsky's approach. In the original Russian, however, it is clear that Vygotsky attempted to show a relationship between his views on children's thought and Lenin's epistemology. He spoke of the "unity and struggle of the opposites of thought and fantasy" in cognition.[30] He was, as we will see, critical of the epistemological dualism that he saw in Jean Piaget's theories of language, and in particular, Piaget's description of a child's "autistic" use of language. Vygotsky emphasized that a Marxist approach to language revealed its "external" or "social" origins.

One of the main problems to which Vygotsky addressed himself was

the interrelation of thought and speech. His work on this topic has frequently been compared to that of Piaget. Vygotsky praised Piaget's
work, calling it "revolutionary," but commented that it "suffers from
the duality common to all trailblazing contemporary works in psychology. This cleavage is a concomitant of the crisis that psychology is
undergoing as it develops into a science in the true sense of the word.
The crisis stems from the sharp contradiction between the factual material of science and its methodological and theoretical premises, which
have long been a subject of dispute between materialistic and idealistic
conceptions." [31]

Piaget in his early work postulated three stages in the development of
the modes of thought of a child: first, autism, second, egocentrism, and
last, socialized thought. In the first, or autistic, stage the child's thought
is subconscious and is directed toward self-gratification. The child does
not yet use language and has not yet adjusted to the existence of other
persons, with their desires and needs. He is not susceptible to the concept of truth and error. In the last stage, socialized thought, the child
has adapted to reality, tries to influence it, and can be communicated
with through language. He has recognized laws of experience and of
logic. In the intermediate stage, egocentrism, the child "stands midway
between autism in the strict sense of the word and socialized thought." [32]
He uses language, but only to himself; he is thinking aloud. Thus, all
three stages constitute a scheme of the development of the thought of a
child based on the assumption that "child thought is originally and naturally autistic and changes to realistic thought only under long and sus-
tained social pressure." [33]

Vygotsky accepted much of this description by Piaget of the individual stages of child development, but he rejected the direction of flow of
the underlying genetic sequence. As Vygotsky described it:

The development of thought is, to Piaget, a story of the gradual socialization
of deeply intimate, personal, autistic mental states. Even social speech is
represented as following, not preceding, egocentric speech.

The hypothesis we propose reverses this course. . . . We consider that
the total development runs as follows: The primary function of speech, in
both children and adults, is communication, social contact. The earliest
speech of the child is therefore essentially social. At first it is global and multi-
functional; later its functions become differentiated. At a certain age the social
speech of the child is quite sharply divided into egocentric and communicative

speech. (We prefer to use the term *communicative* for the form of speech that Piaget calls *socialized* as though it had been something else before becoming social. From our point of view, the two forms, communicative and egocentric, are both social, though their functions differ.) Egocentric speech emerges when the child transfers social collaborative forms of behavior to the sphere of inner-personal psychic functions. . . . In our conception, the true direction of the development of thinking is not from the individual to the socialized, but from the social to the individual.[34]

And thus Vygotsky arrived at the concept for which he is best known, the "internalization of speech":

Piaget believes that egocentric speech stems from the insufficient socialization of speech and that its only development is decrease and eventual death. Its culmination lies in the past. Inner speech is something new brought in from the outside along with socialization. We believe that egocentric speech stems from the insufficient individualization of primary social speech. Its culmination lies in the future. It develops into inner speech.[35]

Since Vygotsky believed that egocentric, and ultimately, inner speech stemmed from primary social speech, occurring through a process of internalization, it was necessary for him to explain the source of the mental states in the earliest stage, the autistic stage of Piaget. What about this child who has not yet "internalized" any part of primary social speech, who has not yet learned to speak at all? Can he think? It becomes obvious that if Vygotsky were to grant that such a child thinks, then he must find quite different roots for thought and speech. And this he did. According to Vygotsky, thought and speech have different genetic roots and develop according to different growth curves that "cross and recross," but "always diverge again." There is a "prelinguistic phase in the development of thought and a preintellectual phase in the development of speech." [36] A crucial moment, explored by William Stern, occurs when the curves of development of thought and speech meet for the first time; from this time forward "speech begins to serve intellect, and thoughts begin to be spoken." [37] Vygotsky believed that Stern exaggerated the role of the intellect as a "first cause of meaningful speech," but he did agree that "Stern's basic observation was correct, that there is indeed a moment of discovery" when the child sees the link between word and object. From this point on, thought becomes verbal and speech rational.[38]

The source of prelinguistic thought is, thus, separate from the source of speech. Prelinguistic thought has a source that is similar to the embryonic thought of some species of animals, while speech always has a social origin. Vygotsky saw a clear tie here with Marxist analysis:

The thesis that the roots of human intellect reach down into the animal realm has long been admitted by Marxism; we find its elaboration in Plekhanov. Engels wrote that man and animals have all forms of intellectual activity in common; only the developmental level differs: Animals are able to reason on an elementary level, to analyze (cracking a nut is a beginning of analysis), to experiment when confronted with problems or caught in a difficult situation. . . . It goes without saying that Engels does not credit animals with the ability to think and to speak on the human level. . . .[39]

At this point in Vygotsky's analysis, his critic is likely to chastise him for reductionism, for drawing too crude a similarity between man and animals. But Vygotsky felt that the answer to such criticism lay in emphasizing the qualitatively new characteristics that emerged after the lines of thought and speech crossed, after the child's great discovery referred to by Stern had occurred. According to Vygotsky, the stage that followed this intersection was not a simple continuation of the earlier:

The nature of the development itself changes, from biological to sociohistorical. Verbal thought is not an innate natural form of behavior but is determined by a historical-cultural process and has specific properties and laws that cannot be found in the natural forms of thought and speech. Once we acknowledge the historical character of verbal thought, we must consider it subject to all the premises of historical materialism, which are valid for any historical phenomenon in human society. It is only to be expected that on this level the development of behavior will be governed essentially by the general laws of the historical development of human society.[40]

Thus, Vygotsky developed for the explanation of the interrelation of thought and language a scheme that contained a high degree of inner consistency and arrived eventually at Marxist conceptions of social development. Thought and language have different roots—thought in its prelinguistic stage being tied to the biological development of man, language in its prerational stage being tied to the social milieu of the child. But these two categories become dialectically involved once the link between them occurs, when the child perceives that every object has a name; from this point onward, one cannot speak of the separateness of thought and language. The internalization of language causes thoughts to be expressed in inner speech; the effect of logic on speech results in coherence and order in oral communication.

Several aspects of Vygotsky's scheme remained unclear. For example, he drew a parallel between the prelinguistic thought of a child and the mental activity of animals, such as chimpanzees. Yet Vygotsky of course granted that, physiologically, there are genuine differences between the brain of a child and that of a chimpanzee. Nonetheless, to what extent those differences result in a qualitatively different sort of prelinguistic thought in the child was not clear in his writings. Within his conception the sociohistorical influences conveyed in language surpassed the biological superiorities of the human brain in accounting for the distinctions between man and animal. As a materialist and monist, Vygotsky agreed that the very sociohistorical factors that he emphasized also had their material causal sources, back in the biological development of man. He quoted Engels's descriptions of the influence of the use of tools upon the development of man. In the final analysis, then, the different roots of thought and language were only relative, not absolute. In the life of the individual human, however, the roots were distinct, and it was here that Vygotsky put his emphasis.

Vygotsky's opinion that language and thought have different roots and that "prelinguistic thought" exists in the early life of a child directly conflicted with Stalin's teachings on linguistics. Stalin wrote in *Marxism and Linguistics:*

It is said that thoughts arise without language material, without the language shell, in, so to speak, a naked form. But this is absolutely wrong. Whatever the thoughts that may arise in the mind of man, they can arise and exist only on the basis of language terminology and phrases. Bare thoughts, free of the language material, free of the "natural matter" of language—do not exist. . . . Only idealists can speak of thinking as not connected with the "natural matter" of language, of thinking without language.[41]

A clearer contradiction by highest authority can hardly be imagined, if one remembers not only Vygotsky's identification of separate sources of thought and language, but also his assertion that "there is no clear-cut and constant correlation between them." [42] Consequently, the rebirth of interest in Vygotsky's writings occurred only after Stalin's death. The ruler of the state had dictated an interpretation of Marxism that the Marxist scientists and intellectuals of the country failed to perceive.

Piaget did not become familiar in detail with Vygotsky's criticisms of his early work until 1961 or 1962, when the abridged translation of *Thought and Language* was made available to him. At that time, twenty-eight years after Vygotsky's death, he published his *Comments on Vygotsky's Critical Remarks*.[43] A discussion of this courteous and interesting reply—in which Piaget granted many essential points to Vygotsky, but retained a number for himself—unfortunately can not be included here.

At this point I would like to shift attention to the postwar period and, particularly, to the person of S. L. Rubinshtein, the Soviet scholar who in the last thirty years has exercised the greatest influence in questions concerning philosophical interpretation of psychology and physiology. First, however, it is necessary to describe the ideological pressures upon Soviet psychologists and physiologists in the immediate postwar period. The physiology session of 1950 was one of the most important events of this sorry epoch in Soviet scholarship. From an intellectual standpoint the 1950 conference is much less interesting than that of 1962, when de-Stalinization had revived Soviet physiology and psychology, and consequently, the later session will receive more attention in this chapter. For those persons more interested in the 1950 meeting and its immediate results, an English-language version of the proceedings is available, as well as several other accounts.[44]

In the period immediately following the war Soviet scientists in many fields initially hoped for a relatively relaxed ideological atmosphere. We have already seen something of the efforts made by several scientists in physics and genetics to return to the more tolerant conditions of the 1920's. This effort failed; immediately after A. A. Zhdanov's death in 1948 controls in several scientific fields, including physics, genetics, cosmology, structural chemistry, and physiology, were tightened. The causes of this ideological campaign are very difficult to identify; the personal characteristics of Stalin seemed to be the most important factor, although the strained international situation and the availability of levers of control in Soviet society were also important conditions permitting Stalin to exercise extraordinary influence on science and scholarship. During the years 1948–52 conferences on science and ideology were held in a number of different fields at which political pressure was exerted on scientists; the "Pavlov" session on physiology and psychology occurred June 28–July 4, 1950, and was sponsored jointly by the Academy of Sciences of the U.S.S.R. and the Academy of Medical Sciences of the U.S.S.R. Unfortunately, the English translation of the speeches given at this conference, published in 1951 in Moscow as "Scientific Session on the Physiological Teachings of Academician Pavlov," does not contain the speeches by P. K. Anokhin, I. S. Beritov, L. A. Orbeli, and others. These speeches were critical of the official position.

In the inaugural address Sergei Vavilov, president of the Academy of Sciences and brother of the deceased geneticist Nikolai Vavilov, indicated that the function of the congress was to return to established Pavlovian teachings; he thus implied what was already known, that there would be no genuine effort at the congress to seek new understandings of the difficult problems of physiology and psychology on a materialist basis. Coming almost two years after the genetics conference where Lysenko and his Michurinist school were officially established as the representatives of the only correct approach to genetics, the physiologists and psychologists were quite aware that the outcome of their conference was predetermined. Vavilov gave the official diagnosis of the state of Soviet physiology and psychology when he commented in his opening statement:

There have been attempts—not too frequent, happily—at an erroneous and unwarranted revision of Pavlov's views. But, more frequently, the ideas and work of researches have not kept to the highroad, but wandered into byways and fieldpaths. Strange and surprising though it may seem, the broad Pavlov road has become little frequented, comparatively few have followed it consistently and systematically. Not all our physiologists have been able, or have always been able, to measure up to Pavlov's straightforward materialism. . . . [T]he time has come to sound the alarm. . . . Our people and progressive humanity generally, will not forgive us if we do not put the wealth of Pavlov's legacy to proper use. . . . There can be no doubt that it is only by a return to Pavlov's road that physiology can be most effective, most beneficial to our people and most worthy of the Stalin epoch of the building of Communism.[45]

The Soviet physiologists who came under the heaviest criticism at the conference were P. K. Anokhin, L. A. Orbeli, and I. S. Beritov. I. P. Razenkov, vice president of the Academy of Medical Sciences, charged that Anokhin, one of the Soviet Union's most distinguished physiologists, "has been guilty of many a serious deviation from Pavlov's teachings, has had an infatuation for the fashionable, reactionary theories of Coghill, Weiss and other foreign authors. . . ." [46] The attribution of pejorative meaning to the term "foreign authors" was typical of the chauvinistic temper of these Stalinist years. Anokhin's thoroughly materialist and scholarly approach to physiology will be considered in a separate section of this chapter. After retreating somewhat from his earlier modifications of Pavlov's teachings, which he saw as somewhat obsolescent although fundamentally important in their time, Anokhin returned in the late fifties and early sixties to the views for which he was criticized in 1950. To the extent that dialectical materialism had an intellectually interesting relationship to Anokhin's work, that relationship must be sought in his writings before 1950 and after 1956. The intervening period was a dark one, not only for Soviet physiology and psychology, but for much of Soviet scholarship.

In a history of Soviet psychology published in Moscow in 1967, a work described by an American psychologist as "pioneering" despite its faults,[47] A. V. Petrovskii told of the "dogmatism" in Soviet psychology following the 1950 session.[48] Petrovskii observed that in the early 1950's there was a strong tendency toward the "liquidation" of psychology entirely, replacing it with Pavlovian physiology. This "nihilistic attitude" toward psychology, Petrovskii continued, was reminiscent of the reflexological currents of the early twenties, when the legitimacy of psychology had also been doubted, but:

If in the twenties the negative attitude of the reflexologists and behaviorists toward psychology could be largely explained—though not justified—by the objective need to criticize the vestiges of subjectivism in psychology, by the beginning of the 1950's the idea of "liquidating" psychology could not be based on any principled considerations whatsoever.[49]

Thus, the current major Soviet historian of Soviet psychology condemned Stalinism in his field in strong terms. To be sure, he dodged the issue of the extent to which the events of the fifties were the responsibility, not only of Stalin, but of the system that permitted him to exercise such power.

Despite the political and ideological pressures of these years Soviet physiology and psychology continued to live and to develop. With better conditions after Stalin's death these fields moved forward once again. One of the best illustrations of the survival ability of Soviet scholars and of their continuing intellectual vitality in the face of great obstacles is found in the person of S. L. Rubinshtein.

Sergei Leonidovich Rubinshtein (1889–1960)

One of the lifelong goals of Sergei Leonidovich Rubinshtein was to give a theoretical analysis of the nature of consciousness and thought on the basis of dialectical materialism. His attempt in this direction was obvious in his writings over a period of many years, from his 1934 article entitled "Problems of Psychology in the Works of Karl Marx" to his 1959 book on the principles of psychology, in which he commented that his interpretation was heavily influenced by the "dialectical materialist understanding of the determination of psychic [mental] phenomena." [50]

Rubinshtein was important in the formation of contemporary Soviet attitudes toward psychology. In 1942 he founded the department of psychology at Moscow University. Around him in the psychology section of the Institute of Philosophy of the Academy of Sciences of the U.S.S.R., which he headed from 1945 to 1960, there grew up a whole school of investigators of the relationship of the psychological to the physiological within the theoretical framework of Marxism. His advanced textbook *Foundations of General Psychology,* published in several editions, was the most authoritative voice in the field for Soviet graduate students. The first edition, published in 1940, received a Stalin Prize. His later work, *Being and Consciousness,* is regarded at the present time as a "deeply creative Marxist work" and has been published in many countries and languages, including Chinese; in 1959 it received a Lenin Prize.

From all these official honors one might think that Rubinshtein was an ideological hack, a mere apologist for Marxism. He was not. Possessing a broad-ranging and subtle mind, he produced even in his relatively elementary 1934 article "what is regarded as the first adequate Marxist theory of motivation and ability." [51] During the worst period for scholarship in the Soviet Union, the Stalinist years immediately after World War II, Rubinshtein came under heavy criticism for his "objective, non-

Party" approach to scholarship and for certain of his theoretical formulations. He bent under the pressure, but he did not break. Though he sharpened the point of his ideological pen, he was still the sort of person who, at the 1947 discussion of the ideological failings of Aleksandrov's *History of Western European Philosophy,* would make a plea for the study of formal logic.[52] This was at a time when formal logic was being displaced by dialectical logic, a campaign with much political support from Party followers. In the 1960's, Rubinshtein emerged again as the most prominent theoretical voice on the knotty problems of the nature of consciousness.[53] He published three books on the topic in 1957, 1958, and 1959 respectively. Upon his death in 1960 the editors of *Problems of Philosophy* honored him with a necrology, which observed that his work would long continue to be of value to psychology.[54]

Although certain themes—for example, the definition of "consciousness"—run through almost all of Rubinshtein's works, there was something of an evolution in his views, a slight but perceptible change that does not seem entirely explainable as a result of political pressure. He was a psychologist, not a physiologist, and his first works are deeply psychological in tone. As time went on, however, he moved more and more toward physiology, maintaining that only with a recognition of the material basis of mental activity could one proceed to an analysis of its most difficult problems. This shift from psychology to physiology is interesting when compared to the contrasting case of Anokhin, a dialectical materialist physiologist, but who, as time went on, became more and more aware of the importance of psychology. In Rubinshtein's case, political influences are undoubtedly relevant, since in the fifties he was criticized in a major Party journal for ignoring Pavlovian physiology; in the later revisions of his work he took this criticism into account. But his last works, published in the more relaxed late fifties, continued to stress the importance of physiology in such a way as to suggest that he had sincerely modified his own views, not just bowed to pressure. In the late fifties he developed several new formulations of the psychophysical problem, which he considered more successful than his earlier attempts and which his contemporaries have also judged more successful.

In the 1946 edition of *Foundations of Psychology* Rubinshtein began with an emphasis on the significance of the *subjective* to psychology. Here he was still defending the hard-won ground of the late twenties and thirties, when psychology's very right to exist and the legitimacy of the term "subjective" had been questioned. Rubinshtein noted that there are no such things as "sensations" or "perceptions" alone, but only a *person's* sensations and perceptions. This was to him, in 1946, the very ground on which psychology begins, its first principle: "The characteristic of 'belonging' to an individual, to a subject, is the first characteristic trait of all that is psychic. . . . My experiences are presented to me differently, in another perspective, as it were, than to any other person. The experiences, the thoughts, the feelings of a subject are *his* thoughts, *his* feelings; these are *his* experiences, a piece of his very life, his flesh and blood." [55] The difference between this 1946 view and his later writings is more one of emphasis than contradiction, since in the earlier writing Rubinshtein recognized, in addition, the psyche as a "reflection of objective reality, existing outside and independently from it." [56] Nonetheless, what one says first is important in judging his intellectual priorities; Rubinshtein in the forties spoke first of the individual; in the fifties, he spoke first of objective reality. In the later years his effort to defend psychology as a discipline was replaced more and more in his priorities by an effort to give a theoretical definition of the psyche, one that was materialistic and monistic and yet assumed a permanent place for psychology as well as for physiology. It was a delicate, subtle, and perhaps inherently inconsistent operation.

In 1946 Rubinshtein's theoretical terminology for the relationship of the psychic to the physical was that of a "dual correlation" (a formulation he would later modify). As he observed:

Every psychic phenomenon is, on the one hand, a product, a dependent component, of the organic life of the individual, and on the other hand, a reflection of the external world surrounding it. . . . These two aspects, always produced in the consciousness of man in unity and mutual interpenetration, appear here as experience and knowledge. The element of knowledge in consciousness especially emphasizes the connection to the outside world, which is reflected in the psyche. Experience is, first of all, a psychic fact as a piece of the very life of the individual, a specific manifestation of his individual life.[57]

Rubinshtein attempted here to distinguish between the objective and the subjective, to tie the objective world to "knowledge" and the subjective world to "experience." Experiences were to him the "subjective aspect of the life-course of a personality." [58] Experiences would be the particular realm of study of psychology.

Rubinshtein already realized that his distinction here could not be maintained in a rigorous fashion. Every "experience" was also connected to the objective world, but differed from "knowledge" in the degree to which the subjective exercised influence: "An experience is an experience of something and, it follows, knowledge of something. It is called an experience not because the second aspect—knowledge—is absent, but because the vital, the personal, aspect in it is predominant." [59] Therefore, "experience" and "knowledge" are not pure categories, but are based on the degree to which either the subjective or the objective is governing. Every experience will contain an aspect of objective knowledge, and every new piece of knowledge gained by an individual will contain a bit of subjective experience. But Rubinshtein felt that the distinction was still a valid one. In experience, subjectivity was predominant; in knowledge, objectivity governed. This distinction led him to a definition of "consciousness" as a unity of experience and knowledge, a unity containing both subjective and objective elements.[60]

This analysis has already brought us to the heart of the psychophysical problem and opens up almost endless possibilities for further comment. No problem is more difficult to solve within the terms of either materialism or idealism. But before pausing for these considerations, I would like to proceed to Rubinshtein's criticisms in 1946 of both "psychophysical parallelism" and "psychophysical identity," and to his reformulation of the problem in the later 1950's.

So far, Rubinshtein's analysis reminds one of psychophysical parallelism, with "knowledge" being the series of physical or real events and "experience" being the series of psychological events. But he firmly rejected such parallelism:

The theory of psychophysical parallelism is based on two errors. It is mistaken to oppose dualistically psychic and physical phenomena as two series of events entirely separate from each other; it is also mistaken to propose that between these two series there is a simple correspondence in the spirit of the old localization theories, according to which there is a sharp correspondence of the psychic to the nervous cell. The unsoundness of this view . . . is illustrated by all contemporary evidence from experimental and clinical research.[61]

Not only did Rubinshtein reject the theory of an unconnected parallel series of psychic and physiological events, but he also disavowed what he called the theory of mutual interaction. In the latter interpretation, psychic and physiological events are seen as a parallel series that *do* influence each other. Support for this interpretation may be found in the fact that physiological changes in an organism often are connected with changes in psychic behavior, or in the opposite phenomenon in which psychic processes, such as strong emotions, often have physiological effects. But Rubinshtein observed that the theory of mutual interaction was based on the same erroneous assumption as the theory of psychophysical parallelism—namely, the dualistic belief that the psychic and the physical are two entirely distinct kinds of phenomena. It then follows that these two sets of phenomena either influence each other (mutual interaction) or remain entirely separate (parallelism). Rubinshtein concluded:

The struggle of the supporters of the theory of mutual interaction against the transformation of man's consciousness into an "epiphenomenon" deprived of all real significance . . . in principle may be completely justified, but they then proceed on the basis of the dualistic premises of the theory of interaction to the completely unjustified notion that psychic forces act externally, as it were, on the flow of physiological processes.[62]

It was the identification of specifically "psychic" forces, entirely outside the realm of objective reality, that caused Rubinshtein to continue looking further for a solution of the psychophysical problem.

There would seem to be a logical exit left for Rubinshtein, the "theory of identity"—that is, the reduction of either the psychic to the physical or the reverse, the reduction of the physical to the psychic. The latter was clearly unacceptable. It led directly, he said, to phenomenalism or even open spiritualism. Starting out with an assumption of the existence of an objective, physical, and material reality, Rubinshtein could hardly end up with a reduction of the physical world to psychic events.

Rubinshtein's theoretical exits seemed to be closing down rapidly. Within the theory of identity one path remained, the reduction of the psychic to the physical. It is the solution that many people outside the Soviet Union assume has been adopted there. It fits well with the popular conception of Pavlovianism. Yet Rubinshtein found it no more palatable. Its goal of explaining the psychic in terms of physics and chemistry was to him an unrealistic and impossible one. It was an example of vulgar, reductionist materialism, a failure to recognize the qualitative differences of matter on different levels. Further, it seemed to leave little or no room for psychology, a discipline that had struggled for its existence in the Soviet Union and that Rubinshtein firmly supported.

What was left? Rubinshtein advanced what he called the principle of psychophysical unity:

The principle of psychophysical unity is the basic principle of Soviet psychology. Within this unity the materialistic bases of the psyche are determining, but the psyche retains its qualitative specificity; it is not reduced to the physical properties of matter and is not converted into an ineffective epiphenomenon.[63]

Rubinshtein attempted, therefore, to work out a position in which there is between the psychic and the physical a unity that allows each to retain its specific characteristics. Consciousness is neither one nor the other, but both. This unity is one of contradictions, resulting in a sort of complementarity between the psychic and physical properties of consciousness, a complementarity that parallels the wavelike and corpusclelike properties of light particles. (This analogy was not used until later, however, since complementarity was having its own troubles in Soviet physics.)

Rubinshtein's 1946 formulation was not successful, and reading his writings of that time, one thinks that he realized it. It papered over apparently irreconcilable differences. When physics attempted to escape the dilemma of quantum mechanics by simultaneously attributing wavelike and corpusclelike properties to light, it was not quite destroying itself in the process (though it might have seemed that way). Materialists could (and did) adjust to this strange concept of physics by speaking of relativistic "matter-energy" instead of matter alone, and observing that both waves *and* particles, and all combinations thereof, would be matter-energy. But a materialist could hardly say that consciousness is both "psychic" and "physical" and leave it there. He needed an equivalence principle here, too, but he did not have one conveniently at hand. If "psychic" is a category, either it must in some way be equated with matter-energy, or one must abandon the view that only matter-energy exists, thereby destroying a fundamental assumption of dialectical materialism. Rubinshtein knew that the only solution lay in linking the psychic to some form of matter—hence his statement above that "the material bases of the psyche are determining"—but he was extremely vague on this linkage. He recognized his solution as "unfinished," and called for further attacks on this "difficult assignment." [64]

Although Rubinshtein's 1946 position was vulnerable, the criticisms that were made of him by V. Kolbanovskii in the Party journal *Bol'shevik* in September 1947 were lightweight intellectually. Most of the criticism was based on Rubinshtein's alleged insufficient political militancy, his failure to criticize adequately psychological theories advanced in western Europe and North America, and his lack of Party spirit. [65]

One of the theoretical criticisms advanced against Rubinshtein stuck, however, and in 1952 he revised his position, abandoning in the process his principle of psychophysical unity, which supposedly described the "dual correlation" of the psychical and the physical. [66] As he wrote:

Materialistic monism means not a unity of two sources—the psychical and the physical—but the presence of a single source, a material source in relation to which the psychical is a product. [67]

This was the beginning of Rubinshtein's revision of his theoretical position, the final result of which was his 1957 monograph, *Existence and Consciousness: Concerning the Place of the Psychical in the Universal Inter-Connections of the Material World*. In the latter work he arrived at a stronger materialistic formulation of the psychophysical problem. Thus, Rubinshtein's revisions seemed to have been more a result of his own awareness of the inadequacy of his earlier position than of the superficial criticism he received during Stalin's last years.

Existence and Consciousness was a book in which Rubinshtein attempted a more systematic and complete analysis of psychic activity than in his earlier text. In order to understand this analysis, it is necessary first to see some of the assumptions on which it was based and then proceed to its details, including the extensions of materialism that it contained.

Rubinshtein's approach was based on a rejection of the "classic" argument for cognitive idealism. As he wrote:

The basic argument of idealism is the following: In the process of cognition there is no way for us to "jump out" of our sensations, perceptions, and thoughts; this means that we can not attain the sphere of real things; therefore, we are obligated to recognize that the very sensations and perceptions themselves are the only possible objects of cognition. At the basis of this classical argument of idealism lies the thought that in order to attain the sphere of real things, it is necessary to "leap out" of the sphere of sensations, perceptions, and thoughts—and that, of course, is impossible.

This line of argument assumes what it is trying to prove. It assumes that sensations and perceptions are only subjective constructs, external to things themselves, to objective reality. But actually objects participate in the very origin of sensations; sensations, arising as a result of the influence of objects on the sense organs, on the brain, are connected with objects in their very origin.[68]

Rubinshtein was correct in stating that this standard argument for idealism assumes what it is trying to prove—that is, it assumes that sensations and perceptions are something other than material reality and therefore must be escaped from in order to approach reality. Since that escape cannot be accomplished, the argument goes, one must accept sensations and perceptions as objects of cognition themselves, and can define them as ideal forms if one wishes. But what Rubinshtein did not explicitly say (although it was implicit in his argument) was that *his* argument also assumed what it was trying to prove. The person who believes that sensations and perceptions are meaningful only as forms of material reality has also made an unprovable assumption. He has as much right to his assumption as the idealist does to his, but he can not justifiably maintain that he has "proved" his case while the idealist has merely assumed what he pretended to prove. All of this merely restates the writer's earlier opinion that the option between materialism and idealism is a matter of philosophic choice, not a matter of logical proof. As the parallel postulate in geometry is the starting point from which several geometries can be constructed, depending on the assumptions made, the mind-body problem is the point from which several philosophies can be built, depending on the assumptions made. The genuinely difficult problem in the case of epistemology is not what can be proved and what can not, but the dilemma presented by the occasional grounds for choice among philosophic assumptions, recognized as such. If science had to wait at every point for rigorous proofs, it would not proceed far. The best form of materialism could be constructed on a few principles openly recognized as unprovable assumptions for which there are, nonetheless, persuasive arguments. Rubinshtein never stated that he was proceeding on such a basis, but what he did say was perfectly reconcilable with such a position.

To Rubinshtein, then, sensations and perceptions do provide an entree into the real material world. He described an epistemology of interaction, of praxis. In 1946, he had rejected the theory of mutual interaction on the basis that it assumed separate interacting series of psychic and physical events. He still, in 1957, rejected such a theory; his new epistemology of interaction was based on the premise of the inter-action of an internal material brain with reflections of external material objects. Two totally separate series of events were not interacting, since both series were based on matter. Thus, what he earlier called experi-ence, or subjective, was to be unpeeled in layers, like an onion, and revealed as also based on objective reality, on matter.

The general philosophic framework from which he approached the problem was one in which the universe is an interconnected material whole. It is an age-old concept, one with similarities to many older sys-tems. As Rubinshtein described this universe:

All phenomena in the world are interconnected. Every action is an interac-tion; every change of one entity is reflected in all the others and is itself an answer to the change of still other phenomena acting upon it. Every external influence is refracted by the internal properties of that body, of that phe-nomenon to which it is subjected. . . . It was not for nothing that Lenin wrote: ". . . it is logical to propose that all matter possesses a property essentially similar to sensation, to the property of reflection. . . ."

This property of reflection is expressed in the fact that every thing is affected by those external influences to which it is subjected. External influ-ences condition even the very internal nature of phenomena and are, so to speak, laid up in it, preserved in it. On the strength of this, all incident influ-ences, all influencing objects, are "represented" or reflected in all other ob-jects. Each phenomenon is in a certain degree "a mirror and echo of the universe." At the same time, the result of this or that influence on any entity is conditioned by the very nature of the latter; the internal nature of phe-nomena is that "prism" through which single objects and phenomena are reflected in others.

This expresses the fundamental property of existence. On this conception is based the dialectical materialist understanding of the determination of phenomena in their interaction and interdependence.[69]

In this interesting and ambitious passage Rubinshtein based himself on concepts already existing within Soviet dialectical materialism, but he presented them in a more complete and speculative form than usually found. The statement that every phenomenon is in a certain degree "a mirror and echo of the universe" derives directly from Marx; commenting on the physiological function of eyes and ears, Marx commented that "these are the organs that tear man away from his individuality, converting him into a mirror and an echo of the universe." [70] However, whether Marx would have extended this limited statement concerning man's sense organs to the broader generalizations of Rubinshtein is by no means clear. The concept of "reflection" that runs through the passage is, of course, derived from Lenin, as Rubinshtein quoted to indicate. Here again, a small comment was expanded into broader meaning. Lastly, a principle of determinism, of universal causation, is also obvious in the passage.

This formulation of causation in an interconnected world became important to Soviet psychology, however. To Soviet theorists as well as to others it has a certain speculative persuasiveness. Rubinshtein's "prism," the internal state of an object through which external influences are refracted, was frequently cited. The editors of *Problems of Philosophy* commented in 1960: "The position defended in *Existence and Consciousness,* in which external causes act through internal conditions, has an essential meaning for the whole system of scientific knowledge." [71]

In the section of his book immediately following, Rubinshtein addressed himself to the problem of the prism more directly. Here he tried to assess the relative weight of "internal factors" and "external factors" in the process of reflection, that property inherent in all matter. The higher the level of the evolution of matter, the more weight the internal factors have: "The 'higher' we rise—from living organisms to man—the more complicated is the internal constitution of phenomena and the greater is the share of the internal conditions compared with the external." [72] Consciousness in man is that form of material reflection in which internal factors play a greater role than in any other form of reflection. Rubinshtein's position on the nature of psychic activity now unfolded. Psychic activity, he said, is both an activity of the brain and a reflection of the external world. Therefore, psychic activity has two different aspects—the ontological and the epistemological. [73] The ontological aspect of the brain is its existence as a nervous system, a material object of great complexity currently being studied by physiologists. The epistemological aspect of psychic activity derives from the cognitive re-lationship of psychic phenomena to objective reality. Rubinshtein be-lieved the distinction between these two aspects to be relative rather than absolute. While the epistemological aspect of psychic activity is dominated by connections with the outside world and the ontological aspect is primarily determined from within, it should always be remem-bered that the brain itself is also, in the end, a result of the influence of the external world. Thus, there is a difference of causal time scales here. The brain is a product of the external environment acting over the en-tire period of natural history. It is a material brain of great organiza-tional complexity formed by natural selection from matter of simpler organization. But ontologically, it exists as a completed entity at any given point in time that cognition takes place, and its internal constitu-tion "refracts" the reflection of external reality, which is also material. Thus, the interaction that occurs is material in origin on both sides, and both sides are products of objective reality, but being formed at differ-ent times, in different places, and in different ways, they interact.

The ontological statement that psychic activity is a function of the brain must never be taken, Rubinshtein warned, to mean that it is a function determined entirely from within, by its cell structure. If such a mistake is made and psychic activity is looked upon as entirely a func-tion of the brain or of the sense organs, then this function will inevitably come to be considered an expression of the states of these organs and lose in that fashion its links to the outside world. This result, said Ru-binshtein, would be physiological idealism: "The brain is only an organ of psychic activity, not its source." [74]

The epistemological aspect of psychic activity, as advanced by Ru-binshtein, was a form of Leninist reflection theory:

The theory of reflection can be roughly described as follows: One senses and perceives not sensations and perceptions, but things and phenomena of the material world. By means of sensations and perceptions one cognizes things themselves, but sensations and perceptions are not these things but only their forms; sensations and perceptions cannot be directly posed in place of things. One can not speak—as frequently is done—about sensations or perceptions as ideal things existing apart from all material reality in the ideal world of consciousness, similar to the way things exist in the material world. The sen-sations and perceptions are forms of the object. Their epistemological content does not exist independently of the object. In this way the dialectical materialist theory of reflection firmly excludes a subjective understanding of the psyche.[75]

An element of Rubinshtein's arguments that now emerges in a crucial way is the nature of sensations and perceptions, which he called "forms of objects existing in the material world." We know from his previous argument that these forms are influenced both by the brain ("refracted by internal conditions") and by objective reality ("reflected within the brain"). We also know that in the case of the brain, the "highest" form of matter, the internal factors exercise more influence than in any other example of reflection we can point to in the universe. Does this mean that the internal factors exercise *more* influence than the external ones? How can we ever know anything certain about the external world? How do Rubinshtein's forms differ from Plekhanov's "hieroglyphs," criticized by Lenin, and do they differ from the forms of *Bildtheorie* psychology?

It is clear already that Rubinshtein did not agree with the presentational epistemology favored by some simple Marxists (and the interpretation that many people, Marxist and non-Marxist, have put upon Lenin's theory of reflection). Sensations and perceptions are not direct "pictures" of external objects, but "forms" of them that are refracted—that is, modified—in the process of transmission. But Rubinshtein also said that he was opposed to representational epistemology, to *Bildtheorie,* in which sensations are considered to be not faithful pictures of reality, but merely representations of it. This position is interesting, since Rubinshtein's "refraction" or "reflection" does not seem far, on the face of it, from representational epistemology. The root of his opposition to representational epistemology seems to have been his belief that it would not allow man to get to the "essence" or "truth" of things, or even to know that he was making approaches to it; at most it would give him some obscure "indicators" of reality.[76] Rubinshtein wanted more. He thought that not only must one recognize that the external objects influence the mental reflection of reality, but that this influence is the most important factor affecting that reflection. As he commented, "In the process of perception the form depends first and foremost on the [external] object, the form of which it is. But the form is not a dead imprint of the object." [77]

Rubinshtein even rejected the "two-factor theory," a thoroughgoing praxis, saying that dialectical materialism

dismisses at its very base the theory of two factors, according to which reflection is determined on the one hand by the subject and on the other by the object. Reflection—psychic activity and its product—is not determined by the object by itself, nor by the subject by itself, nor by one plus the other. Psychic activity is defined by the reflected object.[78]

When Rubinshtein said that psychic activity was not determined by the "subject plus the object," he apparently meant that subjective and objective factors were not of entirely unknown or equal relative influence. Thus, he assigned weights to the internal and external factors of the cognitive process, which resulted in the following delicate and speculative balance of influences: Internal factors are very important in the formation of perceptual forms of reality, more important than in any other case of the universal phenomenon of reflection, but they are still less important than the external factors, the links to objective reality existing outside the brain. Thus, man's knowledge is still in an important sense a faithful reproduction of external reality. Man confirms the truth about that objective reality in practice, when his formulations are either proved or disproved by actual results.

The causal sequence involving consciousness is not, Rubinshtein believed, from consciousness to external reality, but from external reality to consciousness: Therefore the question, How do perceptions make the transition from forms to things? is an incorrectly posed question. "Man does not exist because he thinks, as Descartes put it; he thinks because he exists." [79]

In his assignment of relative weights to internal and external factors Rubinshtein had added another unprovable though fruitful assumption to his system. A strict materialist could accept an explanation of the process of cognition in which the internal factors played *more* of a role than the external ones without contradicting himself, so long as he added to this explanation the belief that the internal factors were also material and had, in their turn, been caused by external influences during the process of evolution. The verification of truth through practice could still play the same role as in Rubinshtein's scheme. Rubinshtein, as we have seen, accepted this evolutionary understanding of the brain.

His addition of a weighting scheme was gratuitous, but reassuring within a tradition that preferred an epistemology in which there occurred as faithful a transmission of information from objective reality to consciousness as possible.

Rubinshtein still believed that the term "subjective" is a legitimate one. "Subjective" was to him a term used to indicate that every aspect of psychic activity displays characteristics unique to the person concerned. Every sensation, every thought, was subjective in this sense. The value of the word "subjective" was not destroyed, in Rubinshtein's opinion, by the fact that these subjective qualities of the individual had, in turn, their objective origins. To deny these origins would be to absolutize subjectivity, to fall prey to a thoroughgoing "subjectivism": "The major way to overcome subjectivism is not in its denial, but in a correct understanding of the subjective as a form of manifestation of the objective." [80] At any one point in time, every person is influenced by both subjective and objective factors, although the human race over the period of its whole history has been influenced only by objective ones. Out of this evolutionary process is created two causal chains that interact with each other. The product of the interaction is consciousness. Psychology studies this interaction. Physiology studies the brain as an organ of the human body.

To a person who is willing to pay the price of some speculation in order to arrive at a conception of consciousness, Rubinshtein's scheme possessed strong points. It advanced a more sophisticated conception of consciousness than previously found within the tradition of materialism. To be sure, it possessed weaknesses, the most obvious of which was connected with the oldest problem in philosophy, pushed back into a more remote recess, but still there. The mind-body problem emerged now around Rubinshtein's "forms" (obrazy) of reality. He defined sensations and perceptions as forms of external material reality. Were these forms material themselves? Is psychic activity, consciousness itself, material? Rubinshtein said no; the forms were "reflections" of objects, not objects themselves.[81] In this way, he said in 1959, "psychic activity is ideal as a cognitive activity of man, and the term 'form' of an object (or phenomenon) is an expression resulting from this recognition." Thus, the dialectical materialism of Rubinshtein contained a category of phenomena called "ideal" that was different from "material." Was this not a surrender of his assumption of monistic materialism? Not at all, he maintained. "The key to the solution of the problem is the fact that, to use a phrase of Hegel's that was specially noted by Lenin, one and the same thing is both it itself and something else, since it appears in different systems of connections and relations." [82] Relying on this Hegelian principle, Rubinshtein maintained that in an epistemological sense the psychic is ideal, while in an ontological sense it is material. The ideal element is precisely the "forms" of reality. Rubinshtein stoutly affirmed, "We are convinced that the recognition of the idealness of psychic activity does not convert it into something spiritual, does not withdraw it from the material world." [83] Many of his critics remained unconvinced. As Rubinshtein observed shortly before his death, "more and more frequently people are affirming that the psychic is material. The partisans of this point of view, which has recently received a certain currency in our philosophic literature, are shutting themselves up inside the ontological aspect of the problem and do not take the trouble to correlate it with the epistemological aspect." [84] We shall hear more from these "partisans of the materiality of consciousness," who were indeed speaking loudly in the philosophic literature of the late fifties and the sixties.

The 1962 Conference

The most interesting event concerning Soviet Marxist philosophy and psychology in the sixties was a great conference on the subject held in May 1962 in Moscow. This all-Union Conference on Philosophic Questions of Higher Nervous Activity and Psychology was convened jointly by the Academy of Sciences of the U.S.S.R., the Academy of Pedagogical Sciences of the Russian Republic (R.S.F.S.R.), and the Ministries of Higher Education of both the U.S.S.R. and the R.S.F.S.R. More than one thousand physiologists, psychologists, philosophers, and psychiatrists participated, coming from all over the Soviet Union. The reports and debates of this conference were published in a volume of 771 pages.[85] Buried in this record are many sharp differences of opinion. It is the best single source for an understanding of the philosophic issues in Soviet physiology and psychology since the passing of the Stalinist era.

The resolution that was approved at the conclusion of the conference inevitably involved such compromises among various points of view that it reveals much less than the debates themselves, which will be discussed below. Nonetheless, the resolution did reproduce the general tone of the conference. The statement noted that the physiology of the nervous system, like many other facets of biology and psychology, was going through a special period of development in which it was coming ever closer to the physical and mathematical sciences. New methods of experimental research—electrophysiology of brain structures and nerve formations on the cellular and subcellular levels; use of computers; statistical methods; the theory of information; and cybernetics—all were leading to new understandings of physiology and psychology. Cybernetics in particular seemed promising. The task of Marxist psychologists and physiologists was to find a way of incorporating these new and valuable sources of knowledge into their disciplines without falling prey to crass materialism on the one hand or idealism on the other.

In the conference reports and debates themselves, two issues emerged as those of the greatest importance. The first was: In view of all this remarkable new knowledge about physiology and information systems, is the reflex approach advocated by Pavlov still valid? The second question was the old one: How now must we define the term "consciousness"? Most of the debates centered around these two questions—legitimate issues of interest to psychologists and physiologists in all countries. On both issues the conference broke into contrasting points of view.

The Validity of the Reflex Approach

The question of the validity of the reflex approach arose in several different forms: around discussions of the significance of Pavlov, the usefulness of the term "reflex arc," and the meaning of the phrase "higher nervous activity." The most energetic critic of the Pavlovian concept of reflexes was N. A. Bernshtein. Bernshtein's opinion that the Pavlovian teaching had become quite obsolescent in the light of modern science was supported by, among others, N. I. Grashchenkov, L. P. Latash, I. M. Feigenberg, M. M. Bongard, and, more indirectly, P. K. Anokhin. Opposing these speakers, in the most direct fashion, were E. A. Asratian, L. G. Voronin, Iu. P. Frolov, A. I. Dolin, N. A. Shustin, A. A. Zubkov and V. N. Chernigovskii.

All these speakers acknowledged Pavlov's immense stature in the history of psychology and physiology. The difference of opinion centered not on his past significance, but on the continuing fruitfulness of his approach. Some of the disagreements were semantic: The defenders of Pavlov tended to describe his views in a very broad, methodological fashion; the new critics looked upon Pavlovianism in a way similar to most non-Soviet physiologists and psychologists—that is, as a stimulus-response approach to nervous behavior. Yet underneath the misunderstandings and heated arguments was a real issue: Was the dialectical materialist understanding of physiology to be tied to the name of Pavlov, or would it acquire other means of identification? This was a question of authentic concern. As one of the supporters of the traditional point of view, V. N. Chernigovskii, observed: "We know that there is a whole group of young people who are skeptical about a whole series of principles of [Pavlov's] teaching on higher nervous activity. . . . I call this group the Young Turks." [86] Yet the Young Turks were in many cases not so young.

N. A. Bernshtein thought that a revolution had occurred in physiology since the beginning of the second quarter of the twentieth century, requiring the modification of many traditional physiological concepts but, given the necessary attention, permitting a new and superior interpretation of life within the traditions of dialectical materialism. The most important element of this revolution, he maintained, was cybernetics. He agreed that cybernetics had some dangers, particularly in the form in which its foreign founders expressed it, but he thought that if this new subject were placed on "the correct methodological rails," it could bring invaluable assistance to the study of biology in general and physiology in particular.

The most important contribution of cybernetics to the problems concerning Bernshtein was the possibility that it presented of explaining on a materialist basis the process of goal-seeking. An organism, seen from the cybernetic point of view, has a definite goal of action; Bernshtein spoke of a "physiology of activity" to distinguish it from the "simply reactive" physiology portrayed in Pavlovian reflex theory. This goal of action must be analyzed carefully, he said.

The goal of action—in other words, the result that the organism is striving to attain—is something that must be achieved but that still does not exist. Therefore, the goal of action is a reflection or model of the necessary future, coded in one way or another in the brain. . . . We should notice that the concept of the reality of such a coded brain model—the extrapolation of the probable future—creates the possibility of a strictly materialist interpretation of such concepts as purposefulness, advisability, etc. . . .

To speak in a metaphor, we can say that the organism constantly plays a game with nature surrounding it—a game whose rules are not defined and whose course of progress, "conceived" by the opponent, is unknown.[87]

In contrast to Pavlovian theory, which Bernshtein characterized as assuming "an equilibrium between the organism and its surrounding milieu," the new conception of life processes assumed "an overcoming of that milieu," a surmounting of the environment. The activity of organisms was directed, he thought, not at simple self-preservation or homeostasis, but at movement toward "a specific program of development." [88]

Bernshtein was quite aware of the dangers of his formulations, which soon resulted in his being criticized at the conference as a teleologist, but he believed his critics were simply ignorant of modern science. He thought that physiologists had been very slow in adjusting to the full implications of the concepts of probabilistic laws in nature. Many of the physiologists of the Pavlovian school, he implied, still dreamed of explaining the human body as a "reactive automat" with its actions rigidly determined in a way that was thoroughly predictable once sufficient facts had been collected. But these orthodox determinists were actually crippling modern materialism, Bernshtein indicated, by tying it to outdated concepts:

Of course, the form of behavior of a reactive automat is more obviously deterministic than the behavior of an organism that is constantly forced to make active choices in stochastic conditions. But the discarding of the concept of an organism as a reactive automat, existing "because" of the stimuli that affect it, is by no means a retreat from scientific determinism in the broad sense; that this is so should be clear from the fact that the shift from describing phenomena by means of single-valued functions to its description by means of the theory of probability does not mean a retreat from a position of strict science.[89]

The possibility of many-valued functions in biological phenomena was also very attractive to Grashchenkov, Latash, and Feigenberg. Their report, "Dialectical Materialism and Several Problems of Modern

Neurophysiology," centered on this issue. They, too, believed that the old conception of the structure of the reflex was "incapable of explaining the observed physical facts." [90] But they thought that Pavlov's system contained a great deal more flexibility than some of his critics believed. The key to many-valued functions in physiology was in the past experience of the organism, as they said Pavlov himself had indicated in his writings on reinforcement. This emphasis on the past, on the genetic approach, was a traditional characteristic of Marxism and should be easily accepted.

Grashchenkov, Latash, and Feigenberg described how many-valued functions distinguished a biological organism from a "simple automat": "One and the same physical stimulus can evoke different reactions, and in turn, one and the same reaction can evoke different stimuli. This phenomenon depends not so much on the inborn structure of the nervous system as on prior individual experience." [91]

Grashchenkov and his colleagues were of the opinion that the concept of prediction on the basis of past experience was the premise of many schemes being currently proposed by Soviet physiologists; Anokhin's "acceptor of action," Bernshtein's "physiology of activity," E. N. Sokolov's "nerve model of a stimulus," and several of I. S. Beritov's views on the physiological structure of behavior all stemmed from such a premise. The characteristic feature of all these hypothetical "predictive structures" was the probabilistic nature of prediction: "Of all possible predicted results the one is chosen which has the highest probability." And Grashchenkov and his friends observed, "The fact that in the process of evolution organisms have developed a mechanism for probabilistic prediction should not be surprising." Such an ability was essential to survival. Furthermore, rather than contradicting determinism, it broadens it by showing that "the final result of dynamic reactions is determined by the information flowing into the brain and by the past experience of the organism." [92]

V. S. Merlin, of the Perm Pedagogical Institute, also thought that the new concepts of probability were very fruitful in physiology. He maintained that a given "nervous-physiological process" does not necessarily give rise to a given "psychic process." In the old days, Merlin continued, such a view would have seemed unacceptable for a materialist, but now that the full significance of quantum mechanics has been realized by materialists, it becomes apparent that laws of probability are fully acceptable as "causal," and therefore there is no reason that a similar approach cannot be taken in psychology and physiology.[93] This would obviously leave room for a much less strictly determined psychology.

But there was still such a thing as "too much room." Grashchenkov, Latash, and Feigenberg guarded against carrying the new concept of probability all the way to a belief in complete spontaneity in psychic phenomena. The distinguished Australian neurophysiologist J. C. Eccles was frequently criticized at the conference for using the uncertainty of quantum mechanics as a means for postulating a realm of action for "mind" as distinct from matter. In his *The Neurophysiological Basis of Mind* (1952) Eccles had given considerable credence to the view that "mind could control the behavior of matter within the limits imposed by Heisenberg's Principle of Uncertainty." [94] The dialectical materialists rejected such a view as being based on the assumption of mind-body duality.

The issue of spontaneity is one that has emerged in several of the discussions described in this book. It also appeared in Grashchenkov's analysis of electroencephalograms (EEG's). He and his colleagues were a little disturbed that some researchers described the EEG patterns as "random and spontaneous." Several physiologists had even said that there is no physiological significance that can be attributed to the EEG. Grashchenkov felt that such an observation would imply a discarding of even probabilistic determinism for a complete, crass indeterminism. There was, he and his co-authors said, a discernible "functional organization" to the EEG, which was obvious in the differences in EEG patterns of states of sleep, narcosis, asphyxia, attention, alarm, and so on. Rather than being "spontaneous," EEG characteristics probably derived from one of two possibilities: Either these rhythms were a direct reflection of the processes of self-regulation, including the vegetative ones; or they were the manifestations of a sort of "undirected searching operation, an oscillation through a range of values, a sounding out of the possible initial displacements of functions subject to regulation." [95]

In sum, then, Grashchenkov, Latash, and Feigenberg agreed with Bernshtein in attempting to modify traditional Pavlovian conceptions, but they were more careful in observing the pitfalls of such an approach.

Another issue very much debated was the continuing validity of the term "reflex arc." Bernshtein believed that the concept of a reflex arc was a part of obsolescent "classical" reflex theory of the first quarter of the twentieth century; he suggested the term "reflex circle." [96] Grashchenkov, Latash, and Feigenberg were equally unhappy with the concept of an open-order reflex arc, but suggested instead what they called a "cyclical innervational structure." [97] Still another speaker, V. N. Miasishchev, of the Bekhterev Psychoneurological Institute in Leningrad, proposed that the reflex be considered a "spiral." He maintained that this model "quite obviously follows the Leninist formula of development. It is a concept that is correct both philosophically and scientifically." [98] He was referring here to Lenin's statement that the approach of the human mind to reality is not a "mirroring," but instead an approach that is "split in two, zigzaglike." [99]

Each of these proposed modifications had the same goal: to indicate that nervous activity is based on a continuous flow of feedback (afferent) signals that serves as the source of information for constant corrective signals. This inward flow of information also changes the very structural nature of the corrective mechanism itself by increasing the store of past experience "deposited" in it. In a sense, the corrective mechanism "manufactures itself" from this store of information. The Soviet interpreters saw in this approach a way of uniting social history (the past history of the individual) and natural history (the inherited characteristics of the species) in a single materialist explanation of behavior.

These critics of the traditional reflex approach were soon themselves the objects of considerable disapproval. The defenders of Pavlov accused the Young Turks of simplifying Pavlov's views by equating his concept of the reflex with the mechanistic one of Descartes. E. V. Shorokhova and V. M. Kaganov, for example, said that Bernshtein regarded reflexes as purely physiological phenomena, ignoring Pavlov's view that they were both physiological and psychological. Shorokhova and Kaganov continued that Bernshtein's concept of the "physiology of activity" contained a definition of reflex that had not changed since the days of Sechenov and was limited in the same sense as the one current in "modern west European physiology." They wished to discard this "atomistic" view but retain the term "reflex" as a description of a phenomenon that was "internally cybernetic." [100]

Other speakers were even more skeptical of Bernshtein and his supporters in their comments. Ia. B. Lekhtman of the Lesgaft Institute of Physical Culture declared:

The archaic interpretation of N. A. Bernshtein of arbitrary actions as spontaneous acts of the nervous system automatically raises our doubts. We have before us an obvious and unequivocal indeterminist interpretation of arbitrary actions, . . . which was long ago refuted by materialist science.[101]

Such interpretations strengthen the ancient fable of free will, he continued, and "if the ideas of cybernetics nourish this fable, then all the worse for cybernetics."

A. A. Zubkov of the Kishenev Medical Institute carried the critique farther. Referring to Bernshtein's statement about the way in which an organism overcomes its environment in accordance with a specific program of development, Zubkov asked, "Just what is this 'specific program of development'? Who drew up this program and put it into living matter as one would put it into a cybernetic machine? There's something here that smells of Aristotelian entelechy." [102]

The approach of L. G. Voronin, Iu. P. Frolov, and E. A. Asratian, old supporters of the Pavlovian school, was to deprecate the originality of people like Bernshtein, Grashchenkov, and Anokhin. Asratian maintained that these three men put great store in novel terms that actually describe phenomena long ago known. He believed, for example, that feedback was described in physiological terms by such people as Bernard, Pavlov, and Sechenov.[103] Frolov, who described himself as the oldest pupil of Pavlov still working, similarly doubted the originality of cybernetics and said it had no philosophy of its own and could be used by people of different schools; much was currently being made of it, he continued, by neopositivists and Gestaltists. Voronin, of Moscow University, maintained that the new criticism of Pavlovianism was not so much based on new scientific facts or proof that Pavlovianism had "aged" as it was simply a revealing of positions that these critics had long wanted to take.[104] The Young Turks were actually "Old Turks." Grashchenkov and Anokhin, he continued, were insisting on modish terms primarily in order to bring in concepts that they long ago favored but did not have cybernetic vocabularies to back up.

There was a ring of truth in this criticism. Grashchenkov had, indeed, characterized Pavlovianism in the thirties as "mechanistic";[105] Bernshtein had called for a replacement of "reflex arc" with "reflex circle" as early as 1935;[106] Anokhin—a biographer of Pavlov, and usually very respectful of him—had criticized his teaching before World War II rather sharply and had, in turn, been the subject of strictures at the 1950 Pavlov session.[107] But to see the controversy of 1962 over Pavlovianism entirely as a reflection of disputes of the thirties would be quite inaccurate. The new developments in neurophysiology and information theory, as exemplified by the works of such people as W. Ross Ashby and Arturo Rosenblueth, were by 1962 exercising great influence on Soviet physiologists and psychologists.[108] These developments seemed to promise new successes in the explanation of decision-making and goal-directed biological development on the basis of materialist assumptions. Since the materialist tradition in physiology was particularly strong in the Soviet Union, it was only natural that these two streams of thought would come together and had, indeed, been anticipated by certain Soviet scholars of the thirties. These older leaders spoke out in 1962 as the most prestigious members of the "cybernetic school," but they were supported by many younger workers. Although this point of view obviously did not go unopposed at the conference, the emphasis in the final resolution on the importance of cybernetics to physiology and the prominent place given to the reports of the cyberneticists indicate that they had achieved an advantageous position in their discussions with the more traditional members of the Pavlovian school.

The Definition of "Consciousness"

If the debate over the validity of the reflex concept was one in which physiologists were the most active participants, the definition of "consciousness" was the issue on which psychologists and philosophers spoke out most frequently. This debate was carried to incredibly fine degrees of detail.

In order to avoid the trap of dualism that the total separation of physiology from psychology presented, the philosophers around Rubinshtein had devised in the late fifties a formula that said that reflex activity is *both* physiological and psychological. Since at this time reflex activity was considered synonymous with psychic activity, the product of this analysis was the position that psychic (reflex) activity is studied in two different aspects by two different kinds of specialists. The physiologist studies psychic (reflex) activity in its ontological aspect, which is material, and he concerns himself with neurophysiology. The psychologist studies psychic (reflex) activity in its epistemological aspect, which is ideal, being based on the ideal forms (*obrazy*), and he concerns himself with cognition ("the refraction of external reality by internal conditions").

This formulation began to break down in the late fifties and early sixties when certain physiologists (Bernshtein and others) began to say that reflex activity is *not* synonymous with psychic activity because the reflex concept is too simple to explain psychic activity; if physiologists would go beyond the reflex approach, they could identify physiological mechanisms (the physiology of activity, the acceptor of action, and so on) that would explain many phenomena earlier thought reserved for psychologists. This approach alarmed many psychologists who feared that the more aggressive physiologists were trying to "swallow their field," as one speaker at the 1962 conference phrased it.

Furthermore, the compromise position of the late 1950's was being undermined from a different quarter. Certain philosophers and psychologists (V. V. Orlov and others) were also arguing that psychic activity was not the same as reflex activity. But while the physiologists like Bernshtein argued this so that physiology could shed its shackles to reflex theory and then proceed more successfully into the realm of psychology, Orlov had other consequences in mind: He wanted to wed physiology to "reflex activity" in order to leave "psychic activity" free for psychologists. He thought that psychic activity should be defined as the "ideal (spiritual) activity of the material brain" and should be the province of psychologists.[109] The physiologists would study the "material brain" itself, and if they wished to describe its functions as "reflex activity," that seemed perfectly natural to Orlov, who saw the study of reflexes as in the tradition of Sechenov and Pavlov, both physiologists. So the situation in 1962 was paradoxical, and can be sharpened in the following way: Both the aggressive physiologists and the aggressive psychologists denied the premise that reflex activity is the same as psychic activity, but for contrasting reasons. The aggressive physiologists denied it because they thought that reflexes were not a sophisticated enough weapon for their continuing campaign to explain psychic activity on a physiological basis; the aggressive psychologists denied it because they looked upon the field of psychic activity as their own domain and did not wish constantly to have to assure their audiences that ontologically everything they were talking about had a material, reflex base. They were not much cheered that the aggressive physiologists no longer thought all psychic activity had a reflex base, since they felt that the intrusion of physiologists belonging to the new cybernetic tradition was no more pleasant a prospect than the intrusion of the physiologists of the old Pavlovian school.

These debates were related to worldwide discussions in physiology and psychology at this time, but they took a different tone and a somewhat different path in the Soviet Union because all concerned—physiologists, psychologists, philosophers—could openly call themselves only materialists. Understandably, therefore, the psychologists felt somewhat more insecure than elsewhere.

In terms of the theoretical definition of consciousness or of the psyche, the biggest difference of opinion was between F. F. Kal'sin and V. V. Orlov. The end of the spectrum toward which Kal'sin's views tended was called by his critics "vulgar materialism." The opposite end, with which Orlov's opinions were associated by those critical of him, was predictably termed "dualism."

Kal'sin, who held a position in the Gorky Pedagogical Institute, was outspoken in his defense of materialism. He believed that the leading professional philosophers had misinterpreted Lenin when they had allowed discussions of ideal forms to creep into Soviet psychology. Consciousness must be considered, he continued, not as a unity of the material and the ideal, but as a purely material phenomenon. "Thought" itself was to Kal'sin merely a particular type of movement of matter. And Kal'sin quoted several sections of Engels's writings that were usually left in decent oblivion by the professional philosophers. Engels, for example, had observed, "The movement of matter is not only crude mechanical movement, not only a change in position; it is also heat and light, electrical and magnetic potential; chemical composition and decomposition, life, and, finally, consciousness." [110]

Kal'sin was highly critical of Kursanov and Kolbanovskii's statement "None of the founders of Marxism-Leninism looked upon thought as a form of movement of matter." How can they maintain such a position in the light of Engels's statement? he asked. And even if Engels had not expressed himself so clearly, he continued, it is obvious that a materialist must recognize thought and consciousness as material phenomena: "Everybody recognizes that thought is a process, a movement. . . . The natural question arises, Just what is it that is moving here?" [111] To Kal'sin, there was only one reply: matter. His simplified approach led several professional philosophers to observe that Kal'sin had not yet learned to distinguish between "a chair and the idea of a chair."

The problem of the nature of consciousness was probably the most serious and divisive issue facing Soviet philosophy of science in the 1960's. On all other questions—quantum mechanics, relativity physics, genetics, and the rest—coherent and defensible positions had been found, positions that gave viable theoretical statements of the problems at hand, yet allowed room for disagreement and further development of science. But the problem of consciousness seemed intractable. Soviet philosophers could not avoid the problem by calling the question of the definition of consciousness "meaningless" in the fashion of many positivistic non-Soviet scholars; they were committed to the constant improvement of an intellectual scheme that included explanations of the stages of development of all matter, and conscious man was ultimately material in that framework.

Solace could be found in the tradition of the mind-body problem as one of the most difficult of all intellectual problems, perhaps the most difficult. The issue was one for all countries and all philosophies. As Herbert Feigl, director of the Minnesota Center for Philosophy of Science, wrote in 1958:

The question arises inevitably: how are the raw feels related to behavioral (or neurophysiological) states? Or, if we prefer the formal mode of speech to the material mode, what are the *logical* relations of raw-feel-talk (phenomenal terms, if not phenomenal language) to the terms and statements in the language of behavior (or of neurophysiology)?

No matter how sophisticated we may be in logical analysis or epistemology, the old perplexities center precisely around this point and they will not down. . . . Schopenhauer rightly viewed the mind-body problem as the "Weltknoten" (world knot). It is truly a cluster of intricate puzzles—some scientific, some epistemological, some syntactical, some semantical, and some pragmatic. Closely related to these are the equally sensitive and controversial issues regarding teleology, purpose, intentionality, and free will.[112]

Within the framework of dialectical materialism it was also reasonable that the problem should be of great complexity, since dialectical materialists recognize principles of qualitative distinction and of varying difficulty on different levels of the evolution of matter. Dialectical materialists opposed the view that such phenomena as the formation of mental abstractions could ever be explained on the basis of physicochemical principles. The human nervous system, including the brain, represents the ultimate and most subtle stage of the evolution of matter; it was natural that the elaboration of its principles would be man's most difficult task.

A thorough description of the shades of opinion in Soviet writings in the late fifties and the sixties would require a book in itself. Here in broad strokes are some of the more identifiable positions.[113]

A few Soviet authors, such as V. M. Arkhipov and I. G. Eroshkin, continued to affirm that psychic activity itself was material; they identified consciousness with nervous processes.[114] They represented the extreme materialist wing of the authors who published on the subject. Close to them were F. F. Kal'sin, and less outspokenly, N. V. Medvedev, B. M. Kedrov, and A. N. Riakin, who characterized psychic activity and thought as a special form of the movement of matter—a form of movement that is no doubt extremely complex, but nonetheless a movement of matter.[115] The scholars so far named were criticized for leaning toward vulgar materialism by still other Soviet writers, such as M. P. Lebedev.[116] The scholars in Rubinshtein's old circle in the Institute of Philosophy affirmed that psychic activity is both physiological and psychological, and continued to maintain that the term "ideal" is perfectly legitimate when used with reference to epistemology (the "reflected" is ideal; the "reflecting" is material).[117] Still other writers openly denied that the term "material" can be applied to psychic activity at all.[118] And V. V. Orlov, as we have seen, did not hesitate to speak of "the spiritual (*dukhovnyi*) activity" of the material brain.[119] The position of the professional philosophers closest to the institutional seats of power—particularly those in the Institute of Philosophy—was in the middle. Their earlier compromise was breaking down, but they hesitated to insist on a new formulation in view of the current effort not to intervene in scientific discussions. But for them dialectical materialism continued to be a middle way between, on the one hand, those scholars —particularly the psychologists—who separated psychic activity entirely from its material substratum, and on the other hand, vulgar materialists like the behaviorists and ultracyberneticists, who questioned the very validity of the term "consciousness."

A subject of considerable concern to Soviet physiologists and psychologists in the 1960's, both at the 1962 conference and elsewhere, was what was called "the philosophical problems of the physiology of the reticular formation." The reticular formation (*formatio reticularis*) is a structure about the size of one's little finger, situated on the upper part of the spinal cord and in the central core of the brainstem. It derives its name from its appearance as that of a reticulum, or fishnet, when viewed under a microscope. It is a part of what is called the "phylogenetically primitive nervous system." [120] In the process of phylogenetic development most of the undifferentiated masses of nerve cells—but not those, apparently, in the reticular formation—became more and more organized around cellular groupings and therefore functionally rather distinct. In contrast to these parts, the reticular formation remained relatively nonspecific and unorganized.[121] Already one can see that the discovery of a part of the brain system of significant importance in mental activity and with such apparently close links to the distant evolutionary past raises questions of an interesting order.

The reticular formation seems to play a rather important role in the process of perception. Studies have indicated that the reticular formation (and some other subcortical systems as well, such as the basal ganglia, brainstem nuclei, and spinal motor neurons) acts as a way station for external sensory information *before* it reaches the primary cortical projection areas. Furthermore, this way station seems to play an active or "distorting" role that modifies these impulses. Such an understanding of the process of perception forces psychologists and physiologists to be even more skeptical than they previously were of defending a close correlation between the world mentally perceived by man and the world of objects external to him. But the "active" or "integrating" role of the reticular formation varies considerably with circumstances. When the sensory signal is novel, the reticular formation seems to have the least influence.

The reticular formation raises some rather significant questions, not only for Soviet scholars, but for others as well. Does the reticular formation merely act as a barrier, screening out information, or does it play a positive, contributory role? If it actually adds information, what are the sources of this information? If a phylogenetically primitive organ, one rather undifferentiated and therefore relatively unapproachable by anatomical or physiological analysis, plays a major role in perception and mental activity, can man explain it through research? What implications does the reticular formation have for the relative weights of rationality and irrationality in mental activity? Can one defend a "copy-theory of knowledge," as most dialectical materialists have done, when the full significance of the reticular formation is taken into account? Can the Pavlovian concept of reflexes, which attributed the major role in perception to the cerebral cortex, be retained?

Soviet philosophers frequently saw the reticular formation as a foundation on which physiological idealism was being rebuilt, as they thought it had once rested, at moments, on Descartes's view of the pineal gland or Müller's concept of specific nerve energies. They were therefore rather critical of those non-Soviet scholars who put particular emphasis on the reticular formation in their explanations of mental activity, such as W. Penfield, H. H. Jasper and A. E. Fessard. The "centrencephalic theory" most closely associated with Penfield was a particular object of their criticisms. In the process of advancing arguments against these attempts to explain the action of the reticular formation the Soviet authors frequently simplified the points of view of their foreign colleagues. Underneath the debate, however, there were several interesting issues. The Soviet authors did not deny the significance of the reticular formation in mental activity; they resisted, however, any interpretation of it that would give it a "dominating role," and they repeated frequently that the reticular formation had its material, evolutionary origins and should not be isolated as a mysterious source of mental activity.

Several Soviet scholars maintained that these issues came to a head in 1953 at an international conference on brain mechanisms and consciousness in Ste.-Marguerite, Canada.[122] At that meeting the French neurophysiologist A. E. Fessard presented a summary report, a very scholarly and well-received one, in which he made several comments to which Soviet writers later took philosophic exception. One of these remarks was Fessard's opinion that subcortical structures "play an important if not predominant role in integrative processes that organize states of consciousness";[123] the other was Fessard's statement that the reticular formation possessed the property of operating on the basis of the principle of "spontaneous activities (output without corresponding input)." [124]

The Soviet writers seemed to fear that the attribution of major or predominant influence in mental activity to subcortical organs would become the basis for idealistic and religious viewpoints; E. A. Asratian criticized Penfield's centrencephalic theory as an attempt to explain man's highest psychic functions on the basis of the most ancient and primitive parts of the brain.[125] And a group of three other Soviet authors objected to Fessard's description of the reticular formation:

The reticular formation is not an isolated source of energy, functioning according to Fessard's principle of "output without input," but instead a specific kind of intermediate transformer, a "transmission station" for afferent . . . stimuli connecting the cerebral cortex with the energy of the external and internal environment of the organism. . . .

The authors of the "reticular" theory of emotions have not taken into consideration that the nervous substrata of the emotions also arose in the process of evolution, in the process of adaptation of the organism to the external environment; . . . one must not make absolute the relative dependence of emotions on this or that nervous center.[126]

Other Soviet scholars cautioned their colleagues, however, not to press the case against men such as Penfield and Fessard too hard, since much that they wrote could be easily accommodated with a materialist interpretation of mental activity; so long as the reticular formation was not directly linked to religion, not made a new "seat of the soul"—and neither Penfield nor Fessard did this—it could be explained on a materialist basis. M. A. Logvin warned the scholars attending the 1962 Moscow conference not to pin labels on non-Soviet physiologists and psychologists, whose philosophical views, he continued, are quite diverse. He particularly objected to lumping together as "idealists" such materialistically oriented scientists as Penfield and H. W. Magoun with other openly idealistic scientists as Charles Sherrington and J. C. Eccles. Logvin noted that Magoun, one of the important researchers of the reticular formation, had defended the value of some Soviet work based on materialistic assumptions and had further admitted that much Western work still displays vestiges of dualism.[127] A similar conciliatory effort was made by A. F. Makarchenko, P. G. Kostiuk, and A. E. Khil'chenko at a 1960 Kiev conference, where they observed that Penfield praised Soviet work on the reticular formation.[128] These latter comments are illustrations, which could be amplified, that the Soviet discussion of the reticular formation and its epistemological implications remained fairly sophisticated and tentative in its assertions. The reticular formation and the question of the relative importance of the subcortical region to mental activity continues to be a subject of interest in Soviet philosophy of science at the present time.

Peter Kuzmich Anokhin

One of the most prominent physiologists in the Soviet Union during the past decades has been Peter Kuzmich Anokhin (1898–   ). In the 1920's as a student and young lecturer Anokhin worked in the laboratories of Pavlov and Bekhterev, and much of his life has been devoted to an evaluation and extension of the Pavlovian tradition. His first significant publications appeared in the early thirties; since that time he has published many works.[129] He attended physiological congresses abroad and is well known outside the Soviet Union; his biography appears in such standard references as the *International Who's Who*. After 1955 he was head of the faculty of the First Moscow Medical Institute and since 1966 a full member of the Academy of Sciences of the U.S.S.R. His research has concerned primarily the central nervous system and embryoneurology.

Anokhin has frequently praised dialectical materialism as a philosophy of science. Throughout his career one of his primary motivations, by his own account, has been the effort to elaborate a materialist and determinist explanation of nervous activity; he has attempted to discover physiological mechanisms underlying forms of human behavior previously described by such indefinite terms as "intention," "choice," "creativity," and "decision-making."

In 1962 Anokhin stated:

The methodology of dialectical materialism is strong precisely because it permits one to rise to a higher level of generalizations and to direct scientific research along more effective routes leading to the most rapid solution of problems.[130]

Anokhin continued that dialectical materialism frequently warns a researcher against falling into interpretations that are ideologically "unacceptable for us." But he also saw a danger in this warning function: It is possible, he observed, to have a science that is philosophically correct but scientifically stagnant. The "enormous motive force hidden in the dialectical materialist methodology" would be fully revealed only if one combined the admonitory function of dialectical materialism with the "logic of scientific progress"—that is, the constant checking, elimination, and confirmation of working hypotheses by experimental facts.[131] Thus, Anokhin hoped for a synthesis of dialectical materialism and rigorous experimental science. There could be no contradiction between the two because the principles of dialectical materialism are developed *by* science. To be sure, dialectical materialism contained the a priori assumptions of materiality and lawfulness, but these were the assumptions of science itself. He commented in 1949, "Nature develops according to the laws of the materialist dialectic. These laws are an absolutely real phenomenon of the objective world." [132]

In one of his earliest works, published in 1935, Anokhin advanced several of the ideas that were in modified forms to play an important role in his understanding of nervous activity. Future historians of neurophysiology and of cybernetic concepts in physiology will need to turn to this early source to evaluate Anokhin's claims to have anticipated such concepts as "feedback" with his "sanctioning afferentiation." Anokhin of course had no knowledge in 1935 of the mathematical foundation of information theory. Furthermore, discussions of "integrated nervous activity" were common in physiology at this time. Charles Sherrington's seminal *The Integrative Action of the Nervous System* had appeared long before, in 1906. Nonetheless, when a person now reads Anokhin's work of 1935, the vocabulary and the concepts do have a ring similar to that found in the subsequent literature of neurocybernetics. He spoke, for example, of neurophysiology in terms of "functional systems" in which the execution of functions is based largely on the set of incoming signals "that direct and correct" the process.[133]

Throughout his life Anokhin was convinced that a physiologist should be both loyal to the Pavlovian school yet simultaneously critical of it. Always proudly calling himself a student of Pavlov, Anokhin nonetheless questioned some of his teacher's concepts. Even in his most critical moments, however, he stoutly defended the materialist assumptions underlying Pavlovianism. In the period immediately after World War II Anokhin declared that he had made errors in several of his earlier writings in which he had criticized Pavlov's method or pointed to predecessors of Pavlov in certain lines of work. Although it cannot be proved, there is considerable reason to believe that these corrections were in response to the changing political scene after the war, when efforts to establish Russian priority in science and ideological factors became more prominent than earlier. Thus, in 1949 Anokhin commented that in his survey of the history of reflex theory from Descartes to Pavlov, published in 1945, he had given too much attention to eighteenth-century materialists and had thus detracted from Pavlov's eminence.[134] Also in 1949 he corrected a criticism of Pavlov that he had published in 1936; in the 1936 publication he had advanced the opinion that it was not correct to say that Pavlov always studied the "whole organism." This synthetic approach was more true of the early Pavlov, who studied blood circulation and digestive processes, than it was of the later Pavlov examining reflex activity.[135] Anokhin in 1936, therefore, implied that the mature Pavlov was something of a reductionist.

In the late 1940's and early 1950's Anokhin became more orthodox while under the shadow of the criticism advanced against him at the 1950 physiology conference, but in the late fifties and sixties he returned to his earlier innovative, even speculative approach and advanced ideas concerning a new architecture of the reflex arc, the use of cybernetics in neurophysiology, and the reliance on more concepts from psychology (as compared with physiology) than earlier. In this later period he clearly believed that the Pavlovian concept of the reflex arc needed modifying, however much he continued in debt to his teacher. He commented in 1962:

Scientific results and theories should be judged according to whether they correspond with reality. . . . [But] some people completely disregard this elementary critical approach and ask only, "Does this new thing concur with what Pavlov said?" And if it does not concur, then it is automatically proclaimed to be a "revision of Pavlov." On the basis of such comparisons we eliminate all possibility of finding anything new. I am not worrying about whether my interpretation will depart from the interpretation of my teacher Pavlov. This is quite natural; we live in a different epoch.[136]

Despite the slight variations in his viewpoints, Anokhin followed a fairly consistent approach to the Pavlovian tradition throughout his life.

This approach can be described as one between two extremes. In his 1949 biography of Pavlov, Anokhin saw dual dangers of opposite nature facing the followers of the great physiologist. On the one hand was the danger that the guiding ideas of Pavlov would be dissipated; on the other was the possibility of turning his teaching into dogma. Anokhin correctly predicted that the greatest danger was that of "canonization." [137]

In 1949, before the physiological congress at which Anokhin was criticized for deviating from Pavlov's principles, Anokhin published a long article surveying what he called "The Main Problems of the Study of Higher Nervous Activity." This article, together with similar surveys published in 1955 and 1963, contains a summary of the work of Anokhin and his school.

Anokhin made it quite clear in 1949 that for twenty years he and his co-workers had attempted to modify the classical method used by Pavlov in studying conditioned reflexes. In several instances, he observed, they used "totally new approaches." [138] They had gone beyond Pavlov's reliance on easily observable secretory glands and muscular reflexes to encephalographic investigations of conditioned reactions, to embryo-physiological studies of higher nervous activity, and to morphophysiological correlations (studying in a parallel fashion both conditioned reflexes and the architectonic features of the cerebral cortex). On the bases of these new approaches they had concluded that the Pavlovian concept of conditioned reflexes was too simplified, particularly its model of the reflex arc with its three links. But Anokhin did not go so far as some other researchers, such as the Soviet academician I. S. Beritashvili (Beritov) in believing that psychic and nervous activity is a unique and unitary form of activity for which the conditioned reflex is not even characteristic. Beritashvili recognized the conditioned reflex as the basic law of activity of the *brain,* but not of psychic functions in general.[139] Anokhin believed Beritashvili to be contradicting not only Pavlov but scientific evidence as well. He continued to defend the conditioned reflex as the key to psychic activity, but he wished to modify the understanding of the conditioned reflex as inherited from Pavlov.

Anokhin believed that the very approach utilized in the classical Pavlovian method blinded researchers to important processes intermediate between the conditioned stimulus and the response. He asked, "Is not the secretory indicator only an organic part of the external expression of the integrated conditioned reaction of the animal, the general form of which took shape long before the stimulation reached the effector mechanisms of the salivary gland?" [140] In other words, Anokhin was turning attention to the internal nervous structure of the conditioned reflex and implying that it was far more complex than simply a direct transmission from a receptor (sense organ) along a nerve to the cortex, a jump along the arc to a neighboring cluster of neurons, and then a transmission of a motor stimulus to an appropriate gland or muscle. He criticized physiologists for falling into the habit of speaking of the passage of conditioned and unconditioned stimuli to the cortex of the brain even though very little was actually known of the physiology of these pathways. And as a matter of fact, it had been found by the 1940's that these stimuli pass through complex structures *before* they reach the cortex. Therefore, said Anokhin, we must be careful about speaking of the "closing" of the reflex arc when it seems so much may be happening before such a closing could occur in the cortical regions. [141]

Pavlov had believed that the arc is closed in the cerebral cortex. He thought that the cerebral cortex contained a multitude of different centers, a "mosaic" as he called it, among which connections could be made. Any one center, made up of a group of cells, was related to a certain activity of the organism. But Anokhin noted that in recent years most neurophysiologists had moved away from this concept of localized centers. It had been demonstrated, for example, that a dog with most of its cerebral cortex removed could still form conditioned reflexes of a diverse character. To be sure, said Anokhin, the cortex plays an important function in the formation of most conditioned reflexes, but its role is much more flexible than earlier thought. The nervous system appears to be an integrated whole in which the parts have a remarkable ability to replace each other. Anokhin called for further study of the means by which this replacement occurs and for investigation of the gathering evidence that some rudimentary conditioned reflexes can be formed in dogs in the subcortical regions. [142]

In the course of his endeavors to explain nervous activity on the basis of material physiological systems Anokhin utilized several terms that have come to be closely linked to his name. These included "return afferentiation"; "sanctioning afferentiation"; "acceptor of action"; and "anticipatory reflection." In recent years Anokhin has usually dropped the phrase "sanctioning afferentiation," first used by him in 1935, but still retains "return afferentiation" (*obratnaia afferentatsiia*) and "acceptor of action" (*aktseptor deistviia*). "Anticipatory reflection" (*operezhaiushchee otrazhenie*) seems to be a more recent development. Each of these phrases described a part of conditioned reflex activity, which Anokhin considered to be characteristic of all organisms of the globe, a means of "entering into temporary adaptive relations with the surrounding world."

In studying the two-way flow of nervous signals that occurs in reflex activity, Anokhin became convinced that the inward, or afferent, flow was the more crucial determinant of subsequent action. His teacher Pavlov had agreed with this observation, although it did not play so important a role in his research as in Anokhin's. In 1911 Pavlov had remarked:

If one divides all the central nervous system into only two halves—the afferent and efferent—then it seems to me that the cerebral cortex is an isolated afferent part. In this section there takes place exclusively the higher analysis and synthesis of the nervous signals that are carried to it, and from there are directed already-prepared combinations of nervous signals to the efferent section. In other words, only the afferent part is active, creative, so to speak, while the efferent is only a passive section for the execution of actions.[143]

Anokhin noticed that this attribution of greater significance to the afferent system than to the efferent one distinguished the Pavlovian reflex system from the Cartesian reflex arc, in which the afferent impulse played the role only of an initiator or impetus. To be sure, Descartes deserved great credit in the history of the "progress of materialistic knowledge" by first introducing the concept of the "external stimulus" and by thus opening the door to the idea of the deterministic dependence of the behavior of animals on the changes occurring in the external world. But Descartes, being a dualist, did not take notice of the question that, for a materialist, becomes pressing: Why is the response usually advantageous? Or why, if the response is not advantageous, does the organism correct its error? For Descartes, this question could be answered in terms of the "higher reasoning" of the organism. Anokhin observed in 1955 that for twenty-five years he had been attempting to answer these questions without any such appeal. Anokhin wished to penetrate "that complicated process of the fitting of reflex activity to the interests of the whole organism—that is, the phenomenon that, practically speaking, deserves the description 'creative.' " [144]

Those physiologists who followed the views of Descartes, Anokhin continued, believed that reflex activity is adaptive or goal-directed from the very beginning of the process. Consequently, they concentrated on discovering already-prepared reflex responses. But with Pavlov's discovery of conditioned reflexes and the phenomenon of "reinforcement," it became clear that a creative, adjusting process lies at the base of reflex acivity. The inadequacy of classical reflex theory became even more clear as a result of experiments in which reflex functions were at first eliminated by vivisection, then restored by compensation. It was through such experiments that Anokhin approached these problems for the first time.

Anokhin soon came to the view that the organism could not begin the process of compensation without signals from the periphery telling of the presence of a defect. But the question still arises, How does the organism "know" that compensation is needed? Anokhin maintained that without what he called "return afferentiation" an answer to this question could not be attempted. By this term he meant "the constant correction of the process of compensation from the periphery." [145] Schematically he represented return afferentiation in the following way:[146]

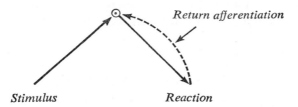

Stimulus   *Reaction*

Anokhin considered this return link in the reflex arc to be intrinsic to reflex activity: "It is difficult to imagine any kind of reflex act of an intact animal that would end with only the effector link of the 'reflex arc,' as is called for in the traditional Cartesian scheme." [147] Every act is, instead, accompanied by an entire integral of afferentiations, greatly varied in terms of strength, localization, time of origin, and speed of transmission. These afferentiations are visual details; temperature, aural, and olfactory sensations; and kinesthetic sensations. The total variety of combinations is infinite. Together they make up one process: "In the presence of constant return afferentiation accompanying, like

an echo, every reflex act, all the natural behavioral acts of an intact
animal may arise, cease, and be transformed into other acts, making up
as a whole an organized chain of effective adaptations to surrounding
conditions." [148]

As a simplified schematic diagram, Anokhin would represent this
"organized chain" in the following way:[149]

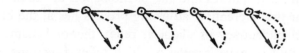

In this chain, return afferentiation serves as an "additional or fourth
link of the reflex." (Anokhin was sharply disputed on this point,
whether a fourth link is a necessary and legitimate addition to Pavlovian
teaching.) In the final step, the desired result has been obtained, so
there is no further effector action. If the process is one of compensation
for a previously destroyed function (for example, by brain-slicing), the
desired compensation has occurred in the final step. If the process is a
more normal one, such as simply picking a glass off the table, that particular goal has also been attained in the final step.

Anokhin guarded against his conception being understood simply as
the belief that "the end of one action is the beginning of the next." Such
an incorrect understanding of what Anokhin was describing would result in a different diagram, one that Anokhin rejected:[150]

Anokhin meant, instead, that the end of one action is a source of return
afferentiation that is transmitted to the nervous center where it is processed before it serves as the cause of a new action. It is in this central
point that the "decision" as to whether the desired result has yet been
obtained is made. This mechanism was called by Anokhin the "acceptor
of action" and deserves a special treatment. It is the acceptor of action
that controls the whole process.

In his discussion of the acceptor of action Anokhin made an attempt
to study intention and will from a physiological and deterministic standpoint. He initially asked, "How does the organism know when it has
reached its goal?" And he replied, "If we stand on a strictly determinis-

tic position, then essentially all the neurophysiological material that we have in our arsenal fails to give us an answer to this question. For the fact of the matter is that for the central nervous system of an animal, all return afferentiations, including sanctioning [that which corresponds to the desired goal] afferentiation, are only complexes of afferent impulses; from the normal point of view of causation there is no obvious reason why one of these stimulates the central nervous system to the further mobilization of reflexive, adaptive acts and another, on the contrary, halts adaptive actions." [151]

There is only one way out, thought Anokhin, and that is the view that there exists in the organism some sort of prepared pattern of nervous impulses with which return afferentiation can be compared. This pattern had to exist before the reflex act itself occurred. If the afferent information coincides with the prepared pattern, then the desired goal has been reached. If it does not, then further effector action is necessary. The whole question then becomes, of course, What is the physiological mechanism containing this pattern, and how was the pattern originally produced?

In order to explain the way in which Anokhin attacked this problem, it is necessary to review several features of classical Pavlovian reflex theory, particularly the relationship between a conditioned reflex stimulation and an unconditioned one. It will be remembered that Pavlov believed that every conditioned reflex is formed on the basis of an unconditioned one. Thus, an unconditioned stimulus—such as food in the mouth—will automatically cause the flow of saliva on the first or nearly first occasion, evoking strong activity in the brain. Such an unconditioned stimulus is usually also accompanied by other stimuli that may become conditioned through training—visual or olfactory sensations and so on. A "temporary" connection is formed between these points, and one can, after a little training, henceforth stimulate the secretory or motor centers of the brain merely by the conditioned stimulus.[152] However, this temporary connection will not be maintained unless it is periodically reinforced by stimulation of the unconditioned center. That is, in the classic experiment saliva will not flow on the strength of the bell signal alone unless periodically it is followed by the presence of food in the mouth of the dog and the stimulation of the unconditioned salivary reflex on which the conditioned reflexes are based.

Anokhin now incorporated his return afferentiation into this scheme. He believed that every conditioned stimulation is sent through the sense organs to the center of the brain that *in the past* had been stimulated many times by the unconditioned stimulus, and that *shortly afterward* the center will again be stimulated by the unconditioned stimulus. Therefore, there arises the possibility of a "matching" or "mismatching" of the *representation* of an unconditioned response that the conditioned stimulus evokes and the actual conditioned response itself, following in a short period of time. Schematically this is represented as follows:[153]

Three successive stages in the development of a conditioned reflex

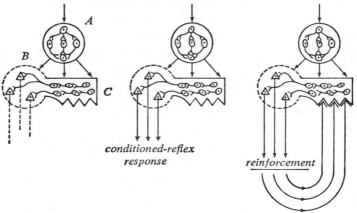

conditioned-reflex
response

reinforcement

In stage one the conditioned stimulus falls upon the appropriate sense organ. In stage two it causes a conditioned reflex response based on a "representation" of an unconditioned reflex, a step that has occurred frequently in the past, but has not yet occurred in this sequence. In stage three, the unconditioned stimulus itself (food in the mouth, for instance) has occurred; the unconditioned response turns out to "match" the conditioned representation of it, and reinforcement occurs.

Anokhin now pushed farther in order to ascertain how powerful the "matching" or "controlling" mechanism was. He found that it was very powerful indeed, as one of his experiments illustrated. He conditioned a dog to go to a feeding box on the left side of a training box in response to the sound "la," and to a feeding box on the right side in response to the sound "fa." In all cases the food was sugared bread. The dog soon became thoroughly conditioned and would immediately go to the correct side. On one day (and only for one time) he introduced a change, however. He placed not bread but meat in the box on the left side and gave the appropriate signal. The dog went to the left box as was its custom, but was obviously surprised to find meat instead of bread there. It demonstrated what animal psychologists call an "orienting-research reaction," but after this moment's hesitation, devoured the meat.

From this point onward, and for twenty days thereafter, the dog's actions were governed by this one event. No matter whether the sound "fa" or "la" was sounded, the dog would always bound to the left box. The experimenters continued to run the experiment as if the exception had never occurred, using sugared bread for food, and placing it in a box on the left side if the signal "la" was to be used and on the right if "fa" was to follow. Yet the dog long persisted in disregarding its old conditioning and searching only the left box for meat. If it found bread there (which it always did if the signal had been "la"), it refused to eat. Only after a long period of the total lack of reinforcement was the old pattern restored, so strong an impression had the one occasion made.[154]

Anokhin believed that this experiment provided further evidence for the presence in the nervous system of a mechanism called the acceptor of action, which is based on very strong, inherited unconditioned reflexes that, in turn, can be linked to conditioned stimuli. In the case of the dog, the unconditioned reflex was the food reflex of carnivorous animals. Anokhin observed that "the acceptor of action" was an abbreviated term for the more accurate but cumbersome phrase "the acceptor of the afferent results of a completed reflex act." The word "acceptor" was a key term conveying both the ideas "to receive" and "to approve" that are contained in the Latin *acceptare*.[155]

According to this scheme, if the nervous system of an animal is acted upon by a conditioned stimulus that has in the past been reinforced by meat, then the acceptor of action will define to what degree the received information corresponds to the earlier afferent experience of the animal. Anokhin represented schematically two cases: the one in which the match is correct and strong reinforcement occurs; the other in which there is a mismatch between the conditioned and the unconditioned stimuli.[156]

Anokhin believed that this approach could help in an understanding of the way in which the nervous system repairs itself after it has been damaged. Let us imagine a form of reflex activity that was schematically

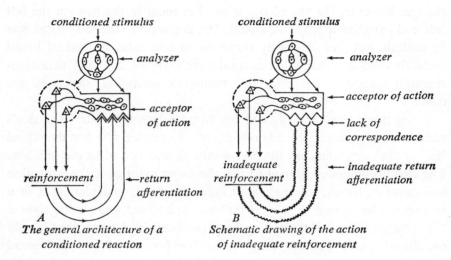

A
The general architecture of a
conditioned reaction

B
Schematic drawing of the action
of inadequate reinforcement

represented by stage *A* in the above drawing, but was destroyed by the existence of some defect caused by surgery or disease. Anokhin would show in the subsequent stages the way in which the components of the appropriate nervous subsystem rearrange themselves until they arrive at an arrangement that yields the proper reinforcement as determined by the afferent signals of the past history of the organism. This reorganization effort is graphically represented by drawing the subsystem in radically different shapes, one after the other, until a proper arrangement is found, shown in the last stage:[157]

Consecutive stages of compensatory adaptation

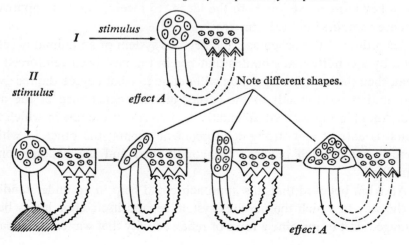

Thus, Anokhin with his acceptor of action was trying to arrive at a physiological explanation for the goal-oriented activity of the nervous system. To be sure, he had not solved the problem of teleology, which has plagued biology throughout its history. On a rather speculative basis he had pushed the boundaries of the riddle back a little farther. Furthermore, despite his effort to identify physiological mechanisms, he had made no effort to localize the acceptor of action in the body itself. Accepting his arguments for a moment, however, one could now give something of an answer to the question, How does the organism know when it has reached its goal? He illustrated that the problems of intention and goal-seeking are amenable to physiological investigation, even if many problems remained. Many neurophysiologists of his time were beginning to think in similar terms. His concepts of reinforcement and acceptor of action bore some resemblances to much earlier research, such as Lynn Thorndike's and Lloyd Morgan's law of effect. Thorndike had shown that when he placed animals in labyrinth-pens, there occurred a curious causal phenomenon that some people incorrectly described as retroactive. It appeared that when the animal made a correct movement toward its eventual escape, it learned to repeat that movement in a similar situation later even though the success always occurred after the movement. Thorndike believed that the success, as an effect of the movement, "stamped in" the movement that had been the cause of it.[158] Anokhin acknowledged the similarity of Thorndike's law of effect to his own viewpoint, but pointed out that Thorndike was not so interested as he in the physiological mechanisms behind the phenomenon. Thorndike identified success with pleasure or satisfaction and described the law of effect primarily in psychological or subjective terms. Anokhin attempted to describe reinforcement and goal-seeking in terms of physiology: the relationship between conditioned and unconditioned stimuli, and the means by which past experience could provide a pattern contained in the acceptor of action, against which future afferent information could be checked.

Anokhin refined his views in 1962 in a speech that he gave to the conference on the philosophic questions of higher nervous activity. In this speech, later published as a long article, he closely tied his opinions about the development of nervous systems to general principles within dialectical materialism, such as space and time as forms of existence of matter and reflection as a property of all matter.

Anokhin had long been fascinated by the related concepts of *time* and *signalization*, believing that they lie at the root of the intellectual problem of the origin of nervous systems and, ultimately, of reflex activity. Nervous systems are, in his opinion, mechanisms for the anticipation of the future. The importance of anticipation is obvious in Pavlov's classic experiments in which the dog anticipates the coming of food on the basis of signals. But in a more general, theoretical sense, one can say, with Anokhin, that anticipation of the future is the basis of most, if not all, nervous activity. The brain itself is sometimes defined as a physiological mechanism for the purpose of utilizing information from the distance receptors (eyes, ears, nose) and therefore giving the animal a time interval for preparatory reactive steps. This is, indeed, one of the standard definitions of the "brain." [159] But if one regards the brain from the evolutionary standpoint, extending its history of development all the way back to the origin of life itself, then it becomes clear that some way of accounting for this property of anticipation must be found that is comprehensible against the background of an initial lifeless, inorganic milieu. Here again Anokhin was wrestling with the problem of teleology, but he now wished to begin, not with the end-products of evolution, but its origins.

In the section of his speech entitled "The Conditioned Reflex as a Reflection of the Spatio-Temporal Structure of the World," Anokhin began with a recognition of the relevance of dialectical materialism to his approach. Not only does dialectical materialism cite space and time as the forms of existence and movement of matter, but it characterizes all matter as possessing the property of reflection. Therefore, the conditioned reflex must be described in terms of space and time, and it must be a reflection within living matter of material conditions external to living matter.

Since living matter developed in a world in which nonliving matter already existed, it was, said Anokhin, "inserted" so to speak, in a world in which spatial and temporal relations were already defined. The relationship of inorganic matter to temporal parameters was entirely passive; no anticipation of events occurred. In terms of the over-all evolution of matter, qualitative new relationships, much more active ones, emerged only with the appearance of living matter. The problem at hand is the way in which these relationships emerged and their essential characteristics.

Anokhin's primary concern here, then, was not the origin of life it-self, but the origin of rudimentary nervous systems. The two are obvi-ously intimately connected—Anokhin believed no life possible without the nervous function of "anticipatory reflection" in some form—but the slight distinction permitted Anokhin to take a different approach from his colleague A. I. Oparin, whose ideas were, nonetheless, important to his own. Oparin, also an outspoken dialectical materialist, is discussed in Chapter VII in this study.

What sort of temporal parameters of the movement of matter existed in the inorganic world at the time of the origin of life? These parameters would be the ones within which the first bit of living matter would have to survive and to which it would have to adapt. These basic characteris-tics of the time structure of the material world Anokhin considered to be a sort of "categorical imperative" for the development of life.[160] Anokhin believed that the first bit of life would encounter three differ-ent types of temporal influences (different in their relationship to the bit of life, not in an absolute sense), all existing in the inorganic world around it:[161]

(1) the action of relatively constant factors;
(2) the action of sequential series of repeating factors (rhythmic or aperiodic);
(3) the action of sequential series of factors that never repeat.

An example of the first type would be the presence or absence of nitrogen or water in the milieu. An example of the second type would be the sequence of day, evening, night, morning, or summer, autumn, winter, spring. An example of the third type would be a solar eclipse, an extremely violent storm or hurricane, or the passing nearby of a comet.

The most interesting category from the standpoint of Anokhin's anal-ysis is the middle one, the action of sequential series of repeating fac-tors. He considered the presence of this category an absolute necessity for the evolution of rudimentary nervous systems, for without it no method of prediction, of anticipatory reflection, could ever be devel-oped, for there would be nothing to predict and no basis for that rhythmic interchange of matter on which life is based. To be sure, the examples already given for this category (diurnal and annual cycles) are not entirely appropriate for the first bit of life, since it is commonly supposed that life originated in deep ocean waters where these influences would be slight; nonetheless, Anokhin reasonably maintained that in such a milieu there would be repeating, perhaps aperiodic, factors (flow of current, change in temperature, presence of various substances). These changing factors were absolutely essential for the development of life, and they had on the internal constitution of the first bit of living matter effects that Anokhin described in terms of dialectical materialist "reflection":

The basis of the development of life and of its relations to the external inorganic world is the repeating influences on the organism. It is precisely these influences—the result of the original properties of the spatio-temporal structure of the inorganic world—that conditioned the whole anatomic organization and adaptive functions of the first living substances. In this sense the organization of living substances is in every sense of the word a reflection of the spatio-temporal parameters of the concrete milieu of their existence.[162]

Some of these repeating factors are harmful to the existence of the organism, and some are helpful. The organism to survive must be adapted in differing ways to these contradictory factors. Anokhin then asked the question that for a physiologist becomes so important: "By means of what sort of concrete mechanisms do these repeating and vital influences cause the organism to be adapted?" Notice that the question here was not, in Anokhin's mind, incorrectly inverted. He did not wish to ask how the organism adapts itself, but how the organism is caused to be adapted by the external factors. At this most primitive level the cause must come from the outside if one is not to attribute a life principle to a closed (and very small) bit of matter itself. This, of course, would be vitalism.

Every external influence on the primitive organism will have effects inside it, said Anokhin. In some cases, these effects can be described as chains of chemical reactions making up "open systems." If we can represent a series of repeating external factors as $(ABCD)$ $(ABCD)$ $(ABCD)$, then the series of resulting internal reactions can be designated as $(abcd)$ $(abcd)$ $(abcd)$. As Anokhin observed: "If in the external world several specific events develop sequentially (for example, seasonal rhythms, changes in temperature, currents in the oceans), then the organism must reflect each of these in specific chemical reconstructions of its protoplasm as long as they attain a given threshold of effect." [163]

According to Anokhin, the crucial step in the evolution of rudimentary nervous systems is now at hand. That step is the origin of mechanisms by which the time intervals separating the effects (*abcd*) (*abcd*) (*abcd*) become *shorter* than the time intervals separating the original causes (*ABCD*) (*ABCD*) (*ABCD*). This will be the first case of "anticipatory reflection." Many such cases are necessary for the evolution of life, Anokhin believed, forming a "struggle for rapidity." [164]

Anokhin believed that this acceleration of reactions would occur on the basis of natural selection.[165] Many of the external influences already discussed could be initiators of whole series of chemical chain reactions. The great majority of the resulting chain reactions would bear no adaptive relationship to subsequent events in the external world. Some, however, would coincidentally result in chemical states inside the primitive organism that would have a survival value when a later external event occurred. If the external events occurred according to a repeating pattern, and if the organism reproduced itself in forms capable of diverse and rapid chain responses to these external events, then those forms producing responses that proved *later* to be advantageous would survive and reproduce. Thus, macrointervals of time (*ABCD*) (*ABCD*) (*ABCD*) produced changes (*abcd*) (*abcd*) (*abcd*), which were originally also separated by macrointervals of time but gradually become separated by microintervals. The latter accelerated series of effects Anokhin designated ($a{\rightarrow}b{\rightarrow}c{\rightarrow}d$) ($a{\rightarrow}b{\rightarrow}c{\rightarrow}d$) ($a{\rightarrow}b{\rightarrow}c{\rightarrow}d$). He pointed to the presence of certain catalysts as a very important factor in causing the acceleration of such reactions.

Anokhin described the process as follows:

If in a small clump of living matter, making up an open system of many molecules, diverse and very rapid reactions become possible, then several of the sequentially repeating external influences, even those separated by large time intervals, will receive the possibility of reflecting themselves in rapid chemical transformations of this substance. . . .

.   .   .   .

Thus, there was created an ability of primitive organisms to reflect the external inorganic world not passively but actively, with the anticipation in their protoplasm of sequential and repetitive phenomena of the external world.

As a matter of fact, since a chain of chemical reactions was established in the protoplasm of the living organism . . . , it is now possible for the action of only the first factor A to initiate the whole subsequent chain of chemical reactions.[166]

The fact that A alone could serve as a "signal" for the whole subsequent chain brought Anokhin closer to his subject of reflexes. There obviously is no true first member of any series of external influences, Anokhin noted. Even A in our example was actually merely an intermediate member of an infinite series stretching back in time. Therefore, the possibilities of signalization are infinite, and many "temporary connections" based on such signalization are established:

Consequently the very fact of the appearance of "signalization" and "temporary connections" must be recognized as one of the oldest principles of the development of living matter. It is precisely in this way that one should interpret I. P. Pavlov's statement, "the temporary nervous connection is the most universal physiological phenomenon in the animal world and in us ourselves." This principle was a result of natural selection and has been driven firmly and ever more deeply into the stabilized structures of protoplasm. . . . One really should not be surprised, then, that this function of the organism, offering it the broadest possibilities of adaptation and progress, rapidly became refined in a special substratum. . . . And it is the nervous system that is this substratum.[167]

This then is the philosophical and evolutionary background against which Anokhin viewed the nervous system. He then proceeded to discuss the conditioned reflex from this standpoint—that is, as a form on a high level of the evolutionary ladder of anticipatory reflection. While stating that dialectical materialism was of methodological assistance in developing his view, Anokhin also seemed to be suggesting an interesting addition to its principles. While dialectical materialism already contained the concept that all matter possesses the property of reflection, Anokhin was suggesting that the additional principle be added that all *living* matter possesses the property of anticipatory reflection. And the latter property was, to Anokhin, a step toward a rudimentary nervous system and conditioned reflexes themselves.[168]

In the 1960's Anokhin's attention turned strongly toward cybernetics, which he recognized as a "revolution in thought, in our very approach to the phenomena of nature and our method of analysis." [169] His attitude to cybernetics was, however, not untinged with the feeling that he had never been properly credited for his anticipation of its approach as early as 1935, when he began applying the term "functional system" to the nervous system. And his term "return afferentiation" was obviously another name for cybernetic feedback. In articles in 1966 and 1968 he attempted to defend the interpretation given above within the framework of cybernetics.[170]

Suggestive as Anokhin's evolutionary interpretation of nervous systems was, it should be noticed that it suffers from the same defects in terms of definition and logic as Oparin's view of "natural selection among prebiological systems." Apparently the only way in which it can be defended against the criticism that Mora advanced against Oparin (see p. 292) is by saying (as Bernal did) that the origin of life or of nervous systems has a "logic of its own." This logic is one of "emergent evolution" or "qualitative leaps" in which different autonomous realms of regularities are successively attained. Yet one must admit that this concept of dialectical logic may be merely a way of dodging, or living with, the ancient problems in biology of teleology and reductionism; these problems seem to be with us still, and they appear particularly clearly in discussions of origins in biology, whether of the origin of life or of nervous systems. Anokhin has pointed to the intimate connection between life and nervous systems, and believes that systems analysis or cybernetics holds out the most fruitful approach, at the present time, to these questions within the tradition of dialectical materialism.

As this survey of philosophical issues in recent Soviet psychology and physiology draws to a close, certain limitations will be evident. The foregoing exposition and discussion have drawn heavily on the physiological aspects of psychology, a concentration deriving in large part from the characteristic stress on physiology pervading Soviet philosophical discussion of psychology. In the description of the network of issues in physiology and psychology a number of prominent Soviet authors have been neglected—among them, A. R. Luria, D. N. Uznadze, B. M. Teplov, and A. N. Leontiev. A more serious limitation, perhaps, is the omission of full discussion of Soviet attitudes toward Freudian psychology. A brief note on this complicated issue will have to suffice.

Soviet psychologists have seriously underestimated the influence of the subconscious on mental activity. Freudianism, after a period of popularity in the Soviet Union in the twenties, became a prohibited subject. In recent years Soviet psychologists have begun to recognize their in-

adequacies in this area, although they have by no means become enthusiastic about psychoanalysis. A concern in the last ten or fifteen years
among many Soviet psychologists, if one judges by the literature, has
been the fear that they have surrendered to Freudianism the whole
realm of the subconscious; they have wanted to make it clear that this is
not so, or at least should not be so. Consequently, there have been a
number of efforts in the Soviet literature to show that Freud was by no
means the first person to point to the importance of the subconscious
realm.[171] This attempt was, no doubt, an attempt to relativize Freud, to
make possible a turning of real attention to the phenomena usually associated with the name of Freud without appearing to embrace Freudianism after years of denying its legitimacy. They have criticized the
"monopoly" of Freudianism abroad, particularly in the United States
(and perhaps with good reason). They have engaged in rather detailed
and controversial semantic discussions of the relative validities of the
terms "nonconscious" (*neosoznavaemyi*), "unconscious" (*bessoznate-
l'nyi*), and "subconscious" (*podsoznatel'nyi*), with several scholars preferring "nonconscious" to the other two terms, which they saw as closer
to Freudianism.[172] But on the whole, they have been moving more and
more toward a recognition of Freud. A. M. Sviadoshch, of the Medical
Institute in Karaganda, commented in 1962 to an audience of psychologists, physiologists, and philosophers:

Without any question S. Freud performed a service for science. He attracted attention to the problem of the "unconscious." He pointed to several
concrete manifestations of the "unconscious," such as its influence on slips
of the pen or of the tongue. However, he also introduced much that was improbable or fantastic to the subject of the "unconscious," such as his assertion of the sexuality of the small child. He created a mistaken psychoanalytic
theory, which we deny.[173]

The work of the Georgian psychologist D. N. Uznadze (1886–1950)
has often been cited as a Soviet alternative to Freud, although it has also
been criticized in the Soviet Union as ideologically suspect.[174] In recent
years it has become less sensitive. F. V. Bassin promoted Uznadze's
theory of "set" (*ustanovka*) as a basis for building a general theory of
nonconscious nervous activity that he considered superior on methodological grounds to that of Freud.

Uznadze worked on his conception from the early twenties, when

Freudianism attracted many Soviet thinkers, until his death. His work has been continued, in particular, by the Institute of Psychology of the Georgian Academy of Sciences. Uznadze, in his classical experiments, asked a subject to compare the size of two small balls by the sense of touch, one ball being placed in each of his hands. At first, during a long series of tests, the balls (of equal weight) were always of different volume, and the smaller ball was always placed in the same hand. Then the subject was given balls of the same volume (and weight). In reply to the question "Which ball is the larger?" the subject answered that the larger ball was in the hand that had always before received the smaller ball. Uznadze explained this illusion in terms of the "internal state" of the subject. On the basis of simple experiments such as this he built a rather elaborate theory of "set," in which the set is formed by past experience.

Several of Uznadze's defenders in the Soviet Union have differed with him on the details of the set theory. Uznadze thought the set remained nonconscious, or at least he was interested in it only when it was. Bassin, on the contrary, thought that much of the merit of Uznadze's approach consisted in the fact that it explained nonconscious factors in terms of a "functional displacement" and thus opened up this realm to objective experiment. Bassin commented, "The nonconscious set fulfills the role, in this way, of an invisible 'bridge' between definite forms of realized experiences and objective behavior, a bridge that, according to Freud, must be fulfilled by the subject's 'unconscious.' " [175]

Even in the sixties, however, Uznadze's views met criticisms linked with older Soviet attitudes. I. I. Korotkin, of the Pavlov Physiological Institute, maintained that the set had to be interpreted either as an epiphenomenon—and therefore was unacceptable in science—or as a material phenomenon, which would then differ little, he thought, from the dynamic stereotype of the Pavlovian school.[176] And M. S. Lebedinskii, from the Institute of Psychiatry of the Academy of Medical Sciences, commented that presenting Uznadze as an alternative to Freud was an unsuccessful attempt, since a Freudian analyst would have no trouble accepting Uznadze's views and still remaining a loyal Freudian.[177]

Despite the continuing differences, Soviet psychologists and physiologists are closer at the present time to their non-Soviet colleagues in terms of their research interests and methods than they have been at any time in the last twenty years. Much of this movement has come from the Soviet side, but by no means all. In recent years there has been a quickening interest among British and American scholars in Soviet psychological and physiological research. And even the earlier hostility of non-Soviet scholars to references to dialectical materialism shows some signs of lessening, while the Soviet writers make fewer such references. Jeffrey A. Gray, of the Institute of Experimental Psychology of the University of Oxford, wrote an introductory article for a 1966 symposium on Soviet psychology that deserves to be quoted at length because of its unusual perceptivity:

The official status of Marxism-Leninism has had the result that philosophy intrudes in scientific research and writing in a way which is totally unexpected for a Western scientist. It is a shock to discover in a text ostensibly concerned with the empirical investigation of psychology that there are frequent references to Marx, Engels, Lenin and even—not so long ago—Stalin. One's first reaction is to dismiss this as a necessary obeisance in the direction of the political powers-that-be—as no doubt, in part, it is; and, in any case, one feels that philosophy has no place in the conduct of scientific research. However, a more sympathetic consideration of the use to which these philosophers are put reveals that there is something of more importance, and perhaps even of real value, going on. In the first place, the Russian habit of making the philosophical background plain for all to see is not such a bad one; above all, it becomes clear that, with different philosophical assumptions, there would be different research and different favoured forms of expression—and this connection is not to be broken simply by keeping the philosophical assumptions out of sight (and out of mind) as the Anglo-Saxon psychologist tries to do. Secondly, there is a good case to be made for the particular assumptions of Marxist philosophy as a reasonable starting point for a scientific psychology—provided, of course, that we are ready to abandon them if our data suggest that other assumptions would make a better starting point. In particular, it can be argued that Marxist assumptions are more consistent with the results of recent psychological and neurophysiological research than are the assumptions contained in the extreme associationist-behaviourist point of view identified with the names of J. B. Watson and C. L. Hull; and that the recent retreat from this extreme position in Anglo-American psychology has made it possible to attempt a *rapprochement* between the views of human nature held in the East and the West.[178]

It is out of a combination of the point of view advanced above by Gray and the recent effort by Soviet scholars to be less zealous in ideological criticisms of their non-Soviet colleagues that some of the most interesting, even exciting, developments in contemporary psychology and physiology may flow. And it should be added that the possibilities for a *rapprochement* between Soviet and non-Soviet intellectual conceptions are by no means limited to physiology and psychology; there are similar opportunities in the other fields of natural science as well.

Concluding Remarks

Contemporary Soviet dialectical materialism is an impressive intellectual achievement. The elaboration and refinement of the early suggestions of Engels, Plekhanov, and Lenin into a systematic interpretation of nature is the most original creation of Soviet Marxism. In the hands of its most able advocates, there is no question but that dialectical materialism is a sincere and legitimate attempt to understand and explain nature. In terms of universality and degree of development, the dialectical materialist explanation of nature has no competitors among modern systems of thought. Indeed, one would have to jump centuries, to the Aristotelian scheme of a natural order or to Cartesian mechanical philosophy, to find a system based on nature that could rival dialectical materialism in the refinement of its development and the wholeness of its fabric.

The most significant function of dialectical materialism in the Soviet Union derives from the comprehensiveness of its conception and the intimacy of its connection with current scientific theory. As a system of thought it is not of immediate utilitarian value to scientists in their work —in fact, converted into dogma it has been a serious hindrance in several cases, although it may have indirectly helped in others—but it does have an important educational or heuristic value. Not only professional Soviet philosophers but many scholars and students in other fields as well have a concept of a unifying principle of human knowledge, the materialist assumption that lies at the base of dialectical materialism. It is not a provable principle, but then neither is it absurd. Soviet scientists as a group have, in fact, faced more openly the implications of their philosophic assumptions than have scientists in those countries—such as the United States and Great Britain—where the fashion is to maintain that philosophy has nothing to do with science. Perhaps out of a meeting of the approaches of Soviet and non-Soviet scholars a way can be found to admit that philosophy does indeed influence science (and vice versa) without allowing that admission to become exaggerated or subject to political manipulation in the way in which it frequently was in the Soviet Union.

In terms of improving the intellectual position of a materialistic explanation of nature, it is clear that Soviet dialectical materialists have made genuine progress in certain fields, progress that to a degree offsets the damaging effects of their failure in genetics. Thirty or forty years ago the crucial questions that dialectical materialists faced were in the area of physics. The new ideas contained in quantum mechanics and the theory of relativity were upsetting to Soviet materialists, as they were to many other traditional thinkers. The dialectical materialists worried about the effects the new physics would have on assumptions that they had previously considered secure: belief in the existence of objective reality, the principle of causality, and the material foundation of reality. Today it is clear that this phase of anxiety has passed. No one knows what the future of physics will bring—and the nature of science is to bring crises—but at the present moment the philosophical problems in physics are much less difficult for dialectical materialists than they were thirty years ago. It is simply not true, as was frequently maintained several decades ago, that relativity physics and quantum mechanics "destroy materialism at its base." These areas of physics no longer contain unique threats to the assumptions of objective reality, causality, and the primary significance of matter. In terms of the new materialist interpretations of nature, relativity and quantum physics support a more satisfying explanation of natural phenomena than Newtonian physics can yield.

The works of such men as Fock, Blokhintsev, Omel'ianovskii, and Aleksandrov are important in this respect, although somewhat different in content. They pointed out that in the light of modern physics, thinkers should give up determinism in the Laplacian sense, but not causality in general. If there were no causality in nature, all possible outcomes of a given physical state would be equally probable. We know very well that according to quantum mechanics the real situation is far from such absolute indeterminism. A concept of causality, based on probability, can still be maintained. Many thinkers continue to find such a causal concept necessary for an understanding of nature. Once accustomed to it, they usually consider it vastly superior to the rigid Laplacian view.

Other Soviet scholars have turned attention to the fact that in general relativity theory the role of matter is not smaller than in classical (Newtonian) physics, but greater. The density of matter in the universe, according to general relativity, determines the configuration of space-time. Matter therefore acquires a significance of which eighteenth- and nineteenth-century materialists could not have dreamed. It is true that the principle of the interchangeability of matter and energy contained in relativity theory seems to demote matter in status (why, for example, should it be considered more important than energy?), but there is another side to the coin that strengthens the convictions of dialectical materialists. Having accepted this interchangeability, they now consider matter and energy synonymous (matter-energy), and proceeding on the basis of the primacy of matter-energy, they do not encounter the ancient problem of the void faced by classical materialists. All voids apparently contain fields of some kind, at least a gravitational one, and therefore a form of matter-energy. The very concept of the void as something that can exist in the real world is therefore thrown into question.

The originality of Soviet dialectical materialism compared to other areas of Soviet thought is not only a result of the talent of a number of Soviet dialectical materialists; it also derives from the nature of classical Marxism and the breathtaking rapidity of development of science itself. Marx wrote a great deal about society, but very little on natural science. The brilliance of his original statements on society overshadowed all subsequent efforts by his supporters in political and economic theory. Before 1917, the system of historical materialism was much more developed than the system of dialectical materialism. Engels, of course, wrote extensively on philosophy of nature and in that sense launched dialectical materialism, but his efforts were rather quickly rendered inadequate by the revolutionary development of scientific theory at the turn of the nineteenth and twentieth centuries. After 1917, Soviet dialectical materialists were forced to seek new paths toward an understanding of nature, because scientific theory itself was already flowing along a new path. Deprived of an adequate Marxist explanation of nature and faced by a revolution in science, Soviet dialectical materialists during the past fifty years have made in philosophy of science an innovative effort that stands out in sharp contrast to other Soviet intellectual efforts.

Perhaps even more significant as a reason for the Soviet achievement in philosophy of nature as compared to other areas of thought in the U.S.S.R. was the system of Communist Party controls over intellectual life, a system that left more room for initiative in scientific subjects than in political ones. The best minds went into scientific subjects, and some of them, naturally enough, were attracted to the philosophic aspects of their work. In the peculiar Soviet environment the esoteric nature of the discussions of dialectical materialism has been something of a boon to writers, screening away censors. Among those authors who write on dialectical materialism, the sections of their works of the highest quality are frequently buried in rather technical discussions. Those scientists who came to dialectical materialism in the late forties for the purpose of defending their science found that interesting work was possible in the philosophy of science. The Communist Party officials continued to consider themselves experts on theories of society, and still crudely intervene today in such discussions. They learned, after several very injurious experiences, that intervention in the knotty problems of scientific interpretation is an exceedingly risky business. Their relative tolerance in this particular field in recent years has had two oddly contrasting effects: Some scientists have abandoned conscious consideration of philosophy now that they are no longer constantly forced to show the relevance of dialectical materialism to science; others have turned to it with new interest now that there is an area of wide dimensions, intellectually speaking, in Soviet philosophy where some innovation is possible.

In retrospect, it is clear that the controversies discussed in this volume reveal very different, even contradictory, aspects of Soviet society. If one is interested primarily in the way in which the Soviet system of political controls over intellectuals created a situation in which unprincipled careerists could gain extraordinary influence in a few scientific fields, then one should turn to the Lysenko affair. Here one can find abundant evidence of the damage done to science by a centralized political system in which the principle of control was extended to scientific theory itself. The Lysenko affair was one of the most flagrant denials of the right of scientists to judge the validity of scientific theory to occur in modern history. The intensely political character that science assumes in all countries is no justification for the intrusion of controls on the judgment of the adequacies of rival scientific explanations. That decision must belong to scientists.

My major goal in writing this volume was not, however, the recounting of this repressive side of the story of Soviet science; for those readers seeking this aspect, I would recommend the books on Lysenko by David Joravsky and Zhores Medvedev. Instead, I have sought to emphasize the more philosophically interesting, rather than the politically dramatic, aspects of the relation between Soviet philosophy and science. In looking through the literature of the past thirty years, I have tried to center my attention on the publications of the best intellectual quality, not the worst. I would do the same if I were studying the relation of Cartesianism to science in the seventeenth and eighteenth centuries, or of Aristotelianism to science in the Middle Ages.

In a commentary on one of my articles on Soviet science the Soviet physicist V. A. Fock wrote that I paid my "main attention to that part of the discussion which proceeded at a higher scientific level and not to articles and viewpoints rightly described by him as 'offensive parodies of intellectual investigations' (particularly numerous in the dark period from 1947 to the early 1950's)." This emphasis meant that I rarely found pertinent items of interest in newspapers, popular political journals, or textbooks of Marxism-Leninism; instead, I looked to the serious monographs and journal articles of Soviet scholars and, wherever possible, to those of natural scientists.

In the eyes of certain of my readers this approach may not seem justified. Those persons who are convinced that all references to dialectical materialism by Soviet scientists have been nothing more than responses to political pressures will doubt that anything worthwhile can be accomplished by studying the works in which these scientists have attempted to illustrate the relevance of Marxist philosophy to their work. I am convinced, however, that quite a few prominent Soviet scientists believe that dialectical materialism is a helpful approach to a study of nature. They have examined many of the same problems of the interpretation of nature that philosophers and scientists in other countries and periods have also examined, and they have slowly developed and refined a philosophy of nature that would almost certainly continue to survive and evolve even if it were no longer propped up by the Communist Party. It is a philosophy of nature that is tied very closely to science itself, and it now depends much more on science for sustenance than on Party ideology. Only by concentrating on scientific literature, rather than on political or ideological sources, can this independent, unofficial side of dialectical materialism be revealed. By recognizing the intellectual sources of much Soviet writing on dialectical materialism, one can begin to understand why there have been such wide disagreements among Soviet scientists on philosophical interpretations of such issues as the physiology of perception, the nature of the universe, and the uncertainty relation of quantum mechanics.

Before turning to a few final observations about the controversies discussed in this volume, it is appropriate to observe that contemporary dialectical materialism possesses several very serious weaknesses. Aside from the political obstacles it still faces in the Soviet Union, the most critical of these, one that seems to be a systemic failing, is the weakness of its defenses against critics standing on the position of philosophic realism. Many scientists and philosophers around the world completely agree with the position of dialectical materialism on the existence of objective reality, but simultaneously decline to consider themselves dialectical materialists. One of the best ways to illustrate their reasons for this refusal is to analyze Lenin's definition of matter:

Matter is a philosophic category for the designation of objective reality which is presented to man in his sensations, an objective reality which is copied, photographed, or reflected by our sensations but which exists independently from these sensations.

A philosophic realist can reply to Lenin that he agrees completely that objective reality exists, that there also exist "objects outside the mind," and that there is nothing supernatural in the world. But where, he can continue, did this word "matter" come from? The realist will then observe that he prefers the term "objective reality" to "matter," since it is clear to him that such phenomena as "consciousness" and "abstraction" are real, but it is far from clear to him that they are material.

The criticism of the philosophic realist reveals that the Leninist definition of matter characterizes its relation to the subject and does not contain a definition of matter itself. There is good reason to believe that Lenin realized this very well, but adopted a "relational definition" of matter nonetheless, since all alternatives to it were much more susceptible to attack. The point is that Lenin's definition of matter is both the *strength* and the *weakness* of dialectical materialism. The definition is strong because it does not depend on the level of development of natural science and thereby acquires much greater permanence. If dialectical materialists attempted to define matter itself—that is, in terms of the totality of its properties—this would mean that eventually the definition would become obsolescent as our knowledge of those properties changed, just as all definitions of matter given by previous materialists have become obsolescent (for example, those of the Greek atomists, who thought of matter in terms of indivisible units). Lenin considered matter to be inexhaustible in its properties and therefore undefinable in terms of them. This belief is one of the most significant differences between dialectical materialism and mechanistic materialism. The Leninist position avoided the built-in obsolescence of previous definitions of matter, but at the same time opened up dialectical materialists to the criticism that they have no clear way of demonstrating the superiority of the term "matter" to "objective reality."

Strictly speaking, of course, Lenin did not give a definition of matter. The principle of the materiality of the universe does not flow out of the Leninist position on epistemology quoted above. It constitutes, instead, a separate assumption. The assumptive character of materialism is not quite the fatal flaw that its opponents might try to make of it, since all conceptual systems contain some assumptions. Man cannot pursue his most human goal of trying to understand without paying the price of making some assumptions. It is a question of choosing one's assumptions carefully and remaining open to other possibilities. One can argue (for example, on the grounds of economy) that some assumptions are more justified than others; furthermore, one can defend materialism against realism on precisely this basis. But the assumptive character of materialism does mean that dialectical materialism does not uniquely flow out of the facts of science, as some of its defenders would maintain.

Turning now to the chapters of this volume and the problems of interpretation of nature described there, we can easily see that they contain many falsely inflated disputes and many examples of attempts at manipulation by political ideologues of issues that belonged to philosophers and scientists. Nonetheless, in all the issues except the Lysenko controversy and the resonance chemistry dispute, genuine intellectual questions were contained within the frameworks of the over-all debates.

And even in the chemistry discussion—prompted by a crude emulator of Lysenko—there was the very real problem of the significance and meaning of models in nature. The Lysenko affair alone was totally artificial from an intellectual standpoint; the few legitimate scientific and philosophic issues raised there either were outdated or were misunderstood by the supporters of Lysenko. I think that this assessment will stand even if the inheritance of acquired characters—by far not the only issue—should become accepted in future biology. Lysenko was incapable of understanding the biology of his time. It would be both ahistorical and inaccurate to try to defend him in the name of future biology. Such a defense of the theory of inheritance of acquired characters would be another matter.

In the other disputes, however, we frequently found talented scientists who fully understood modern science making arguments that are in many cases at least plausible while connecting them with dialectical materialism in ways that do not seem to have been merely the results of political pressure. Some of these opinions were "wrong," as judged by contemporary science, to be sure, but they are frequently still understood by contemporary scientists as legitimate and reasonable points of view in the context of their times. Some of them are quite relevant even yet, and more are being developed.

Among the Soviet scholars' viewpoints that have been recognized as valuable, either in their time or at the present time, and in which dialectical materialism may have played a role are: V. A. Fock's and A. D. Aleksandrov's criticisms of certain interpretations of quantum mechanics and relativity; D. I. Blokhintsev's philosophical interpretation of quantum mechanics; O. Iu. Schmidt's analysis of planetary cosmogony; V. M. Ambartsumian's views of star formation and criticisms of certain cosmological theories; G. I. Naan's "quasi-closed" cosmogonical models; A. L. Zel'manov's view of a "manifold universe"; many Soviet criticisms of the concept of an absolute beginning of the universe or of a nondevelopmental cyclical one; A. I. Oparin's views of the origins of life and his criticism of mechanism in biology; a number of Soviet philosophers' and scientists' views on cybernetic evolution of matter; L. S. Vygotsky's opinions on thought and language; S. L. Rubinshtein's concepts of perception and consciousness; and P. K. Anokhin's revision and extension of Pavlovian physiology.

Yet one should be very careful about assuming that the scientific views of any specific one of the above Soviet scientists were, in fact, importantly influenced by dialectical materialism. In the sentence at the beginning of the above paragraph I said that dialectical materialism "may" have played a role in the intellectual development of these scientists, not that it had in any identifiable case. There is, indeed, no way of demonstrating beyond a doubt that the views of a particular scientist were importantly influenced by intellectual Marxism. Such demonstrations are not in the nature of intellectual history in general, completely aside from the question of Marxism. We can show that a scientist held idea x, as evidenced in his writings, and we often can show that he stated in print that there was a connection between x and concept y in Marxism as he interpreted it. But there is no way that we can prove an actual causal link between x and y. There are, in fact, many possible explanations besides that of true intellectual stimulation. The scientist may have come upon idea x independently and then used y merely as a supporting argument. He may have created such linkages for no other reason than the political pressure being exerted upon him. He may actually have used such linkages merely for purposes of his career, aware that a scientific interpretation that could be called Marxist would have a better chance of receiving official favor.

What justification do I have, then, for advancing the interpretation that Marxism was important as an intellectual influence in Soviet science? My interpretation derives primarily from the impression given by reading enormous amounts of Soviet literature on philosophy and science written at times of greatly varying political pressures. I have tried to describe the characteristics of that literature in the previous chapters. It is my opinion that when one looks not at any one scientist and his views, but at the total corpus of the writings of the scientists mentioned above, one is justified in observing that their interpretations of science, and even, in some instances, their scientific research itself, demonstrate characteristics that one can persuasively argue derive in some degree from dialectical materialism. Furthermore, there is no clear relation between political pressure and the moments when the Soviet scholars wrote on dialectical materialism. Many Soviet scientists continue to write on dialectical materialism today, while many more never do. It is possible in the Soviet Union for a scientist to ignore totally dialectical materialism in his publications, a fact that should cause us to take more seriously those who continue to devote attention to it.

In the final analysis, the problem of causation in this study of science and Soviet Marxism is not fundamentally different from that in other areas of intellectual history and in other countries. Philosophy and politics influence scientists in all countries. The filiation of ideas is an extremely difficult process to study, but the attempt is worthwhile. Furthermore, dialectical materialism deserves particular consideration by historians and philosophers of science by virtue of its more intimate interaction with science than any other current in contemporary philosophy.

A sophisticated materialism that is open to criticism and debate, of which dialectical materialism might some day become a true form, is a philosophic point of view that can be helpful to scientists. It is most valuable to the scientist when his research approaches the outermost limit of knowledge, the area where speculation necessarily plays the greatest role, the approach to the cosmic, the infinite, or the origins or essence of forms of being. It is least valuable, and quite capable of being crudely used with harmful results, when applied to the immediate, the next stage of research. Dialectical materialism could not help a scientist with the details of laboratory work. It would never predict the result of a specific experiment. It certainly would never tell him how to raise crops or treat mental illness. But it might warn him not to fall prey to mysticism in the face of the sometimes overwhelming mystery and awe of the unknown. Through its nonreductionism it might remind him how contradictory and difficult the explication of nature is and how dangerous it is to reduce the complex phenomena on one level to combinations of simple mechanisms on a lower level. It might warn him that the sudden appearance of anomalies in research is not reason for discarding a realist epistemology or a commitment to the existence of at least some regularities in nature, whether probabilistic or strictly deterministic. It might remind him, through its emphasis on the interconnectedness of nature, of the importance of an ecological approach to the biological world and of the significance of the historical view for an understanding of the development of matter. It might encourage him to erect temporary explanatory schemes larger than any one science, but ones that do not pretend to possess final answers. At the same time, it might also assure him that the retention of commitments to epistemological realism and natural order is by no means a renunciation of art or mystery in nature. Nothing is more baffling than the ingenuity of man and his creations and the beauty of nature, of which man is a part. A sophisticated materialism could handle such considerations equally as well as a sophisticated idealism, and would start from assumptions that are more consonant with the naturalism implicit in most science.

Appendixes

Appendixes

Appendix I   Lysenko and Zhdanov

In Chapter I of this book, entitled "Introduction: Background of the Discussions," I commented that there is serious reason to question the existing interpretation by many historians of the role of Andrei Zhdanov in the ideological campaign in the sciences after the Second World War.[1] In this appendix I would like to give evidence for my doubts. I would also like to repeat my statement that without access to Party archives it is impossible to resolve definitely this issue.

One of the textbook commonplaces of scholarship on the Soviet Union is the view that the ideological campaign in the arts and sciences after the war was directed by Zhdanov, a member of the Central Committee of the Communist Party and one of Stalin's most favored assistants. The name that has been given to this campaign is the Zhdanovshchina. Many authors have considered the session of the Lenin Academy of Agricultural Sciences from July 31 to August 7, 1948, when Lysenko received explicit Party approval, the culmination of the Zhdanovshchina. Yet the role of Zhdanov is by no means clear. Recent evidence points more and more to Stalin himself; Zhdanov may actually have opposed the campaign in the sciences.

Some scholars would maintain that until the archives are opened, nothing meaningful can be said about the relations among the top leaders in the Soviet Union. To attempt to evaluate the roles of Stalin and Zhdanov, they would say, is an exercise in Kremlinology, a dubious enterprise involving the use of unreliable sources. This point of view is, in many ways, a persuasive one; Kremlinological analyses are frequently based on the slenderest reeds of evidence. I can not accept it in this particular case, however, for the following reason: To resist inquiry into the politics of the period is to abandon the field to the existing Kremlinological interpretation, one that is widely accepted even though the evidence for it is quite weak.

I shall cite a few well-known non-Soviet texts favoring the existing interpretation, not because I consider these authors to be more convinced of its validity than others, but merely to illustrate its currency. Leonard

443

Schapiro in his history of the Communist Party spoke of "the series of assaults conducted by Zhdanov between 1946 and 1948 on artistic, academic and scientific activity." [2] Georg von Rauch observed that "the climax of the Zhdanovshchina was reached in the discussion arising out of Lysenko's biological theories, which in the summer of 1948 led to the Academy of Sciences announcing its official position." [3] Donald Treadgold in his textbook on Soviet history described the session on biology in the following fashion: "In August 1948 Zhdanov carried his attack into natural science. . . . Lysenko's contentions were imposed upon the assembled Soviet biologists. . . . After crowning his work with this ideological coup, Zhdanov died suddenly on the last day of August 1948." [4] These three examples could be multiplied easily. Indeed, I accepted this understanding of Zhdanov's role in science myself until six or seven years ago.

If one looks for direct evidence to support the existing interpretation, it turns out to be exceedingly scarce. I have not been able to locate a single mention by Zhdanov of Lysenko, or even a statement by Zhdanov of the necessity to introduce ideological considerations into biology. Zhdanov's notorious speech of June 24, 1947, in which he criticized G. F. Aleksandrov's *History of West European Philosophy,* contains only one paragraph on specific issues in science, and nothing whatsoever on biology.[5] In his other major speeches and articles after the war Zhdanov was similarly silent.[6] The record of the biology conference itself, which occurred while Zhdanov was still alive, does not contain a single reference to him.[7] If Zhdanov were the proud architect of the conference, would he have gone unmentioned in the proceedings? Possibly, perhaps, but Lysenko said nothing on Zhdanov's role. Even in his eulogy after Zhdanov's death Lysenko made no attempt to portray Zhdanov as the defender of his view.[8] These omissions are sufficient to cause the first doubts about the current interpretation, as they did in my case. Zhdanov, the ideological spokesman of the Party, could hardly lecture on ideology and scholarship and omit reference to Lysenko only accidentally. In these years the phrase "ideology and science" made everyone think, rightly or wrongly, of Lysenko—everyone, it seems, except Zhdanov.

The most important piece of the puzzle is Iurii A. Zhdanov, son of Andrei Zhdanov and from 1949 to 1953 husband of Stalin's daughter Svetlana; Iurii Zhdanov after the war became the head of the science section of the Central Committee, no doubt partly as a result of his father's influence. Young Zhdanov was a chemist who in 1948 completed a thesis at the Institute of Philosophy of the Academy of Sciences on the philosophic significance of the concept of homologous compounds, a topic that once interested Friedrich Engels.[9] In later years he moved more directly into chemistry, eventually defending a dissertation at Moscow State University on the chemical synthesis of certain types of carbohydrates.[10] He also continued to publish on the philosophy of science.[11] Young Zhdanov in 1947 and 1948 definitely opposed Lysenko. He publicly criticized both Lysenko's leadership of the Lenin Academy of Agricultural Sciences and his views on intraspecific competition and mechanism of transmission of hereditable characteristics. In a May 1948 article on scientific creativity Iurii Zhdanov censured persons who try to make fiefdoms of certain areas of science:

A proprietary attitude toward science—when a scientist looks upon other people trying to conduct research in the same area with unhealthy jealousy, when he tries to use his position and authority to hinder the development of other directions in his area of science—cannot and must not take place in our science. We recall the fervent, even startling, words of I. V. Michurin: "My followers must outstrip me, contradict me, even demolish my work, while at the same time continuing it. Progress is created only by means of such a consistent demolishing of [earlier] work." [12]

The reference to the Russian horticulturist Michurin leaves little doubt that Zhdanov was thinking particularly of biology and Trofim Lysenko. It is difficult to imagine that Andrei Zhdanov would promote his son as a specialist on questions of ideology and science in the Central Committee if he did not agree with him on the most controversial question in this area, that of biology.

On the final day of the historic session of the Lenin Academy of Agricultural Sciences, and still slightly over three weeks before Zhdanov's death, *Pravda* carried a letter from Iurii Zhdanov addressed to Stalin. From this rather long letter I will cite only enough to suggest its spirit:[13]

To Comrade Stalin:

In a paper on controversial questions of contemporary Darwinism given at a school for lecturers, I certainly made quite a number of serious mistakes.

1. The very attitude of this paper was mistaken. I obviously underestimated my new position as a member of the Central Committee's staff, underestimated my responsibility, did not realize that my statement would be appraised as the official view of the C.C.

· · · · · · · · · · · · · ·

3. My sharp and public criticism of Academician Lysenko was a mistake. . . . I do not agree with certain theoretical propositions of Academician Lysenko . . . but criticism of all these deficiencies should not be done in the way that I did in my paper.

. . . .

I consider it my duty to assure you, Comrade Stalin, and in your person the Central Committee of the CPSU(b), that I have been and remain an ardent Michurinist. My mistakes derive from an insufficient preparation for the struggle for Michurin's teachings. All this is because of inexperience and immaturity. I will repair my mistakes in my work.

July 10, 1948   Iurii Zhdanov

Since Zhdanov Senior did not publish his own views on specific issues in science, any attempt to describe them is necessarily speculative. Nevertheless, certain events in the year before and the year after his death provide clues to his opinions. One of the results of Andrei Zhdanov's criticisms of Soviet philosophers in his June twenty-fourth speech was the creation of a new journal in philosophy, *Problems of Philosophy*. The relationship of the speech to the founding of the journal was indicated by the fact that the entire first issue was devoted to the speech and the ensuing discussion of Soviet philosophy of which Andrei Zhdanov was chairman.[14] Three issues of the journal appeared before Andrei Zhdanov's death; it is almost certain that he and his son, leaders of the Party's ideological concerns, played a large role in determining the policy of the journal. The striking aspect of these three issues is that they contained the most unorthodox articles on the philosophy of science during the period from the end of the war until Stalin's death. Rather than being rigidly ideological—as the current interpretation of the Zhdanovshchina would suppose—they were, in the Soviet context, genuinely innovative. An article in the second issue by the theoretical physicist M. A. Markov is even today probably the strongest defense of the Copenhagen position in quantum mechanics published in the Soviet Union. I discussed this article at length in the quantum mechanics chapter.[15] In the same issue Academician I. I. Schmalhausen, who was later forced to recant his views, published an article on biology that was clearly directed against Lysenko.[16]

On September 7, 1948, one week after Zhdanov's death, *Pravda* carried an article criticizing the editorial position of *Problems of Philosophy* and a series of articles on philosophy of science.[17] The first issue of *Problems of Philosophy* after Zhdanov's death contained a new and ominous tone: a letter to Lysenko congratulating him on his fiftieth birthday, and three articles defending Michurinist biology.[18] The next issue announced the replacement of the editor, B. M. Kedrov, by D. I. Chesnokov. Among the criticisms advanced against Kedrov was his sponsoring of Schmalhausen's article on biology.[19]

An obvious possibility issuing from this evidence is that Andrei Zhdanov disagreed with the ideological campaign in the sciences. It is impossible to know just what his attitude would have been toward many of the ideological controversies in the sciences of the postwar period, since most of these debates occurred after his death. Conferences similar in tone, if not in result, to that for the biologists were held in a number of other sciences: astronomy in December 1948 and April 1951; cellular biology in March 1950; physiology in June and July 1950; and resonance chemistry in June 1951. Zhdanov may well have disagreed strongly with Lysenko in biology but at the same time have been committed to the need for a Marxist interpretation of the sciences. Andrei Zhdanov's comments on science in the June twenty-fourth speech indicate that he was particularly hostile toward any attempt to justify religion or epistemological idealism through science. Here he was entirely in the tradition of materialism as a conceptual scheme and was, further, closely following Lenin, who directed his criticisms in *Materialism and Empirio-Criticism* in the same direction. There is little reason to believe that Zhdanov saw either the inheritance of acquired characters or any other of Lysenko's theories, such as the phasic development of plants, as required by dialectical materialism. And he would have been correct in this interpretation of dialectical materialism. On the other hand, Zhdanov was one of the most important supporters of the postwar ideological campaign in music, art, and literature. I am not, therefore, making an attempt to "rehabilitate" Zhdanov in terms of non-Soviet scholarship. I merely question the logic that says that since Zhdanov favored some of the intrusions in scholarship that took place after the war, he favored them all.

In the years 1948–51 ideological restrictions upon science were at a maximum in the Soviet Union. This was the period when even such prominent scientists as V. A. Fock and A. I. Oparin yielded to Stalinist pressures; as I have described in previous chapters, Fock at this time wavered in his support of the concept of complementarity in physics, and Oparin, who had frequently spoken favorably of Lysenko, temporarily supported the scientifically worthless views of Olga Lepeshinskaia in cytology. In 1951 Iurii Zhdanov published an article entitled "On Criticism and Self-Criticism in Scientific Work," in which he lauded Lysenko's theory of stages of plant growth and the practical results of his field work.[20] If one wishes to maintain that the evidence on Iurii Zhdanov's attitude toward Lysenko is so contradictory that nothing meaningful can be learned from the sources of the period, this article is important evidence of that contradiction. In this article Iurii spoke of the "significant contribution" to biological science made by Lysenko. It is my opinion, however, that this article has to be seen in the context of its time. In later works, when the political controls became more lenient, Iurii Zhdanov began moving away from Lysenko again. Furthermore, recent evidence from the books of Zhores Medvedev and Svetlana Allilueva, which will be discussed subsequently, supports the view that in 1951 Iurii Zhdanov was writing under heavy constraints.

On January 16, 1953, more than four years after his apologetic letter and shortly before Stalin's death, Iurii Zhdanov published in *Pravda* an article entitled "Against Subjective Distortions of Natural Science," which was interpreted outside the Soviet Union as marking a possible resurgence of "neo-Zhdanovism." [21] The article was, indeed, highly ideological in tone and discussed a number of the past controversies concerning dialectical materialism and natural science. A careful reading of the article leads one to believe, however, that it was directed *against* the view that is usually considered the zenith of the Zhdanovshchina, Lysenkoism. The subtlety of this article should not be missed: Iurii Zhdanov was in guarded Aesopian language trying to explain that he was a dialectical materialist of the first order, but that he did not see Lysenkoism as having anything to do with dialectical materialism. Taking as his point of departure Stalin's recent *Economic Problems of Socialism in the U.S.S.R.,* which contained a criticism of those people who did not recognize the "objective" laws of science, Iurii Zhdanov discussed all the recent important ideological issues in the sciences—physiology, resonance chemistry, quantum mechanics, relativity physics—except biology.[22] His neglect even to mention biology seems to indicate that he thought Lysenko's theories were irrelevant to dialectical materialism. Iurii considered the importance of ideology to science to be the need to recognize objective reality and the necessity to avoid subjective (anthropomorphic, teleological, idealistic) elements. And as a matter of fact, there was a great deal to be said for Iurii's position in view of Lysenko's anthropomorphic and teleological conceptions involving such oddities as "marriage for love" and "co-operation for the good of the species" in the plant world.

In 1959 Iurii Zhdanov published a little book entitled *Lenin and Natural Science,* in which he made many comments on contemporary science, including a criticism of the opponents of cybernetics. Not once did he mention the name of Lysenko. But his goal of separating Michurinism from Lysenkoism seems fairly clear. He maintained that the supporters of "formal genetics" had made mistakes; in particular, they minimized the importance of the external environment in determining heredity. He also felt that they refused to recognize "the facts" of vegetative hybridization. But his strongest criticism was reserved for the unnamed (surely Lysenko and his supporters) Soviet scholars who attacked genetics wholesale:

The critics of Mendelism swept aside the whole concept of heredity according to formal genetics and did not take the trouble to explain a series of findings of the defenders of formal genetics. . . . Individual scholars used the pretext of struggling against anti-Michurinist tendencies to ignore or discard without the slightest analysis the rich factual material gathered by cytologists, biochemists, and embryologists proceeding on the basis of formal genetics. This attitude was against the interests of materialist biology and hindered the development of several branches of knowledge, particularly biochemical genetics, chemical embryology, and radiation genetics.[23]

Iurii Zhdanov favored the view that in the final analysis internal hereditary factors were determined by the external environment. This was a reasonable opinion when organisms are being considered over the entire evolutionary scale from the origin of life itself onward. This was the essence of Michurinism to Zhdanov.

During the last few years new evidence on the question of the Zhdanovs and Stalin has come from several different directions. The most striking information came from a history of the genetics controversy published in 1969 in the United States; the author was Zhores A. Medvedev, a Soviet biologist. According to Medvedev, Iurii Zhdanov, as head of the department of science of the Central Committee, "came out against Lysenko in a number of speeches." Medvedev also thought that Andrei Zhdanov agreed with his son; according to him in the spring of 1948 Andrei "subjected Lysenko to severe criticism" at a meeting of the Central Executive Committee.[24] Further evidence came from yet another source. Over two years after I wrote the first draft of this appendix, Svetlana Allilueva, Stalin's daughter and Iurii's former wife, published her *Twenty Letters to a Friend*. Miss Allilueva recalled that shortly after their marriage her husband "was suffering on account of his work in the Central Committee. He didn't know, he couldn't imagine, what he had gotten himself into." [25] At this point in the text Miss Allilueva's translator inserted a footnote relating Iurii Zhdanov's criticism of Lysenko and of his apology to Stalin for his "mistake." [26]

Miss Allilueva returned to the topic in her second book, *Only One Year*. She wrote:

. . . my father gave his support to Trofim Lysenko, flattered by the false "practical conclusions" of that climber, who played cunningly on my father's weakness for everything "practical." I recall how in 1948 Yuri Zhdanov, who was then working in the Central Committee's Department of Science, came out against Lysenko, who, in turn, was instantly defended by my father. "Now genetics are finished!" said Yuri at the time. In obedience to Party discipline, Yuri Zhdanov had to "acknowledge his errors" and write a repentant letter to Stalin, which was published in *Pravda*. And then he was made to present the refutation of the chromosome theory "from a Marxist position." [27]

We see, then, that there is strong evidence for the supposition that persons who considered themselves loyal dialectical materialists—Andrei and Iurii

Zhdanov—were by no means convinced of the validity of Lysenkoism. They were bureaucratic protectors of the faith, to be sure, and not independent intellectuals, but they may have correctly seen nonetheless that the support by the Party of a charlatan could only, in the end, cause great damage both to their doctrine and to their nation. Their approach reminds one somewhat of a statement of St. Augustine, another defender of a doctrine, who wrote:

In points obscure and remote from our sight if we come to read anything in Holy Scripture that is, in keeping with the faith in which we are steeped, capable of several meanings, we must not, by obstinately rushing in, so commit ourselves to any one of them that, when perhaps the truth is more thoroughly investigated, it rightly falls to the ground and we with it.[28]

Pope Urban VIII forgot Augustine's sage advice when considering the case of Galileo; similarly, Stalin apparently ignored the advice of the Zhdanovs.

There is a more important aspect to the case of the Zhdanovs than what is specifically contained in the events just recounted. Unless the non-Soviet observers can learn to distinguish between dialectical materialism as an intellectual approach to nature and the vulgar "applications" of it found in Lysenkoism, there is little hope of understanding how distinguished and honest scientists in the Soviet Union (and other countries as well) could have sincerely defended dialectical materialism for decades and how some of them still defend it. Still less can one understand how some philosophers in the Soviet Union wish to elaborate dialectical materialist analyses of nature in spite of disapproval of their approaches by Party-controlled editorial boards. The concept of a dialectical materialism *in spite of* the Communist Party is a very difficult one for most Americans, since they see the one as being wholly the creation of the other.

Appendix II   H. J. Muller on Lenin and Genetics

The following article by the noted American biologist Hermann J. Muller (1890–1967) was published in 1934 in the Soviet Union in a book entitled *To the Memory of V. I. Lenin*. The article appeared in both the English and the Russian languages. The following pages are a complete and unedited reproduction of the English language version in the original publication. At the time it was published, Muller was a senior geneticist of the Institute of Genetics of the Academy of Sciences in Moscow.

The decision to include the article as an appendix was prompted by the realization that to protect Muller from his own philosophical beliefs during the most important part of his scientific career would be to repeat the mistake that has already been made far too often by historians of science. Great scientists, like all great men and women, are human beings whose views are conditioned by their times in very complex ways. In 1934 Muller considered Marxism quite important to his scientific work. Furthermore, it is clear that his Marxism was no short-lived affair. When Muller wrote this article, he was a mature scientist of forty-four years; seven years before, he had presented the paper on the induction of mutations in *Drosophila* by means of X rays, for which in 1946 he won the Nobel Prize. Furthermore, we know that Muller had been attracted to Marxism for many years, and he believed—erroneously or not—that it was a viewpoint relevant not only to politics, but also to science. In this article Muller even maintains that Marxism had influenced his views of biology twenty-four years earlier, at the time of the birth of genetics at Columbia University; he says, "It is interesting to note that in the rise of this so-called 'Morgan school,' better called 'the *Drosophila* school,' there was a strong direct Marxian influence. . . ." (See p. 462, and the subsequent section where Muller describes Morgan's young students influencing their teacher's research.) Historians of biology are certain to question this interpretation, and may even totally reject it, but it is quite difficult to contradict the fact that

Muller thought that it was true. To hide Muller's Marxist assumptions would be a distortion of his intellectual development.

Even if one accepts that Marxism was for many years an important part of Muller's scientific conceptions, what justification is there for including this article in a study of Soviet science and philosophy? First of all, Muller's obviously sincere interest in dialectical materialism is further evidence that quite a few distinguished scientists, Soviet and non-Soviet, have at one time or another considered dialectical materialism a valuable approach to science. I have maintained in this study that a number of the Soviet scientists described here were not hypocritically praising Marxism; Muller's praise of similar ideas while living in the Soviet Union is another example of the impact of Marxism on scientists.

Muller's article also supports the opinion that the leading Marxist geneticists in the Soviet Union believed from an early point that the inheritance of acquired characters and the denial of material carriers of heredity—views to be advanced by Lysenko—were contrary to dialectical materialism. No matter how disconcerting Muller's polemical style and political preferences may be to non-Marxist readers today, it should be noticed that if the views contained in this article had prevailed in the Soviet Union—and there is good reason to believe that they coincided with those of many Soviet Marxist geneticists, such as I. I. Agol and S. G. Levit—Lysenko could never have won control over Soviet biology. Muller delivers here a devastating attack on the concept of the inheritance of acquired characters, based on his understanding of materialism.

The article also deserves republication by virtue of the anonymity into which it fell; indeed, it almost disappeared entirely, as a result of its later intentional disregard by *both* its American author and its Soviet audience. In the Soviet Union, after the rise of Lysenko, Muller was condemned as a "Mendelist-Morganist," and his writings were suppressed. Muller returned to the United States, where he became quite critical of the Soviet Union after many of his Soviet Marxist friends and fellow geneticists, such as Agol and Levit, had died in the purges. The list of Muller's publications that I obtained from Indiana University after Muller's death does not include the following article, although the list seems to be exhaustive, containing 372 items published between 1914 and 1967.[1] The list does contain his popular articles critical of the Soviet Union, which were published after the Second World War. These observations are not meant as personal criticisms of Muller. After the war he was summoned before the House Un-American Activities Committee for a public hearing; Muller, along with many other people, was defending himself against political attack, and quite naturally, he did not point to the evidence that could be best used against him. Muller was, nonetheless, a very courageous man who warned repeatedly after the war of the dangers of weapons-testing and the genetic damage of radiation. As his colleague Tracy M. Sonneborn remarked, Muller never lost his faith in socialism despite his criticism of the Soviet regime.[2] And he was never a narrow ideologue or a believer in "scientism," as distinguished from science. In his book *Out of the Night,* published in New York in 1935 after Soviet authorities refused publication in the U.S.S.R., Muller wrote:

Man . . . can win the ability to push up the great avenue of happiness and power only by the continued and consciously fostered growth of that intelligence and cooperation which have brought him to his present status. . . . Intelligence, in the form of technical knowledge and ability in respect to material things, has of late grown astonishingly; but, without a corresponding growth in social motivation and in the means of carrying it out, man's great new tools—so much more dangerous and more easily misdirected on a large scale than were the primitive instruments of the past—may work only misery and even destruction. *Love must balance knowledge, or we fail.*[3]

In recent years interest in the Soviet Union in the analysis of genetics by Muller has grown. The Soviet biologist Zhores A. Medvedev (who, like Muller, could not gain permission for publication in the U.S.S.R.) in his 1969 history of Lysenkoism called the article "a very deep and serious dialectic-philosophic treatment of genetic problems." [4] I. T. Frolov, the current editor of the Soviet journal *Problems of Philosophy,* in his 1968 work on genetics and philosophy did not refer to this particular article of Muller's, but he quoted frequently from Muller's later criticisms of Lysenko.[5]

Muller's article is, of course, a product of its time and will seem dated in some of its scientific particulars and political emotions. But it deserves attention as an indicator of the attitudes of an important group of Marxist geneticists in the period in which Lysenko was beginning his campaign for power.

Lenin's Doctrines in Relation to Genetics

Hermann J. Muller

If geneticists had been able to realize and correctly to apply the general principles advocated by Lenin with regard to the basis and the development of scientific thought, and if they had followed the methods which he used in his own thinking on scientific matters, the science of genetics would have developed much more directly and rapidly in the past, free from the chief deviations to which it has been subject. Instead, it has learned slowly, through costly mistakes, and is still much encumbered with the psychological impedimenta of the past age. On the other hand, the great advances which genetics has made, in spite of these hindrances, furnish additional confirmation of the correctness of the principles on which Lenin laid stress in his discussions of scientific matters. The clear formulation of the essentials of Lenin's standpoint and procedure in so far as they are pertinent to natural science, should be correspondingly invaluable in the guiding of the course of the future development of our science.

The principle of natural science with which Lenin concerned himself most was that of materialism. Most of Lenin's book "Materialism and Empirio-criticism" (the work in which he has most to say about natural science) was devoted to bringing out the issue of materialism versus idealism in a clear cut fashion: to showing that true science could only be unequivocally materialistic, to analyzing in detail various recently advertised substitutes for materialism which claimed to be not idealistic either, and to exposing these fabrications as inconsistent muddles of the two points of view. So long as any traces of feudal and of decadent bourgeois attitudes remain extant among intellectuals—and that means for a considerable time yet to come—it will be necessary to continue vigilantly in this work of detecting, exposing, and extracting from scientific theory these growths and tendencies which are the products of veiled anti-materialism, masquerading in the costume of "empiricism" and in various other disguises. Genetics has been and is no exception to this rule. But since Lenin, of course, did not concern himself directly with this field, it is necessary for geneticists to be all the more assiduous in understanding Lenin's principles, so that they themselves may make proper application of them.

Perhaps the most basic and obvious application of materialistic thinking to genetics is the primary recognition of the materiality of that most fundamental object about which all genetics revolves, namely the gene. To a so-called "naive" scientist, the Mendelian rules of heredity afford a direct argument for the existence of definite hereditary materials, or "genes," which are subject to these laws of distribution. Thus it was to have been expected that some anti-materialists, following the re-discovery of Mendelism, would tend to align themselves against it, and to oppose its extension and generalization. It was only natural that Karl Pearson, the frank idealist with whom Lenin has so frequently taken issue, and the developer of a purely conceptual, scholastic and empirical treatment of heredity, in elaborate mathematical form, should have been the most outstanding figure in this reaction against Mendelism. His opposition diverted much valuable scientific energy from the correct path of development of genetics, and it has only been in the past few years that a group of mathematicians (notably Wright,

Haldane and Fisher) has grown up within the ranks of the Mendelians, to develop the mathematical side of genetic theory in a way in which it could and should have been done thirty years before. We may note in addition that some other branches of science for which Pearson had largely formed the connecting link with biology proper (for instance, much of anthropology and sociology), were kept from utilizing true genetic principles in their development through this reactionary stand of Pearson in genetics, due ultimately to his idealistic preconceptions. Physical anthropology to this day remains largely "biometrical," and is barely beginning to feel, internally, the effects of the genetic revolution which must ultimately cause its entire reorganization.

The above opposition to Mendelism itself, on the part of Pearson and various others (for instance, among the botanists, Harper, among the paleontologists, Osborn) represented only the first anti-materialist wave which the young science of genetics had to stem. For many of those who accepted the Mendelian rules as such adopted a second line of anti-materialist defence, in that they endeavored to put them upon as vague as possible a basis. It was no doubt the feeling that the identification of the Mendelian units, the genes, with the chromosomes was merely a materialist vulgarity, which, as much as anything else, held Bateson and, following him, almost the whole British school of Mendelians (excepting Lock and Doncaster) and much of continental Europe as well, aloof from taking part in this so fertile liaison. And yet it was evident very early, from the extraordinary parallelism between the methods of chromosome and gene distribution, that the former must (barring almost a miracle of scientific coincidence) constitute the visible material basis of the latter. Not to recognize this was but one way of staving off the advance of materialism longer, and it is no accident that we find the center of this reaction in England, the land where idealism probably has its firmest roots in intellectual circles (note the parallel movement in physics now centered there). This resistance to further genetic development even on the part of Bateson—one of those who had at first done the most in helping to establish Mendel's laws as such—probably impeded the progress of genetics more than did the opposition of the non-Mendelians, since it led away from fruitful lines more of those who could otherwise have taken a real part in genetic advance.

Of course the refusal of Bateson and his school to accept the chromosomes as the basis of Mendelism was not alleged to be any opposition to materialism, but pretended to be founded in a kind of empiricism, as has so often been true in other similar cases, as, for example, in the case of the Machian "empirio-criticists" whom Lenin attacked. Most of the other reactionaries in genetics also have avoided committing themselves openly to an anti-materialist position. Their background of anti-materialism has usually shown through, however, in their refusal to recognize concrete materialistic explanations when the evidence for the latter had become so strong that it should have carried conviction. Confronted by such situations, they continued to interpose ill-conceived objections and to offer, in place of the concrete materialistic theories of the advancing section of scientists, alternatives which on analysis could be shown to rest upon vagaries of thought not expressible in concrete terms. It is common among some of the thinkers of this type in biological sciences, especially among the physiologists, to hide behind the words "physiological" and "dynamic," as opposed to "morphological" and "static." They make a pretense here of using the real truth that all matter and structure imply motion, but they do this simply in order to avoid having to give a concrete idea of these material structures, therewith conveniently forgetting the complementary truth that, contrariwise, structure must also be involved in all motion, in all that is "dynamic." Thus their show of subtlety is in fact but a means of escaping into a realm where materialism is at least half dissolved. Accordingly intellectuals of this type are to be classed with that group of compromisers, of evaders and corrupters of materialism whom Lenin deemed it even more important to expose and to combat than the frank out-and-out idealists themselves. All this by no means implies that the materialist must have a ready explanation for all phenomena of nature, or that each process must have as its basis some one object or thing, in one-to-one correspondence. He must, however, be ever on the alert to discover concrete processes in operation, definable in exact ways, that is, in terms of the movements of given materials. The way of science is to look for such explanations and then to test their truth, to proceed further in like manner. This is very far from vague talk about "tendencies," about "dynamic" relationships, etc.

In addition to the above groups of compromisers and sabotagers of materialism, there have been some biologists who, while accepting Mendelism and genes, and in some cases even some sort of connection of the genes with the chromosomes, nevertheless, adopting a third line of defense, have stated outright that these genes must be regarded purely as "concepts," that is, as mental abstractions, not as real things. Chief among these were Johannsen and East. They had to guide their course, however, within such a difficult no-man's-land of paradox (like the Oswald school of non-atomists in physical chemistry) that it was not possible for their following to grow very large, before the falsity of their position became too manifest.

The above discussion has concerned itself chiefly with that phase of genetics which deals with the processes of heredity, whereby the biological bases (the genes) for the traits of one generation become transmitted to and apportioned among the next generation. In this field, despite the opposition of the reactionaries, the concrete materialistic chromosome and gene theories have at last emerged definitely victorious. Another equally important side of genetics is that dealing with the negation of heredity, in other words, with variations of the genes—with the process whereby genes of one generation come to be in some measure different from those of the preceding generation. It is by a kind of synthesis of these two opposed processes, of heredity and variation, that evolution occurs; to have a proper understanding of evolution we must therefore have a correct conception not only of heredity, but also of variation. But it appears that in the latter field, even more than in the former, progress has been hindered and much confusion made through the workings of anti-materialist ideologies.

The true materialist immediately realizes that the nature of the variations which occur in the hereditary material can not be determined by the needs of the organism. For that would imply some kind of purposive guidance, inner or outer, of these variations and would necessitate an ultimately teleological and therefore idealistic basis of living things. The big stumbling block to the acceptance of the generality of non-purposive variation was, however, the very elaborate and obvious adaptiveness of all living things, including that of practically all their parts. As Haeckel pointed out in his "Riddle of the Universe," this apparent purposiveness remained till the middle of the nineteenth century the weightiest argument of the teleologically minded. It was Darwin's greatest contribution that he pointed out that this adaptiveness was merely the long-term resultant of non-purposive variations—variations which were accidental so far as their having any relation to their possible later usefulness was concerned—since the mechanism of multiplication of organisms involved the "natural selection" of those variations which we term useful or adaptive, and the dying out of the others. The value of this great contribution of Darwin's as an indispensable basis for materialism in general was of course recognized by Marx and Engels. And Lenin delights in the dismay of all veiled as well as open anti-materialists which was later caused by the popular work of Haeckel above mentioned, which had, as its chief thesis, the setting forth of the ateleological nature of this Darwinian basis of biological phenomena. Despite all of Haeckel's faults, his exceptionally strong side is, says Lenin, his exposition of the triumph of naturo-historical materialism. In this exposition, the doctrine of the basically non-adaptive nature of genetic variation necessarily occupies one of the two key-positions; the other key doctrine is the related one of the dependence of the mind upon the brain.

As has been stated above, the general non-adaptiveness of variations is a necessary doctrine for materialism; it will readily be seen, further, that, if we grant the non-adaptiveness of variations (except as adaptiveness may occur accidentally), then the doctrine of natural selection remains as the only possible materialistic explanation of the emergence of adaptive organisms, as resultants of this series of non-adaptive variations. Therefore the denial either of natural selection, or of the generality of variation of a kind that is accidental, in the above sense, is tantamount to a denial of materialism, even though such attempts may and very often do hide under the cloak of a seemingly materialistic method of explanation. We find accordingly that ever since the publication of Darwin's "Origin of Species," attempts, sometimes isolated and sometimes in powerful groups, have been renewed, with the object of tearing down one or the other, or both, of these two fundaments of scientific biology. Following Lenin's example in his treatment of the Machian school in physics, it is the duty of clear-minded biologists to see through the pretensions of those espousing all such allegedly new-fangled doctrines, and to reveal them ever again for the well-known asses in lion's skins that they are. Marvellous are the names with which these awesome garments have been provided by their occupants—ranging from the "Mneme" of Semon, down through to the "Aristogenesis" of Henry Fairfield Osborn, but nearly all of them can be comprised under the old categories of directive germinal variation (Orthogenesis) or inheritance of acquired characters (Lamarckism).

When from out of the more general biological thought and activity of the past century the concrete science of genetics began to crystallize, and to show itself to lie at the core of the problems concerned with the origination and elaboration of all biological phenomena, it was only to be expected that it would now furnish the chief stage on which the anti-materialists would seek to give a demonstration of these claims of theirs. As a matter of fact, unfortunately for these aspirations, few of them had the mental acuteness for real genetics; however, they constituted themselves a more or less invading fringe around and to some extent within this field. And their activities did much to hold back its development, besides creating for outsiders to genetics proper an impression of disorder, from which they might select that which suited them. Time and again the claims were made, by people who seemed to outsiders to be responsible geneticists, that it could be proved that variations were directive, in that in the presence of given environments or external agencies, definite sorts of changes, of evolutionary significance, were regularly produced. If this were really the usual method of evolution, the scope of natural selection would become so greatly restricted that it could no longer serve as an explanation of adaptation. And since, in evolution, a given group does become adjusted in a complicated way to the kind of environment surrounding it, this method of evolution would have to mean that, in the main, the variations were adaptive, i.e., purposive, i.e., non-materialistic in basis. It is easy to see, then, that all such claims really involve the idea of orthogenesis and that this in turn implies an Aristotelian perfecting process, that is, purpose in nature. Of course all these attempts have eventually proved ill-founded, but much energy has thereby been diverted into useless channels. And the battle is always being started afresh—witness, most recently, the widely advertised claims of Jollos, which are being eagerly seized upon by those wide circles of anti-Darwinian taxonomists, paleontologists, physiologists, ecologists, etc., who are ever hungry for such stimulants.

The Lamarckian variety of anti-selectionism is less obviously teleological than the orthogenetic variety, and yet it is easy to show that it, too, must have an anti-materialistic basis. For while it is sometimes postulated that the inheritance of acquired characters is effected through some kind of a mechanism (though at other times some sort of inner or outer spirit has frankly been held responsible), nevertheless the existence of such a magic, incredibly versatile mechanism within the basic organization of all living things, including even the simplest (involving, at the same time, the existence of the various mechanisms whereby the somatic adaptations themselves were produced, and also of those whereby the latter were then transmitted to the next generation), would be a fact which in itself could only be "explained" through some act of purposive creation. This is much more evident now, in the light of our knowledge of some of the mechanisms of somatic adaptation, of reproduction, and of inheritance, than it was in the days of Darwin, who still thought he could admit some Lamarckism, in a secondary role.

A somewhat similar criticism holds also for those views concerning the relation of genetics and evolution according to which the latter has taken place principally by means of the formation of new combinations of existing genes through hybridization, or by the loss of pre-existing genes (or gene parts), or both. For here we would be driven back to a conception of the primordial living matter as having already pre-formed within itself all the germinal organization necessary for its most complicated later evolutionary developments, and the original creation of such an elaborately pre-determined vehicle of fate would be an unparalleled and inexplicable mystery, surely requiring the direct intervention of a creator. These latter ideas (unlike those of Lamarckism, which never gained much foothold in genetics) have played an important role in determining the trends of a considerable part of genetic work, even in recent years—work which otherwise could have been directed more usefully. It will be seen, however, that before such direction could be effective, ruthless analysis of the basic ideas, or confusion of ideas, involved in the viewpoints in question, carried on in the same spirit as that in which Lenin criticized the empirio-criticists, would have been necessary.

To those scientists who would protest that we should not make such pre-judgements regarding scientific possibilities, on the basis of a prior "philoso-phical" assumption of materialism, but should rather follow in any direction in which the empirical facts of the case seem to be leading, we may retort, with Lenin, that all the facts of daily life, as well as those of science, together form an over-whelming body of evidence for the materialistic point of view—one which we need not here digress to outline—-and that therefore we are justified, in our further scientific work, in taking this principle as a founda-tion for our higher constructions. It too is ultimately empirical, in the better sense of the word, and it has the overwhelming advantage of being founded upon the evidence as a whole, rather than upon just some restricted portion of the latter. By this it is not implied, however, that the deviations in evolutionary and genetic theory above referred to really had the em-pirical evidence even of their own field in their favor; the latter had to be greatly warped, and looked at purblindly, in order that it should seem to furnish evidence for these views which were themselves in truth the products of idealistic predilections.

The influence of these tendencies of idealistic origin in genetics, as in the fields with which Lenin chiefly dealt, extended a good deal further than merely to the works of these persons who were themselves idealists or even semi-idealists in their philosophy, for there were of course many naive natural scientists, and even would be materialists suffering from confusion of thought or not sufficiently analytical to see the philosophical implications of some of the doctrines being pressed upon them. (Examples of this sort of thing in fields ancestral to genetics are furnished, in fact, by some of the utterances of Haeckel himself, as Lenin has pointed out.) And in this category certainly belonged T. H. Morgan also, even though he is now regarded by many of those unfamiliar with the actual development of his points of view as the leader of the frankly materialistic movement in genetics. For although Morgan did, from the time of his first genetic work, have a general standpoint in science which should certainly be classed as material-istic, although, be it noted, not Marxistic, nevertheless he failed to grasp the role of natural selection as the sole factor determining the adaptive course of evolution, and so he was influenced by the then prevalent reactionary views into attempting to explain the direction of evolution through the character of the environmental agencies which supposedly produced the variations, and through an additional (also borrowed) hypothesis of variation tending to continue further in any direction in which it has once begun (see his article on the subject in "Science" in 1910). He tried to find still a third, and this time original, escape from natural selection in the fact of dominance, which he erroneously thought would cause a given type automatically to increase itself at the expense of the rest of the population (see his contribution in the series of lectures epitomizing the results of scientific progress in different fields, given at Columbia University in 1909). He opposed, moreover, the idea of the noncontamination of the gene in crosses, a basic principle of Mendelism which is necessary for the efficacy of the natural selection process, and the lack of which constituted the chief weakness of Darwinism in the nineteenth century (see his scheme of gene contamination in the "American Naturalist," 1910). And in the face of more and better evidence than had ever confronted anyone else before (we must remember that he was in the same laboratory with Wilson, the soundest cytologist and general biologist of the time) he long resisted the acceptance of a generalized chromosome theory of heredity.

It was in fact in the hope of gaining evidence for his various substitutes for Darwinism that Morgan began in 1909 his experiments with *Drosophila,* which had been shown by the prior work of Castle, Lutz, Stevens, and Moenkhaus to be well adapted for such a study. The results proved, however, to be glaringly at variance with his views, and at the same time he soon found himself pressed in his interpretations, by a small group of younger co-workers occupying the official positions of "students," whose ideas, despite their officially subordinate position, Morgan realized that he should take seriously. These "students" had been influenced greatly by their studies under Wilson, and even more by Lock's remarkably prophetic book ("Recent Progress in the Study of Variation, Heredity and Evolution," 1906, 1909), which has precisely the modern standpoint upon all essential questions of heredity (role of the chromosomes and their interchange of linked genes, universality of Mendelism, multiple factors, and even something of "balanced lethals"). Slowly and against his will Morgan was forced to give way to this double pressure of facts and arguments. So far as concerns the role played here by the winning of the facts themselves, it may be remarked that the earlier of these, in 1909–1911, were mostly contributed by himself, while the great bulk of the facts of real significance, subsequent to 1911, and practically all after 1913, were found by the younger workers quite independently of any guidance from him, in experiments which they had planned on the basis of their own more advanced viewpoints. Their results and interpretations were, however, later accepted by Morgan and presented chiefly by him to the scientific and lay public, so that these developments have sometimes been referred to, especially in circles farthest removed from contact with the original work, as "Morganism." In this way did the central trend of modern materialist genetics have its origin.

It is interesting to note that in the rise of this so-called "Morgan school," better called "the *Drosophila* school," there was a strong direct Marxian influence, which played a part whose importance it is difficult to measure quantitatively. This is not the place to go into personal details; in this case these details belonged in that category of accidents which are so dependent upon the conditions of the time that they are eventually bound to happen repeatedly, in one form or another, and so in time produce predictable consequences in the manner recognized by Marx and Lenin. It need only be said that a portion of the group of younger *Drosophila* workers had working class connections, were class-conscious, and had absorbed a Marxian, materialist viewpoint, which had in fact played a role in the choice and direction of their occupation. As has been explained, there is on each of the questions mentioned only one stand which, when analyzed, is really consistent with materialism. It is not strange, then, that it was exactly this portion of the group, and others in proportion to their nearness to this portion, who were the most militant in the formative years of 1911 to 1915, in pressing forward the acceptance of a clear-cut, generalized gene and chromosome theory of inheritance and of mutations, and a correlative ateleological conception of evolution, as occurring through the natural selection of rare accidental mutations in these otherwise stable Mendelian genes. The plentiful results obtained by the others (some of whom were given more time and opportunity for experimental work and writing), when regarded in this light, were found increasingly to provide the desired direct empirical evidence for these views, and this led to the planning, chiefly under the instigation of the more outspoken materialists, of still further experiments, which continued to solidify and extend this entire materialistic theory of genetics. We see, then, that although it was without knowledge of Lenin, that section of genetics which has proved itself to be nearest the truth was nevertheless carried forward, at least in part, in the spirit of Lenin, because it worked, partly unconsciously, but partly even consciously, in the spirit of Marx. If we can make this spirit still more conscious in the future, it should be correspondingly more effective in its results.

There is yet another way in which the method of mental procedure of Lenin, in common with that of Marx, is applicable to genetics, and for the lack of which much scientific energy has been and is even yet being misspent. I refer to that subtle, more critical and effective way of thinking known as the dialectic method, in distinction to the crudities of what Lenin terms "vulgar materialism" or "vulgar mechanism." In the early stages of the rise of Mendelism most geneticists, and conspicuously the Bateson school, fell into a naive fallacy with regard to the manner in which the hereditary units, the genes, were related to the visible characters of the organisms in which they were contained. Each gene was assumed to determine a particular physical character in almost one-to-one correspondence, through some unanalyzed and presumably direct mechanistic relationship. This mistake constitutes a typical example of over-simplification in science, combined with a static, scholastic, cut-and-dried attitude towards the objects being studied, a view neglecting the complicated processes ("movements" in the Marxian sense) whereby these objects are interrelated to one another and undergo their development. A realization of the complex realities of matter, especially of living matter, of its inter-connectedness, of the determination of qualitative by quantitative differences and *vice versa,* would have shown the untenability of this "gene-equals-character" point of view from the start. On the whole, genetics had to educate itself in this respect through a long and weary course of experience, although in the *Drosophila* laboratory the essentials of this criticism (pointing out the necessary dependence of one and the same character on many genes and of many characters on one and the same gene, the unpredictability of the results in untried gene combinations, the impropriety of deducing presence of a gene from presence of a character or from dominance, etc.) were urged by the more advanced section in 1911–1912 on general theoretical grounds (*vide* manuscripts of that date). The fact that the results in genetics failed to bear out the above too naive expectations was then taken by some geneticists to mean that the Mendelian laws were wrong, or required modification, and a false antithesis was likewise set up, in some quarters, between Mendelism and Darwinism—an antithesis which has not yet been everywhere eradicated from the minds of biologists. But, as genetics has progressed, it has been painfully forced, by its data, finally to the necessary dialectic position in regard to this question of the gene-character relationship. There are still backslidings, however, as when "sub-genes" were recently postulated which, like the genes before, are now supposed to have the above simple relationship to the finished characters of the organism. It is therefore still necessary to remain alert in this respect.

Another very grave fallacy, of what may be called a dialectic nature, to which the great majority of geneticists who have at all concerned themselves with the questions at issue have been subject, is the bodily taking over of the biological principles (which we might now more specifically call the genetic principles) concerning evolution through the struggle for existence, and the quite uncritical transference of them to human affairs, with the simple substitution, in the biological formula, of nations, races, classes, etc. for biological species, and of the social struggle for the biological struggle. From this our biologists and geneticists of course draw the conclusion that a continuation of the competitive system of business, of a class society, and of race divisions necessarily constitutes the path of biological betterment (whatever that may be defined or felt to mean), and therefore, too, of betterment in general. Thus the natural scientists throw overboard all of social science, forgetting the special laws which apply to social processes and structures, that put the latter on a different level, as it were, from the simpler biological relationships of non-social, non-intelligent organisms.

It may be instructive at this point to show by a more concrete analysis just what a large combination of serious fallacies at once is involved in the above false theorization. In the first place, our biologists neglect to realize the purely biological fact that natural selection can not be trusted to invariably bring about the improvements in adaptation that may be necessary for a species, since in fact most species become extinguished only a comparatively few remaining, which branch out and compensate for the loss of the others. These others sometimes actually become, before their final extinction, less well adjusted than they were, through the natural survival of characteristics that, while temporarily helping the individual and his immediate offspring, do so only at the expense of the species as a whole. Secondly, it is conveniently forgotten that biological "progress," in the sense of increase in numbers, is by no means always synonymous with the sort of progress which we can in any sense consider desirable. Thirdly, the social groups under consideration fail, in important ways, to form real biological groups such as a zoologist deals with in his studies of evolution, and (4) the social struggle is at the same time very far from being identifiable with the biological "struggle for existence," since economic success does not necessarily mean more effective biological multiplication and it may mean the reverse. The further assumption (5) is made that the social battles of today, dependent on our present-day economic system, can continue indefinitely, just as if there were no principles of economic and social evolution *per se,* which will, as Marx showed, change the conditions of social life entirely (and therefore too, of life considered from the point of view of the individual). (6) In accordance with the above pre-conceptions these "social Darwinians" must necessarily regard the groups and individuals who happen to be economically dominant at the present time as genetically superior in regard to those traits which would be philosophically considered as desirable, in spite of the fact that neither natural nor social science has been able to bring forward any valid evidence for such a point of view, and in spite of contrary evidence from history. How is it possible for people who are considered scientists to commit so many fallacies at once? It is of course because of their class bias, because this fabrication of theirs forms a necessary part of the apologetics of the contradictory system upon which the continued existence of their class as such depends. A knowledge of Marxism and Leninism very quickly leads to a dissolution of the above superficial system of rationalization. Lenin expressed himself quite definitely with regard to the specific problem in question, in his comments upon the biologico-sociology of Bogdanov and of Lange. To quote a few words from him in this connection: ". . . there is no available application of the ideas of 'selection,' . . . and so forth, in the province of the social sciences. They are simply shallow phrases . . . Nothing is easier than to tack on the labels of 'energetics' or 'biologico-sociology' to the phenomena, say, of crises, revolutions, class struggles, etc., but there is nothing more sterile, more scholastic and deadly than an occupation of this sort . . . the application of biologic ideas in general to the domain of social sciences can only result in a meaningless phrase . . . the 'social energetics' of Bogdanov, his modification of Marxism through the doctrine of social selection, exemplifies that sort of phraseology" (V. I. Lenin, "Materialism and Empirio-criticism," Collected works [Engl. ed.], Vol. XIII, pp. 283–284).

Since the general biologists have perforce handed over to their more specialized group, the geneticists, the major tasks concerned with selection and its effects, the above criticism now has a special pertinence for the latter. The public naturally looks to them as the authorities in this field, and tends to believe their pronouncements, yet most of the geneticists in capitalist countries are shamefully ignorant and prejudiced in regard to social affairs, and allow themselves to be the instruments of propagation of the above bourgeois philosophy. Going now beyond the existing system as it is, the majority of the group known as "eugenists," which includes many semigeneticists and pseudo-geneticists and also not a few who have made real contributions in technical genetics, are advocating the interference of man in the processes of biological selection with the object of artificially increasing the numbers of those types which today tend to obtain economic dominance. In other words, basing their argument upon the above fallacies they turn from defense to aggression and, as the events in Germany have shown, their movement is readily adaptable to the aims even of the most extreme fascism, serving, in fact, as one of the most conspicuous instruments of the latter. As yet geneticists as a body have taken no stand against these atrocities committed by some of their representatives, and very few individual geneticists have raised their voices in protest. Here, then, in this pretended application of genetics, we have the most obvious and disastrous example of perversion of the science through lack of use of the principles which Lenin espoused.

But while Lenin was opposed to the crude "vulgarly mechanistic" carrying over of physical or biological principles to social, he was heart and soul in favor of the further development of science at all "levels," and of the rational application of its findings in the service of man, for he realized what a profound influence upon social possibilities advances in theory and practice of the more basic sciences could exert—witness, for instance, his drive for electrification. And his references to Darwinism show that he appreciated the profound significance of biological evolution and of the processes of fortuitous variation and biological selection which produce it. While, therefore, Leninism must stand diametrically opposed to any pretended developments of genetics of the type discussed in the preceding paragraph, it must actively support the progress of true genetics, the study of the relation of genetics to social life, and the socially directed application of genetics in the betterment of life. Lenin took obvious delight in quoting Marx's maxim that "philosophers seek to understand the world; our task is to change it." And, as his whole life showed, he was not afraid of the application of this principle throughout every sphere of human existence, regardless of past superstitions and of outworn practice that had subserved the interests of the feudal or bourgeois systems. It is therefore up to geneticists now, in the temper of Leninism, not only to work more effectively forward in developing the theory of our science, freer than before from anti-materialistic and anti-dialectic deviations, but also to re-examine thoroughly the whole field of the practical application of genetics to human affairs, and to remodel accordingly much of our research in applied genetics.

For it is first possible in a socialist society to make an adequate attack upon the problems of improving our breeds of cultivated plants and domesticated animals, since only there can the work be carried out upon an adequate cooperative scale, with the backing of sufficient critical research, and with the possibility of using sufficiently long-term methods. In this field the necessary preliminary expenditures are so great, and the profits are in some cases so long postponed, as well as, in individual cases, so uncertain, that even the greatest capitalist may not dare to venture upon it, despite the overwhelming profits to be eventually gained, by the work as a whole. The check on the capitalist in this field is further strengthened by the fact that improved varieties can not long be "cornered," owing to the power of self-multiplication of living things, so that any profits must usually accrue in large measure to the community, rather than to the producer as such. This makes it necessary that the community itself be its own producer in this instance. As a result of these and other circumstances, we do in fact find by far the greatest enterprises in applied genetics in the world already located in the USSR. Our responsibility for correctly developing the principles thereof in the spirit of Leninism, as outlined above, becomes correspondingly increased.

In the realm of the study of genetics in relation to man we also find that the most serious work, in proportion to its means, is now being carried out in the USSR, as a result of similar circumstances. I have outlined in another article (" 'Rassenpflege' and Genetics") the tasks confronting the science of human genetics in a socialized society which wishes to provide its human material with as good as possible an environment for its development and for its efficient work. As has previously been pointed out by Levit, who has laid a solid basis for such work in his own institute, a most intricate and far-reaching program of investigation is called for here, in which the labors of physicians, professional geneticists, psychologists, educators and administrators must be harmoniously combined in mutual knowledge and action. For the hereditary units, the genes, of man, as of the higher animals, are myriad and manifold and complex in their interactions with one another and with different environmental situations, and in the possibilities of their operation. Stupendous enough are the problems which confront the scientist in the pioneer work of mapping out the genetic structure of a population, and of the individuals comprising it. But the resourcefulness and cooperativeness of science will be even further taxed in the work of learning how best to adapt methods of medical treatment, of general mode of living, type of work, kind of education, etc., to the specific genetic compositions which the human beings concerned will then have been found to have, so as to increase their health, happiness and productivity, not merely negatively, by doing away, so far as possible, with the diseases and with the less pronounced inefficiencies and disharmonies that are so common, but also positively, so as to develop most actively all especially valuable potentialities. Lenin was prominent among those who pointed out that the natural differences between men made an identity of treatment of each not the ideal condition, but that, on the contrary, as higher stages of communism were approached, each individual should be taken into consideration in accordance with his own need and ability, in such a way that he and society would receive the maximum benefit. An elaboration of this idea, in terms of modern genetic knowledge, brings out as a necessary implication the principle that existing genetic differences must be recognized, in order that optimal material adjustments can be made. This recognition involves us in the great program of work just mentioned, inasmuch as the detection of genetic differences and their disentanglement from environmental effects is, as all the true genetic findings of the last years have shown us, a matter for highly technical research and not one to be arrived at by the simplified methods of the social Darwinians and eugenists who, as above pointed out, assume all obvious social differences to be genetic in basis. Throwing overboard their simplifications and perversions, then, we are confronted with the serious work of construction on the basis of the realistic recognition of both genetic and social truths at once.

With the progress of socialized research, the reality of the genetic differences between one individual and another, differences extending to traits of the most diverse kinds, will become ever more convincingly and detailedly demonstrated, without the results being vitiated and corrupted by the bias of bourgeois prejudices. In accordance with these newer and surer findings it will then become increasingly evident that not only should the environment (training, treatment, etc.) of the individual be adjusted so as better to fit his own genetic equipment, but also that conditions be increasingly controlled, so as to allow the individuals of later generations to receive as advantageous as possible a genetic equipment. This brings us before another vast body of work in the future, which involves not only the co-operation of those in all the fields of activity mentioned previously, but also of those in social sciences in general and in philosophy, to aid in the working out of optimal methods and objectives. At the present time, however, few beside the general biologist and geneticist, who have vividly in mind the stupendous results of the biological evolution of the past, which brought living matter upward from microbe to man, and who are aware of how much we can now artificially improve upon nature in the speed and scope of biological evolution, can realize the vastness of the potentialities thus raised. Capitalist society, with its idealist and religious background, would naturally shrink before such revolutionary implications, but for Marxist-Leninist dialectic materialism, in consideration of the facts of genetics, it is one of the logical phases of the world movement, and only in a society freed from the prejudices and conflicts of class, race and sex, and from the encumbrances of religious superstitions and customs, can an effectual attack be made upon this hitherto impregnable recess of nature, this last stronghold in which the gods of the past still find some refuge. With socialist enthusiasm, however, this hidden way can be opened up to furnish a new avenue of unending victory for the triumphant workers.

Lenin, along with Marx and Engels, emphasized the continuance of the process of evolution, the transient nature of things that are, the fact that no general end stage had yet been reached, and that there was no reason to suppose that it would be reached. As this applies with regard to social affairs, so also it applies to biology and to the march both of artificial physical technique, and of physical processes in general. Considering this now on the biological field, we must rid ourselves of the ignorant and religious preconceptions that our own biological nature, that is, our own genetic material, any more than our own social structure, represents the last possible word in perfection. It is neither made in the image of god nor is it, according to a modern variation of that idea, in itself a diety. Lenin sharply criticized Lunacharsky for his evasion of materialism by the deification of human nature itself, since not even in this final refuge of the religious spirit is the wished-for deity to be found. Our present genetic composition is therefore simply one among the material things of life, although in a sense the most precious one with which we are provided, and it is therefore up to us to change it, and to continue to change it, in all such ways as will best further the harmonious and effective development of the worker's society. In this way, it will eventually become the privilege of every individual to reap the benefits of the biological fruits of socialism, as well as of the other fruits, and to enjoy increasingly that world conquest, on the path of which Lenin helped so much to set us.

Notes

I: Introduction: Background of the Discussions

1. Materialism and atheism are related but not synonymous concepts. Certain materialists, notably in the seventeenth century, combined materialism and theism. Hence I have said here that materialists *avoid* religious elements in scientific explanations, not that they necessarily deny the existence of God. Materialism in recent times, however, has usually been, explicitly or implicitly, atheistic. Soviet dialectical materialism is, of course, explicitly atheistic.

2. One might maintain that the aspect of dialectical materialism that was relevant in the biology controversy was the principle of the unity of theory and practice; according to this interpretation, Lysenko was much more willing than the classical geneticists to apply his theory to the betterment of Soviet agriculture. This point will be discussed in the genetics chapter. In the meantime, one might note that the principle of the unity of theory and practice is based on an unstated concept of time. Any theoretical development in science *should* be quickly applied, said the dialectical materialists, but how quickly was not specified. Obviously the application of a theory cannot in every case be simultaneous with its development. Premature widespread application would result in great waste. Therefore, the whole question of applying theory becomes subject to discussion. In a rational atmosphere this discussion would revolve around criteria such as completeness of the theory, expenses and risks involved in attempts to apply it, and gains to be obtained from application. From the standpoint of such criteria, the Soviet geneticists of the thirties were not noticeably guilty of divorcing theory from practice. Indeed, Nikolai Vavilov, Lysenko's opponent, was devoted to the union of theory and practice in the best Marxist sense: He wished to combine the highest scientific principles with a commitment to the betterment of society through science. Lysenko, on the contrary, caused great harm to Soviet agriculture.

3. See, for example, Conway Zirkle, *Evolution, Marxian Biology and the Social Scene* (Philadelphia, 1959).

4. As L. C. Dunn commented, belief in the inheritance of acquired characteris-

tics "solaced most of the biologists of the nineteenth century." *A Short History of Genetics* (New York, 1965), p. x.

5. Joravsky's full-length history of the Lysenko affair appeared too late to be utilized in this study, but he published numerous articles concerning the episode before his book appeared. See, for example, his "The First Stage of Michurinism," in J. S. Curtiss, ed., *Essays in Russian and Soviet History* (New York, 1963), pp. 120–32; "The Vavilov Brothers," *Slavic Review* (September 1965), pp. 381–94; also pp. 233–71 of his *Soviet Marxism and Natural Science* (New York, 1961). See also his *The Lysenko Affair* (Cambridge, Mass., 1970).

6. See pp. 94–5.

7. De Broglie would find causality by replacing current quantum theory by a theory (pilot-wave or, later, double-solution) that would restore classical concepts. Nagel would consider existing quantum theory "causal." See the latter's "The Causal Character of Modern Physical Theory," in S. W. Baron, E. Nagel, and K. S. Pinson, eds., *Freedom and Reason: Studies in Philosophy and Jewish Culture* (Glencoe, Ill., 1951), pp. 244–68; and *The Structure of Science: Problems in the Logic of Scientific Explanation* (New York, 1961), pp. 316–24. These issues will be discussed at much greater length in the chapter on quantum mechanics.

8. For a discussion of the nonreductiveness of dialectical materialism, see pp. 54–5.

9. See p. 263.

10. Leonard Schapiro, *The Communist Party of the Soviet Union* (New York, 1960), p. 343.

11. See Chap. 6, note 27.

12. See my *The Soviet Academy of Sciences and the Communist Party, 1927–1932*, Princeton, 1967, particularly Chaps. IV and V.

13. For Soviet criticism, see I. T. Frolov, *Genetika i dialektika*, Moscow, 1968, especially pp. 10–16 and 61–68.

14. V. P. Egorshin, "Estestvoznanie i klassovaia bor'ba," *Pod znamenem marksizma* (No. 6, 1926), p. 135.

15. Quoted in Frolov, *op. cit.*, p. 68.

16. *Ibid.*, p. 66.

17. See Raymond A. Bauer, Alex Inkeles, and Clyde Kluckhohn, *How the Soviet System Works: Cultural, Psychological and Social Themes* (Cambridge, Mass., 1956), pp. 114, 116–17, and 118–19.

18. See, in particular, the Intro. and Chaps. 1, 2, 3, and 5 of his *Historical Materialism*. Nikolai Bukharin, *Historical Materialism* (Ann Arbor, Mich., 1969), pp. 9–83 and 104–29.

19. The literature on the origins of the Cold War has been growing rapidly, and no attempt can be made to list or summarize it here. For an example of the revisionist literature, see Gar Alperovitz, *Atomic Diplomacy: Hiroshima and Potsdam; the use of the atomic bomb and the American confrontation with Soviet power* (New York, 1965). Discussions of the issue, from somewhat different viewpoints, are in Arthur Schlesinger, Jr., "Origins of the Cold War," *Foreign Affairs*, October 1967, pp. 22–52; and Hans J. Morgenthau, "Arguing About the Cold War," *Encounter*, May 1967, pp. 37–41.

20. See, for example, Thomas Kuhn, *The Copernican Revolution: Planetary Astronomy in the Development of Western Thought* (Cambridge, Mass., 1957), pp. 198–9.
21. See Appendix I.
22. An example of the co-operation of philosophers with scientists in the defense of science can be found in the second issue of the new journal *Problems of Philosophy*, created in 1947, when the ideological scene was already becoming strained. This issue contained an article by the theoretical physicist M. A. Markov strongly defending quantum mechanics and one by the biologist I. I. Schmalhausen clearly directed against Lysenko. After Zhdanov's death the editorial board of the journal was criticized by *Pravda* for publishing these articles, and the editor was replaced. See pp. 74–5 and 79. Also see Chap. 3, note 21.
23. See Chap. 8, p. 321 in particular.
24. Maksimov, who received his education as a physicist, might be an exception here, since he was considered by some people to be a philosopher of science. David Joravsky described him as "a physicist to philosophers, a philosopher to physicists." David Joravsky, *Soviet Marxism and Natural Science, 1917–1932* (New York, 1961), p. 185.

II: Dialectical Materialism: The Soviet Marxist Philosophy of Science

1. "Dialectical materialism" is believed to appear for the first time in the following passage from Plekhanov: "He [Hegel] showed that we are free only to the degree that we know the laws of nature and sociohistorical development and to the degree that *we, submitting to them,* rely upon them. This was a great gain both in the field of philosophy and in the field of social science—a gain that, however, only modern, dialectical, materialism has exploited in full measure." Plekhanov did not in any way indicate that he was coining a phrase here; it is possible that there was an earlier usage of which Plekhanov was aware. Later in the same article Plekhanov repeated the phrase "dialectical materialism," and without a comma separating the words. G. V. Plekhanov, *Izbrannye filosofskie proizvedeniia,* I (Moscow, 1956), pp. 443, 445.
2. The place where Engels perhaps came closest to saying "dialectical materialism" was in his general introduction to *Anti-Dühring.* There he talked of "modern materialism" in both the organic and inorganic realms. He then said, according to an English translation, "In both cases modern materialism is essentially dialectic. . . ." In the original German, the approach to the term "dialectical materialism" is not quite so close, since in the above sentence a pronoun is used for "modern materialism." After referring to *der moderne Materialismus* in an earlier sentence, Engels said, *In beiden Fällen ist er wesentlich dialektisch. . . ." Anti-Dühring* (Berlin, 1962), p. 24. But the thought is clear enough, and Plekhanov and Lenin were entirely within the spirit of Engels's passage when they used the term "dialectical material-

ism." For the English translation given here, see *Anti-Dühring* (Moscow, 1959), p. 35.

3. The best single source for Marx's and Engels's writings, and for dates of composition, is *Karl Marx, Friedrich Engels, Werke* (or *Marx-Engels-Werke*) (Berlin, 1956–67), 40 vols. in 41. Lenin's works are available in several editions, the latest of which is the fifth: V. I. Lenin, *Polnoe sobranie sochinenii* (Moscow, 1958–66), 55 vols. plus index.

4. See K. Marx, *Matematicheskie rukopisi*, Moscow, 1968. See also the special edition of *Voprosy istorii estestvoznaniia i tekhniki* (*Vypusk* 25), 1968, dedicated to the 150th anniversary of Marx's birth; this issue contained a previously unpublished manuscript on technology and an interesting discussion by Ernst Kol'man of the mathematics manuscripts.

5. This statement is in the preface to the second edition of *Anti-Dühring*, written in 1885. The translation is from *Anti-Dühring* (in English) (Moscow, 1959), pp. 16–17.

6. Frederick Engels, *Dialectics of Nature* (New York, 1940), p. 243.

7. George Lichtheim, *Marxism: An Historical and Critical Study* (New York, 1961), p. 245.

8. Z. A. Jordan, *The Evolution of Dialectical Materialism: A Philosophical and Sociological Analysis* (New York, 1967), p. 15.

9. "Razlichie mezhdu naturfilosofiei Demokrita i naturfilosofiei Epikura," in *K. Marks i F. Engel's: iz rannikh proizvedenii* (Moscow, 1956), pp. 23–98.

10. *Ibid.,* pp. 46–54.

11. Marx, *Capital* (New York, 1967), I, p. 309.

12. Jordan, *op. cit.,* p. 26.

13. There is a distinction, of course, between "revision" in the sense of modification of a scientific theory and "revisionism" as a political stance. Political revisionism has usually meant the abandonment by Marxists of a commitment to revolution; it has been prompted by a spectrum of motivations, ranging from genuine political conservatism and personal fear of violent change to the intellectual conviction that theoretical Marxism was incorrect in insisting, in every case, upon the necessity for violent revolution as a means to achieve fundamental social change. But there is a connection, obviously, between "conceptual revision" and "political revisionism"; if Marxism is to claim that it is scientific, it must in principle be willing to question the universal commitment to political revolution if empirical data seem to justify such questioning.

14. This statement in no way overlooks the fact that in science there is also often a great reluctance to revise a conception. Thomas Kuhn even implied that some dogma is necessary for scientific research. But scientists are avowedly in favor of revision and expect it to occur even if they may have difficulty in adjusting to it. To revise is to progress, in the vocabulary of scientists; to revise has traditionally been to deviate from truth, in the vocabulary of orthodox Marxism. For a discussion of resistance to discovery in scientific research, see Thomas Kuhn, "The Function of Dogma in Scientific Research," in A. C. Crombie, ed., *Scientific Change* (New York, 1963), pp. 347–69; and Bernard Barber, "Resistance by Scientists to Scientific Discovery," *Science,* CXXXIV (1961), 596–602.

15. There are both similarities and differences between Renaissance humanism and the new Marxist humanism. Both were rebellions against forms of scholasticism (the one Aristotelian, the other dogmatic Marxist), but the former was largely based on the rediscovery of the classical world and a new preoccupation with man instead of God, while the latter is an effort to counteract the old pretensions of Marxism to science with a new attention to man. Both contained a certain disdain for natural philosophy, which may have been healthy in terms of preceding history but which ignored important facets of intellectual endeavor. Among those historians of science who have been most critical of Renaissance humanism are Lynn Thorndike, Pierre Duhem, and Thomas Kuhn. The last remarked, "If humanism had been the only intellectual movement of the Renaissance, the Copernican Revolution might have been long postponed. . . . The humanists did not, however, succeed in stopping science." Thomas Kuhn, *The Copernican Revolution* (New York, 1959), p. 127. A rather successful compromise view was advanced by Marie Boas, who maintained that the Renaissance humanists never attacked science as such. Marie Boas, *The Scientific Renaissance 1450–1630* (New York, 1962), p. 27.

16. Jean-Paul Sartre, *Search for a Method* (New York, 1963), p. 175.

17. Engels, *Anti-Dühring* (Moscow, 1959), p. 17.

18. *Ibid.,* pp. 171–2. Also, Marx, *Capital*, p. 309.

19. See, for example, the discussion in Ernest Nagel, *The Structure of Science: Problems in the Logic of Scientific Explanation* (New York, 1961), pp. 29–78. Nagel commented, "We are certainly free to designate as a law of nature any statement we please. There is often little consistency in the way we apply the label, and whether or not a statement is *called* a law makes little difference in the way in which the statement may be used in scientific inquiry. Nevertheless, members of the scientific community agree fairly well on the applicability of the term for a considerable though vaguely delimited class of universal statements. Accordingly, there is some basis for the conjecture that the predication of the label, at least in those cases where the consensus is unmistakable, is controlled by a felt difference in the 'objective' status and function of that class of statements. It would indeed be futile to attempt an ironclad and rigorously exclusive definition of 'natural law.' " *Ibid.,* pp. 49–50.

20. E. Dühring, *Cursus der Philosophie als streng wissenschaftlicher Weltanschauung und Lebensgestaltung* (Leipzig, 1875).

21. Engels, *Anti-Dühring* (Moscow, 1959), p. 54.

22. *Ibid.*

23. Engels's positivism in this passage is noted by David Joravsky, *Soviet Marxism and Natural Science, 1917–1932* (New York, 1961), p. 9.

24. His opposition to crude materialism in *Anti-Dühring* would have been clearer if he had printed the original preface to the first edition, which he wrote in May 1878. In this preface Engels diluted the positivistic element with an emphasis on dialectics: "It is precisely dialectics that constitutes the most important form of thinking for present-day natural science, for it alone offers the analogue for, and thereby the method of explaining, the evolutionary processes occurring in nature, interconnections in general, and transi-

tions from one field of investigation to another." Engels later substituted another preface for this one; the original did not appear in print until forty-seven years later as a part of *Dialectics of Nature*. See Engels, *Anti-Dühring* (Moscow, 1959), p. 455 and editor's note, p. 451.

25. Engels, *Dialectics of Nature* (Moscow, 1954), pp. 151–227.
26. See Haldane's introduction to *Dialectics of Nature* (New York, 1940), p. xiv.
27. John Keosian, in his *The Origin of Life*, a book currently much used in universities in the United States, wrote, "Engels, in his *Dialectics of Nature*, was among the first to consider the spontaneous generation and the vitalistic theories from a materialist viewpoint. He condemned them both and maintained that life could have resulted only from a continuous evolution of matter, the origin of life being merely a rung in the long ladder of development." (New York, 1968), p. 11. See also J. D. Bernal, *The Physical Basis of Life* (London, 1951), p. 39, and his more recent *The Origin of Life* (London, 1967), pp. 4 and 131.
28. Howard Selsam and Harry Martel, eds., *Reader in Marxist Philosophy* (New York, 1963), pp. 326–7. Another American philosopher, Paul K. Feyerabend, commented, "There are not many writers in the field today who are as well acquainted with contemporary science as was Lenin with the science of his time, and no one can match the philosophical intuition of that astounding author." To this sentence Feyerabend added the following footnote remark: "I am here thinking mainly of Lenin's comments on Hegel's *Logik* and *Geschichte der Philosophie. Materialism and Empiriocriticism* is a different story." Lenin's comments on Hegel to which Feyerabend referred are in the *Philosophical Notebooks*. See Paul K. Feyerabend, "Dialectical Materialism and Quantum Theory," *Slavic Review*, XXV (September 1966), 414.
29. This is not the place to give a full discussion of the Soviet Marxist philosophy of science, with all of the varying interpretations of the decades between Marx and Engels's period and the early 1930's. It is necessary to discuss only those aspects of that philosophy that seemed likely to influence scientific interpretations. Those aspects—materialism and epistemology, the laws of the dialectic, the union of theory and practice, and the categories—are discussed below. Engels, especially, wrote a great deal on these subjects. After his death, the center of attention shifts to Russia, where new trends in the philosophy of science centered around Plekhanov, Bogdanov, and Lenin. Then from 1917 to 1931 no less than four "schools" in the philosophy of science successively became influential. These were the antiphilosophical "vulgar materialists," represented by such men as Emmanuel Enchmen and O. Minin; the "mechanists," including I. Stepanov and L. I. Axel'rod; the followers of A. M. Deborin, called Deborinites (or, by their opponents, Menshevik idealists); and the officially approved "dialectical materialists," headed by M. B. Mitin. Mitin became a member of the editorial board of the chief Soviet philosophy journal, *Under the Banner of Marxism*, in 1931; later, for many years he was chief editor of the current leading Soviet philosophy journal, *Problems of Philosophy*.
30. Engels, *Anti-Dühring* (Moscow, 1959), p. 76.

31. *Ibid.*, p. 86. Engels's statement here on matter and motion will be relevant in the controversy over the theory of chemical bonds. See p. 303. Engels's view is also very similar to the Aristotelian conception of motion. In Galileo's noted dialogue the role of the Aristotelian was taken by Simplicio, who at one point said that ". . . no motion can either exist or even be imagined except as inhering in its subject." Galileo Galilei, *Dialogue Concerning the Two Chief World Systems—Ptolemaic & Copernican,* trans. Stillman Drake (Berkeley and Los Angeles, 1962), p. 121. A difference here, of course, is that between "subject" and "matter," a difference a realist (but not a materialist) might find significant.

32. Engels, *Dialectics of Nature* (New York, 1940), pp. 322–3.

33. Engels, *Dialectics of Nature* (Moscow, 1954), p. 328.

34. Engels, *Ludwig Feuerbach* (New York, 1935), p. 30.

35. Quoted by Gustav Wetter, *Dialectical Materialism: A Historical and Systematic Survey of Philosophy in the Soviet Union* (New York, 1958), p. 281. The passage is in *Ludwig Feuerbach,* p. 31.

36. Wetter, *op. cit.,* p. 281.

37. Quoted in *ibid.,* p. 283.

38. Engels, *Dialectics of Nature,* pp. 310–11.

39. G. V. Plekhanov, *Izbrannye filosofskie proizvedeniia,* Vol. I (Moscow, 1956), p. 501.

40. For a helpful and clear discussion of the distinction between "presentational" and "representational" theories of perception, see Joseph G. Brennan, *The Meaning of Philosophy* (New York, 1967), pp. 121–2.

41. G. V. Plekhanov, *Protiv filosofskogo revizionizma* (Moscow, 1935), pp. 168–9. One may notice the similarity of this analogy to Plato's famous parable of the cave, in which an imprisoned man tries to determine the nature of reality from its shadows on the walls of his prison. Plekhanov, however, considered his viewpoint materialist, in contrast to the idealism of Plato.

42. G. V. Plekhanov, *Izbrannye filosofskie proizvedeniia,* Vol. I, pp. 475 and *passim.*

43. Ernst Mach, *The Analysis of Sensations and the Relation of the Physical to the Psychical* (New York, 1959), p. 12; originally published as *Beiträge zur Analyse der Empfindungen,* Jena, 1886.

44. *Ibid.,* p. 8.

45. The principle of economy was not essentially an original idea. From early Greek times the view was rather frequently expressed that simplicity is a desirable characteristic of scientific explanation; this opinion is sometimes summarized as the principle of Occam's razor. Despite this rather traditional aspect of Mach's principle of economy, he was unkindly criticized for it. The expression "economy of thought" led to the unfair but clever comment that the best way of economizing thought is not to think. Sir Harold Jeffreys, *Scientific Inference* (Cambridge, England, 1957), p. 15.

46. Mach's opinions helped break the way for such attitudes as the "principle of complementarity" of modern physics, considered in some detail in the chapter on quantum mechanics.

47. Bogdanov discussed the two realms of sensations in his *Empiriomonizm* (Moscow, 1905), pp. 15 ff.

48. *Ibid.,* p. 25.
49. *Ibid.,* p. 41.
50. Joravsky, *op. cit.,* pp. 27, 33, and *passim.*
51. The works that particularly irritated Lenin were *Studies in* (Lenin said it would have been more proper to say "against") *the Philosophy of Marxism* (St. Petersburg, 1908), a symposium by Bazarov, Bogdanov, Lunacharskii, Berman, Helfond, Iushkevich, and Suvorov; Iushkevich's *Materialism and Critical Realism;* Berman's *Dialectics in the Light of the Modern Theory of Knowledge;* and Valentinov's *The Philosophical Constructions of Marxism.* Lenin attacked these publications in his preface to the first edition of *Materialism and Empirio-Criticism: Critical Comments on a Reactionary Philosophy* (Moscow, 1952), pp. 9–11.
52. Lenin, *Materialism and Empirio-Criticism,* p. 11.
53. Even today, in the last third of that century, many intellectuals still believe materialism to be discredited by science, although it is clear that such a conclusion is entirely unwarranted.
54. Cited by Lenin, *ibid.,* p. 267. See Louis Houllevigue, *L'Évolution des Sciences* (Paris, 1914), pp. 87–8.
55. For Lenin on Poincaré, see *Materialism and Empirio-Criticism,* pp. 260–2, 265.
56. *Ibid.,* p. 271.
57. *Ibid.,* p. 130.
58. See p. 460.
59. Lenin, *Materialism and Empirio-Criticism,* p. 235.
60. *Ibid.,* p. 271.
61. *Ibid.,* pp. 269–70.
62. Selsam and Martel, eds., *op. cit.,* p. 331.
63. *Ibid.,* pp. 362–3.
64. Willard Van Orman Quine, *From a Logical Point of View: Logico-Philosophical Essays* (New York, 1963), p. 44.
65. *Ibid.,* p. 18.
66. As discussed on p. 347, some Soviet enthusiasts for cybernetics in the early sixties discussed rephrasing the laws of the dialectic in cybernetic terms. See Chap. 9, note 63.
67. G. W. F. Hegel, *Science of Logic,* trans. W. H. Johnston and L. G. Struthers (London, 1951), p. 473.
68. *Ibid.,* pp. 477–8.
69. *Ibid.,* p. 475. See also J. M. E. McTaggart, *A Commentary on Hegel's Logic* (Cambridge, 1910), pp. 3–4.
70. Hegel, *Encyclopedia of Philosophy,* trans. G. A. Mueller (New York, 1959), p. 77.
71. Engels, *Anti-Dühring* (Moscow, 1959), p. 34.
72. H. B. Acton, *The Illusion of the Epoch: Marxism-Leninism as a Philosophical Creed* (Boston, 1957), p. 101.
73. *The Logic of Hegel,* trans. William Wallace (Oxford, 1892), p. 205.
74. Engels, *Anti-Dühring* (Moscow, 1959), pp. 175, 176. Since each of these concepts has equal validity in economics as well as science, according to Soviet dialecticians, the transitions from capitalism to socialism to commu-

nism occur according to qualitative leaps resulting from sufficient quantitative changes in modes of production and social organization.

75. Engels then compared this need for a definite, but varying, minimum of cavalry in order to defeat the Mamelukes to Marx's economic principle that a definite, though varying, minimum sum of exchange-values is needed to make possible the transformation of these values into capital. *Ibid.,* pp. 176–7.

76. See, for example, Karl Marx and Frederick Engels, *Selected Works,* Vol. II (Moscow, 1958), pp. 388–9.

77. The concept of dialectical levels of natural laws was particularly important in the thought of A. I. Oparin, the noted writer on the origin of life. See Chap. 7. It was also important in the discussions of physiology and psychology described in Chap. 10.

78. See, in particular, pp. 254 ff.

79. N. Berdyaev, *Wahrheit und Lüge des Kommunismus* (Lucerne, 1934), p. 84. Quoted in Wetter, *op. cit.,* p. 551.

80. *The Logic of Hegel,* p. 22.

81. The principle of the struggle of opposites is ancient in natural philosophy. Fire and water were the most important of the two pairs of Aristotelian elements and were seen as opposites. Some medieval alchemists incorporated parts of Aristotelian philosophy into an essentially materialistic view of nature in which simple forms of matter were changed into superior forms by natural processes that could be, at least potentially, duplicated by man.

82. Engels was drawing here, of course, on early-nineteenth-century nature philosophy. For an interesting effort to describe this nature philosophy as of fundamental importance to the development of field theory, see L. Pearce Williams, *The Origins of Field Theory* (New York, 1966). Of particular interest is Williams's discussion of Hans Christian Oersted, many of whose ideas on the polarities in nature were similar to Engels's. *Ibid.,* pp. 51 ff.

83. Quoted from Hegel, *Science of Logic,* trans. Johnston and Struthers, Vol. II (London, 1929), pp. 66 ff., in Wetter, *op. cit.,* p. 335.

84. Engels, *Anti-Dühring* (New York, 1939), p. 155.

85. Engels, *Anti-Dühring* (Moscow, 1959), p. 193.

86. *Ibid.,* pp. 178–96.

87. *Ibid.,* p. 188.

88. Engels admitted the attack, but did not identify the mathematician. *Ibid.,* p. 17.

89. V. I. Lenin, *Filosofskie tetradi* (Moscow, 1938), p. 212. Stalin in 1938 omitted the Law of the Negation of the Negation altogether and cast the basic laws of the dialectic in a different mold; see "O dialekticheskom i istoricheskom materializme," in *Istoriia vsesoiuznoi kommunisticheskoi partii (Bol'shevikov): Kratkii kurs* (Moscow, 1938), pp. 101–4.

90. *Kratkii slovar' po filosofii* (Moscow, 1966), p. 119.

91. *Ibid.,* p. 120.

92. *Ibid.,* p. 119.

93. *Ibid.*

94. V. I. Lenin, *Filosofskie tetradi* (Moscow, 1965), p. 229.

95. *Kratkii slovar' po filosofii* (Moscow, 1966), p. 120.

96. The *Short Philosophical Dictionary* published in Moscow in 1955 listed the categories as matter, motion, time, space, quality, quantity, reciprocal connection, contradiction, causality, necessity, form and content, essence and appearance, possibility and actuality, etc. See Gustav A. Wetter, *Soviet Ideology Today* (New York, 1966), p. 65.

97. This is the eleventh thesis of Marx's *Theses on Feuerbach*, originally printed in 1888 as an appendix to Engels's *Ludwig Feuerbach and the Outcome of Classical Philosophy*. The reference may be found in C. P. Dutt, ed., *Ludwig Feuerbach* (New York, 1935), p. 75.

98. Engels, *Ludwig Feuerbach* (New York, 1935), pp. 32–3.

99. For a Soviet discussion of the importance of practice as a criterion of truth, published near the end of the Stalin period, see M. N. Rutkevich, *Praktika— osnova poznaniia i kriterii istiny* (Moscow, 1952).

100. But it did, in his mind, provide a beneficial prod to scientific development. As Engels observed, "If a technical demand appears in a society, then it will move science ahead more than ten universities." K. Marx and F. Engels, *Izbrannye pis'ma* (Moscow, 1947), p. 469.

101. Benjamin Farrington, *Greek Science* (Baltimore, 1961), pp. 94–5.

102. See Alexandre Koyré, "Galileo and Plato," in his *Metaphysics and Measurement* (London, 1968), pp. 16–43; and his *Études Galiléennes* (Paris, 1966), pp. 277–91. For criticism of Koyré's view, see Thomas P. McTighe, "Galileo's 'Platonism': A Reconsideration," in Ernan McMullin, ed., *Galileo: Man of Science* (New York, 1968), pp. 365–87.

103. L. Pearce Williams, *The Origins of Field Theory* (New York, 1966), p. 47.

104. A. Einstein, *Sobranie nauchnykh trudov*, IV (Moscow, 1967), p. 248.

III: Quantum Mechanics

1. Two valuable collections of articles indicating the diversity of the views expressed are S. Körner, ed., *Observation and Interpretation in the Philosophy of Physics with Special Reference to Quantum Mechanics* (New York, 1962); and R. G. Colodny, ed., *Beyond the Edge of Certainty* (Englewood Cliffs, N. J., 1965).

2. Expressed mathematically as $\Delta x \Delta p_x \geqq \hbar/2$, where Δx and Δp_x are the limits of precision within which the value of a coordinate and of momentum, respectively, can be simultaneously determined and $\hbar = $ Planck's constant divided by 2π.

3. It should perhaps be noted that from the time of classical science to the present there have been many debates over the physical significance of mathematical formalisms. The author of the preface to Copernicus's *De Revolutionibus*, the Lutheran theologian Andreas Osiander, tried (contrary to the astronomer's wishes) to urge Copernicus's readers to regard his system as a useful mathematical fiction not representative of physical truth. One could also cite Newton, who resolutely refused to say that his gravitational theory proved that matter has an innate power of attraction; he insisted that the

mathematics were the same *as if* this were so, but that he could not state that such attraction did in fact exist. The cases of Copernicus and Newton as described are well known; for easy reference, see Thomas S. Kuhn, *The Copernican Revolution* (New York, 1959), p. 187; and Alexandre Koyré, *From the Closed World to the Infinite Universe* (Baltimore, 1957), pp. 178–9.

4. See Hilary Putnam, "A Philosopher Looks at Quantum Mechanics," in Colodny, *op. cit.*, p. 78. To say that de Broglie "originally" proposed the undulatory theory means only within the framework of the modern mathematical apparatus; wave interpretations of light extend, of course, back to Fresnel and Young in the early nineteenth century and beyond. Similarly, the statement that Born "originally" suggested the corpuscular theory does not deny Newton's (or the early atomists') theories of light. See Vasco Ronchi, *Histoire de la lumière,* trans. Jean Taton (Paris, 1956).

5. The explanation for the spot imprint given by de Broglie was that of the "reduction of the wave packet."

6. Quoted in Max Jammer, *The Conceptual Development of Quantum Mechanics* (New York, 1966), p. 329.

7. J. Robert Oppenheimer, *The Open Mind* (New York, 1955), p. 82.

8. The first person to give a precise definition of complementarity was not Bohr but Pauli, and it turned out that Bohr did not quite agree with Pauli's formulation. Such differences have continued to plague interpreters of quantum mechanics. See Jammer, *op. cit.*, pp. 355–6, and particularly the difference between what von Weizsäcker called "parallel complementarity" and "circular complementarity."

9. A summary of the early warnings is in David Joravsky, *Soviet Marxism and Natural Science, 1917–1932* (New York, 1961), *passim,* esp. pp. 285–6. Fock's name will be spelled with a "c," as a result of his preference for this transliterated form, although the transliteration system used elsewhere in this book would dictate "Fok."

10. Bohr indicated that the concept of complementarity might be applied to such areas as physiology, psychology, biology, and sociology in his *Atomtheorie und Naturbeschreibung* (Berlin, 1931), and "Causality and Complementarity," *Dialectica,* II (No. 3–4, 1948), 312–319. This issue of *Dialectica* was devoted entirely to the concept of complementarity and included one article in which the author advanced the thesis that complementarity is potentially valid in all areas of systematic study: F. Gonseth, "Remarque sur l'idée de complémentarité," pp. 413–20.

11. *Otchet o deiatel'nosti akademii nauk SSSR za 1929 g.* (Leningrad, 1930), Vol. I (Appendix).

12. K. V. Nikol'skii, "Printsipy kvantovoi mekhaniki," *Uspekhi fizicheskikh nauk,* XVI (No. 5, 1936), 537–65. Nikol'skii later published a book setting forth the same view: *Kvantovye protsessy* (1940). Nikol'skii's 1936 article indicated his agreement with the position of Einstein, Podolsky, and Rosen in their debate with Bohr. See A. Einstein, B. Podolsky, and N. Rosen, "Can Quantum-Mechanical Description of Physical Reality Be Considered Complete?" *Physical Review,* XLVII (No. 10, May 15, 1935), 777–80; and Niels Bohr, "Can Quantum-Mechanical Description of Physical Reality Be

Considered Complete?" *Physical Review,* XLVIII (No. 8, October 15, 1935), 696–702.

13. Nikol'skii, "Otvet V. A. Foku," *Uspekhi fizicheskikh nauk,* XVII (No. 4, 1937), 555. In his criticism of Nikol'skii, Fock maintained that quantum mechanics described the action of an individual micro-object as well as statistical groups: "K stat'e Nikol'skogo 'Printsipy kvantovoi mekhaniki,'" *ibid.,* pp. 553–4.

14. Fock, "Mozhno li schitat', chto kvantomekhanicheskoe opisanie fizicheskoi real'nosti iavliaetsia polnym?" *ibid.,* XVI (No. 4, 1936), 437. In his introduction Fock clearly indicated that he considered Bohr the victor in the debate.

15. Fock also engaged in a debate before the war with A. A. Maksimov, another important participant in the later controversy. See Fock, "K diskussii po voprosam fiziki," *Pod znamenem marksizma,* No. 1 (1938), pp. 149–59. In 1937 and 1938 *Pod znamenem marksizma* contained a number of articles on the philosophic interpretation of quantum mechanics, including contributions by Maksimov, E. Kol'man, P. Langevin, and Nikol'skii.

16. See also Omel'ianovskii's defense of relativity theory in this period in his "Lenin o prostranstve i vremeni i teoriia otnositel'nosti Einshteina," *Izvestiia akademii nauk SSSR (Seriia istorii i filosofii),* No. 4 (1946), pp. 297–308.

17. M. E. Omel'ianovskii, *V. I. Lenin i fizika XX veka* (Moscow, 1947), *passim,* esp. p. 77. Omel'ianovskii accepted the relativity of simultaneity and of spatial and temporal intervals, concepts that were to be severely criticized in Soviet philosophical journals in the coming months.

18. *Ibid.,* p. 95.

19. For critical reviews of Omel'ianovskii, see M. Karasev and V. Nozdrev, "O knige M. E. Omel'ianovskogo 'V. I. Lenin i fizika XX veka,'" *Voprosy filosofii,* No. 1 (1949), pp. 338–42; V. V. Perfil'ev, "O knige M. E. Omel'ianovskogo 'V. I. Lenin i fizika XX veka,'" *Voprosy filosofii,* No. 1 (1948), pp. 311–12. The second edition was published in Ukrainian, *Borot'ba materiializmu proty idealizmu v suchasnii fizytsi* (Kiev, 1947).

20. A. A. Zhdanov, *Vystuplenie na diskussii po knige G. F. Aleksandrova "Istoriia zapadnoevropeiskoi filosofii," 24 iiunia 1947 g.* (Moscow, 1951), p. 43.

21. The first four issues were under the editorship of B. M. Kedrov, who was replaced by D. I. Chesnokov after Kedrov had sponsored a series of controversial articles. Kedrov obviously supported the Markov article and was held responsible for the criticism it incurred. Five articles in the first issues of *Voprosy filosofii,* including Markov's, were criticized in an article in *Pravda,* "Za boevoi filosofskii zhurnal," September 7, 1949.

22. M. A. Markov, "O prirode fizicheskogo znaniia," *Voprosy filosofii,* No. 2 (1947), pp. 140–76.

23. Maksimov charged that around Fock in the P. N. Lebedev Physics Institute of the Academy of Sciences there was a group of scientists who refused to admit dialectical materialism into science. A. A. Maksimov, "Bor'ba za materializm v sovremennoi fizike," *Voprosy filosofii,* No. 1 (1953), p. 178. When Markov's viewpoint was discussed in this institute, very little substantive criticism was expressed; see L. L. Potkov, "Obsuzhdenie raboty M. A.

Markova 'O mikromire,' " *Voprosy filosofii*, No. 2 (1947), pp. 381–2. The criticism came later.

24. Markov, p. 150. The "hidden parameter" theories have been promoted in recent years by David Bohm in particular. See his *Causality and Chance in Modern Physics* (New York, 1961), esp. pp. 79–81, 106–9, 111–16.
25. Markov, p. 146.
26. *Ibid.*, p. 163.
27. E. Schrödinger, "Die gegenwärtige Situation in der Quantenmechanik," *Die Naturwissenschaften*, XXIII, No. 48 (November 29, 1935), 812.
28. Putnam, pp. 94 ff. Hans Reichenbach also analyzed the cat paradox in "The Principle of Anomaly," *Dialectica*, II, No. 3–4 (1948), 344.
29. A. A. Maksimov, "Ob odnoi filosofskom kentavre," *Literaturnaia gazeta*, April 10, 1948, p. 3.
30. "K diskussii po stat'e M. A. Markova," *Voprosy filosofii*, No. 1 (1948), p. 225.
31. Maksimov, "Ob odnoi filosofskom kentavre," p. 3. One of the characteristics of Maksimov's article was its inaccuracies, as many critics in letters to the editor of *Literaturnaia gazeta* pointed out. In the quotation cited, for example, Maksimov stated that Markov had said that microreality "does not exist" before measurement, a statement that Markov never made, although he did say that the state of a system is "prepared" by measurement. In addition, Maksimov described Markov as saying that there existed a sharp division between the microlevel and the macrolevel of physical reality, a statement that Markov not only did not make but specifically denied.
32. See "Diskussiia o prirode fizicheskogo znaniia: Obsuzhdenie stat'i M. A. Markova," *Voprosy filosofii*, No. 1 (1948), pp. 203–32. Among the other contributors were B. G. Kuznetsov and S. A. Petrushevskii.
33. A. A. Maksimov, "Diskussiia o prirode fizicheskogo znaniia," *Voprosy filosofii*, No. 3 (1948), p. 228.
34. "Ot redaktsii," *ibid.*, pp. 231–2.
35. Soviet philosophers were quite straightforward in recognizing the discrediting of complementarity. Thus, Storchak observed, "In the course of the discussion of M. A. Markov's article it was established that the principle of complementarity was contrived as an idealistic distortion of the foundations of quantum mechanics." "Za materialisticheskoe osveshchenie osnov kvantovoi mekhaniki," *Voprosy filosofii*, No. 3 (1951), p. 202.
36. Ia. P. Terletskii, "Obsuzhdenie stat'i M. A. Markova," *Voprosy filosofii*, No. 3 (1948), p. 229.
37. He seems to have played a role in this controversy similar to Chelintsev's in the theory of resonance dispute. See Chap. 8.
38. A. A. Maksimov, "Marksistskii filosofskii materializm i sovremennaia fizika," *Voprosy filosofii*, No. 3 (1948), p. 114.
39. See, for example, V. P. Tugarinov, "Sootnoshenie kategorii dialekticheskogo materializma," *Voprosy filosofii*, No. 3 (1956), p. 155.
40. A. I. Uemov, *Veshchi, svoistva, otnosheniia* (Moscow, 1963). I learned of the importance of this book in conversations with Soviet philosophers in 1970 at the Institute of Philosophy of the Academy of Sciences.

41. As an example of one of the newer articles on the topic discussed here, see L. G. Antipenko, "Razvitie poniatiia material'nogo ob' 'ekta v fizike mikromira," *Voprosy filosofii*, No. 1 (1967), pp. 104–13.
42. D. I. Blokhintsev, *Vvedenie v kvantovuiu mekhaniku* (Moscow and Leningrad, 1944).
43. D. I. Blokhintsev, *Osnovy kvantovoi mekhaniki* (Moscow and Leningrad, 1949).
44. *Ibid.*, p. 8.
45. *Vvedenie v kvantovuiu mekhaniku*, p. 34.
46. *Ibid.*, p. 42.
47. See Reichenbach, *op. cit.*, p. 345.
48. *Vvedenie v kvantovuiu mekhaniku*, p. 42.
49. *Ibid.*, pp. 52, 58.
50. In a laudatory review of Blokhintsev's second edition, Storchak observed that the book would serve well as a dialectical materialist statement of quantum mechanics. "Za materialisticheskoe osveshchenie osnov kvantovoi mekhaniki," *Voprosy filosofii*, No. 3 (1951), p. 202.
51. *Vvedenie v kvantovuiu mekhaniku*, p. 34, and *Osnovy kvantovoi mekhaniki*, p. 45; italics added.
52. Blokhintsev drew his references from Bohr's *Atomtheorie und Naturbeschreibung* (Berlin, 1931) and P. Jordan's *Physics of the Twentieth Century* (New York, 1944).
53. Compare *Osnovy kvantovoi mekhaniki*, p. 547, lines 17–21, with *Vvedenie v kvantovuiu mekhaniku*, p. 34, lines 6–7.
54. *Osnovy kvantovoi mekhaniki*, p. 57.
55. D. I. Blokhintsev, "Kritika idealisticheskogo ponimaniia kvantovoi teorii," *Uspekhi fizicheskikh nauk*, XLV, No. 2 (October 1951), 195–228. Reprinted with two pages of preface as "Kritika filosofskikh vozzrenii tak nazyvaemoi 'kopengagenskoi shkoly' v fizike," in A. A. Maksimov *et al.*, eds., *Filosofskie voprosy sovremennoi fiziki* (Moscow, 1952), pp. 358–95.
56. "Kritika idealisticheskogo ponimaniia kvantovoi teorii," p. 209.
57. *Ibid.*, p. 210.
58. Blokhintsev now defined the ensemble as the microsystem plus the macroinstrument. *Ibid.*, p. 212.
59. *Istoriia vsesoiuznoi kommunisticheskoi partii (bol'shevikov): Kratkii kurs* (Moscow, 1945), p. 101.
60. "Kritika idealisticheskogo ponimaniia kvantovoi teorii," p. 213.
61. Many contemporary analysts of quantum mechanics agree with Blokhintsev on this point. See, for example, P. K. Feyerabend, "Problems of Microphysics," in R. G. Colodny, ed., *Frontiers of Science and Philosophy* (Pittsburgh, 1962), p. 207.
62. This position of Blokhintsev's illustrates that he was not in complete agreement with the interpretation of Nikol'skii before World War II, as has often been said. Nikol'skii agreed with Einstein, Podolsky, and Rosen. See Nikol'skii, "Printsipy kvantovoi mekhaniki."
63. "Kritika idealisticheskogo ponimaniia kvantovoi teorii," p. 211.
64. V. A. Fock, "O tak nazyvaemykh ansambliakh v kvantovoi mekhanike," *Voprosy filosofii*, No. 4 (1952), p. 170.

65. *Ibid.,* p. 173.
66. "Otvet akademiku V. A. Foku," *Voprosy filosofii,* No. 6 (1952), pp. 172–3. In articles in the sixties Blokhintsev was less concerned with the physical significance of the wave function than with relativistic quantum mechanics, quantum field theory, and attempts to find a system for the rational arrangement of elementary particles. See, for example, his "Problema struktury elementarnykh chastits," in I. V. Kuznetsov and M. E. Omel'ianovskii, eds., *Filosofskie problemy fiziki elementarnykh chastits* (Moscow, 1964), pp. 47–59.
67. D. I. Blokhintsev, *Printsipal'nye voprosy kvantovoi mekhaniki* (Moscow, 1966).
68. D. I. Blokhintsev, *The Philosophy of Quantum Mechanics* (Dordrecht-Holland and New York, 1968).
69. *Ibid.,* p. v.
70. Quoted in *ibid.,* p. 1.
71. *Ibid.,* p. 2.
72. *Ibid.,* p. 34.
73. *Ibid.,* p. 11.
74. *Ibid.,* p. 14.
75. *Ibid.,* p. 33.
76. *Ibid.,* p. 35.
77. *Ibid.,* p. 22.
78. *Ibid.*
79. *Ibid.*
80. *Ibid.,* p. 100.
81. *Ibid.*
82. *Ibid.*
83. *Ibid.,* p. 25.
84. *Ibid.,* p. 41.
85. *Ibid.,* p. 110.
86. N. R. Hanson, "Five Cautions for the Copenhagen Interpretation's Critics," *Philosophy of Science,* October 1959, p. 327.
87. V. A. Fock, "Nil's Bor v moei zhizni," *Nauka i chelovechestvo 1963,* Vol. II (Moscow, 1963), pp. 518–19.
88. *Ibid.,* p. 519.
89. V. A. Fock, "Ob interpretatsii kvantovoi mekhaniki," in P. N. Fedoseev *et al.,* eds., *Filosofskie problemy sovremennogo estestvoznaniia* (Moscow, 1959), p. 235. In 1965 Fock wrote in the following way of his approving but nonetheless critical approach to Bohr's interpretation: *Le mérite d'une nouvelle position du problème de la description des phénomènes à l'échelle atomique appartient à Niels Bohr; le point de vue adopté dans le présent article est le résultat de nos recherches et méditations ayant pour but d'approfondir, de préciser—et si nécessaire de critiquer et de corriger—les idées de Bohr.* V. Fock, "La physique quantique et les idéalisations classiques," *Dialectica,* No. 3–4 (1965), p. 223.
90. See, for example, Shirokov, "Filosofskie voprosy teorii otnositel'nosti," in V. N. Kolbanovskii *et al.,* eds., *Dialekticheskii materializm i sovremennoe estestvoznanie* (Moscow, 1964), pp. 58–80.

91. Fock, "K diskussii po voprosam fiziki," *Pod znamenem marksizma*, No. 1 (1938), p. 159.

92. See p. 73 and note 14 above.

93. Fock, "K diskussii po voprosam fiziki"; and "Protiv nevezhestvennoi kritiki sovremennykh fizicheskikh teorii," *Voprosy filosofii*, No. 1 (1953), pp. 168–74; Maksimov, "O filosofskikh vozzreniiakh akad. V. F. Mitkevich i o putiakh razvitiia sovetskoi fiziki," *Pod znamenem marksizma*, No. 7 (1937), pp. 25–55; and "Bor'ba za materializm v sovremennoi fizike," *Voprosy filosofii*, No. 1 (1953), pp. 175–94.

94. Fock, "Osnovnye zakony fiziki v svete dialekticheskogo materializma," *Vestnik Leningradskogo universiteta*, No. 4 (1949), p. 39; and M. E. Omel'ianovskii, *Filosofskie voprosy kvantovoi mekhaniki* (Moscow, 1956), p. 35.

95. V. A. Fock and A. B. Migdal, in N. S. Krylov, *Raboty po obosnovaniiu statisticheskoi fiziki* (Moscow and Leningrad, 1950), p. 8.

96. Even if Fock's hypothesis were to be granted, the existence of objective reality would not necessarily be denied, since there is no reason why such reality has to be defined in terms of certain parameters, such as position and momentum. Nevertheless, such an interpretation would require a more sophisticated view of reality than is often granted it.

97. Omel'ianovskii, "Dialekticheskii materializm i tak nazyvaemyi printsip dopolnitel'nosti Bora," in A. A. Maksimov *et al.*, *Filosofskie voprosy sovremennoi fiziki* (Moscow, 1952), pp. 404–5.

98. Fock, "O tak nazyvaemykh ansambliakh v kvantovoi mekhanike," *op. cit.*, p. 172.

99. Fock, "Kritika vzgliadov Bora na kvantovuiu mekhaniku," *Uspekhi fizicheskikh nauk*, XLV No. 1 (September 1951), 13.

100. Fock, "Ob interpretatsii kvantovoi mekhaniki," *op. cit.*, pp. 212–36.

101. In 1952 de Broglie, after defending the Copenhagen Interpretation for over twenty years, returned to his earlier belief in its replacement by a theory based on the "instinctive" position of a physicist, that of realism. Louis de Boglie, "La Physique quantique restera-t-elle indéterministe?" *Revue d'Histoire des Sciences et de leurs applications*, V, No. 4 (October–December 1952), 309.

102. See David Bohm, *op. cit.*

103. Vigier remarked, "A particle is thus considered as an average organized excitation of a chaotic subquantum-mechanical level of matter, similar in a sense to a sound wave propagation in the chaos of molecular agitation." In this same article Vigier credited Blokhintsev with providing the essential ideas for his model. J.-P. Vigier, "Probability in the Probabilistic and Causal Interpretation of Quantum Mechanics," in Körner, *op. cit.*, pp. 75, 76.

104. Fock, "Ob interpretatsii kvantovoi mekhaniki," p. 215.

105. *Ibid.*, p. 218.

106. The intermediate form, said Fock, would be a case when wavelike and corpusclelike properties appear simultaneously (although not sharply), such as when an electron is partially localized (corpusclelike property) and at the same time displays wave properties (wave function has the character of a

standing wave with an amplitude rapidly decreasing with increasing distance from the center of the atom).

107. *Ibid.,* p. 219.
108. *Ibid.,* p. 222.
109. See, for example, his "Lenin o prichinnosti i kvantovaia mekhanika," *Vestnik akademii nauk SSSR,* No. 4 (1958), pp. 3–12; "Filosofskaia evoliutsiia kopengagenskoi shkoly fizikov," *Vestnik akademii nauk SSSR,* No. 9 (1962), pp. 86–96; "Problema elementarnosti chastits v kvantovoi fizike," in *Filosofskie problemy fiziki elementarnykh chastits* (Moscow, 1964), pp. 60–73; and his "Lenin i filosofskie problemy sovremennoi fiziki" (Moscow, 1968).
110. See the 1946 article and 1947 book discussed on pp. 73–4 and 484.
111. "Zadachi razrabotki problemy 'Dialekticheskii materializm i sovremennoe estestvoznanie,' " *Vestnik akademii nauk SSSR,* No. 10 (1956), pp. 3–11.
112. *Lenin i sovremennaia nauka* (Moscow, 1970). Omel'ianovskii published an article in the same volume, "Sovremennye filosofskie problemy fiziki i dialekticheskii materializm," pp. 226–52.
113. *Filosofskie voprosy kvantovoi mekhaniki* (Moscow, 1956).
114. *Ibid.,* pp. 21–2.
115. *Ibid.,* pp. 253, 254.
116. *Ibid.,* p. 74. See note 2 above.
117. *Ibid.,* p. 71.
118. *Ibid.,* p. 32.
119. The record of the conference was published in P. N. Fedoseev, *et al.,* eds., *Filosofskie problemy sovremennogo estestvoznaniia* (Moscow, 1959). For the Chesnokov reference, see p. 650.
120. Omel'ianovskii, "The Concept of Dialectical Contradiction in Quantum Physics," in *Philosophy, Science and Man: The Soviet Delegation Reports for the XIII World Congress of Philosophy* (Moscow, 1963), p. 77.
121. *Ibid.,* p. 75.
122. Omel'ianovskii, "Filosofskie aspekty teorii izmereniia," in *Materialisticheskaia dialektika i metody estestvennykh nauk* (Moscow, 1968), pp. 207–55.
123. *Ibid.,* p. 248.
124. See N. Bohr, "Kvantovaia fizika i filosofiia," *Uspekhi fizicheskikh nauk,* No. 1 (1959), p. 39.
125. An interesting discussion of Soviet reactions to the double solution approach is in Edwin Levy, Jr., "Interpretations of Quantum Theory and Soviet Thought" (unpublished Ph.D. dissertation, Indiana University, 1969). Also see the discussion of quantum mechanics by two men who have worked closely with de Broglie in J. Andrade e Silva and G. Lochak, *Quanta* (New York and Toronto, 1969).
126. Kol'man "Sovremennaia fizika v poiskakh dal'neishei fundamental'noi teorii," *Voprosy filosofii,* No. 2 (1965), p. 122. Kol'man, a Czech who has spent long periods of time in Moscow, has played a very interesting role in disputes over the philosophy of science. Among Czech scientists he is generally known as an ideologue, but in the Soviet Union he has often been a "liberal" in the various controversies, although he favored Lysenko in the

early genetics controversy. As early as 1938 he was praised by Fock for his views on relativity physics. In cybernetics he was the first person to plead with Party officials for a recognition of the value of the new field. Kol'man, "Chto takoe kibernetika," *Voprosy filosofii,* No. 4 (1955), pp. 148–59. The article on physics cited above is definitely within this liberal tradition.

127. See A. A. Tiapkin, "K razvitiiu statisticheskoi interpretatsii kvantovoi mekhaniki na osnove sovmestnogo koordinatno-impul'snogo predstavleniia," in *Filosofskie voprosy kvantovoi fiziki* (Moscow, 1970), pp. 139–80. I am grateful to one of the editors of this book, L. G. Antipenko, for showing it to me in page proofs shortly before publication.

128. *Ibid.,* p. 152.

129. *Ibid.,* p. 144.

130. Tiapkin cited Omel'ianovskii as an example of Soviet philosophers who accepted the Copenhagen School too uncritically. *Ibid.,* pp. 144–5.

131. *Ibid.,* pp. 153–4.

132. *Ibid.*

133. *Ibid.,* p. 178.

134. *Ibid.,* p. 153.

135. See my comments in the introduction, with comparison of Soviet views on quantum mechanics to those of Paul Feyerabend, David Bohm, Louis de Broglie, and Ernest Nagel, pp. 7–8.

136. See p. 98 above.

137. The Copenhagen Interpretation remains very strong. Max Jammer said in 1966, "As is well known, this interpretation is still espoused today by the majority of theoreticians and practicing physicists. Though not necessarily the only logically possible interpretation of quantum phenomena, it is *de facto* the only existing fully articulated consistent scheme of conceptions that brings order into an otherwise chaotic cluster of facts and makes it comprehensible." Max Jammer, *The Conceptual Development of Quantum Mechanics* (New York, 1966), p. vii.

138. See especially pp. 394–6.

139. Comments by Fock and the American philosopher Paul K. Feyerabend on an earlier version of this chapter, as well as my reply, may be found in the *Slavic Review,* September 1966, pp. 411–20.

IV: Relativity Theory

1. The relationship between x, y, z, t and x', y', z', t' in the two reference frames S and S' are given by the following equations. The transformation based on these equations is called a Galilean transformation:

$$x = x' + ut$$
$$y = y'$$
$$z = z'$$
$$t = t'$$

where u is the relative velocity of S and S'.

2. The equations are:
$$x = (x' + ut')\,\gamma \qquad t = (t' + \frac{x'u}{c^2})\Big/\gamma$$
$$y = y' \qquad\qquad \gamma = \sqrt{1-\beta^2}$$
$$z = z' \qquad\qquad \beta = u/c$$
$$c = \text{velocity of light}$$

3. See the discussions in Maxim William Mikulak, "Relativity Theory and So-viet Communist Philosophy (1922–1960)" (unpublished Ph.D. dissertation, Columbia University, 1965), and in David Joravsky's "The 'Crisis' in Physics," in his *Soviet Marxism and Natural Science, 1917–1932* (New York, 1961), pp. 275–95. Dialectical materialist philosophers were aware of the problems of interpretation presented by new developments in physical theory and occasionally in the twenties pointed to the dangers of "Machism" in physics. In 1930 A. M. Deborin gave an official speech in the Academy of Sciences of the U.S.S.R. entitled "Lenin and the Crisis of Contemporary Physics." The physicists seemed undisturbed, however. V. P. Volgin, ed., *Otchet o deiatel'nosti akademii nauk SSSR za 1929 g.,* Vol. I, Appendix (Leningrad, 1931).

4. An established physicist opposing relativity in the name of dialectical mate-rialism was A. K. Timiriazev; some of those scientists who came to its de-fense in the same name, at least briefly, were A. F. Ioffe, I. E. Tamm, and O. Iu. Schmidt, all men of impressive scientific talent.

5. S. Iu. Semkovskii, *Dialekticheskii materializm i printsip otnositel'nosti* (Moscow and Leningrad, 1926), pp. 9, 11.

6. *Ibid.,* p. 54.

7. Joravsky, *op. cit.,* pp. 275–6.

8. Quoted in A. A. Semenov, "Ob itogakh obsuzhdeniia filosofskikh vozrenii akademika Mandel'shtama," *Voprosy filosofii,* No. 3 (1953), p. 200, from L. I. Mandel'shtam, *Polnoe sobranie trudov* (Moscow, 1950), Vol. V, p. 178.

9. N. D. Papaleksi, *et al., Kurs fiziki* (Moscow and Leningrad, 1948), Vol. II, p. 539. Aleksandrov criticized this view in his "Teoriia otnositel'nosti kak teoriia absoliutnogo prostranstva-vremeni," *Filosofskie voprosy sovremennoi fiziki* (Moscow, 1959), p. 284.

10. A. A. Zhdanov, *Vystuplenie na diskussii po knige G. F. Aleksandrova 'Istoriia zapadnoevropeiskoi filosofii' 24 iiunia 1947 g.* (Moscow, 1951).

11. For an example of the limits of vulgarity in the criticism of relativity theory, see A. A. Maksimov, "Protiv reaktsionnogo einshteiniantsva v fizike," *Krasnyi flot,* June 23, 1952. Maksimov had once been considerably more positive about Einstein and relativity theory, although he granted the need to rebuild the philosophic base of relativity. See his "Teoriia otnositel'nosti i materializm," *Pod znamenem marksizma,* No. 4–5 (1923), pp. 140–56. Another example of simplistic opposition was I. V. Kuznetsov's statement, "The unmasking of reactionary Einsteinism in the area of physical science is one of the most pressing tasks of Soviet physicists and philosophers," in A. A. Maksimov, *et al.,* eds., *Filosofskie voprosy sovremennoi fiziki* (Moscow, 1952), p. 47.

12. The relatively objective view was G. I. Naan, "K voprosu o printsipe otnosi-

tel'nosti v fizike," *Voprosy filosofii*, No. 2 (1951), pp. 57–77. Naan's view was criticized by a host of authors, as will be discussed. The editorial criticism of Naan is in "K itogam diskussii po teorii otnositel'nosti," *Voprosy filosofii*, No. 1 (1955), p. 138.

13. A. A. Maksimov, "Bor'ba za materializm v sovremennoi fizike," *Voprosy filosofii*, No. 1 (1953), p. 194.

14. G. I. Naan, "Sovremennyi 'fizicheskii' idealizm v SShA i Anglii na sluzhbe popovshchiny i reaktsii," *Voprosy filosofii*, No. 2 (1948), pp. 290 ff. Naan based his criticism of Frank on the latter's *Foundations of Physics* (Chicago, 1946) and articles. Naan criticized Bertrand Russell's *History of Western Philosophy* for its sensationalism and James Jeans's comments about the "disappearance" of matter.

15. M. E. Omel'ianovskii, "Falsifikatory nauki: ob idealizme v sovremennoi fizike"; A. A. Maksimov, "Marksistskoi filosofskii materializm i sovremennaia fizika"; R. Ia. Shteinman, "O reaktsionnoi roli idealizma v fizike"; in *Voprosy filosofii*, No. 3 (1958), pp. 105–24, 143–62, 163–73. Omel'ianovskii directed his critique of Carnap at the latter's *The Logical Syntax of Language*. Omel'ianovskii considered Einstein to be a "materialist," but not a "dialectical materialist," a view for which he was criticized by M. M. Karpov, who considered Einstein to be a thoroughgoing idealist. M. M. Karpov, "O filosofskikh vzgliadakh A. Einshteina," *Voprosy filosofii*, No. 1 (1951), pp. 130–41.

16. Omel'ianovskii, *ibid.*, pp. 144, 155. Frank included a chapter on "Logical Empiricism and Philosophy in the Soviet Union" in his *Between Physics and Philosophy* (Cambridge, Mass., 1941).

17. A. A. Maksimov, "Marksistskoi filosofskii materializm i sovremennaia fizika," *Voprosy filosofii*, No. 3 (1948), p. 114.

18. *Ibid.*

19. G. A. Kursanov, "Dialekticheskii materializm o prostranstve i vremeni," *Voprosy filosofii*, No. 3 (1950), p. 186 and *passim*.

20. Einstein commented in his 1916 obituary of Mach: "He [Mach] conceived every science as the task of bringing order into the elementary single observations which he described as 'sensations.' This denotation was probably responsible for the fact that this sober and cautious thinker was called a philosophical idealist or solipsist by people who had not studied his work thoroughly. . . . I can say with certainty that the study of Mach and Hume has been directly and indirectly a great help in my work. . . . Mach recognized the weak spots of classical mechanics and was not very far from requiring a general theory of relativity half a century ago. . . . It is not improbable that Mach would have discovered the theory of relativity, if, at the time when his mind was still young and susceptible, the problem of the constancy of the speed of light had been discussed among physicists." Quoted in Philipp Frank, "Einstein, Mach, and Logical Positivism," in Edward H. Madden, *The Structure of Scientific Thought* (Boston, 1960), p. 85, from *Physikalische Zeitschrift*, XVII (1916), 101 ff.

21. L. I. Storchak, "Znachenie idei Lobachevskogo v razvitii predstavlenii o prostranstve i vremeni," *Voprosy filosofii*, No. 1 (1951), pp. 142–8.

22. Another attempt to avoid Mach by way of Lobachevskii is N. V. Markov,

"Znachenie geometrii Lobachevskogo dlia fiziki," in A. A. Maksimov *et al.*, eds., *Filosofskie voprosy sovremennoi fiziki*, pp. 186–215.

23. Quoted in David Dinsmore Comey, "Soviet Controversies Over Relativity," in *The State of Soviet Science* (Cambridge, Mass., 1965), p. 191.

24. V. Shtern, "K voprosu o filosofskoi storone teorii otnositel'nosti"; D. I. Blokhintsev, "Za leninskoe uchenie o dvizhenii"; G. A. Kursanov, "K kriticheskoi otsenke teorii otnositel'nosti"; *Voprosy filosofii*, No. 1 (1952), pp. 175–81, 181–3, 169–74. Shtern's views were stated in more detail in his *Erkenntnistheoretische Probleme der Modernen Physik* (Berlin, 1952).

25. Kursanov, "K kriticheskoi otsenke . . . ," p. 170.

26. Shtern's simple view of relativity was later criticized thoroughly by P. G. Kard, "O teorii otnositel'nosti," *Voprosy filosofii*, No. 5 (1952), pp. 240–7, but Kard simultaneously spoke positively of Blokhintsev's effort to preserve a concept of absolute space. See Blokhintsev, *op. cit.*

27. Blokhintsev, *op. cit.* Blokhintsev believed that each larger and more "inertial" frame of reference was an improvement over the previous one as a result of its possession of a relative grain of truth. Compare this view with Lenin's statement in *Materialism and Empirio-criticism* on relative and absolute truth: "Human thought by its very nature is able to give and does give absolute truth, which is accumulated as the sum total of relative truths, but the limits of the truth of each scientific proposition are relative, now expanding, now shrinking with the growth of knowledge." V. I. Lenin, *Sochineniia,* 4th ed., Vol. XIV (Moscow, 1947), p. 122.

28. I. V. Kuznetsov, "Sovetskaia fizika i dialekticheskii materializm," in A. A. Maksimov, *et al.*, eds., *Filosofskie voprosy sovremennoi fiziki*, pp. 31–86; and R. Ia. Shteinman, *op. cit.*, pp. 234–98.

29. Kuznetsov, *op. cit.*, p. 72.

30. V. A. Fock, "Protiv nevezhestvennoi kritiki sovremennykh fizicheskikh teorii," *Voprosy filosofii*, No. 1 (1953), p. 174.

31. *Ibid.*, p. 168.

32. See p. 484, footnote 15.

33. Thus, Fock briefly mentioned that the statement that the "speeds of light to and fro are equal" is a "natural" assumption. But the attribution to ϵ of any value between 0 and 1 in the celebrated equation $t_2 = t_1 + \epsilon (t_3 - t_1)$ is conventional. On the general topic of congruency Fock agreed with Aleksandrov that the precise meaning of x, y, z, and t can be derived *from* the law of wave-front propagation. They are not given a priori. See Fock, *The Theory of Space, Time and Gravitation* (New York, 1959), pp. 24, 147–8.

34. For Russian and German editions, see A. D. Aleksandrov, *Vnutrennaia geometriia vypuklykh poverkhnostei* (Moscow-Leningrad, 1948); and *Die innere Geometrie der konvexen Flächen* (Berlin, 1955).

35. A. D. Aleksandrov, "Leninskaia dialektika i matematika," *Priroda*, No. 1 (1951), pp. 5–15.

36. A. D. Aleksandrov, "Ob idealizme v matematike," *Priroda*, No. 7 (1951), pp. 3–11; and No. 8 (1951), pp. 3–9.

37. *Pravda*, October 4, 1966, p. 2. I am grateful to David D. Comey for pointing out this quotation to me and for the translation.

38. Albert Einstein, *Sobranie nauchnykh trudov,* Vols. I–IV (Moscow,

1965–67). A few small collections of Einstein's scientific works had been published earlier, and in 1922–24 a four-volume collection of his scientific publications through 1922 was published in Japanese; nonetheless, the Soviet editors prepared the first edition covering all of Einstein's lifetime. In preparing this significant publication, the editors corresponded frequently with Einstein's secretary, Miss Helen Dukas. Conversation with Miss Dukas, Princeton, New Jersey, January 16, 1970. See Herbert S. Klickstein, "A Cumulative Review of Bibliographies of the Published Writings of Albert Einstein," *Journal of the Albert Einstein Medical Center,* July 1962, pp. 141–9.

39. A. D. Aleksandrov, "Teoriia otnositel'nosti kak teoriia absoliutnogo pro-stranstva-vremeni," *Filosofskie voprosy sovremennoi fiziki* (Moscow, 1959), pp. 273–4.

40. *Ibid.,* p. 279.

41. In a well-known paper of 1908 on space and time, Minkowski spoke of the "Relativity Postulate," and then commented, *Indem der Sinn des Postulats wird, das durch die Erscheinungen nur die in Raum und Zeit vier-dimensionale Welt gegeben ist, aber die Projektion in Raum und in Zeit noch mit einer gewissen Freiheit vorgenommen werden kann, möchte ich dieser Behauptung eher den Namen* Postulat der absoluten Welt *(oder kurz Welt-postulat) geben. Gesammelte Abhandlungen von Hermann Minkowski* (Leipzig and Berlin, 1911), p. 437.

42. Aleksandrov, "Teoriia otnositel'nosti . . . ," p. 282.

43. Adolf Grünbaum, "Geometry, Chronometry, and Empiricism," in Herbert Feigl and Grover Maxwell, eds., *Minnesota Studies in the Philosophy of Science,* Vol. III, *Scientific Explanation, Space, and Time* (Minneapolis, 1962), p. 522.

44. Aleksandrov, "Teoriia otnositel'nosti . . . ," p. 283.

45. *Ibid.,* p. 284.

46. A. D. Aleksandrov and V. V. Ovchinnikov, "Zamechaniia k osnovam teorii otnositel'nosti," *Vestnik leningradskogo universiteta,* No. 11 (1953), pp. 95–109.

47. Aleksandrov, "Teoriia otnositel'nosti . . . ," p. 303.

48. The difficulties of such an opinion will be discussed later.

49. Aleksandrov, "Teoriia otnositel'nosti . . . ," p. 301.

50. See Alfred A. Robb, *The Absolute Relations of Time and Space* (Cambridge, England, 1921). For the historically minded, one can point out that cosmologies based on light are very old in the history of science, although considerable differences existed among the various systems. One of the oldest and most extensive cosmologies of light was that of Robert Grosseteste of the late twelfth and early thirteenth century. Grosseteste believed light was the first effective principle of movement by which the operations or "becoming" of natural things were caused. See A. C. Crombie, *Robert Grosseteste and the Origins of Experimental Science, 1100–1700* (Oxford, 1953), esp. pp. 91–124. In many earlier writings, such as those of the Spanish Jew Avice-bron, St. Augustine, Pseudo-Dionysius, and St. Basil, the idea was presented that light is a form that actualizes the potentiality of matter as a universal *continuum.*

51. Aleksandrov, "Teoriia otnositel'nosti . . . ," p. 274. Even though Aleksandrov agreed with much of Robb's interpretation, he expressed his wish to dissociate himself from Robb's remark that the Einsteinian relativity of simultaneity converts the universe into a kind of "nightmare." See Robb, *op. cit.,* p. v.

52. V. Fock, *The Theory of Space, Time and Gravitation,* p. xviii.

53. V. A. Fock, "Comments," *Slavic Review,* September 1966, p. 412. The above English version of Fock's comments is the one approved by Fock. The phrasing is a little awkward in spots, particularly in the sentence beginning "Even such statements. . . ."

54. By the term "Galilean space," Fock meant space of maximum uniformity. As he commented, in such space:
    "(a) All points in space and instants in time are equivalent.
    (b) All directions are equivalent, and
    (c) All inertial systems, moving uniformly and in a straight line relative to one another, are equivalent (Galilean principle of relativity)."
    Fock, *The Theory of Space, Time and Gravitation,* p. xiii.

55. See the discussion of Fock in P. S. Dyshlevyi, *V. I. Lenin i filosofskie problemy reliativistskoi fiziki* (Kiev, 1969), pp. 148 ff.

56. V. A. Fock, "Les principes mécaniques de Galilée et la théorie d'Einstein," *Atti del convegno sulla relatività generale: problemi dell'energia e onde gravitazionali* (Florence, 1965), p. 12.

57. See Hans Reichenbach, *The Philosophy of Space and Time* (New York, 1958), p. 223.

58. *Ibid.,* p. 226.

59. V. A. Fock, "O roli printsipov otnositel'nosti i ekvivalentnosti v teorii tiagotenii Einshteina," *Voprosy filosofii,* No. 12 (1961), p. 51.

60. Fock, *The Theory of Space, Time and Gravitation,* p. xv.

61. *Ibid.,* p. xvi.

62. It should be noted that a somewhat different argument from Fock's can be made against the equivalence of the Copernican and Ptolemaic systems. Norwood Russell Hanson, an American philosopher of science, argued "that what has been called 'geometrical equivalence,' 'mathematical equivalence,' and even 'absolute identity' is actually no more than *observational equivalence.*" See his "Contra-Equivalence: A Defense of the Originality of Copernicus," *ISIS,* LV (September 1964), 308–25.

63. Fock, *The Theory of Space, Time and Gravitation,* p. xvi.

64. *Ibid.,* pp. 351–2.

65. *Ibid.,* p. 351. See also V. A. Fock, "Poniatiia odnorodnosti, kovariantnosti i otnositel'nosti," *Voprosy filosofii,* No. 4 (1955), p. 133.

66. See p. 135.

67. Fock, *The Theory of Space, Time and Gravitation,* pp. 351–2.

68. The most general covariant equation for an interval of space-time is $ds^2 = g_{\mu\nu}\, dx_\mu\, dx_\tau$. Here $g_{\mu\nu}$ is a tensor, that is, a magnitude that transforms according to well-defined rules whenever a transformation to a new coordinate system occurs. In Galilean space-time the coefficient $g_{\mu\nu}$ remains unchanged, but in Riemannian space-time $g_{\mu\nu}$ is a function of the coordinates.

69. Many physicists would emphasize at this point that one of the principles of GTR is that *no* auxiliary functions are to be introduced.
70. Fock, *The Theory of Space, Time and Gravitation*, p. xviii.
71. Fock, "Les principes . . . ," p. 11.
72. *Ibid.*, pp. 9, 11.
73. P. S. Dyshlevyi, *V I. Lenin i filosofskie problemy reliativistskoi fiziki* (Kiev, 1969), p. 143.
74. *Ibid.*
75. See, for example, J. M. Bochenski, ed., *Bibliographie der sowjetischen Philosophie*, Vols. I–V, (Dordrecht-Holland, 1959–64).
76. M. F. Shirokov, "Filosofskie voprosy teorii otnositel'nosti," in V. N. Kolbanovskii, *et al.,* eds., *Dialekticheskii materializm i sovremennoe estestvoznanie* (Moscow, 1964), p. 59.
77. M. F. Shirokov, "O materialisticheskoi sushchnosti teorii otnositel'nosti," in I. V. Kuznetsov and M. E. Omel'ianovskii, eds., *Filosofskie voprosy sovremennoi fiziki* (Moscow, 1959), pp. 325 ff.
78. M. F. Shirokov, "O preimushchestvennykh sistemakh otscheta v n'iutonovskoi mekhanike i teorii otnositel'nosti," *Voprosy filosofii*, No. 3 (1952), pp. 128–39.
79. See the discussion in P. S. Dyshlevyi, *op. cit.*, pp. 137–9.

V: Cosmology and Cosmogony

1. H. Bondi, "The Steady-State Theory of the Universe," in H. Bondi, W. B. Bonnor, R. A. Lyttleton, and G. J. Whitrow, *Rival Theories of Cosmology* (London, 1960), pp. 17–18.
2. William Bonnor, *The Mystery of the Expanding Universe* (New York, 1964); Fred Hoyle, *The Nature of the Universe* (New York, 1960); G. C. McVittie, *Fact and Theory in Cosmology* (New York, 1961); George Gamow, *The Creation of the Universe* (New York, 1955); Otto Struve, *The Universe* (Cambridge, Mass., 1962); Gerard de Vaucouleurs, *Discovery of the Universe* (New York, 1957); and G. J. Whitrow, *The Structure and Evolution of the Universe* (London, 1959). A very interesting recent summary is W. H. McCrea, "Cosmology After Half a Century," *Science*, CLX (June 21, 1968), 1295–1299. McCrea's discussion is very relevant to this chapter since it cites recent evidence supporting the view that the universe is isotropic and homogeneous, questions of much dispute among Soviet astronomers such as Ambartsumian. It should be kept in mind, however, that McCrea's evidence postdated much of the discussion described in this chapter. Further, some Soviet astronomers would use the term "metagalaxy" where he uses "universe."
3. Categories I, II, III, and IV are "relativistic" in the sense of accepting both special and general relativity; category V accepts special relativity but rejects general relativity; category VI is a substantial adaptation of relativity involving abandonment of the conservation laws. Models IIa, IIb, IIc, and IIId can be called big-bang models, although IIb might be best described as "multiple-big-bang." Models IIIa, IIIb, IIIc, and IIIe are not of the big-bang type; IIIa

and IIc start with an infinite length of time in the static Einstein state; IIIe contains infinite contraction and expansion phases without a singular state between. IV, in contrast to the other models, is based on a rejection of the cosmological principle.

4. The cosmological term (λ) was originally introduced by Einstein to provide a force of repulsion resisting gravitational collapse in a static model (Ib). He later abandoned it after shifting to expanding models and after seeing, as a result of Friedmann's work, that expanding models could be constructed without it. The cosmological term was retained by other cosmologists (III); its effect, appreciable only when enormous distances are involved, is to speed up the rate of expansion.

5. One should not forget that in the twenties and thirties, before Stalinism deeply affected Soviet intellectual life, there was a more sophisticated body of literature on philosophic aspects of cosmology and cosmogony. Scientists in those years frequently did not have a deep knowledge of dialectical materialism, but even the great A. A. Friedmann made some effort to connect his views of the universe with materialism. See A. A. Friedmann (Fridman), *Mir kak prostranstvo i vremia* (Moscow, 1965 [1923]), esp. p. 32. Also, B. P. Gerasimovich, *Vselennaia pri svete teorii otnositel'nosti* (Khar'kov, 1925). The work of M. A. Bronshtein is also relevant.

6. Soviet writers were, of course, not the only critics of the later writings of James Jeans. The American physicist Freeman Dyson commented, ". . . he went from bad to worse, becoming a successful popular writer and radio broadcaster, accepting a knighthood and ruining his professional reputation with suave and shallow speculations on religion and philosophy." Freeman J. Dyson, "Mathematics in the Physical Sciences," *Scientific American,* September 1964, p. 129.

7. E. T. Whittaker, *The Beginning and End of the World* (London, 1943), p. 63.

8. Bonnor, *The Mystery of the Expanding Universe,* p. 119.

9. Alexandre Koyré, *From the Closed World to the Infinite Universe* (Baltimore, 1957).

10. E. J. Dijksterhuis commented, "The strong influence which Newton's religious ideas exercised on his scientific thought is revealed, among other things, in his belief in the existence of absolute space and absolute time. The former to him symbolized God's omnipresence, the latter His eternity." Dijksterhuis, *The Mechanization of the World Picture* (London, 1961), p. 487. In the *Scholium Generale* of his *Principia* Newton observed, "This most beautiful system of the sun, planets, and comets, could only proceed from the counsel and dominion of an intelligent and powerful Being." *Principia,* trans. Florian Cajori (Berkeley and Los Angeles, 1966), Vol. II, p. 544.

11. Quoted in Wetter, *Dialectical Materialism: A Historical and Systematic Survey of Philosophy in the Soviet Union,* trans. Peter Heath (New York, 1958), p. 436.

12. R. A. Alpher, H. Bethe, and G. Gamow, "The Origin of the Chemical Elements," *Physical Review,* LXXIII (April 1, 1948), 803–4.

13. G. A. Kursanov, "Dialekticheskii materializm o prostranstve i vremeni," *Voprosy filosofii,* No. 3 (1950), pp. 173–91.

14. He returned from the ice escape by way of Alaska and the United States, where he became a member of the New York Explorers' Club.
15. "Otto Iul'evich Schmidt (1891–1956)," in I. V. Kuznetsov, *Liudi russkoi nauki: ocherki o vydaiushchikhsia deiateliakh estestvoznaniia i tekhniki, matematika, mekhanika, astronomiia, fizika, khimiia* (Moscow, 1961), p. 404.
16. Akademik O. Iu. Schmidt, *Proiskhozhdenie zemli i planet* (Moscow, 1949, 1957, 1962). Also available in English as *A Theory of Earth's Origin* (Moscow, 1958).
17. Schmidt, *A Theory of Earth's Origin*, p. 11.
18. H. N. Russell, *The Solar System and Its Origin* (New York, 1935), pp. 95–6.
19. Schmidt, *A Theory of Earth's Origin*, p. 17.
20. The sun is more than four light-years from the closest star; just traveling that distance with a relative velocity of twenty kilometers per second would require approximately 10^5 years. Otto Struve and Velta Zebergs, *Astronomy of the 20th Century* (New York, 1962), p. 173.
21. Another feature of Schmidt's system was his belief that the earth was originally cold and gained heat later as a result of the breakdown of radioactive elements.
22. Schmidt, *A Theory of Earth's Origin*, pp. 83–4.
23. *Ibid.*, pp. 84–5.
24. O. Iu. Schmidt, "O vozmozhnosti zakhvata v nebesnoi mekhanike," *Doklady akademii nauk SSSR*, LVII (1947), 213–16; O. Iu. Schmidt and G. F. Khil'mi, "Problema zakhvata v zadache o trekh telakh," *Uspekhi matematicheskikh nauk*, No. 26 (1948), pp. 157–9.
25. G. F. Khil'mi, "O vozmozhnosti zakhvata v probleme trekh tel," *Doklady akademii nauk SSSR*, LXII, No. 1 (1948), 39–42; also see the bibliography given in Schmidt, *Proiskhozhdenie zemli i planet* (1962), p. 93.
26. See the work of V. V. Radzievskii and T. A. Agekian cited by Schmidt in *A Theory of Earth's Origin*, pp. 92, 132, 137.
27. Schmidt, *A Theory of Earth's Origin*, pp. 93–4.
28. Von Weizsäcker first presented his theory in "Über die Entstehung des Planetensystems," *Zeitschrift für Astrophysik*, XXII (1943), 319–55.
29. V. S. Safronov, "Problema proiskhozhdeniia zemli i planet: soveshchanie po voprosam kosmogonii solnechnoi sistemy," *Vestnik akademii nauk SSSR*, No. 10 (1951), pp. 94–102.
30. Schmidt, *A Theory of Earth's Origin*, p. 15.
31. Faye (1819–1902) is best known for his view that the sun can be considered a "thermal machine" that radiates heat from its surface. See Giorgio Abetti, *The History of Astronomy* (London and New York, 1952), p. 199. O. Iu. Schmidt, "Problema proiskhozhdeniia zemli i planet," *Voprosy filosofii*, No. 4 (1951), p. 121.
32. *Ibid.*, p. 122.
33. *Ibid.*
34. *Ibid.*
35. *Ibid.*, p. 124.
36. V. G. Fesenkov, "O kosmogonicheskoi gipoteze akademika O. Iu. Shmidta i

o sovremennom sostoianii kosmogonicheskoi problemy," *Voprosy filosofii*, No. 4 (1951), p. 136.

37. "Ot redaktsii," *Voprosy filosofii*, April 1951, p. 150.

38. V. G. Fesenkov, "Evoliutsiia i vozniknovenie zvezd v sovremennoi galaktike," *Voprosy filosofii*, No. 4 (1952), p. 115.

39. See A. S. Arsen'ev, "Nekotorye metodologicheskie voprosy kosmogonii," *Voprosy filosofii*, No. 3 (1955), pp. 38, 41. Also see V. G. Fesenkov, "Problema kosmogonii solnechnoi sistemy," *Priroda*, No. 4 (1940), p. 14.

40. For a scientific biography of Ambartsumian, see A. B. Severnyi and V. V. Sobolev, "Viktor Amazaspovich Ambartsumian (K shestidesiatiletiiu so dnia rozhdeniia)," *Uspekhi fizicheskikh nauk*, September 1968, pp. 181–3.

41. V. A. Ambartsumian, "Nekotorye voprosy kosmogonicheskoi nauki," *Kommunist*, No. 8 (1959), p. 86.

42. V. A. Ambartsumian, "Nekotorye metodologicheskie voprosy kosmogonii," in P. N. Fedoseev *et al.*, eds., *Filosofskie problemy sovremennogo estestvoznaniia* (Moscow, 1959), p. 290.

43. V. A. Ambartsumian, "Problema proiskhozhdeniia zvezd," *Priroda*, No. 9 (1952), pp. 9–10.

44. V. A. Ambartsumian, "Stars of T Tauri and UV Ceti Types and the Phenomenon of Continuous Emission," in George H. Herbig, ed., *Non-Stable Stars*, International Astronomical Union Symposium No. 3 (Cambridge, Eng., 1957), pp. 177–85.

45. Ambartsumian, "Problema proiskhozhdeniia zvezd," p. 10.

46. He is frequently attributed this achievement in ordinary biographical accounts, outside the Soviet Union as well as within that country. See, for example, John Turkevich, *Soviet Men of Science* (Princeton, 1963), pp. 15–16.

47. Ambartsumian quite sensibly agreed with the uniformitarianism so well expressed by Lyell in the following way: "It may be necessary in the present state of science to supply some part of the assumed course of nature hypothetically; but, if so, this must be done without violation of probability, and always consistently with the analogy of what is known both of the past and the present economy of our system." Charles Lyell, *Principles of Geology*, 12th ed., Vol. I (London, 1875), p. 299.

48. I can not resist giving the mnemonic device for remembering the types: "Oh, Be A Fine Girl, Give Me a Kiss Right Now, Smack!"

49. Ambartsumian, "Problema proiskhozhdeniia zvezd," p. 14.

50. *Ibid.*, p. 18.

51. Jordan was very frequently criticized by Soviet scientists and philosophers. See his *Die Herkunft der Sterne* (Stuttgart, 1947).

52. Ambartsumian, "Problema proiskhozhdeniia zvezd," pp. 12–13.

53. See, for example, the article "Interstellar Matter" in the *McGraw-Hill Encyclopedia of Science and Technology* (New York, 1960), Vol. VII, p. 222: "V. A. Ambartsumian first pointed out that superluminous stars of high temperature, which cannot be very old because of the tremendous rate at which mass is converted into energy, are always found in clouds of gas and interstellar particles. Such associations are clear proof that stars must form from this material."

54. Just as dialectical materialists did not believe it correct to speak of the "birth" of the universe as a whole, so they also considered it incorrect to speak of its "death." They criticized those non-Soviet writers who spoke of the white-dwarf stage in stellar evolution as the "cemetery of celestial matter," or who used the term "white death of the universe." The Soviet philosophers frequently said that the white-dwarf stage of stellar evolution is simply another "state" of matter, not an end point of the universe. See, for example, the same G. A. Kursanov as above in his "O mirovozzrencheskom znachenii dostizhenii sovremennoi astronomii," *Voprosy filosofii*, No. 3 (1960), p. 64. A more extended discussion of a similar nature revolved around the Second Law of Thermodynamics and the "heat-death" of the universe. A number of different attempts were made by Soviet authors to refute this theory, most of them based on the belief that it is incorrect to extend the realm of the second law from closed systems to infinite ones. Other writers (S. T. Meliukhin and G. I. Naan) believed that there must exist in the universe "anti-entropic" processes counteracting the processes described by the second law. This view also arose in the analysis of cybernetics. Still others (L. D. Landau and E. M. Lifshits) opposed the heat-death interpretation on the basis of relativistic thermodynamics. See "Oproverzhenie 'teorii' teplovoi smerti," in *Filosofiia estestvoznaniia* (Moscow, 1966), pp. 130–6, and references contained therein. Also, E. A. Sedov, "K voprosu o sootnoshenii entropii informatsionnykh protsessov i fizicheskoi entropii," *Voprosy filosofii*, No. 1 (1965), pp. 135–45.

55. V. A. Ambartsumian, "Zakliuchitel'noe slovo," in P. N. Fedoseev *et al.*, eds., *op. cit.*, pp. 575–6.

56. The term "metagalaxy" here refers only to a part of the universe as a whole, that part about which man has direct evidence. The term "metagalaxy" was first used by Harlow Shapley. See, for example, his *The Inner Metagalaxy* (New Haven, 1957). The Soviet philosopher G. I. Naan remarked that although Shapley saw the need for a distinction between "universe" and "metagalaxy," he was not sufficiently careful in using it. See his "Gravitatsiia i beskonechnost'," in P. S. Dyshlevyi and A. Z. Petrov, eds., *Filosofskie problemy teorii tiagoteniia Einshteina i relativistskoi kosmologii* (Kiev, 1965), pp. 275, 278.

57. Ambartsumian, "Nekotorye metodologicheskie voprosy kosmogonii," in P. N. Fedoseev, *et al.*, eds., *op. cit.*, p. 271.

58. *Ibid.*

59. *Ibid.*, p. 272.

60. *Ibid.*, p. 273.

61. The term "parsec" is a contraction of "parallax second"; in terms of distance one parsec equals 19.2 trillion (19.2×10^{12}) miles. When the parallax of a star, as measured from the earth, is one second of arc, the distance to the star is defined as one parsec.

62. Ambartsumian, "Metodologicheskie voprosy kosmogonii," in P. N. Fedoseev, *et al.*, eds., *op. cit.*, p. 286.

63. *Ibid.*, pp. 287–8.

64. "Rech' Akad. AN Estonskoi SSR G. I. Naan," in Fedoseev, *et al.*, eds., *op. cit.*, p. 419.

65. Ambartsumian, "Vvodnyi doklad na simpoziume po evoliutsii zvezd," in Ambartsumian, *Nauchnye trudy,* Vol. II (Erevan, 1960), pp. 143–63; "Nekotorye osobennosti sovremennogo razvitiia astrofiziki," in *Oktiabr' i nauchnyi progress,* Vol. I (Moscow, 1967), pp. 73–85; "Contemporary Natural Science and Philosophy," in *Papers for XIV International Wien Philosophical Congress, Section 7* (Moscow, 1968), pp. 41–72, also published in Russian in *Uspekhi fizicheskikh nauk,* September 1968, pp. 1–19; "Ob evoliutsii galaktik," "Mir galaktik," and "Perspektivy razvitiia astronomii," in Ambartsumian, ed., *Problemy evoliutsii Vselennoi* (Erevan, 1968), pp. 85–127, 176–94, and 232–5; "Kosmos, kosmogoniia, kosmologiia," *Nauka i religiia,* No. 12 (1968), pp. 2–37; "Nestatsionarnye ob' 'ekty vo Vselennoi i ikh znachenie dlia kosmogonii," in Ambartsumian, ed., *Problemy sovremennoi kosmogonii* (Moscow, 1969), pp. 5–18; Ambartsumian and V. V. Kaziutinskii, "Revoliutsiia v sovremennoi astronomii," *Priroda,* No. 4 (1970), pp. 16–26.
66. Ambartsumian, "Mir galaktik," pp. 188–9.
67. Ambartsumian, "Contemporary Natural Science and Philosophy," pp. 26–9, 60–3.
68. *Ibid.,* p. 62.
69. Ambartsumian, "Vvodnyi doklad na simpoziume po evoliutsii zvezd," pp. 145–6.
70. Ambartsumian, "Nestatsionarnye ob' 'ekty vo Vselennoi i ikh znachenie dlia kosmogonii," pp. 5–18.
71. See B. V. Kukarkin and A. G. Masevich, "Sovetskie astronomy na VIII s' 'ezde mezhdunarodnogo astronomicheskogo soiuza v Rime," *Voprosy filosofii,* No. 1 (1953), p. 226. Also, see Struve and Zebergs, *Astronomy of the 20th Century,* pp. 32–3.
72. Jordan, *op. cit.*
73. H. Bondi, *Cosmology* (Cambridge, Eng., 1961), p. 164.
74. A similar use of the law of the unity and struggle of opposites in astronomy is in A. S. Arsen'ev, "Nekotorye metodologicheskie voprosy kosmogonii," *Voprosy filosofii,* No. 3 (1955), pp. 32–44.
75. A. S. Arsen'ev, "O sub' 'ektivizme v sovremennoi kosmogonii," *Priroda,* No. 6 (1954), pp. 47–56.
76. In the early nineteenth century the Alps and the Andes seemed as difficult to explain in terms of probable, gradualist causes as the planetary system did in the 1940's. The similarity of problem does not necessarily lead, of course, to a similarity in solution. See Frank Dawson Adams, *The Birth and Development of the Geological Sciences,* Dover reprint edition (New York, 1954), esp. pp. 210–76. A discussion that points out that catastrophism and uniformitarianism did not play simple "negative" and "positive" roles, respectively, vis-à-vis later developments, especially in biology, is in Walter F. Cannon, "The Uniformitarian-Catastrophist Debate," *ISIS,* LI (March 1960), 38–55.
77. It hardly needs to be pointed out that "catastrophism" is a purely relative term. Glaciers, earthquakes, and volcanic eruptions have obviously occurred in the earth's history on a scale far greater than any presently being witnessed. The explosions of supernovae could hardly be more cataclysmic. To

the extent that Arsen'ev's arguments here have strength, that strength derives more from the anthropomorphism of certain very rare events connected with the earth's history ("birth," "death," etc.), which has been inserted by some astronomers, rather than from catastrophism itself. And even that argument is an open one. Arsen'ev's opposition to catastrophism was expressed even more strongly in an article that he published in 1955; here he referred to the thought contained in the system of Kant and Laplace that the solar system was created as a result of a long and gradual process, and commented: "The whole development of modern cosmogony confirms this idea. . . ." A. S. Arsen'ev, "Nekotorye metodologicheskie voprosy kosmogonii," *Voprosy filosofii,* No. 3 (1955), pp. 32–44.

78. Arsen'ev, "O sub' 'ektivizme v sovremennoi kosmogonii," p. 52.
79. O. S. Gevorkian, "Bor'ba osnovnykh filosofskikh napravlenii v sovremennoi kosmogonii," in P. N. Gapochka, B. M. Kedrov, and M. N. Maslina, *Voprosy dialekticheskogo i istoricheskogo materializma* (Moscow, 1956), pp. 244–82.
80. V. I. Sviderskii, *Filosofskoe znachenie prostranstvenno-vremmenykh pred- stavlenii v fizike* (Leningrad, 1956).
81. *Ibid.,* p. 262.
82. S. T. Meliukhin, *Problema konechnogo i beskonechnogo* (Moscow, 1958).
83. In 1959, G. A. Kursanov developed the following interpretation: Einstein's world view was a mixture of natural-science materialism and Machist philos- ophy. Although one could find many points on which Mach and the Vien- nese group negatively influenced Einstein, in the final analysis, the natural- science materialism of Einstein was the more important influence and was responsible for the greatness of his work. G. A. Kursanov, "K otsenke filo- sofskikh vzgliadov A. Einshteina na prirodu geometricheskikh poniatii," in I. V. Kuznetsov and M. E. Omel'ianovskii, eds., *Filosofskie voprosy sovre- mennoi fiziki* (Moscow, 1959), pp. 393–410. A similar approach resulting in an even more positive evaluation of Einstein is in B. G. Kuznetsov's biog- raphy *Einstein;* Kuznetsov spoke of the "absolute spiritual purity" of Ein- stein's work as a "struggle for the sovereignty of the mind against all forms of mystical antiintellectualism." See Kuznetsov, *Einstein* (Moscow, 1962), and my review in *ISIS,* June 1964, pp. 251–2.
84. Meliukhin, *Problema konechnogo i beskonechnogo,* p. 75.
85. *Ibid.,* p. 178.
86. See note 54 above.
87. Meliukhin, *Problema konechnogo i beskonechnogo,* p. 189.
88. *Ibid.,* p. 194. Reginald Kapp stated the relationship between space and mat- ter in relativistic cosmology in the following way: "Bearing in mind that matter is not so much *in* space as *of* space, the most accurate description, and a noncommittal one, of the uttermost elementary constituent of matter may be *a bit of differentiated* space." Reginald O. Kapp, *Towards a Unified Cosmology* (London, 1960), p. 52.
89. Meliukhin, *Problema konechnogo i beskonechnogo,* p. 195.
90. *Ibid.,* p. 196.
91. I. S. Shklovskii and Carl Sagan, *Intelligent Life in the Universe* (San Fran- cisco, 1966), p. 135. By means of an unusual denotation system, it is pos-

sible to tell which sentences, and even which phrases, were written by each of the two authors, one a Soviet scientist, the other an American. The two scholars were not, however, in disagreement on basic issues.

92. An example of an exception was I. P. Plotkin, who wrote that the view that physics did not touch upon the problem of systems that are infinitely large "completely denied cosmology as a science." "A refusal to consider such problems," he continued, "would deal a damaging blow to science and philosophy." Among Soviet physicists whom he criticized for avoiding the issue of infinitely large systems were the well-known L. D. Landau and E. M. Lifshits, who in their text *Statistical Physics* commented: ". . . if we try to apply statistics to the universe as a whole, regarding it as a single, closed system, then we immediately run into a sharp contradiction between theory and experimental evidence." See I. P. Plotkin, "O fluktuatsionnoi gipoteze Bol'tsmana," *Voprosy filosofii*, No. 4 (1959), pp. 138–40; and L. D. Landau and E. M. Lifshits, *Statisticheskaia fizika, klassicheskaia i kvantovaia* (Moscow, 1951), pp. 43–4.

93. A. S. Arsen'ev, "O gipoteze rasshireniia metagalaktiki i krasnom smeshchenii," *Voprosy filosofii*, No. 8 (1958), p. 190.

94. "Rech' Akad. G. I. Naana," in *Filosofskie problemy sovremennogo estestvoznaniia*, p. 420.

95. An edited report of the conference was published: I. Z. Shtokalo, *et al.*, eds., *Filosofskie voprosy sovremennoi fiziki* (Kiev, 1964).

96. P. S. Dyshlevyi, "Prostranstvo-vremennye predstavleniia obshchei teorii otnositel'nosti," in Shtokalo *et al.*, eds., *op. cit.*, pp. 71, 81.

97. P. K. Kobushkin, "Nekotorye filosofskie problemy reliativistskoi kosmologii," in Shtokalo *et al.*, eds., *op. cit.*, pp. 116–48.

98. Kobushkin cited, in particular, Naan as a source for the concept of a "quasi-closed metagalaxy." G. I. Naan, *Trudy shestogo soveshchaniia po voprosam kosmogonii* (Moscow, 1959), pp. 247 *passim;* and "O sovremennom sostoianii kosmologicheskoi nauki," *Voprosy kosmogonii*, Vol. VI (Moscow, 1958), pp. 277–329.

99. Such a star would, apparently, violate the mass-radius relation of dwarf stars, which says that the larger the mass, the smaller the radius. According to this relation a dwarf greater than 1.2 solar masses would be reduced to a point. See, for example, W. S. Krogdahl, *The Astronomical Universe* (New York, 1962), p. 371. Also see J. R. Oppenheimer and G. M. Volkoff, "On Massive Neutron Cores," and the accompanying article by Richard C. Tolman, "Static Solutions of Einstein's Field Equations for Spheres of Fluid," *Physical Review*, LV (February 15, 1939), 374–81 and 364–73.

100. Kobushkin, *op. cit.*, p. 125.

101. *Ibid.*, p. 126.

102. Kobushkin was not committed to the Lemaître model, however. He was also very interested in inhomogeneous, anisotropic models such as expanding and rotating ones (Gödel, Heckmann, *et al.;* category IV), and spent considerable time discussing them. See *op. cit.*, pp. 131–9. He was also fascinated by the possibility of connecting the micro-, macro-, and megaworlds through physical constants, in a fashion similar to that once favored by Arthur Eddington, a person who like Lemaître had once been severely criticized in the

Soviet Union. Kobushkin began his work on this topic before World War II, and it is possible that not until the sixties was the ideological scene relaxed enough for him to publish his findings.

103. P. V. Kopnin, "Razvitie kategorii dialekticheskogo materializma—vazhnei-shee uslovie ukrepleniia soiuza filosofii i estestvoznaniia," in P. S. Dyshlevyi and A. Z. Petrov, eds., *op. cit.,* p. 5. I consider this essay interesting and unorthodox in terms of its implications for Soviet cosmology, despite the classification by certain non-Soviet authors of Kopnin among the "old guard" for certain works on dialectical logic. See Thomas J. Blakeley, *Soviet Theory of Knowledge* (Dordrecht-Holland, 1964), p. 6.

104. Kopnin, *op. cit.,* p. 8.

105. *Ibid.,* p. 9. This view, although somewhat controversial in the Soviet context, had been expressed as early as 1931: "In the materialist dialectic there is no closed or complete system of categories, nor can there be such a thing. All that can be spoken of is a systematic account of the laws and categories which truly reflect the dialectic of reality, a further working-out of these, and the exploration of new categories, representing hitherto unknown forms of material motion." Translated and quoted in Wetter, *op. cit.,* from G. Obich-kin, *Osnovnye momenty dialekticheskogo protsessa poznaniia* (Moscow and Leningrad, 1933), p. 80.

106. Kopnin, *op. cit.,* p. 13.

107. *Ibid.*

108. V. I. Sviderskii, "O dialektiko-materialisticheskom ponimanii konechnogo i beskonechnogo," in Dyshlevyi and Petrov, eds., *op. cit.,* p. 261, *passim;* also see S. T. Meliukhin's comments and objections in "O filosofskom ponimanii beskonechnosti prostranstva i vremeni," *ibid.,* p. 295.

109. Sviderskii, "O dialektiko-materialisticheskom ponimanii konechnogo i beskonechnogo," in Dyshlevyi and Petrov, eds., *op. cit.,* p. 267.

110. G. I. Naan, "Gravitatsiia i beskonechnost'," in Dyshlevyi and Petrov, eds., *op. cit.,* p. 284.

111. See his "O beskonechnosti vselennoi," *Voprosy filosofii,* No. 6 (1961), pp. 93–105.

112. Naan, "Gravitatsiia i beskonechnost'," p. 269.

113. *Ibid.,* p. 271.

114. *Ibid.*

115. A. Einstein and E. G. Straus, "The Influence of the Expansion of Space on the Gravitation Fields Surrounding the Individual Stars," *Reviews of Modern Physics,* XVII, No. 2–3 (1945), 120–4.

116. I. D. Novikov, "O povedenii sfericheski-simmetrichnykh raspredelenii mass v obshchei teorii otnositel'nosti," *Vestnik MGU,* Seriia III, No. 6 (1962), pp. 66–72.

117. He called the combination "dialectical" in a different spot; see "Gravitatsiia i beskonechnost'," p. 275.

118. A Schwarzschild shell is a stage in the life history of a star; see Bart J. Bok, *The Astronomer's Universe* (Cambridge, Eng., 1958), pp. 86–7; and Otto Struve, *Stellar Evolution* (Princeton, 1950), p. 149.

119. See A. L. Zel'manov, "Nereliativistskii gravitatsionnyi paradoks i obshchaia teoriia otnositel'nosti," *Nauchnye doklady vysshei shkoly: fiziko-matemati-*

cheskie nauki, No. 2 (1958), pp. 124–7; "K postanovke kosmologicheskoi problemy," *Trudy vtorogo s' 'ezda vsesoiuznogo astronomo-geogezicheskogo obshchestva 25–31 Ianvaria 1955 g.* (Moscow, 1960), pp. 72–84; "Metagalaktika i Vselennaia," in *Nauka i chelovechestvo 1962* (Moscow, 1963), pp. 383–405; "Kosmos, kosmogoniia, kosmologiia," *Nauka i religiia,* No. 12 (1968), pp. 2–37; "Mnogoobrazie material'nogo mira i problema beskonechnosti Vselennoi," in *Beskonechnost' i Vselennaia* (Moscow, 1969), pp. 274–324.

120. Zel'manov, "Mnogoobrazie material'nogo mira . . . ," p. 278.
121. Zel'manov, "K postanovke . . . ," pp. 73–4.
122. "Rech' A. L. Zel'manova," in P. N. Fedoseev, *et al.,* eds., *op. cit.,* pp. 434–41.
123. A. L. Zel'manov, "O beskonechnosti material'nogo mira," in M. E. Omel'ianovskii and I. V. Kuznetsov, eds., *Dialektika v naukakh o nezhivoi prirode* (Moscow, 1964), p. 260. Beginning in the late forties a number of scientists in various countries had begun to work on anisotropic, inhomogeneous models. See, for example, K. Gödel, "An Example of a New Type of Cosmological Solutions of Einstein's Field Equations of Gravitation," *Reviews of Modern Physics,* XXI (July 1949), 447–50; and Amalkumar Raychaudhuri, "Relativistic Cosmology I," *Physical Review,* IIC (May 15, 1955), 1123–6. For previous work by Zel'manov on the same subject, see A. L. Zel'manov, "Khronometricheskie invarianty i soputstvyiushchie koordinaty v obshchei teorii otnositel'nosti," *Doklady akademii nauk SSSR,* CVII, No. 6 (1956), 815–18; and "K reliativistskoi teorii anizotropnoi neodnorodnoi Vselennoi," *Trudy shestogo soveshchaniia po voprosam kosmogonii* (Moscow, 1959), pp. 144–73.
124. Zel'manov, "O beskonechnosti material'nogo mira."
125. *Ibid.,* pp. 263–4.
126. According to the view of the history of science advanced by Thomas Kuhn, relativity physics was, however, not a simple addition, or modification, to classical physics, but a paradigm that inherently contradicted classical physics. See the discussions of this interesting and crucial issue in Thomas Kuhn, *The Structure of Scientific Revolutions* (Chicago and London, 1962), pp. 100–1, and the penetrating review of Kuhn's book by Dudley Shapere, "The Structure of Scientific Revolutions," *The Philosophical Review,* LXXIII (July 1964), 383–94, esp. the discussion of the existence or nonexistence of an inherent contradiction, pp. 389–90.
127. Zel'manov, "O beskonechnosti . . . ," p. 268.
128. Zel'manov, "Mnogoobrazie material'nogo mira . . . ," p. 280 and *passim.*
129. Zel'manov, "Metagalaktika i Vselennaia," p. 390.
130. *Beskonechnost' i Vselennaia* (Moscow, 1960). For another interesting recent book, see K. P. Staniukovich, S. M. Kolesnikov, V. M. Moskovkin, *Problemy teorii prostranstva, vremeni i materii* (Moscow, 1968).
131. See E. M. Chudinov, "Logicheskie aspekty problemy beskonechnosti Vselennoi v reliativistskoi kosmologii," in *Beskonechnost' i Vselennaia,* pp. 181–218; E. M. Chudinov, "Filosofskie problemy sovremennoi fiziki i astronomii" (Moscow, 1969); V. V. Kaziutinskii, "Astronomiia i dialektika (k 100-letiiu so dnia rozhdeniia V. I. Lenina)," in *Astronomicheskii kalen-*

dar' 1970 (Moscow, 1969), pp. 138–70; G. I. Naan, "Poniatie beskonech-nosti v matematike i kosmologii," in *Beskonechnost' i Vselennaia*, pp. 7–77; A. L. Zel'manov, "Mnogoobrazie material'nogo mira i problema beskonech-nosti Vselennoi," in *Beskonechnost' i Vselennaia*, pp. 274–324.

132. V. A. Ambartsumian and V. V. Kaziutinskii, "Revoliutsiia v sovremennoi astronomii," *Priroda*, No. 4 (1970), pp. 16–26. Ambartsumian also approved of Kaziutinskii's approach in a short foreword he wrote to the latter's "Revoliutsiia v astronomii" (Moscow, 1968).

133. Quoted by Kaziutinskii in "Astronomiia i dialektika," *Astronomicheskii kalendar' 1970* (Moscow, 1969), pp. 140–1, from V. L. Ginzburg, "Kak ustroena Vselennaia i kak ona razvivaetsia vo vremia" (Moscow, 1968), p. 51.

134. V. V. Kaziutinskii, "Revoliutsiia v astronomii" (Moscow, 1968), p. 33.

135. See, for example, E. M. Chudinov, "Logicheskie aspekty problemy beskonechnosti Vselennoi v reliativistskoi kosmologii," p. 218. The Estonian philosopher G. I. Naan said in 1969 that the infinitude of the universe is a postulate, rather than something that can be "proved" or "disproved." But, Naan continued, without this postulate in one form or another man can not correctly understand the world as something which exists external to him, his will, and his consciousness. See G. I. Naan, "Poniatie beskonechnosti v matematike i kosmologii," pp. 76–7.

136. See L. B. Bazhenov and N. N. Nutsubidze, "K diskussii o probleme beskonechnosti Vselennoi," in *Beskonechnost' i Vselennaia*, p. 130.

137. They included M. Kaufman, F. Hoyle, J. V. Narlikar, N. C. Wickramasinghe, and D. Layzer. See R. B. Partridge, "The Primeval Fireball Today," *American Scientist*, LVII, No. 1 (1969), 37–74, esp. pp. 42–6. Partridge himself, however, preferred the fireball interpretation; he commented, "It is only fair for me to announce my personal bias in advance: I believe the fireball picture to be consistent with all the experimental data, and to be the simplest theoretical explanation of these data." *Ibid.*, p. 43.

138. *Ibid.*, p. 72.

139. Scriven quoted in Kapp, *Towards a Unified Cosmology*, p. 49; Bonnor, *The Mystery of the Expanding Universe*, p. 204.

VI: Genetics

1. Both the professional biologists and the professional philosophers had to be ordered to follow Lysenkoism. There was even a "Morganist school" among Soviet Marxist biologists in the twenties. See David Joravsky, *Soviet Marxism and Natural Science* (New York, 1961), p. 300. The official pressure exerted upon the Academy of Sciences can be readily seen as early as 1938. In May of that year the Council of People's Commissars (headed by Stalin's assistant V. M. Molotov) refused to approve the Academy's work plan. Lysenko shared the podium with Molotov in criticizing the Academy. See "V akademii nauk SSSR," *Vestnik akademii nauk SSSR*, No. 5 (1938), pp. 72–73. Shortly afterward, the Presidium of the Academy criticized the Institute of Genetics for refusing to recognize Lysenko's work. "Khronika," *Vestnik*

akademii nauk SSSR, No. 6 (1938), p. 75. The philosophers, who as ideologists might be expected by non-Soviet observers to have supported Lysenko, also had to be forced into line. In 1948, after the victory of Lysenko at the session of the Lenin Academy of Agricultural Sciences, the Presidium of the Academy of Sciences criticized the Institute of Philosophy for not giving "the necessary support to the Michurinist, materialist direction in biology." "Postanovlenie prezidiuma akademii nauk SSSR ot 26 avgusta 1948 g. po voprosu o sostoianii i zadachakh biologicheskoi nauki v institutakh i uchrezhdeniiakh akademii nauk SSSR," *Pravda,* August 27, 1948, p. 1.

2. Conway Zirkle maintained that there existed a peculiarly Marxist form of biology from the days of Marx and Engels onward. With the advent of Marxism to Russia this view supposedly gained much strength there. See his *Evolution, Marxian Biology, and the Social Scene* (Philadelphia, 1959). As I shortly explain above, I disagree with Zirkle's thesis that a peculiarly "Marxist biology" existed.

3. Darwinism attracted much attention among the populists; at first it was accepted with open arms as a symbol of materialism and scientific rationalism. Typical of such a reception was that of D. I. Pisarev. Later, however, at the hand of V. A. Zaitsev, a Russian Proudhonist, Darwinism received a racist interpretation that alarmed Zaitsev's fellow radicals. Zaitsev's close friend N. D. Nozhin attempted to reinterpret Darwinism within the spirit of the Proudhonist ideal of *mutualité.* The noted populist N. G. Chernyshevskii was hostile to Darwinism in general and criticized sharply Darwin's comparison of selection of domestic animals with selection in the wild state. Chernyshevskii's superficially brilliant critique was based entirely on the concept of the inheritance of acquired characters. Another criticism of Darwinism from a Russian radical came from Prince Peter Kropotkin, whose identification of co-operation as well as competition in the organic world was valuable in a scientific sense and, contrary to some opinion, could be included within a Darwinist framework. See D. I. Pisarev, *Selected Philosophical, Social and Political Essays* (Moscow, 1958), pp. 303–9, 344–452; V. A. Zaitsev, *Izbrannye sochineniia* (Moscow, 1934), Vol. I, pp. 26, 228–37, 429–37; N. D. Nozhin, "Nasha nauka i uchenye," *Knizhnyi vestnik,* April 15, 1866, pp. 175–8; N. D. Nozhin, "Po povodu statei 'Russkago Slova' o nevol' 'nichestvo," *Iskra,* No. 8 (1865), pp. 115–17; N. G. Chernyshevskii, *Polnoe sobranie sochinenii* (Moscow, 1939), Vol. X, pp. 737–72, esp. 758–9; P. A. Kropotkin, *Mutual Aid, a Factor of Evolution* (London, 1902); James Allen Rogers, "Darwinism, Scientism and Nihilism," *Russian Review,* XIX (1960), 10–23; and James Allen Rogers, "Charles Darwin and Russian Scientists," *Russian Review,* XIX (1960), 371–83. I am grateful to Miss Barbara Dafoe of Columbia University and Mr. Joseph Fuhrmann of Indiana University for help on this topic.

4. Here L. C. Dunn's comment is appropriate; belief in the inheritance of acquired characteristics, he said, "solaced most of the biologists of the nineteenth century." *A Short History of Genetics* (New York, 1965), p. x.

5. See Mark B. Adams, "The Founding of Population Genetics: Contributions of the Chetverikov School 1924–1934," *Journal of the History of Biology,* Spring, 1968, pp. 23–39.

6. On Michurin see David Joravsky, "The First Stage of Michurinism," in J. S. Curtiss, ed., *Essays in Russian and Soviet History* (New York, 1963), pp. 120–32. Also see T. D. Lysenko, ed., *I. V. Michurin: Sochineniia v chetyrekh tomakh*, Vols. I–IV (Moscow, 1948).

7. Hudson and Richens commented: "In his firm belief in the importance of the environment on genetic constitution Burbank anticipated the later theories of Lysenko. Several of his remarks on the power of the environment to modify genetical constitution presaged Lysenko's theory of 'shattering,' while his conclusion that 'heredity is nothing but stored environment' heralds Lysenko's dictum that 'hereditary constitution is as it were a concentrate of the environmental conditions assimilated by the plant organism in a number of preceding generations.' His tentative hypothesis of sap hybridization may be the antecedent of Lysenko's theory of graft hybridization. . . ." Hudson and Richens, *op. cit.*, p. 13.

8. A view supported in classical genetics by the belief that wild alleles are usually recessive.

9. An example is N. P. Dubinin, "I. V. Michurin i sovremennaia genetika," *Voprosy filosofii*, No. 6 (1966), pp. 59–70. Michurin's biological views were fully developed before the Revolution; the few changes that did occur in his opinions after 1917 were moves toward Mendelism rather than away from it; see pp. 197 and 253, and P. S. Hudson and R. H. Richens, *The New Genetics in the Soviet Union* (Cambridge, Eng., 1946), p. 12.

10. N. P. Dubinin, *op. cit.*, p. 64.

11. A useful short biography of Lysenko in English is Maxim W. Mikulak, "Trofim Denisovich Lysenko," in George W. Simmonds, ed., *Soviet Leaders* (New York, 1967), pp. 248–59; Soviet sources are M. S. Voinov, *Akademik T. D. Lysenko* (Moscow, 1950) and *T. D. Lysenko* (Moscow, 1953). An interesting short description of Lysenko appeared in *Pravda* in August 1927. The visiting journalist described the young agronomist in the following way: "If one is to judge a man by first impression, Lysenko gives one the feeling of a toothache; God give him health, he has a dejected mien. Stingy of words and insignificant of face is he; all one remembers is his sullen look creeping along the earth as if, at very least, he were ready to do someone in. Only once did this barefoot scientist let a smile pass, and that was at mention of Poltava cherry dumplings with sugar and sour cream. . . . The barefoot Professor Lysenko now has followers, pupils, an experimental field. He is visited in the winter by agronomic luminaries who stand before the green fields of the experiment station, gratefully shaking his hand. . . ." Quoted in Zhores A. Medvedev, *The Rise and Fall of T. D. Lysenko* (New York and London, 1969), pp. 11–12.

12. A bibliography of Lysenko's works from 1923 to 1951 is in T. D. Lysenko, *Agrobiologiia* (Moscow, 1952).

13. T. D. Lysenko, *Agrobiology* (Moscow, 1954), p. 17.

14. *Ibid.*, p. 18.

15. *Ibid.*, p. 21

16. T. D. Lysenko, "Vliianie termicheskogo faktora na prodolzhitel'nost' faz razvitiia rastenii," *Trudy azerbaidzhanskoi tsentral'noi opytnoselektsionnoi stantsii*, No. 3 (1928), pp. 1–169.

17. A. L. Shatskii, "K voprosu o summe temperatur, kak sel'skokhoziaistvenno-klimaticheskom indekse," *Trudy po sel'skokhoziaistvennoi meteorologii,* XXI, No. 6 (1930), 261–3. When the eminent plant physiologist N. A. Maksimov first discussed this formula, he commented that it was of "great interest" but was based on "too few experiments and must be further checked." Maksimov was a critic of Lysenko in later years but was forced to recognize him, as were many others. See N. A. Maksimov, "Fiziologicheskie faktory, opredeliaiushchie dlinu vegetatsionnogo perioda," *Trudy po prikladnoi botanike, genetike i selektsii,* XX (1929), 169–212; and N. A. Maksimov and M. A. Krotkina, "Issledovaniia nad posledstviem ponizhennoi temperatury na dlinu vegetatsionnogo perioda," *Trudy po prikladnoi botanike, genetike i selektsii,* XXIII (1929–30), vyp. 2, 427–73. The latter article contains genuine criticism of Lysenko's unclear terms.

18. See Hudson and Richens, *op. cit.,* p. 28, for references to Lysenko's debate over mathematics with the eminent A. N. Kolmogorov. Conway Zirkle suggested that Lysenko was the victim of a frustration complex: "Unable to handle the simplest mathematics, Lysenko resents it violently and denounces any application of mathematics to a biological problem. This puts Mendelism beyond his reach. As he equates all genetics with the 3 to 1 ratio, it is evident that he comprehends practically nothing of the modern developments in this field, and his complex makes him resent the very existence of a science which frustrates him." *Death of a Science in Russia* (Philadelphia, 1949), p. 96.

19. T. D. Lysenko, *Agrobiology,* p. 23.

20. The journal *Agrobiologiia* ceased publication in 1966.

21. *Biulleten' iarovizatsii,* No. 1 (January 1932), p. 63.

22. The most interesting and complete discussion of vernalization I have found is O. N. Purvis, "The Physiological Analysis of Vernalization," in W. H. Ruhland, ed., *Encyclopedia of Plant Physiology,* Vol. XVI (Berlin, 1961), pp. 76–117. Interestingly enough, vernalization has been found to have a genetic basis, a fact that Lysenko, if he knew of it, would undoubtedly have found unpleasant: "Up to six genes have been found responsible for the cold requirement of vernalizable plants. In the case of three of these genes, Hh, Ii, and Kk which are found in wheat, barley, rye, and *Arabidopsis,* the cold requirement is caused by the recessive alleles; in the case of the other genes, Ss, Tt, and Uu, found in wheat, barley, *Hyoscyamus* and *Arabidopsis,* by the dominant alleles." "Vernalization," *McGraw Hill Encyclopedia of Science and Technology,* Vol. XIV (New York, 1966), p. 305.

23. Julian Huxley, *Heredity East and West* (*Lysenko and World Science*) (London, 1949), p. 71.

24. See Hudson and Richens's discussion of this case, *op. cit.,* p. 39.

25. See, for example, *ibid.,* pp. 32–51.

26. See Joravsky's discussion in Part Four of his *Soviet Marxism and Natural Science,* pp. 233–71.

27. Even where Soviet agriculture could have used recent developments in agronomic science, its extreme backwardness made such applications very difficult. The state of development of Soviet agriculture immediately prior to collectivization has been a topic of much debate in the Soviet Union in re-

cent years, particularly since the criticism of Stalin in 1956. The prevalent interpretation among Soviet historians before 1956 was that the material-technical base of Soviet agriculture had so advanced by 1929 that a "contradiction" between the new productive forces and the old productive relationships had arisen, and this contradiction led to the necessity of collectivization. Newer interpretations, however, question seriously this thesis; the author of the first important study of the problem since 1956 concluded that a new material-technical base had not yet been created by the early thirties. Such a view implies, of course, that an important theoretical justification for rapid collectivization was absent. See V. P. Danilov, *Sozdanie material'no-tekhnicheskikh predposylok kollektivizatsii sel'skogo khoziaistva v SSSR* (Moscow, 1957); and M. L. Bogdenko and I. E. Zelenin, "Osnovnye problemy istorii kollektivizatsii sel'skogo khoziaistva v sovremennoi sovetskoi istoricheskoi literature," in M. P. Kim, *et al.*, eds., *Istoriia sovetskogo krest'ianstva i kolkhoznogo stroitel'stva v SSSR* (Moscow, 1963), pp. 192–222, esp. pp. 194–5.

28. Experiments on *Drosophila melanogaster* were conducted in the famous "fly room" at Columbia University from 1910 to 1928; H. J. Muller, one of T. H. Morgan's students, took the first fruit flies to Soviet Russia in 1922. Hybrid corn was available commercially in the United States after 1933, but the greatest expansion came in the forties; by 1949, 77.6 per cent of the total United States acreage was in hybrid corn. L. C. Dunn, *A Short History of Genetics* (New York, 1965), p. 140; A. H. Sturtevant, *A History of Genetics* (New York, 1965), pp. 45–57; Paul C. Mangelsdorf, "Hybrid Corn," in L. C. Dunn, ed., *Genetics in the 20th Century* (New York, 1951), p. 564. For the impact of Muller's 1922 visit to Russia, see Mark B. Adams, "The Founding of Population Genetics: Contributions of the Chetverikov School 1924–1934," *Journal of the History of Biology*, Spring, 1968, pp. 23–39.

29. A Soviet student discussed the enormous problems in Soviet agriculture after collectivization and the help given by the followers of Michurinist biology in Aleksandr S. Kuroedov, "Rol' sotsialisticheskoi sel'sko-khoziaistvennoi praktiki v razvitii michurinskoi biologii," unpublished dissertation for the degree of *kandidat,* Moscow State University, 1952, esp. pp. 99–105.

30. I am particularly indebted to David Joravsky for an understanding of the importance of Lysenko's nostrums to his ascent in the 1930's.

31. Based on conversations held with Soviet officials in Moscow and Leningrad, May–July 1970.

32. For an article in which Lysenko urged involving "thousands" of collective farm workers in his experiments, see his "Obnovlennye semena: beseda s akademikom T. D. Lysenko," *Sotsialisticheskoe zemledelie,* September 16, 1935, p. 1. The emphasis here seems to be as much on personal involvement as on technical advantage.

33. There is considerable evidence that vernalization did in fact lead to decreased yields. See, for example, O. M. Targul'ian, ed., *Spornye voprosy genetiki i selektsii: raboty IV sessii akademii 19–27 dekabriia 1936 goda* (Moscow and Leningrad, 1937), pp. 189–93, 204–5.

34. Prezent commented, "Genetics gives birth to dialectics." Later he called this

"material for the criticism of the path over which I have traveled." *Pod znamenem marksizma*, No. 11 (1939), pp. 95, 112.

35. Dubinin said at the conference in 1939, "Academician Lysenko is greatly confused about questions of Mendelism. But I think that to a considerable degree this confusion must be attributed, Academician Lysenko, to your helper, Comrade Prezent. (Voice from the floor shouts 'Correct!')" *Pod znamenem marksizma*, No. 11 (1939), p. 186. Hudson and Richens commented, "There is indeed evidence that the full elaboration of the genetical system of Lysenko is principally due to Prezent. . . ." *Op. cit.*, p. 15.

36. T. D. Lysenko, *Agrobiology*, p. 65.

37. *Ibid.*, p. 68.

38. Nazi Germany passed a compulsory sterilization law on July 14, 1933. Hudson and Richens commented, "Although still a matter of controversy, there can be no doubt that genetical research has demonstrated the heterogeneity of the human race, and has therefore provided a potential basis for the development of theories of racial and class distinction. It seems clear that Lysenko and Prezent realized these implications and found in them a serious objection to the theory of social equality. The growing political tension between Russia and Germany no doubt served to inflame these suspicions." *Op. cit.*, p. 27. And J. B. S. Haldane, a communist, bitter foe of fascism, and brilliant geneticist, anticipated the issue in 1932 when he commented, "The test of the devotion of the Union of Socialist Soviet Republics to science will, I think, come when the accumulation of the results of human genetics, demonstrating what I believe to be the fact of innate human inequality, becomes important." *The Inequality of Man* (London, 1932), p. 137.

39. See Filipchenko's numerous articles in *Izvestiia buro po evgenike* and *Izvestiia buro po genetike i evgenike* in the period 1922–6. Also "Spornye voprosy evgeniki," *Vestnik kommunisticheskoi akademii*, No. 20 (1927), pp. 212–54. Corresponding Academician N. P. Dubinin, one of Lysenko's major opponents and director of the Institute of General Genetics of the Academy of Sciences, praised Filipchenko's work of the 1920's and 1930's in an article that appeared after Lysenko's discrediting. "I. V. Michurin i sovremennaia genetika," *Voprosy filosofii*, No. 6 (1966), p. 69.

40. Quoted in Joravsky, *op. cit.*, p. 306, from A. S. Serebrovskii, "Antropogenetika," *Mediko-biologicheskii zhurnal*, No. 5 (1929), p. 18.

41. His brother, Sergei, eventually became president of the Academy of Sciences of the U.S.S.R. See David Joravsky, "The Vavilov Brothers," *Slavic Review*, September 1965, pp. 381–94.

42. See his statement in O. M. Targul'ian, ed., *op. cit.*, p. 462.

43. Th. Dobzhansky, the noted geneticist and also a Russian, commented in 1947: "Vavilov was an ardent Russian patriot. Outside of Russia he was regarded by some as a communist, which he was not. But he did wholeheartedly accept the revolution, because he believed that it opened broader possibilities for the development of the land and of the people of Russia than would have been otherwise. In October, 1930, during a trip to the Sequoia National Park in the company of this writer (and with nobody else present), he said with much emphasis and conviction that, in his opinion, the opportunities for serving mankind which existed in the USSR were so great and so

inspiring that for their sake one must learn to overlook the cruelties of the regime. He asserted that nowhere else in the world was the work of scientists more appreciated than in the USSR." Th. Dobzhansky, "N. I. Vavilov, A Martyr of Genetics, 1887–1942," *Journal of Heredity*, XXXVIII, No. 8 (August 1947), 229–30.

44. See note 28 above.

45. T. M. Sonneborn, "H. J. Muller, Crusader for Human Betterment," *Science*, November 15, 1968, p. 772.

46. H. J. Muller, *Out of the Night: A Biologist's View of the Future* (New York, 1935), p. vii.

47. See Huxley, *op. cit.*, p. 183.

48. Sonneborn, *op. cit.*, p. 774.

49. A. Kol', "Prikladnaia botanika ili leninskoe obnovlenie zemli," *Ekonomicheskaia zhizn'*, January 29, 1931, p. 2.

50. *Ekonomicheskaia zhizn'*, March 13, 1931.

51. *Ibid.*

52. C. D. Darlington, "The Retreat from Science in Soviet Russia," in Zirkle, ed., *Death of a Science in Russia*, pp. 72–3. It is useful to compare the corruption of Lysenko by publicity to that of Luther Burbank. See Walter L. Howard, "Luther Burbank: A Victim of Hero Worship," *Chronica Botanica*, IX, No. 5–6 (1945). Of course, the fact that Burbank was lionized, invited to give sermons, pictured on a postage stamp, and honored by placing Arbor Day on his birthday did not interfere with classical genetics in the United States. But his exploitation by shady companies gave an example of the obvious fact that a capitalist society produces its own peculiar forms of corruption.

53. T. D. Lysenko, "Iarovizatsiia—eto milliony pudov dobavochnogo urozhaia," *Izvestiia*, February 15, 1935, p. 4.

54. *Ibid.* As Joravsky has pointed out, Stalin changed his mind frequently; his support of Lysenko at one time is hardly sufficient to explain Lysenko's long period of dominance.

55. After discussing the principle of the transition of quantity into quality, Stalin remarked that Mendeleev's Periodic Table is an illustration of this principle and that "the same thing is shown in biology by the theory of neo-Lamarckism, which is supplanting neo-Darwinism." *Anarkhizm ili sotsializm?* (Moscow, 1950), p. 25.

56. Targul'ian, ed., *op. cit.*, p. 374.

57. Targul'ian, ed., *Spornye voprosy genetiki i selektsii: raboty IV sessii akademii 19–27 dekabriia 1936 goda* (Moscow and Leningrad), 1937.

58. *Ibid.*, p. 72.

59. *Ibid.*, p. 336.

60. See the importance of the principles of "intelligence" and "co-operation" in his *Out of the Night*, p. 37.

61. Another critic of this sort was A. R. Zhebrak, a well-known Soviet geneticist who lost his teaching job at the Timiriazev Academy in 1948. In 1939 Zhebrak said that classical genetics is a confirmation of the laws of the dialectic, especially of the Law of the Transition of Quantity into Quality. *Pod znamenem marksizma*, No. 11 (1939), p. 98. N. P. Dubinin has also written about

genetics in Marxist philosophy journals. And in 1936 at one point he tried to turn the tables on Prezent by accusing him of following Weismann! Targul'ian, ed., *op. cit.*, p. 339.

62. Targul'ian, ed., *op. cit.*, p. 114.
63. The members of the editorial board of *Pod znamenem marksizma* at this time were: V. V. Adoratskii, M. B. Mitin, E. Kol'man, P. F. Iudin, A. A. Maksimov, A. M Deborin, A. K. Timiriazev, and M. N. Korneev.
64. Vavilov's speech at the 1939 conference has been called "weak" or "ineffective" by non-Soviet commentators, but I find it quite outspoken and carefully grounded on both theoretical and practical arguments. *Pod znamenem marksizma*, No. 11 (1939), pp. 127–40.
65. Lysenko on scientific method needs no comment: ". . . in order to obtain a definite result, one must want to obtain namely that result; if you want to obtain a definite result, you will obtain it. . . . I need only those people who obtain what I need." *Pod znamenem marksizma*, No. 11 (1939), p. 95.
66. Quoted in V. Kolbanovskii, "Spornye voprosy genetiki i selektsii (obshchii obzor soveshchaniia)," *Pod znamenem marksizma*, No. 11 (1939), p. 93.
67. T. D. Lysenko, "Po povodu stat'i akademika N. I. Vavilova," *Iarovizatsiia*, No. 1 (1939), p. 140.
68. Quoted from Iu. Polianskii in *Pod znamenem marksizma*, No. 11 (1939), p. 103.
69. *Pod znamenem marksizma*, No. 11 (1939), p. 125.
70. There is much more information on the arrest, trial, and fate of Vavilov than given here in Zhores A. Medvedev, *The Rise and Fall of T. D. Lysenko* (New York and London, 1969), esp. pp. 67–77.
71. Medvedev, *op. cit.*, p. 110.
72. *The Situation in Biological Science: Proceedings of the Lenin Academy of Agricultural Sciences of the U.S.S.R., July 31–August 7, 1948, Complete Stenographic Report* (New York, 1949); also published in English in the U.S.S.R. (Moscow, 1949); also in Russian: *O polozhenii v biologicheskoi nauke. Stenograficheskii otchet sessii vsesoiuznoi akademii sel'sko-khoziaistvennykh nauk imeni V. I. Lenina, 31 iiulia–7 avgusta 1948 g.*, (Moscow, 1948).
73. Lysenko, nonetheless, put great emphasis on food, especially at certain points in an organism's growth. This emphasis will be particularly clear in his experiments with increasing the butterfat content of dairy milk. See p. 247. The nutrient theory probably has connections with Darwin's belief that "of all the causes which induce variability, excess of food, whether or not changed in nature, is probably the most powerful." Quoted in Hudson and Richens, *op. cit.*, p. 7.
74. T. D. Lysenko, *Agrobiology*, p. 34.
75. *Ibid.*, pp. 34–5. Emphasis in the original.
76. *Ibid.*, p. 64.
77. See O. N. Purvis, "The Physiological Analysis of Vernalization," in Ruhland, ed., *op. cit.*, Vol. XVI, pp. 76–117.
78. The connection between vernalization and heredity is clearly revealed in his statement: ". . . *during the vernalization of seeds or plants an accumulation*

of changes takes place. These changes remain in the cells in which they have taken place and are also transmitted to all the new cells formed by them." *Agrobiology*, p. 50. Emphasis in the original.

79. Lysenko was confident that he could produce new varieties with desirable characteristics in two to three years. In accordance with his theory he eliminated hundreds of varieties without even testing them. He commented, "We have no right, legal or moral, to waste one or two years on phasic analysis in cases where we can dispense with it." *Ibid.*, p. 110.

80. *Ibid.*, p. 83.

81. "Genotype" means the array of genes which an individual receives through heredity. "Phenotype" is the totality of the physical and behavioral characters that an individual displays, and is the result of an interaction between the genotype and the environment.

82. T. D. Lysenko, *Heredity and its Variability*, trans. Th. Dobzhansky (New York, 1946), p. 10. Also notice Lysenko's observation that "a living body qualitatively altered by living conditions always possesses an altered heredity. But it is by no means always the case that the qualitatively changed parts of the body of an organism can establish normal metabolism with many other body parts; hence, the changes do not always become fixed in the sex-cells." P. 24.

83. *Ibid.*, p. 1.

84. T. D. Lysenko, *Agrobiologiia*, p. 436.

85. Lysenko's writings on the gene were very confused. At one point he said, ". . . we deny that the geneticists and cytologists will see genes under the microscope. . . . The heredity basis is not some kind of self-reproducing substance *separate* from the body. The hereditary basis is the cell. . . ." At another time he said, "Academician Serebrovskii is also wrong when he says Lysenko denies the existence of genes. Neither Lysenko nor Prezent have ever denied the existence of genes. We deny the correctness of your conception of the word 'genes.' . . ." *Agrobiology*, pp. 186, 188. Lysenko did admit the reality of chromosomes.

86. "Desired" conditions in the sense that they were those of a locality in which the Soviet government wished to grow certain plants. Thus, the conditions of northern Russia might be "undesirable" for agriculture in an absolute sense, but the government wished to find varieties that could be raised there for very understandable reasons.

87. T. D. Lysenko, *Heredity and Its Variability*, p. 32.

88. Erik Nordenskiöld, in his old but still very valuable *The History of Biology* (New York, 1935), commented on Darwin's theory of pangenesis, "Darwin is here, as so often elsewhere, a speculative natural philosopher, not a natural scientist." This remark, typical of those expressed by inductivist historians of science in the period when Darwinism seemed to be yielding to modern genetics, seems excessively critical; a great part of Darwin's genius was his willingness to build a system that was in part speculation, but all the time checking the system against empirical data. Would that Lysenko had been as willing to discipline himself. If he had, his name would probably be known today only to a few specialists in vernalization. For a discussion of inductivist historians of science, unnecessarily combative but generally sound, see

Joseph Agassi, *Towards an Historiography of Science,* in *History and Theory: Studies in the Philosophy of History,* Beiheft 2 (The Hague, 1963), pp. 1–31. Agassi has, however, too simple a view of Marxist interpretations of the history of science.

89. *Heredity and Its Variability,* pp. 55–65.

90. *Ibid.,* p. 51.

91. Lysenko quoted Michurin's belief that "the farther apart the crossed parental pairs are in respect to place of origin and environmental conditions, the more easily the hybrid seedlings adapt themselves to the environmental conditions of the new locality." *Agrobiology,* p. 85. If applied to man, this view would be, of course, a strong argument for racial mixing. Lysenko did not extend his system to man, here or elsewhere.

92. Hudson and Richens, *op. cit.,* p. 48.

93. Darwin believed in the possibility of true graft hybrids, as did many nineteenth-century biologists and selectionists. Luther Burbank, who, like Lysenko, kept very poor records, advanced a tentative theory of sap hybridization similar in many ways to graft hybridization. Michurin's "mentor theory" postulated the influence of the stock by the scion.

94. For an elementary discussion of blending inheritance see E. W. Sinnott, L. C. Dunn, and Th. Dobzhansky, *Principles of Genetics* (New York, 1950), pp. 97 ff. and 121 ff.

95. Hudson and Richens, *op. cit.,* pp. 42–3.

96. T. D. Lysenko, *Izbrannye sochineniia* (Moscow, 1958), Vol. II, p. 48.

97. Dunn, *A Short History of Genetics,* p. x.

98. Since Darwin ascribed validity to both natural selection and the inheritance of acquired characters, both the neo-Mendelians and Michurinists could call themselves Darwinians.

99. Many writers attribute to Lamarck a vitalistic view of nature. The issue is not nearly that simple; as C. C. Gillispie has commented: "His dichotomy of organic and inorganic nature provides no escape into transcendentalism, and that has always been the door through which vitalists have slipped from science into mystery. Life is a purely physical phenomenon in Lamarck, and it is only because science has (quite rightly) left behind his conception of the physical that he has been systematically misunderstood, and assimilated to a theistic or vitalistic tradition which in fact he held in abhorrence." *The Edge of Objectivity* (Princeton, 1960), p. 276.

100. V. A. Shaumian maintained at the 1948 conference that the influence of the process of milking cows must have a hereditary effect: "Can such a vigorous determining factor of action on an udder applied from generation to generation over an expanse of many years remain without result? We attribute no less importance to the factor of the milking process than to feeding because milking is one of the most important methods and means of exercise for a milch cow." This view is, of course, pure Lamarckism, based on the inheritance of the effects of use, and similar to Lamarck's famous description of the lengthening of a giraffe's neck. Zirkle, ed., *Death of a Science in Russia,* p. 148.

101. Lysenko commented: ". . . no positive results can be obtained from work conducted from the standpoint of Lamarckism. The very fact that we do

succeed in changing the hereditary nature of plants in a definite direction by suitable training shows that we are not Lamarckians. . . ." *Agrobiologiia,* p. 221. Prezent was even more outspoken on the issue. But Lysenko in another spot referred to Lamarck more favorably: "Let us note, by the way, that the Morganists are vainly trying to frighten people with Lamarckism. Lamarck was a wise man. The importance of his work cannot of course be compared to Darwinism. There are serious errors in Lamarck's theory. But in his own time there did not exist a more progressive biologist than Lamarck." *Iarovizatsiia,* No. 5(8) (1936), pp. 45–68.

102. See, for example, V. I. Polianskii and Iu. I. Polianskii, *Sovremennye problemy evoliutsionnoi teorii* (Leningrad, 1967), p. 5 and *passim.*
103. Charles Coulston Gillispie, "Lamarck and Darwin in the History of Science," in Bentley Glass, Owsei Temkin, and William L. Straus, Jr., eds., *Forerunners of Darwin: 1745–1859* (Baltimore, 1959), pp. 268–9.
104. *Ibid.,* p. 277.
105. See Erik Nordenskiöld, *The History of Biology* (New York, 1935), p. 324.
106. Pollen fertilization deserves some comment, since it was one of the most controversial issues in Lysenko's writings. Lysenko believed that the ova of plants would select particular pollen grains (one form of nutrient) at fertilization, those that would result in the best adaptation to local conditions. At one point Lysenko even spoke of "marriage for love" in the plant world. Hudson and Richens, *op. cit.,* p. 58, have tried to show that this belief, although crudely expressed, need not necessarily have been based on anthropomorphic concepts. They say that even when a plant is grown in an environment to which it is not adapted, "it is not inconceivable that natural selection should so refine the selective power that any departure from the norm in the environment should tend to bring about compensating changes in the selection of nutrients at any part of the life cycle." I find the last sentence very dubious; it would, I believe, be improved if "any departure from the norm" were changed to "some departures" (those previously experienced and overcome), and if "any part of the life cycle" were changed to "any part of the life cycle prior to or during the fertile period." Lysenko later abandoned the "marriage for love" phrase and criticized anthropomorphic concepts in biology. The pollen concept has connections with Darwin's ideas on pollen prepotency. For Lysenko's later conservatism, see his statement, "purpose is bound up with consciousness, which is absent in nature." T. D. Lysenko, "Teoreticheskie osnovy napravlennogo izmeneniia nasledstvennosti sel'skokhoziaistvennykh rastenii," *Pravda,* January 29, 1963, pp. 3–4.
107. Theodosius Dobzhansky, *The Biological Basis of Human Freedom* (New York, 1956), p. 10.
108. Dunn, *A Short History of Genetics,* p. 155. Muller's early sympathy for dialectical materialism has already been cited.
109. Sturtevant, *op. cit.,* p. 67.
110. Dunn, *A Short History of Genetics,* p. 215.
111. Targul'ian, ed., *op. cit.,* p. 131.
112. *Ibid.,* p. 137.
113. *Istoriia vsesoiuznoi kommunisticheskoi partii (bol'shevikov): kratkii kurs* (Moscow, 1945), p. 101. The same citation was used against physicists who

emphasized the distinctness of the boundary between the macrolevel and the microlevel. See p. 85.

114. T. D. Lysenko, *Izbrannye sochineniia,* Vol. II (Moscow, 1958), p. 49.

115. As Dobzhansky described it: "Mutations are . . . changes induced ultimately by the environment, but the properties of a mutant are dependent on the nature of the gene that made the change, rather than on the environmental agency which acted as the trigger that set off the change." *The Biological Basis of Human Freedom,* p. 19.

116. Lysenko commented, "It seems to me that what must urgently be found here is not proof of whether Lysenko is right or wrong, but the best and most practical way of achieving the goal the plant breeder has set himself, namely, to breed the best varieties in the shortest time. Every theoretical proposition that helps practical farming will be more useful and, of course, come nearer the truth than all the theoretical propositions that give neither immediate nor future, neither direct nor indirect, guidance to practical action in our socialist agriculture." *Agrobiology,* p. 111.

117. T. D. Lysenko, *Izbrannye sochineniia,* Vol. II, p. 6, quoting from I. V. Michurin, *Sochineniia,* Vol. IV, p. 72. The opening speech at the 1937 genetics conference, given by the supposedly neutral A. I. Muralov, president of the Lenin Academy of Agricultural Sciences, emphasized the bond between theory and practice: "What must all participants at the conference remember during the discussion of issues in selection and genetics? They must first of all keep in mind the assistance which science must render socialist production, arming it with scientific theory." Targul'ian, ed., *op. cit.,* p. 5. Lysenko forced even his opponents to wish him good luck by such statements as: "I think that if the propositions we are advancing, particularizing and developing turn out to be fundamentally wrong, it should be regretted not only by me and my staff, but also by all the opponents of these propositions, for we would be deprived of an effective method of breeding new varieties." *Agrobiology,* p. 112.

118. Schrödinger said that a physicist "would be inclined to call de Vries's mutation theory, figuratively, the quantum theory of biology. We shall see later that this is much more than figurative. The mutations are actually due to quantum jumps in the gene molecule. But quantum theory was but two years old when de Vries first published his discovery, in 1902. Small wonder that it took another generation to discover the intimate connection!" *What is Life? & Other Scientific Essays* (Garden City, N. Y., 1956), p. 35. This section of Schrödinger's book is based on lectures given in Dublin in February 1943.

119. The same issue was discussed from the standpoint of DNA by a conservative but intelligent philosophy graduate student in the U.S.S.R. Ferents Pinter, "Aktual'nye voprosy vzaimootnosheniia marksistskoi filosofii i genetiki," dissertation for the degree of *kandidat* of the philosophical sciences, Moscow State University, 1965.

120. T. D. Lysenko, *Agrobiology,* pp. 551–2.

121. See pp. 72–3, 90–1, 98, 109 and *passim.*

122. Several Soviet reviewers observed in 1965: "It is well known that the formula 'Science is the foe of chance' was proclaimed at the end of the 1940s. This formula is incorrect, for it is based on the confusion of two completely

different concepts of chance. As is known, it has done no small amount of harm to science and to practice, but nowhere have its authors openly renounced it, and to this day it figures as a component part of 'the Michurinist teaching,' although it has nothing in common with the views of Michurin himself." I. L. Knuniants, B. M. Kedrov, L. Ia. Bliakher, "A Book that Does Not Deserve High Appraisal," *Pravda,* January 24, 1965, p. 2, from *Current Digest of the Soviet Press,* XVII, No. 4 (1965), p. 7. Lysenko's view of chance was criticized in the Soviet Union in 1957 by A. L. Takhtadzhan: "From the point of view of mechanistic materialism, statistical laws are only a temporary stage of our knowledge. In reality statistical laws are just as objective laws of nature as are the laws of the unique." *Botanicheskii zhurnal,* No. 4 (1957), p. 596.

123. The knife cuts both ways. If one accepts the inheritance of acquired characters, building a "new man" may seem more possible, but the same argument would lead to racist positions or even belief in the superiority of aristocracies. As Julian Huxley wrote: ". . . it is very fortunate for the human species that acquired characters are *not* readily impressed on the hereditary constitution. For if they were, the conditions of dirt, disease, and malnutrition in which the majority of mankind have lived for thousands of years would have produced a disastrous effect upon the race." Huxley, *op. cit.,* p. 138. Of course, any serious discussion of the implications for man of a hypothetical inheritance of acquired characters would have to involve discussions of time: the number of generations required to fix the heredity of a new trait and the number of generations in new conditions to erase it.

124. As late as 1958, *Pravda* spoke of N. K. Kol'tsov (1872–1940), a prominent Soviet biologist associated with eugenicist views in the twenties, as a "shameless reactionary who is known for his wild theory that preaches 'the improvement of human nature.' " *Pravda,* December 14, 1958. Yet "the improvement of human nature" is precisely the reason given by some writers for the entire Lysenko affair. John Langdon-Davies wrote that the controversy occurred because "a limit was being set to the extent to which environmental change at the hands of the U.S.S.R. planners could be expected to alter human nature permanently for the better." *Russia Puts the Clock Back* (London, 1949), pp. 58–9.

125. A map of the shelter-belt scheme is given in *Ogonek,* No. 10 (March 1949), pp. 4–5.

126. A. Bovin, "Na trassakh gosudarstvennykh lesnykh polos," *Pravda,* May 8, 1950, p. 2.

127. See his "Gnezdovaia kul'tura lesa," *Ogonek,* No. 10 (March 1949), pp. 6–7. Also his "Posev polezashchitnykh lesnykh polos gnezdovym sposobom" (Moscow, 1950).

128. See *Pravda,* April 17, 1943.

129. Kropotkin devoted his attention primarily to animals, including man, rather than to plants. He wrote, "If we resort to an indirect test, and ask Nature: 'Who are the fittest: those who are continually at war with each other, or those who support one another?' we at once see that those animals which acquire habits of mutual aid are undoubtedly the fittest." Kropotkin did not deny the existence of competition among members of the same species, nor

did he actually deny the accuracy of the phrase "survival of the fittest"; he maintained merely that those species that co-operate are "fittest." P. Kropotkin, *Mutual Aid: A Factor of Evolution* (London, 1902), p. 6. See also above, p. 196, and esp. note 3.

130. T. Lysenko, "Novoe v nauke o biologicheskom vide," *Pravda*, November 3, 1950, p. 2.

131. T. D. Lysenko, *Agrobiologiia*, p. 504.

132. Lysenko commented, "Wild plants, particularly forest trees, possess the extremely useful biological ability of self-thinning. . . . This occurs because a given area of tree crown can support only a certain number of plants. Therefore, some of the trees normally die." "Gnezdovaia kul'tura lesa," *Ogonek*, No. 10 (March 1949), p. 7. But Lysenko at another point commented that the example of thousands of tree seedlings crowding each other is actually not a case of intraspecies competition because many seedlings are required to overpower the grass that is trying to crowd them out. See his "Teoreticheskie osnovy napravlennogo izmeneniia nasledstvennosti sel'skokhoziaistvennykh rastenii," *Pravda*, January 29, 1963, pp. 3–4.

133. *Botanicheskii zhurnal*, No. 2 (1955), p. 213. See also V. Ia. Koldanov, "Nekotorye itogi i vyvody po polezashchitnomu lesorazvedeniiu za istekshie piat' let," *Lesnoe khoziaistvo*, No. 3 (1954), pp. 10–18.

134. In "O vnutrividovykh i mezhvidovykh vzaimootnosheniiakh sredi rastenii," *Botanicheskii zhurnal*, XXXVIII, No. 1 (1953), V. N. Sukhachev maintained that Darwin was correct, contrary to Lysenko's views, in saying that intraspecific competition exists and that as a general rule, the closer the structure of organisms the more intense the competition. This phenomenon is important in explaining the progressive separation of characteristics in evolutionary development. Sukhachev observed that one must, however, be careful in talking about "competition" in the plant world, since it can easily be given an anthropomorphic connotation; furthermore, the existence of "competition" does not deny the simultaneous existence of "co-operation." Nonetheless, for lack of a better term, Sukhachev felt the word "competition" is legitimate. Other authors were less critical of Lysenko's position; the discussion even included an entry by Lysenko himself, a reprint of his article on "species" for the second edition of the *Bol'shaia sovetskaia entsiklopediia*. T. D. Lysenko, "Novoe v nauke o biologicheskom vide," *Botanicheskii zhurnal*, XXXVIII, No. 1 (1953), 44–56; see also M. M. Il'in, "Filogenez pokrytosemennykh s pozitsii michurinskoi biologii," *Botanicheskii zhurnal*, Vol. XXXVIII, No. 1 (1953), pp. 97–118.

135. See James M. Swanson, "The Bolshevization of Scientific Societies in the Soviet Union; An Historical Analysis of the Character, Function and Legal Position of Scientific and Scientific Technical Societies in the USSR 1929–1936," unpublished dissertation, Indiana University (Bloomington, Ind., 1967).

136. A. A. Rukhkian, "Ob opisannom S. K. Karapetianom sluchae porozhdeniia leshchiny grabom," *Botanicheskii zhurnal*, XXXVIII, No. 6 (1953), 885–91.

137. Bibliographies of this debate are given in *Botanicheskii zhurnal*, No. 2 (1954), pp. 221–3; and No. 2 (1955), pp. 213–14.

138. Rubashevskii was the author of *Filosofskoe znachenie teoreticheskogo nasledstva I. V. Michurina* (Moscow, 1949). See also the two references above, note 137; and V. N. Sukhachev and N. D. Ivanov, "K voprosam vzaimootnoshenii organizmov i teorii estestvennogo otbora," *Zhurnal obshchei biologii*, XV, No. 4 (July–August 1954), 303–19.

139. "Diskussii: rasshiriat' i uglubliat' tvorcheskuiu diskussiiu po probleme vida i vidoobrazovaniia," *Botanicheskii zhurnal*, XL, No. 2 (1955), 206.

140. *Ibid.*, p. 207, and references at end of article.

141. *Ibid.*

142. *Ibid.*, p. 208.

143. *Ibid.*

144. B. Sokolov, "Ob organizatsii proizvodstva gibridnykh semian kukuruzy," *Izvestiia*, February 2, 1956, p. 2.

145. I. Knuniants and L. Zubkov, "Shkoly v nauke," *Literaturnaia gazeta*, January 11, 1955, p. 1.

146. V. N. Sukhachev and N. D. Ivanov, "Toward Problems of the Mutual Relationships of Organisms and the Theory of Natural Selection," *Current Digest of the Soviet Press*, VII, No. 1 (February 16, 1955), 11, from *Zhurnal obshchei biologii*, July–August 1954, pp. 303–19.

147. "Expand and Intensify Creative Discussion of Species and Species Formation," *Current Digest of the Soviet Press*, VII, No. 24 (July 27, 1955), 5, from *Botanicheskii zhurnal*, No. 2 (1955), pp. 206–13.

148. Lysenko and Maltsev had known each other for more than twenty years and spoke well of each other. They were both delegates to the Second all-Union Congress of Collective Farm Shock Workers in 1935. At the Twentieth Congress of the Communist Party in 1956 Maltsev gave a speech that attracted considerable attention.

149. Medvedev, *op. cit.*, p. 137.

150. See, for example, his "Shire primeniat' v nechernozemnoi polose organomineral'nye smesy," *Izvestiia*, April 27, 1957, p. 2. But Lysenko also took notice of his critics; he accused Sukhachev, the editor of *Botanicheskii zhurnal* and *Biulleten' moskovskogo obshchestva ispytatelei prirody*, of a "direct denial of the entire concept of materialist biology" and of launching a "highly unscientific criticism of my works." T. Lysenko, "Teoreticheskie uspekhi agronomicheskoi biologii," *Izvestiia*, December 8, 1957, p. 5.

151. "Rech' tov. N. S. Khrushcheva na soveshchanii rabotnikov sel'skogo khoziaistva Gor'kovskoi, Arzamasskoi, Kirovskoi oblastei, Mariiskoi, Mordovskoi i Chuvashskoi ASSR 8 Aprelia 1957 goda v gorode Gor'kom," *Pravda*, April 10, 1957, p. 2.

152. "Soveshchanie rabotnikov sel'skogo khoziaistva oblastei iugo-vostoka," *Pravda*, March 19, 1955, p. 2. For a criticism by Lysenko of the method of obtaining hybrid corn, written in 1935, see his *Agrobiology*, p. 123.

153. "Rech' tovarishcha T. D. Lysenko," *Pravda*, February 26, 1956, p. 9.

154. See his "Interesnye raboty po zhivotnovodstvu v Gorkakh Leninskikh," *Pravda*, July 17, 1957, pp. 5–6; and *Agrobiologiia*, No. 4 (1957), pp. 123–7.

155. "Ukaz prezidiuma verkhovnogo soveta SSSR," *Pravda*, September 29, 1958, p. 1.

156. "Ob agrobiologicheskoi nauke i lozhnykh positsiiakh 'Botanicheskogo zhurnala,' " *Pravda,* December 14, 1958, pp. 3–4.

157. The members of the Lysenko school attempted to incorporate their fertilizer schemes into an over-all approach to farming based on a "single plan." In this way they kept a tight control over agronomy. See "Rech' tovarishcha G. A. Cheremisinova," *Pravda,* February 14, 1964, p. 2. On December 23, 1959, the Central Committee officially approved Lysenko's fertilizer methods. See "O dal'neishem razvitii sel'skogo khoziaistva: doklad predsedatelia soveta ministrov RSFSR tovarishcha D. S. Polianskogo," *Pravda,* December 23, 1959, pp. 1–2.

158. *Pravda* and *Izvestiia* announced on April 10, 1956, that the U.S.S.R. Council of Ministers had "complied with the request" of Lysenko that he be released as president of the Academy. In June of the same year, however, he was elected a member of the Presidium of the Academy. In August 1961 he was re-elected president, but in April 1962 he stepped down again for "reasons of health." The man who replaced him, M. Olshanskii, was one of his strong defenders. See, for example, his "Protiv fal'sifikatsii v biologicheskoi nauke," *Sel'skaia zhizn',* August 18, 1963, pp. 2–3.

159. For information on the relationship between cybernetics and genetics in the Soviet Union I am grateful to S. C. McCluskey of Columbia University, who made a short unpublished study of the topic.

160. Schrödinger, *op. cit.,* p. 71. In his *The Double Helix* (New York, 1968), James Watson emphasized the importance of Schrödinger's little book to Francis Crick's decision to leave physics and enter biology. P. 13.

161. Lysenko condemned Schrödinger's essay in his opening address at the 1948 session of the Lenin Academy of Agricultural Sciences. See T. D. Lysenko, *The Situation in Biological Science* (New York, 1949), p. 23.

162. Many of these articles are available in English, as a result of the translation services of the Joint Publications Research Service (JPRS) of the U.S. Department of Commerce. In the following citations, the serial number of the JPRS publication is given in parentheses: R. L. Berg and N. V. Timofeev-Ressovskii, "Paths of Evolution of the Genotype," *Problems of Cybernetics,* No. 5 (1961) (JPRS 10,292, January 1962—1016); A. A. Liapunov and A. G. Malenkov, "Logical Analysis of Hereditary Information," *Problems of Cybernetics,* No. 8 (1962) (JPRS 21,048, November 1963—19,390); Zh. A. Medvedev, "Errors in the Reproduction of Nucleic Acid and Proteins and Their Biological Significance," *Problems of Cybernetics,* No. 9 (1963) (JPRS 21,448, November 1963—19,750); I. I. Schmalhausen, "Hereditary Information and Its Transformations," *Transactions of the Academy of Sciences, USSR,* Biological Sciences Section, 120, 1 (1958), American Institute of Biological Sciences, Washington, D. C.; I. I. Schmalhausen, "Fundamentals of the Evolutionary Process in the Light of Cybernetics," *Problems of Cybernetics,* No. 4 (1960) (JPRS 7403, March 1961—4686); I. I. Schmalhausen, "Natural Selection and Information," *News of the Academy of Sciences, USSR,* Biology Series, Vol. 25–1 (1960) (JPRS 2815, September 1960—14,090); I. I. Schmalhausen, "Evolution in the Light of Cybernetics," *Problems of Cybernetics,* No. 13 (1965) (JPRS 31,567, October 1967—

16,376); N. V. Timofeev-Ressovskii, "Some Problems of Radiation Biogeo-cenology," *Problems of Cybernetics,* No. 12 (1964) (JPRS 31,214, September 1965—14,503); K. S. Trincher, *Biology and Information—Elements of Biological Thermodynamics* (Moscow, 1964) (JPRS 28,969, April 1965—6475).

163. At one point he described the Law of the Life of Biological Species as the fact that "everything in the life of each biological species, and consequently of each living body, is directed . . . at preserving and increasing the numbers of the given species. . . ." T. D. Lysenko, "Teoreticheskie osnovy napravlennogo izmeneniia nasledstvennosti sel'skokhoziaistvennykh rastenii," *Pravda,* January 29, 1963, pp. 3–4.

164. T. D. Lysenko, "Interesnye raboty po zhivotnovodstvu v Gorkakh Leninskikh," *Pravda,* July 17, 1957, pp. 5–6.

165. Prikazy po ministerstvu sel'skogo khoziaistva SSSR ot 5 ianvaria 1961 g. No. 3, "Ob opyte raboty eksperimental'nogo khoziaistva 'Gorki Leninskie' po povysheniiu zhirnomolochnosti korov," i ot 26 iiunia 1963 g. No. 131, "Ob uluchsheniia raboty po sozdaniiu zhirnomolochnogo stada krupnogo rogatogo skota v kolkhozakh i sovkhozakh putem ispol'zovaniia plemennykh zhivotnykh, proiskhodiashchikh s fermy 'Gorki Leninskie,' i ikh potomkov."

166. *Pravda,* January 16, 1961.

167. *Ibid.*

168. *Pravda,* December 15, 1963; also, see *Pravda,* July 12, 1962.

169. The 1961 Party Program, still technically in effect, calls for "developing in breadth and depth the Michurinist teaching in biological science, which derives from the point of view that the conditions of life are the most important in the development of the organic world." *Materialy XXII s' 'ezda KPSS* (Moscow, 1962), p. 416.

170. A very powerful refutation of Lysenko's view, which derived its strength from its illustration of the overwhelming practical utility as well as theoretical beauty of modern genetics, was Zh. Medvedev and V. Kirpichnikov, "Perspektivy sovetskoi genetiki," *Neva,* No. 3 (1963), pp. 165–75. This article elicited a response from the supporter of Lysenko M. A. Olshanskii in "Protiv fal'sifikatsii v biologicheskoi nauke," *Sel'skaia zhizn',* August 18, 1963, pp. 2–3.

171. "O rezul'tatakh proverki deiatel'nosti bazy 'gorki leninskie,' " *Vestnik akademii nauk,* No. 11 (1965), p. 124.

172. Lysenko, of course, never accepted the idea of coexistence of different approaches to biology. For an example of his exclusivity, where he demands that biologists abandon "incorrect" theories, see his *Agrobiology,* p. 135.

173. B. E. Bykhovskii, academic secretary of the Department of General Biology of the Academy of Sciences, commented: "Almost from the very date of the formation of this department we began to receive signals that all was not well in the administration of Lenin Hills farm." *Vestnik akademii nauk,* No. 11 (1965), p. 107. Also see Lysenko's comments on Semenov in *ibid.,* p. 61; N. N. Semenov, "Nauka ne terpit sub' 'ektivizma," *Nauka i zhizn',* No. 4 (1965), pp. 38–43; and my chapter on Semenov's reform plans, "Reorganization of the Academy of Sciences," in Peter H. Juviler and Henry W. Morton, eds., *Soviet Policy-Making* (New York, 1967), pp. 133–61. Semenov

told Walter Sullivan of *The New York Times* in the summer of 1967, "My goal, since 1950, has been to achieve a marriage between biology and chemistry. At first it was slowed by the difficulties of the time—the Lysenko problem. However five years ago I was able to form a new Division of Biophysics, Biochemistry and Physiologically Active Compounds within the Academy. . . . At first it was a mechanical mixture, but now it is nearly a chemical compound." Personal conversation, Walter Sullivan, July 14, 1967.

174. See Medvedev, *op. cit.,* pp. 215–17.

175. *Ibid.,* pp. 198–9.

176. See, for example, V. Dudintsev, "Net, istina," *Komsomol'skaia pravda,* October 23, 1964, for an interesting story of a specialist on polyploidy who was forced out of her job in Vavilov's old institute but who for ten years carried out research in her own garden. Also see V. Bianki and V. Stepanov, "Kto napisala oproverzhenie?" *Komsomol'skaia pravda,* March 16, 1965, p. 2.

177. N. Vorontsov, "Zhizn' toropit: nuzhny sovremennye posobiia po biologii," *Komsomol'skaia pravda,* November 11, 1964, p. 3.

178. Anatolii Agranovskii, "Nauka na veru ne prinimaet," *Literaturnaia gazeta,* January 23, 1965, p. 2. The details of this article were later confirmed by the investigating committee of the Academy of Sciences. See "O rezul'tatakh proverki deiatel'nosti bazy 'gorki leninskie,' " *Vestnik akademii nauk,* No. 11 (1965), p. 33.

179. *Vestnik akademii nauk,* No. 11 (1965).

180. *Ibid.,* p. 93. Notice that there is one exception to the decline in butterfat content, that opposite $\frac{9}{16}$ [$\frac{5}{16}$]. The number $\frac{9}{16}$ must be a typographical error in the report. The correct order in sequence would be $\frac{5}{16}$; this correction would also explain the comment in the report, "Only in one instance, with a difference of $\frac{1}{16}$, was the lower group a little higher."

181. *Ibid.,* p. 108.

182. One of Lysenko's eccentricities may actually have limited the damage he did. He frowned upon artificial insemination. On his own farm it was not practiced. Therefore, each of his bulls was able to impregnate only forty to forty-five cows a year. Through artificial insemination one bull can service on an average two thousand cows annually, and Lysenko's mixed breed bulls could have spoiled much larger numbers of cows. No doubt artificial insemination was used with some bulls originally from Lysenko's farm, since many Soviet farms employed the practice and Lysenko himself spoke favorably of it on occasion. See *ibid.,* p. 15, for evidence that Lysenko did not permit artificial insemination on his farm; see T. D. Lysenko, "Vazhnye rezervy kolkhozov i sovkhozov," *Pravda,* March 14, 1959, pp. 2–3, for statements showing that he recommended it elsewhere.

183. There is evidence that Lysenko or his assistants tried to conceal important evidence, particularly that pertaining to the reasons for eliminating certain numbers of the herd. *Vestnik akademii nauk,* No. 11 (1965), pp. 17, 81.

184. "Raz' 'iasneniia komissii v sviazi s zamechaniiami akademika T. D. Lysenko," *Vestnik akademii nauk,* No. 11 (1965), p. 73.

185. *Vestnik akademii nauk,* No. 11 (1965), pp. 91–2.

186. See pp. 244–5.

187. Walter Sullivan, "The Death and Rebirth of a Science," in Harrison E.

Salisbury, ed., *The Soviet Union: The Fifty Years* (New York, 1967), p. 287.

188. Walter Sullivan, personal conversation, July 14, 1967.

189. N. P. Dubinin, "Sovremennaia genetika v svete marksistskoleninskoi filosofii," in M. E. Omel'ianovskii, ed., *Lenin i sovremennoe estestvoznanie* (Moscow, 1969), pp. 287–311. For information about Dubinin's sufferings during Lysenko's period of greatest influence, see Medvedev, *op. cit.*

190. Zh. Medvedev and V. Kirpichnikov, *op. cit.*, p. 169.

191. Lysenko commented in 1957, "It has been proved, and this can be experimentally verified, that heredity is not a special substance separate from the living body, as 'classical genetics' maintains, but a property of the living body itself. Hence it is futile to search for a specific substance of heredity or an organ of heredity in the living body." This statement shows clearly that those diehards among the followers of Lysenko who believe that the relative supercession of the gene by DNA, or by cytoplasmic inheritance, proves that Lysenko was right after all are simply incorrect. To Lysenko, insistence on DNA, or any identifiable material substance, as the carrier of heredity would be just as unacceptable as the gene. See Lysenko, "Teoreticheskie uspekhi agronomicheskoi biologii," *Izvestiia*, December 8, 1957, p. 4.

192. As Sewall Wright observed: "I am sure that most geneticists would consider the view that heredity is something that can be sucked out of an egg with a micropipette, or shattered with X-rays, with consequences in later generations that exactly parallel the ones that can be seen in the chromosomes, as less idealistic than such popular relics of sympathetic magic as the inheritance of acquired characters, or the usually associated doctrine of maternal impressions." Sewall Wright, "Dogma or Opportunism?" *Bulletin of the Atomic Scientists,* May 1949, p. 141.

193. G. Platonov, "Dogmy starye i dogmy novye," *Oktiabr'*, No. 8 (1965), pp. 149–65. This article, which attempted a sort of "synthesis" of classical genetics and Michurinism, was severely criticized by V. P. Efroimson in a letter to *Voprosy filosofii*, No. 8 (1966), pp. 175–81.

194. Ferents Pinter, *op. cit.* Pinter believed that the advent of DNA research and cytoplasmic inheritance proved that in certain respects the Michurinists were correct in their criticism of Mendelism. This comment ignores the fact that the Michurinists made absolutely no contribution in this direction; on the contrary, this research proceeded directly out of the neo-Mendelian tradition. As Julian Huxley commented, "Even if some new theoretical interpretation proves to be required, it cannot start from far behind the present front of science, as Lysenko does, but must take account of existing knowledge." Huxley, *op. cit.*, p. 218.

195. It is, of course, true that Lysenko twisted Michurin's beliefs. This has been pointed out many times, including the 1936 genetics conference. See, for example, Targul'ian, ed., *op. cit.*, pp. 399–400.

196. N. P. Dubinin, "I. V. Michurin i sovremennaia genetika," *Voprosy filosofii*, No. 6 (1966), pp. 59–70.

197. I. T. Frolov, *Genetika i dialektika* (Moscow, 1968).

198. *Ibid.*, p. 13.

199. *Ibid.*, p. 16 and *passim*.

200. *Ibid.,* p. 253.
201. See, for example, R. C. Buck, "On the Logic of General Behavior Systems Theory," in H. Feigl and M. Scriven, eds., *Minnesota Studies in the Philosophy of Science,* Vol. I (Minneapolis, 1956), pp. 223–38; and Carl G. Hempel, "General System Theory and the Unity of Science," *Human Biology,* XXIII (1951), 313–27.
202. Frolov, *op. cit.,* p. 253.
203. Iu. S. Viatkin and A. S. Mamzin, "Sootnoshenie strukturnofunktsional'nogo i istoricheskogo podkhodov v izuchenii zhivykh sistem," *Voprosy filosofii,* No. 11 (1969), pp. 46–56.
204. See the discussions in *Problemy metodologii sistemnogo issledovaniia* (Moscow, 1970).
205. *Ibid.,* pp. 386–410.
206. *Ibid.*

VII: Origin of Life

1. C. H. Waddington, "That's Life," *New York Review of Books,* February 29, 1968, p. 19.
2. Sidney W. Fox, ed., *The Origins of Prebiological Systems and of Their Molecular Matrices* (New York, 1965), p. 98.
3. H. C. Bastian, *The Beginning of Life* (New York, 1872).
4. Frederick Engels, *Dialectics of Nature* (New York, 1940), p. 189.
5. Quoted in Jacques Nicolle, *Louis Pasteur: A Master of Scientific Enquiry* (London, 1961), p. 67.
6. A. I. Oparin, *Proiskhozhdenie zhizny* (Moscow, 1924).
7. It appeared as Appendix I in J. D. Bernal's *The Origin of Life* (London, 1967), and was translated by Ann Synge. I was unable to obtain the original Russian edition of 1924 in any library in the United States, although a thorough search was made by the Inter-Library Loan Service. Consequently, I was forced to use the 1967 English translation, which is apparently complete.
8. A. I. Oparin, *The Origin and Initial Development of Life,* NASA Technical Translation F-488 (Washington, D. C., 1968).
9. John Keosian, *The Origin of Life,* 2nd ed. (New York, 1968), p. 12.
10. Bernal, *op. cit.,* pp. 240–1.
11. A. I. Oparin, "K voprosu o vozniknovenii zhizni," *Voprosy filosofii,* No. 1 (1953), p. 138.
12. Oparin, *The Origin and Initial Development of Life,* p. 4.
13. A. I. Oparin, "The Origin of Life," in Bernal, *op. cit.,* p. 214.
14. *Ibid.,* p. 217.
15. Bernal, *op. cit.,* p. 251.
16. J. B. S. Haldane, "The Origin of Life," reprinted from *The Rationalist Annual* (1929), in Bernal, *op. cit.,* p. 249.
17. Haldane, *op. cit.,* p. 248
18. See *Aleksandr Ivanovich Oparin* (*Materialy k biobibliografii uchenykh SSSR, Seriia biokhimii, vypusk* 3) (Moscow and Leningrad, 1949), p. 5.

19. J. B. S. Haldane, *The Marxist Philosophy and the Sciences* (New York, 1939), p. 3.
20. Letter to the author from C. H. Waddington, October 16, 1969.
21. A. I. Oparin in Bernal, *op. cit.*, p. 203.
22. Oparin's discussion at this early date of crystals with reference to life was fascinating; it reminds one of Erwin Schrödinger's later discussion in *What is Life?*, a treatment that was of considerable influence on the early thought of several of today's prominent molecular geneticists, as I mentioned previously (p. 521). Bernal wrote of Oparin: "His consideration of crystals, which also have the capacity of growth and replication of form, came very close to modern ideas of self-reproduction, which has been found to be the key to molecular biology, whose ideas were at that time far below the horizon of research." Bernal, *op. cit.*, p. 237.
23. Wöhler actually began with organic substances in his "synthesis" of urea. There is a controversy in the history of science over the significance of Wöhler's accomplishment. For a recent chapter in that controversy, see *The New York Review of Books*, April 9, 1970, pp. 45–6.
24. N. L. Kudriavtseva wrote in 1954, "The carbide theory of Mendeleev, having received a geological foundation, reappears as the most simple and clear theory of the origin of petroleum, explaining processes of the formation of hydrocarbons in the earth's crust without relying on any kind of mysterious living matter and assuming the development of matter in the usual way, from simpler to more complicated forms." "K voprosu o vozniknovenii zhizni," *Voprosy filosofii*, No. 1 (1954), p. 220.
25. The following exchange took place at a conference at Wakulla Springs, Florida, in October 1963: "Dr. Buchanan [J. M. Buchanan, Department of Biology, MIT]: 'At what point did Dr. Oparin decide that the synthesis of complex organic molecules would come from methane, ammonia, water, and hydrogen, and how did he choose these particular compounds?' Dr. Oparin: 'Almost 40 years ago, in 1924, in the book published at that time, I was led to this view by Mendeleev, who has expressed the hypothesis of inorganic origin of oil, which was then subsequently rejected by geologists. Also, very stimulating to us was the discovery of methane in the atmosphere of the large planets.' " Fox, ed., *op. cit.*, p. 97.
26. Oparin in Bernal, *op. cit.*, p. 226.
27. See pp. 148–53.
28. Oparin in Bernal, *op. cit.*, p. 228.
29. Quoted in Bernal, *op. cit.*, p. 21.
30. See the discussion in Fox, ed., *op. cit.*, p. 97.
31. Oparin in Bernal, *op. cit.*, p. 229.
32. *Ibid.*
33. See, for example, the discussion in his *Life: Its Nature, Origin and Development* (Edinburgh and London, 1961), pp. 10–13.
34. Oparin in Bernal, *op. cit.*, p. 211.
35. Oparin, *Life: Its Nature, Origin and Development*, p. 9.
36. Oparin, *The Origin of Life* (New York, 1938), p. 162.
37. Oparin's hypothesis of coacervates as protocells should not be confused with

Olga Lepeshinskaia's later views, which Oparin criticized as a simple belief in spontaneous generation. See pp. 276, 279.

38. Oparin, *The Origin of Life*, p. 174.
39. *Ibid.*, p. 176.
40. *Ibid.*, p. 192.
41. *Ibid.*, pp. 194–5.
42. *Ibid.*, p. 203.
43. *Ibid.*, p. 31.
44. *Ibid.*, p. 32.
45. *Ibid.*. p. 33.
46. *Ibid.*, p. 137.
47. *Ibid.*, p. 136.
48. R. Beutner, *Life's Beginning on the Earth* (Baltimore, 1938).
49. *Ibid.*, p. 89.
50. *Ibid.*, p. 91.
51. *Ibid.*, p. 97
52. *Ibid.*, p. 96.
53. *Ibid.*, p. 97.
54. Oparin, *The Origin of Life*, p. 174.
55. Zhores A. Medvedev, *The Rise and Fall of T. D. Lysenko* (New York and London, 1969), pp. 137–8.
56. For one of his strongest defenses of Lysenko, see his *Znachenie trudov tovarishcha I. V. Stalina po voprosam iazikoznaniia dlia razvitiia sovetskoi biologicheskoi nauki* (Moscow, 1951), pp. 10–15.
57. Medvedev, *op. cit.*, p. 214.
58. Her views were published as an article in 1950, but appeared in a more complete version in her *Kletka: ee zhizn' i proiskhozhdenie* (Moscow, 1952).
59. Lysenko supported Lepeshinskaia in "Novoe v nauke o biologicheskom vide —o rabotakh deisvitel'nogo chlena meditsinskikh nauk SSSR O. B. Lepeshinskoi" (Moscow, 1952).
60. A. I. Oparin, *Znachenie trudov tovarishcha I. V. Stalina po voprosam iazykoznaniia dlia razvitiia sovetskoi biologicheskoi nauki* (Moscow, 1951), pp. 14–15.
61. See the following note.
62. A debate over the origin of life occurred in the Soviet Union in the early fifties, much of it of poor intellectual quality. A description of this debate is in Gustav A. Wetter, *Der Dialektische Materialismus und das Problem der Entstehung des Lebens* (München, 1958). The main disputes concerned whether protein was the substance essential to life and whether life was a molecular or supramolecular phenomenon. Z. N. Nudel'man agreed with A. P. Stukov and S. A. Iakushev that the properties of life can be found on the molecular level of protein (and thereby criticized Oparin's view), but Nudel'man explained those properties in terms of the molecular structure of protein, in contrast to Stukov and Iakushev, who held an amorphous and almost vitalistic view of protein. A. E. Braunshtein, who spoke positively of much of Oparin's approach, saw the importance of protein as the carrier of

life, not in its "chemical structure," but in its "special mechanism for the exchange of matter." Nudel'man thought the qualitative transition from "living" to "nonliving" matter occurred in the transition from the microstructure of the molecule to its macrostructure ("macrostructure" referred to the whole molecule; "microstructure" to its parts). Oparin continued to think the simplest bit of life was supramolecular. Oparin was criticized both by the simple Lysenkoites, including Lepeshinskaia, who often tended toward vitalism, and also by some of the newer molecular biologists, who thought Oparin failed to recognize the significance of their research. For Lepeshinskaia's recognition of her differences with Oparin, see A. I. Ignatov, "Mezhdunarodnyi simpozium po proiskhozhdeniiu zhizni na zemle," *Voprosy filosofii*, No. 1 (1958), p. 154. For an earlier sharp criticism of Oparin from the standpoint of Lepeshinskaia and her supporters, see A. P. Skabichevskii, "Problema vozniknoveniia zhizni na zemli i teoriia akad. A. I. Oparina," *Voprosy filosofii*, No. 2 (1953), pp. 150–5. For other articles in the discussion, see: A. S. Konikova and M. G. Kritsman, "Zhivoi belok v svete sovremennykh issledovanii biokhimii," *Voprosy filosofii*, No. 1 (1953), pp. 143–50; A. P. Stukov and S. A. Iakushev, "O belke kak nositele zhizni," *Voprosy filosofii*, No. 2 (1953), pp. 139–49; N. L. Kudriavtseva, "K voprosu o vozniknovenii zhizni," *Voprosy filosofii*, No. 2 (1954), pp. 218–21; Z. N. Nudel'man, "O probleme belka," *Voprosy filosofii*, No. 2 (1954), pp. 221–6; A. M. Emme, "Neskol'ko zamechanii po voprosu o protsesse vozniknoveniia zhizni," *Voprosy filosofii*, No. 1 (1956), pp. 155–8; Laslo Takach, "K voprosu o vozniknovenii zhizni," *Voprosy filosofii*, No. 3 (1955), pp. 147–50; M. G. Kritsman and A. S. Konikova, "K voprosu o nachal'noi forme proiavleniia zhizni," *Voprosy filosofii*, No. 1 (1954), pp. 210–16; A. F. Sysoev, "Samoobnovlenie belka i svoistvo razdrazhimosti—vazhneishie zakonomernosti zhiznennykh iavlenii," *Voprosy filosofii*, No. 1 (1956), pp. 152–5; and A. V. Kozhevnikov, "O nekotorykh usloviiakh vozniknoveniia zhizni na Zemle," *Voprosy filosofii*, No. 1 (1956), pp. 149–52.

63. A. I. Oparin, "K voprosu o vozniknovenii zhizni," *Voprosy filosofii*, No. 1 (1953), pp. 138–42.

64. Russian edition: A. I. Oparin and V. G. Fesenkov, *Zhizn' vo Vselennoi* (Moscow, 1956); English edition: A. Oparin and V. Fesenkov, *Life in the Universe* (New York, 1961).

65. Oparin and Fesenkov, *op. cit.*, p. 121.

66. See pp. 149 ff.

67. Oparin and Fesenkov, *op. cit.*, p. 239.

68. *Ibid.*, p. 16.

69. *Ibid.*, p. 245.

70. *Ibid.*

71. A. I. Oparin, *The Origin of Life on the Earth*, trans. Ann Synge (London, 1957).

72. *Ibid.*, p. 37. See also Lepeshinskaia, *Kletka: ee zhizn' i proiskhozhdenie.*

73. Oparin, *The Origin of Life on the Earth*, p. 37.

74. Oparin, *The Origin of Life*, pp. 59–60.

75. Oparin, *The Origin of Life on the Earth*, p. 285.

76. *Ibid.*, p. 286.

77. For a relatively popular account, see Chap. 17, "DNA or Protein?" in George and Muriel Beadle, *The Language of Life: An Introduction to the Science of Genetics* (Garden City, N. Y., 1966).
78. Oparin, *The Origin of Life on the Earth,* p. 287.
79. *Ibid.,* pp. 95–102.
80. A. I. Oparin, *Zhizn', ee priroda, proiskhozhdenie i razvitie* (Moscow, 1960); English edition: *Life: Its Nature, Origin and Development.* For the record of the 1957 conference, see: A. I. Oparin, A. G. Pasynskii, A. E. Braunshtein, and T. E. Pavlovskaya, eds., *Proceedings of the First International Symposium on The Origin of Life on the Earth* (New York, London, Paris, and Los Angeles, 1959).
81. H. Fraenkel-Conrat and B. Singer, "The Infective Nucleic Acid from Tobacco Mosaic Virus," in *Proceedings of the First International Symposium on the Origin of Life on the Earth* (New York, 1959), p. 303.
82. Wendell H. Stanley, "On the Nature of Viruses, Genes and Life," in *Proceedings of the First International . . . ,* p. 313.
83. *Ibid.,* p. 316.
84. Oparin, *The Origin of Life on the Earth,* pp. 274–5.
85. Stanley, *op. cit.,* pp. 274–5.
86. *Proceedings of the First International . . . ,* p. 368.
87. Oparin, *Life: Its Nature, Origin and Development,* p. 66.
88. *Ibid.*
89. *Ibid.,* p. 4.
90. *Ibid.,* p. 5.
91. *Ibid.,* p. 6.
92. *Ibid.,* p. 9.
93. *Ibid.,* p. 13.
94. *Ibid.,* p. 11.
95. *Ibid.,* p. 13.
96. *Ibid.,* p. 37.
97. *Ibid.,* p. 18.
98. *Ibid.,* p. 19.
99. *Ibid.,* p. 20.
100. *Ibid.,* pp. 33–5.
101. *Ibid.,* p. 24.
102. Quoted in *ibid.*
103. E. Chargaff, "Nucleic Acids as Carriers of Biological Information," in *Proceedings of the First International . . . ,* p. 299.
104. *Ibid.,* p. 297.
105. The proceedings of this conference were published as Fox, ed., *op. cit.*
106. Peter T. Mora, "The Folly of Probability," in *ibid.,* pp. 39–64.
107. Mora maintained that "physiochemical selectivity" could lead only to "a temporary metastable order or function which will cease and tend to disperse more and more as its complexity increases." *Ibid.,* p. 47.
108. *Ibid.,* p. 57.
109. *Ibid.,* p. 48.
110. *Ibid.,* p. 50.
111. *Ibid.,* p. 57. Mora maintained elsewhere that his approach, like Oparin's, was

materialistic. See P. T. Mora, "Urge and Molecular Biology," *Nature* (July 20, 1963), pp. 212–9.

112. J. D. Bernal, communicating by mail, said that it posed the most fundamental questions of the origin of life that have been raised "at this conference or, as far as I know, elsewhere." N. W. Pirie commented, "Dr. Mora, you have started people thinking." *Ibid.,* pp. 52, 57.

113. Fox, ed., *op. cit.,* pp. 53–5.

114. A. I. Oparin, *The Origin and Initial Development of Life* (Washington, D. C., 1968).

115. *Ibid.,* pp. 101–2.

116. B. I. Chuvashov, "K voprosu o vozniknovenii zhizni na Zemle," *Voprosy filosofii,* No. 8 (1966), pp. 76–83.

117. Walter Sullivan, "Moon Soil Indicates Clue to Life Origin," *The New York Times,* January 7, 1970.

118. The official reports published by NASA were quite inconclusive. See the special issue of *Science* (January 30, 1970) devoted to the analysis of the lunar materials.

VIII: Structural Chemistry

1. J. R. Partington, *A Short History of Chemistry* (London, 1948), p. 255.

2. *Ibid.*

3. *Ibid.,* p. 216.

4. Indeed, today chemists still work primarily by gathering data on chemical reactions rather than approaching, as the physicist attempts to do, the submolecular and subatomic levels. The theory of resonance itself, as Linus Pauling pointed out, was derived largely by the chemists' method. This stress upon the empirical approach of chemists at a more gross level does not ignore, of course, the increasing use by chemists of physical methods of investigation such as spectroscopic, X-ray, and electron diffraction methods, which are valuable supplements to their work. See Linus Pauling, *The Nature of the Chemical Bond* (Ithaca, N. Y., 1960), pp. 219 ff.

5. For Kekulé's exposition of his theory see August Kekulé, "Sur la constitution des substances aromatiques," *Bulletin de la Société Chimique,* III (1865), 98; also August Kekulé, "Untersuchungen über aromatische Verbindungen," *Justus Leibig's Annalen der Chimie,* CXXXVII (1866), 129. Alexander Findlay in *A Hundred Years of Chemistry* (New York, 1937), p. 147, said, "It is probable that Kekulé regarded his theory mainly as an elegant philosophical system into which all the known facts relating to the aromatic compounds could be neatly and satisfactorily grouped together; and the first to regard the theory capable of experimental proof was Kekulé's pupil, Koerner." Kekulé was a thorough chemist who laboriously checked his theories with empirical tests. Nevertheless, he considered speculation one of the most fruitful methods of investigation; according to his own testimony, he received the inspiration for his two most important scientific theories while dozing. The chain structure of aliphatic compounds is supposed to

have come to him while dreaming atop an omnibus on a summer night; the ring structure of aromatic compounds supposedly flashed into his mind while he was dozing in front of the fire.

6. Sidney J. French, *The Drama of Chemistry* (New York, 1937), p. 93.

7. Students often think of these structures as being isomers or tautomers, but they are neither, since the Kekulé molecules do not exist. Isomers are compounds composed of the same elements in the same proportions, but different in properties because of differences in structure. Tautomers are isomers that change into one another rapidly and are usually in equilibrium with one another.

8. E. F. Armstrong, *Chemistry in the Twentieth Century* (London, 1924), p. 121.

9. The resonance theory was anticipated in the 1920's by several German and English chemists, especially C. K. Ingold in England. Ingold called his particular version of essentially the same phenomenon "mesomerism," a more accurate description than "resonance," since it means literally "between the forms." "Resonance," on the other hand, connotes movement, which does not occur in chemical resonance. The term "resonating system," often used by chemists, is even less precise.

10. G. W. Wheland, *Resonance in Organic Chemistry* (New York, 1955), p. 3.

11. These configurations for the resonance structure of carboxylate ions are given in Pauling, *op. cit.*, p. 275.

12. Here, in particular, the five structures should not be thought of as isomers or tautomers. The latter exist, whereas resonance structures do not.

13. Wheland, *op. cit.*, pp. 7–8.

14. *Ibid.*, p. 4.

15. Pauling, *op. cit.*, p. 186

16. More exactly, the C-C bond (single) length in ethane is 1.536 ± 0.016Å, the C-C bond length in benzene is 1.393 ± 0.005Å, and the C-C bond (double) length in ethylene is 1.330 ± 0.005Å. *Tables of Interatomic Distances and Configuration in Molecules and Ions* (London, 1958), pp. M135, M196, M129. An Angstrom unit is equal to one hundred millionths of a centimeter.

17. Use of the theory of resonance is evident in A. N. Nesmeianov, R. Kh. Freidlina, and A. E. Borisova, "O kvazikompleksnikh metalloorganicheskikh soedineniiakh," *Izvestiia akademii nauk: otdelenie khimicheskikh nauk*, Jubilee Bulletin (1945), pp. 239–50. In this article Nesmeianov and his coworkers explained the properties of certain compounds on the basis of the resonance theory, including the concept of superpositioning. They referred to Pauling's 1944 book on resonance.

18. D. N. Kursanov and V. N. Setkina, "O vzaimodeistvii chetvertichnikh solei ammoniia s prostimi efirami," *Doklady akademii nauk SSSR*, LXV (April 21, 1949), 847–55.

19. E. N. Prilezhaeva, Ia. K. Syrkin, and M. V. Volkenshtein, "Ramaneffekt galoidoproizvodnikh etilena i elektronni rezonansa," *Zhurnal fizicheskoi khimii*, XIV (1940), 1396–1418.

20. M. I. Kabachnik, "Orientatsiia v benzolnom kol'tse," *Uspekhi khimii*, XVII (January 1948), 96–131.

21. Ia. K. Syrkin and M. E. Diatkina, *Khimicheskaia sviaz' i stroenie molekul* (Moscow, 1946).

22. "Soviets Blast Pauling, Repudiate Resonance Theory," *Chemical and Engineering News,* XXIX (September 10, 1951), 3713.

23. *The Structure of Molecules and the Chemical Bond* (New York, 1950).

24. G. V. Chelintsev, *Ocherki po teorii organicheskoi khimii* (Moscow, 1949), p. 2.

25. *Ibid.,* pp. 107–8.

26. V. M. Tatevskii and M. I. Shakhparanov, "Ob odnoi makhistskoi teorii v khimii i ee propagandistakh," *Voprosy filosofii,* No. 3 (1949), pp. 176–92.

27. Quoted in *ibid.,* p. 177.

28. *Ibid.*

29. "K 70-letiiu so dnia rozhdeniia I. V. Stalina," *Zhurnal fizicheskoi khimii,* No. 12 (1949), pp. 1385–6; *Pravda,* December 17, 1949, p. 1.

30. A biographical article on Butlerov and additional bibliographical information may be found in *Russkii biograficheskii slovar',* Vol. III (St. Petersburg, 1908), pp. 528–33. For longer but somewhat less reliable articles, see *Bol'shaia sovetskaia entsiklopediia,* Vol. VI (Moscow, 1951), pp. 378–83 and 383–9. A valuable more recent article on Butlerov is L. L. Potkov, "Teoriia khimicheskogo stroeniia A. M. Butlerova," *Zhurnal fizicheskoi khimii,* XXXVI, No. 3 (1962), 417–28. Butlerov was not unknown to scientists outside of Russia. He traveled extensively in Europe and knew Kekulé well. He spent quite a bit of time among German chemists, worked with Liebig, and delivered papers in German. In 1861 at Speyer, Germany, he developed the concept that the chemical structure of molecules determines the reactions which any particular substance undergoes. In 1876 he was made an honorary member of the fledgling American Chemical Society, which still possesses his appreciative letter of acceptance. See Henry M. Leicester, "Alexander Mikhailovich Butlerov," *Journal of Chemical Education,* XVII (May 1940), 208, 209.

31. Leicester, *op. cit.*

32. J. Jacques, "Boutlerov, Couper et la Société chimique de Paris (Notes pour servir à l'histoire des théories de la structure chimique)," *Bulletin de la Société chimique de France,* 1953, pp. 528–30.

33. Pauling, *op. cit.,* p. 4.

34. The earlier editions of Pauling's book were published in 1939 and 1944.

35. I. M. Hunsberger, "Theoretical Chemistry in Russia," *Journal of Chemical Education,* XXXI (1954), p. 506.

36. See note 5 above.

37. Quoted from S. N. Danilov, "A. M. Butlerov—osnovatel' teorii khimicheskogo stroeniia (1861–1951)," *Zhurnal obshchei khimii,* XXI (October 1951), 1740.

38. Quoted in O. A. Reutov, "O nekotorikh voprosakh teorii organicheskoi khimii," *Zhurnal obshchei khimii,* XXI (January 1951), 196. Reutov criticized this statement of Butlerov's "if it is erected into a principle."

39. See "Na uchenom sovete instituta organicheskoi khimii AN SSSR," *Izvestiia akademii nauk: otdelenie khimicheskikh nauk,* No. 4 (1950), pp. 438–44.

40. The committee in charge of writing the report consisted of D. N. Kursanov,

chairman, M. G. Gonikberg, M. M. Dubinin, M. I. Kabachnik, E. D. Kaverzneva, E. N. Prilezhaeva, N. D. Sokolov, and R. Kh. Freidlina.

41. D. N. Kursanov *et al.*, "K voprosu o sovremennom sostoianii teorii khimicheskogo stroeniia," *Uspekhi khimii*, XIX, No. 5 (1950), 532.

42. *Ibid.*, pp. 537 ff.

43. *Ibid.*

44. Reutov, *op. cit.*, p. 187.

45. The verbatim record of this conference was published as *Sostoianie teorii khmicheskogo stroeniia; vsesoiuznoe soveshchanie 11–14 iiunia 1951 g.: stenograficheskii otchet* (Moscow, 1952).

46. *Ibid.*, p. 303. Her defense occurred in a speech at the Institute of Organic Chemistry of the Academy of Sciences of the U.S.S.R. I have not been able to find a copy of this speech.

47. J. B. S. Haldane, *The Marxist Philosophy and the Sciences* (New York, 1939), p. 101. Haldane's and Diatkina's arguments were based on the dialectic, but the criticism of resonance in the Soviet Union centered on the use of multiple fictitious images.

48. *Sostoianie teorii khimicheskogo stroeniia*, p. 67.

49. *Ibid.*, pp. 47 ff.

50. Wheland, *op. cit.*, p. viii.

51. *Sostoianie teorii khimicheskogo stroeniia*, pp. 81, 86.

52. *Ibid.*

53. *Ibid.*, pp. 86–7.

54. *Ibid.*, pp. 79–80.

55. *Ibid.*, p. 87. Chelintsev named Academicians A. N. Nesmeianov, A. N. Terenin, B. A. Kazanskii; Member of the Ukrainian Academy of Sciences A. I. Kiprianov; Corresponding members of the U.S.S.R. Academy of Sciences Ia. K. Syrkin, V. N. Kondratiev, I. L. Knuniants, A. I. Brodskii; professors and doctors of sciences M. V. Volkenshtein, M. I. Kabachnik, D. N. Kursanov, R. Kh. Freidlina, M. E. Diatkina, D. A. Bochvar, B. M. Berkenheim, A. P. Terentiev, B. A. Ismailskii, B. M. Mikhailov, A. Ia. Iakubovich, A. I. Titov, L. I. Smorgonskii, M. G. Gonikberg; "docents" and "candidates" of sciences V. M. Tatevskii, M. I. Shakhparanov, M. D. Sokolov, and O. A. Reutov. Kabachnik tried to demonstrate that although he had mistakenly supported the resonance theory, he had realized his mistake in 1950 and had published an article correcting himself. At this point there was a shout from the floor, "You were forced to do it." *Sostoianie teorii khimicheskogo stroeniia*, p. 270.

56. *Ibid.*, p. 89.

57. *Ibid.*, p. 181.

58. *Ibid.*, p. 350.

59. *Ibid.*, pp. 255–6.

60. *Ibid.*, p. 365.

61. *Ibid.*, p. 370.

62. *Ibid.*, p. 376.

63. These defects were "corrected" in a revised report. See *Sostoianie teorii khimicheskogo stroeniia* (Moscow, 1954).

64. *Sostoianie teorii khimicheskogo stroeniia* (1952), p. 376.

65. *Ibid.*
66. A. N. Nesmeianov, " 'Contact Bonds' and the 'New Structural Theory,' " *Bulletin of the Academy of Sciences of the U.S.S.R.: Division of Chemical Sciences* (Consultants Bureau, N. Y.), No. 1 (1952), pp. 215–21.
67. *Ibid.*, p. 221.
68. G. V. Chelintsev, "O teorii khimicheskogo stroeniia A. M. Butlerova i ee novykh uspekhakh," *Zhurnal obshchei khimii,* XXII (February 1952), 350–60.
69. B. A. Kazanskii and G. V. Bykov, "K voprosu o sostoianii teorii khimicheskogo stroeniia v organicheskoi khimii," *Zhurnal obshchei khimii,* XXIII (January 1953), 175.
70. *Ibid.*
71. S. N. Danilov, "A. M. Butlerov (1828–1886)," *Zhurnal obshchei khimii,* XXIII (October 1953), 1601–12.
72. A. A. Kalandiia, "Otvet na stat'iu G. V. Tsitsishvili po povodu raboty A. A. Kalandiia 'Raschet molekuliarnykh ob' 'emov neorganicheskikh soedinenii tipa AnBmOs,' " *Zhurnal obshchei khimii,* XXIV (January 1954), 193–6.
73. Tsitsishvili criticized Kalandiia in his article "Ob oshibakh A. A. Kalandiia v stat'e 'Raschet molekuliarnikh ob' 'emov neorganicheskikh soedinenii tipa AnBmOs' i ego popytakh ukrepit porognuiu kontseptsiiu rezonansa," *Zhurnal obshchei khimii,* XXII (December 1952), 2240–45. The original article of Kalandiia's appeared in *Zhurnal obshchei khimii,* XIX (September 1949), 1635.
74. Kalandiia, "Otvet . . . ," p. 193.
75. *Ibid.*, p. 199.
76. Arbuzov had a bout with the Russian chemist Vladimir Chelintsev in 1913, the curious nature of which raises the possibility that the Chelintsev family may have been troublemakers in Russian chemistry on more than one occasion. Gennadi Vladimirovich Chelintsev's patronymic indicates that he was the son of a Vladimir Chelintsev, but I have not been able to determine whether this Vladimir is the same as the one who debated Arbuzov in 1913. Arbuzov described the debate as "the most important crisis of my career." The exact nature of the issue is unknown, but a full-scale debate between V. Chelintsev and Arbuzov was scheduled in St. Petersburg. Arbuzov many years later said, "The debate for which I prepared myself with great anxiety never did take place since my opponent did not appear. The meeting was held just the same, and I delivered the report in the presence of all the prominent chemists of St. Petersburg. On the question of my controversy with V. V. Chelintsev, all the chemists rallied to my side, and on the next morning I was pleasantly surprised when 100 printed copies of the detailed proceedings of the meeting were handed to me." These comments of Arbuzov's appeared in an article totally unconnected with the resonance dispute. *Zhurnal obshchei khimii,* XXV (August 1955), 1387.
77. *Sostoianie teorii khimicheskogo stroeniia* (1952), p. 363.
78. A. N. Nesmeianov and M. I. Kabachnik, "Dvoistvennaia reaktsionnaia sposobnost' i tautomeriia," *Zhurnal obshchei khimii,* XXV (January 1955), 41–87
79. *Ibid.*, p. 71.

80. G. V. Chelintsev, "O vtorom izdanii doklada komissii OKhN AN SSSR 'Sostoianie teorii khimicheskogo stroeniia v organicheskoi khimii,'" *Zhurnal obshchei khimii,* XXVII (August 1957), 2308–10.

81. *Sostoianie teorii khimicheskogo stroeniia v organicheskoi khimii* (Moscow, 1954).

82. M. I. Shakhparanov, *Dialekticheskii materializm i nekotorye problemy fiziki i khimii* (Moscow, 1958), p. 86.

83. Note of L. H. Long, translator of Walter Huckel, *Structural Chemistry of Inorganic Compounds* (New York, 1950), p. 437.

84. Ia. K. Syrkin, "Sovremennoe sostoianie problemy valentnosti," *Uspekhi khimii,* XXVIII (August 1959), 903–20.

85. *Ibid.,* p. 919.

86. M. I. Batuev, "K voprosu o sopriazhenii v benzole," *Zhurnal obshchei khimii,* XXIX (November 1958), 3147–54; also Pauling, *op cit.,* p. 233.

87. On March 26, 1962, Peter Kapitsa, the outstanding Soviet physicist, criticized the attitude of Soviet philosophers toward resonance theory, as well as their attitudes toward the theory of relativity, Heisenberg's indeterminacy principle, genetics, and cybernetics. *Ekonomicheskaia gazeta,* March 26, 1962, p. 10.

88. The lecture was later printed in the Soviet Union; see Linus Pauling, "Teoriia rezonansa v khimii," *Zhurnal vsesoiuznogo khimicheskogo obshchestva im. D. I. Mendeleeva,* No. 4 (1962), pp. 462–7. I am grateful to Dr. Pauling for a reprint of this article.

89. G. I. Shelinskii, *Khimicheskaia sviaz' i izuchenie ee v srednei shkole* (Moscow, 1969). See esp. pp. 31, 44–6, and 136.

90. In addition to the sources cited below, see: Iu. A. Zhdanov, "Obrashchenie metoda v organicheskoi khimii" (Rostov, 1963); R. B. Dobrotin, "Khimicheskaia forma dvizheniia materii" (Leningrad, 1967); V. I. Kuznetsov, *Evoliutsiia predstavlenii ob osnovnykh zakonakh khimii* (Moscow, 1967); the last section of V. I. Kuznetsov, *Razvitie ucheniia o katalize* (Moscow, 1964); B. M. Kedrov and D. N. Trifonov, *Zakon periodichnosti i khimicheskie elementy* (Moscow, 1969); R. V. Kucher, "Metodologicheskie problemy razvitii teorii v khimii," *Voprosy filosofii,* No. 6 (1969), pp. 78–85; S. E. Zak, "Kachestvennye izmeneniia i struktura," *Voprosy filosofii,* No. 1 (1967), pp. 50–8; and Iu. A. Zhdanov, "Znachenie leninskikh idei dlia razrabotki metodologicheskikh voprosov khimii," *Filosofskie nauki,* No. 2 (1970), pp. 80–90.

91. Iu. A. Zhdanov, *Ocherki metodologii organicheskoi khimii* (Moscow, 1960).

92. N. A. Budreiko, *Filosofskie voprosy khimii* (Moscow, 1970).

93. B. M. Kedrov, *Tri aspekta atomistiki:* Vol. I, *Paradoks Gibbsa;* Vol. II, *Uchenie Dal'tona;* Vol. III, *Zakon Mendeleeva* (Moscow, 1969).

94. N. N. Semenov, "Marksistsko-leninskaia filosofiia i voprosy estestvoznaniia," *Vestnik akademii nauk,* No. 8 (1968), pp. 24–40.

95. *Ibid.,* p. 24.

96. *Ibid.,* p. 25.

97. *Ibid.,* pp. 38–9.

98. *Ibid.,* p. 39.

99. *Ibid.,* p. 35.
100. See V. Shvyrev, "Materialisticheskaia dialektika i problemy issledovaniia nauchnogo poznaniia," *Kommunist,* XVII (November 1968), 40–51.
101. Quoted in my "A Soviet Marxist View of Structural Chemistry," *ISIS,* LV (March 1964), 30.
102. Gustav Wetter, *Dialectical Materialism: A Historical and Systematic Survey of Philosophy in the Soviet Union* (New York, 1958), pp. 432–6.
103. *Ibid.,* pp. 435–6.
104. For an example of such a misinterpretation of the resonance theory, see "On 'Nonresonance' Between East and West," *Chemical and Engineering News,* XXX (June 16, 1952), 2474, and the correction sent in by G. W. Wheland printed on p. 3160 of the same volume.
105. Mary Hesse, "Models and Analogy in Science," *The Encyclopedia of Philosophy,* Vol. V (New York, 1967), pp. 356–7.
106. Mary B. Hesse, "Models in Physics," *The British Journal for the Philosophy of Science,* November 1953, p. 214.

IX: Cybernetics

1. The following chapter draws heavily from three of my articles that appeared elsewhere: "Cybernetics," in George Fischer, ed., *Science and Ideology in Soviet Society* (New York, 1967), pp. 83–106; "Cybernetics in the Soviet Union," in *The State of Soviet Science* (Cambridge, Mass., 1965), pp. 3–18; and "Cybernetics—Soviet Aspect," in *Enzyklopaedisches Woerterbuch* (Freiburg, forthcoming).
2. Even though the cybernetics discussion is one of the latest of the topics of philosophy and science in the Soviet Union, I have placed this chapter before the one on physiology and psychology since cybernetics is important to the last stages of the discussion of physiology and psychology in the following chapter.
3. Stalin's agreement with Pavlov's approach to physiology may have resulted in unusual promotion of that field in the thirties, forties, and early fifties. Cybernetics falls outside the Stalinist period, however, and is particularly interesting as an indicator of the relationship of philosophy and science in the later period of greater freedom.
4. The formation of regional economic councils to replace the earlier centralized control was announced in "Zakon o dal'neishem sovershenstvovanii organizatsii upravleniia promyshlennost'iu i stroitel'stvom," *Pravda,* May 11, 1957, pp. 1–2.
5. For a description of the intellectual excitement of the early days of the development of cybernetics, particularly at the time of the Josiah Macy Foundation meetings of 1946 and 1947, see Norbert Wiener, *Cybernetics, or Control and Communication in the Animal and the Machine* (Cambridge, Mass., 1962), pp. 1–29.
6. The first chapter of a Soviet pamphlet on information and control theory was entitled "Ways to Overcome Complexity." See A. I. Berg and Iu. I. Cherniak, *Informatsiia i upravlenie* (Moscow, 1966), pp. 6–22. The authors

maintained that the complexity of the national economy in the preceding years experienced a qualitative leap, but believed that cybernetics provided ways to cope with these new intricacies.

V. M. Glushkov, A. A. Dorodnitsyn, and N. Fedorenko wrote that the application of cybernetics to economic planning would "produce a tremendous national economic effect and at least double the rate of development of the national economy." See their "O nekotorykh problemakh kibernetiki," *Izvestiia,* September 6, 1964, p. 4.

7. A surprisingly large number of the important Soviet publications on cybernetics were translated into English by the Joint Publications Research Service of the U.S. Department of Commerce. However, the quality of translation was very poor. Two bibliographies, useful for the earlier debate, are David D. Comey, "Soviet Publications on Cybernetics," and L. R. Kerschner, "Western Translations of Soviet Publications on Cybernetics," both in *Studies in Soviet Thought,* IV, 2 (February 1964), 142–77.

8. A similar view was also expressed frequently by philosophers. Ernst Kol'man, for example, commented, "The goal of our development—a communist society—is, from the cybernetic point of view, an open, dynamic system with ideal autoregulation." G. S. Gurgenidze and A. P. Ogurtsov, "Aktual'nye problemy dialektiki," *Voprosy filosofii,* No. 10 (1965), p. 147.

9. See I. M. Nekrasova, *Leninskii plan elektrifikatsii strany i ego osushchestvlenie v 1921–1931 gg.* (Moscow, 1960); also, Maurice Dobb, *Soviet Economic Development Since 1917* (New York, 1966), p. 339.

10. An interesting person in this sort of activity was Alexei Gastev, a disciple of Taylor, who combined readings of "shock-work poetry" with interests in labor efficiency. Gastev disappeared in the purges; in 1962 he was formally rehabilitated by Aksel I. Berg, Chairman of the Scientific Council on Cybernetics of the U.S.S.R. Academy of Sciences. An Iu. A. Gastev, who was very likely Alexei's son, became active in cybernetics work in the sixties, especially in education. See Iu. A. Gastev, "O metodologicheskikh voprosakh ratsionalizatsii obucheniia," in A. I. Berg *et al.,* eds., *Kibernetika, myshlenie, zhizn'* (Moscow, 1964), pp. 459–72, esp. pp. 466–7. Kendall E. Bailes, Columbia University, has made several unpublished studies of Soviet attitudes toward labor in these years.

11. See *Programma kommunisticheskoi partii Sovetskogo Soiuza* (Moscow, 1961), pp. 71–3.

12. My visit to the Moscow Power Institute in May 1963 and the description of cybernetic research by the pro-rector, D. V. Rosewig, is given in Oliver J. Caldwell and Loren R. Graham, "Moscow in May 1963: Education and Cybernetics," Washington, D. C. (U.S. Office of Education Bulletin), 1964, pp. 39–42.

13. Such a school was described to me in Moscow in 1963; see *ibid.,* p. 20.

14. A. I. Berg, ed., *Kibernetika na sluzhbu kommunizma* (Moscow, 1961).

15. *Ibid.,* p. 8.

16. V. D. Belkin, "Kibernetika i ekonomika," in Berg, ed., *Kibernetika na sluzhbu kommunizma.*

17. *Ibid.,* p. 188. The opposition to cybernetics was connected with the opposition to econometrics. See S. M. Shaliutin, "O kibernetike i sfere ee primene-

niia," in V. A. Il'in, V. N. Kolbanovskii, and E. Kol'man, eds., *Filosofskie voprosy kibernetiki* (Moscow, 1960), p. 55.

18. A. I. Kitov, "Kibernetika i ekonomika," in Berg, ed., *Kibernetika na sluzhbu kommunizma*. A discussion in English of a cybernetic model of a centrally planned economy devised by the Polish Academy of Sciences Econometric Commission was published the year before: Henry Greniewski, *Cybernetics Without Mathematics* (Warsaw, 1960), p. 196.

19. V. A. Venikov, "Primenenie kibernetiki v elektricheskikh sistemakh," and I. Ia. Aksenov, "Transportnye problemy kibernetiki," in Berg, ed., *Kibernetika na sluzhbu kommunizma*. By 1961 there was, of course, an immense literature in many languages on these and other problems concerning the application of cybernetics.

20. During a trip to the Soviet Union in 1963 an incomplete diagnostic machine was demonstrated to me by engineers of the Moscow Power Institute. See Caldwell and Graham, *op. cit.*, p. 42. See also S. N. Braines, "Nevrokibernetika," and A. D. Voskresenskii and A. I. Prokhorov, "Problemy kibernetiki v meditsine," both in Berg, ed., *Kibernetika na sluzhbu kommunizma*. An English translation of an article on the same subject in the same period was E. Khudiakova, "Cybernetics and Medical Diagnosis," *The Soviet Review*, February 1961.

21. Caldwell and Graham, *op. cit.*

22. For a Soviet article in this field, see N. D. Andreev and D. A. Kerimov, "Vozmozhnosti ispolzovaniia kiberneticheskoi tekhniki pri reshenii nekotorykh pravovykh problem," in Berg, ed., *Kibernetika na sluzhbu kommunizma*, pp. 234–41. A superior discussion of some of the same problems, published in approximately the same period, was Lucien Mehl, "Les sciences juridiques devant l'automation," *Cybernetica*, No. 1, 2, (1960), pp. 22–40, 142–70.

23. A skeptical Soviet view of the use of cybernetics in military planning was B. S. Ukraintsev, "O vozmozhnostiakh kibernetiki v svete svoistva otobrazheniia materii," in Il'in *et al.*, eds., *op. cit.*, p. 130.

24. There is a whole separate literature on homeostasis, much too extensive to cite. It has been connected with the Principle of Least Action, variously attributed to Maupertuis, Leibniz, Euler, Lagrange, or Hamilton. In biology, Bentley Glass linked it to the work of Claude Bernard, Le Chatelier, as well as W. B. Cannon. See Bentley Glass, "Maupertuis, Pioneer of Genetics and Evolution," in Bentley Glass, Owsei Temkin, William L. Straus, Jr., eds., *Forerunners of Darwin, 1745–1859* (Baltimore, 1959), pp. 53–5.

25. Roger Nett and Stanley A. Hetzler, *An Introduction to Electronic Data Processing* (New York, 1963), pp. 172–3. I am indebted to Mrs. Helen Powers, a student at Columbia University, for pointing out this reference to me.

26. See the discussions of the position of science in the framework of Soviet Marxist ideology in my *The Soviet Academy of Sciences and the Communist Party, 1927–1932* (Princeton, 1967), pp. 34–8, 192–3; and in my "Reorganization of the Academy of Sciences," in Peter H. Juviler and Henry W. Morton, *Soviet Policy-Making* (New York, 1967), pp. 151–3.

27. These articles were: V. P. Tugarinov and L. E. Maistrov, "Protiv idealizma

v matematicheskoi logike," *Voprosy filosofii,* No. 3 (1950), pp. 331–9; **M.** Iaroshevskii, "Kibernetika—'nauka' mrakobesov," *Literaturnaia gazeta,* April 5, 1952, p. 4; "Materialist" (pseud.), "Komu sluzhit kibernetika?" *Voprosy filosofii,* No. 5 (1953), pp. 210–19. The first of these three attacked cybernetics only indirectly.

28. Tugarinov and Maistrov, *op. cit.,* was an example.
29. Iaroshevskii, *op. cit.,* p. 4. In view of later events in Vietnam, the Soviet critic's comments do not appear so exaggerated, although this motivation can hardly be attributed to such an opponent of war research (after World War II) as Norbert Wiener. See Noam Chomsky's discussion of General William Westmoreland's description of the Vietnam war as a technological "great success" in which machines carry "more and more of the burden." Noam Chomsky, "After Pinkville," *The New York Review of Books* January 1, 1970, p. 4.
30. To reduce all complex forms of the movement of matter to combinations of simple forms would mean subscribing, many Soviet authors said, to vulgar rather than dialectical materialism. See, for example, M. N. Andriushchenko, "Otvet tovarishcham V. B. Borshchevu, V. V. Il'inu, F. Z. Rokhline," *Filosofskie nauki,* III, No. 4 (1960), 108–10. It is clear, however, that the basic "law" underlying this argument—that of the transition of quantity into quality—could be used in favor of the notion of thinking machines as well as against it. A sufficiently sophisticated arrangement of computer components and perhaps even the integration of organic material—as certain Soviet scientists have suggested—could involve a change in "qualitative" relationships.
31. "Nauka i zhizn'," *Kommunist,* No. 5 (1954), pp. 3–13.
32. E. Kol'man, "Chto takoe kibernetika," *Voprosy filosofii,* No. 4 (1955), pp. 148–59.
33. E. Kol'man, "Chuvstvo mery," in A. I. Berg and E. Kol'man, eds., *Vozmozhnoe i nevozmozhnoe v kibernetike* (Moscow, 1964), pp. 51–64. Also, in the same spirit, see Kol'man's "Kibernetika stavit voprosy," *Nauka i zhizn',* XXVIII, No. 5 (1961), pp. 43–5.
34. S. L. Sobolev, A. I. Kitov, and A. A. Liapunov, "Osnovnye cherty kibernetiki," in *Voprosy filosofii,* No. 4 (1955), p. 147.
35. E. Kol'man, "O filosofskikh i sotsial'nykh problemakh kibernetiki," in Il'in *et al.,* eds., *op. cit.,* p. 90; A. Kolmogorov, "Avtomaty i zhizn'," in Berg and Kol'man, eds., *op. cit.,* p. 10.
36. See W. Ross Ashby, *An Introducton to Cybernetics* (London, 1956), pp. 83–5, 270–1; and V. N. Kolbanovskii, "O nekotorykh spornykh voprosakh kibernetiki," in Il'in *et al.,* eds., *op. cit.,* pp. 257–8.
37. Norbert Wiener, *The Human Use of Human Beings: Cybernetics and Society* (New York, 1954), pp. 105–11.
38. See, for example, his *The Nerves of Government: Models of Political Communication and Control* (New York, 1963).
39. A general Soviet criticism of the use of cybernetics by sociologists outside the Soviet Union was E. A. Arab-Ogly, "Sotsiologiia i kibernetika," *Voprosy filosofii,* No. 5 (1958). Another more specific criticism of a cybernetic interpretation of history was I. Ia. Aksenov, "O vtorom mezhdunarodnom kongresse po kibernetike," in Il'in *et al.,* eds., *op. cit.,* p. 367.

40. S. M. Shaliutin, "O kibernetike i sfere ee primeneniia," in Il'in *et al.*, eds., *op. cit.*, pp. 25–7.
41. L. A. Petrushenko, "Filosofskoe znachenie poniatiia 'obratnaia sviaz' v kibernetike," *Vestnik leningradskogo universiteta: seriia ekonomiki, filosofii, i prava*, 17 (1960), 76–86.
42. B. V. Akhlibinskii and N. I. Khralenko, *Chudo nashego vremeni: kibernetika i problemy razvitiia* (Leningrad, 1963).
43. Kolbanovskii, *op. cit.*, p. 248.
44. Berg *et al.*, eds., *op. cit.*; Il'in *et al.*, eds., *op. cit.*
45. Wiener, *The Human Use of Human Beings*, p. 32.
46. A. M. Turing, "Computing Machinery and Intelligence," *Mind*, LIX (October 1950), 434.
47. Kol'man, "Kibernetika stavit voprosy," *Nauka i zhizn'*, XXVIII, No. 5 (1961), 44–5.
48. A. I. Berg, "Nauka velichaishikh vozmozhnostei," *Priroda*, No. 7 (1962), p. 21.
49. Todor Pavlov, "Avtomaty, zhizn', soznanie," *Nauchnye doklady vysshei shkoly: filosofskie nauki*, No. 1 (1963), p. 53.
50. See p. 341.
51. A. Kolmogorov, "Avtomaty i zhizn'," in Berg and Kol'man, eds., *op. cit.*, pp. 22–3.
52. B. Bialik, "Tovarishchi, vy eto ser'ezno?" and S. Sobolev, "Da, eto vpolne ser'ezno!" in Berg and Kol'man, eds., *op. cit.*, pp. 77–88.
53. E. Kol'man, "Chuvstvo mery," in Berg and Kol'man, eds., *op. cit.*, p. 53.
54. N. P. Antonov and A. N. Kochergin, "Priroda myshleniia i problema ego modelirovaniia," *Filosofskie nauki*, VI (1963), 42.
55. "Norbert Viner v redaktsii nashego zhurnala," *Voprosy filosofii*, No. 9 (1960), p. 164.
56. Ashby, like Wiener, at one point said that computers may eventually become so complex that neither the men who constructed them nor the men who operated them would know why they do what they do. W. R. Ashby, "Design for a Brain," *Electronic Engineering*, XX, No. 250, 382–3. This view was sharply criticized by the Soviet author E. A. Arab-Ogly in "Sotsiologiia i kibernetika," *Voprosy filosofii*, No. 5 (1958).
57. Thomas J. Watson, Jr., chairman of the International Business Machines Corporation, commented that computers can help humans enrich the quality of their lives, but they "can never handle our destiny or even make simple decisions of morality." He warned against attributing to computers "more abilities, more potential, than they really have." *The New York Times*, November 12, 1968, p. 15.
58. Ashby, *An Introduction to Cybernetics*, p. 152.
59. Hartley's ideas were presented at the International Congress of Telegraphy and Telephony at Lake Como in September 1927 and published as "Transmission of Information," *The Bell System Technical Journal*, VII (1928), 535–63.
60. Claude Shannon and Warren Weaver, *The Mathematical Theory of Communication* (Urbana, Ill., 1949), pp. 19, 105.

61. Planck described his formulation of this relationship in his *Scientific Autobiography and Other Papers* (New York, 1949), pp. 40–2.
62. Louis de Broglie, "La Cybernétique," *La Nouvelle Nouvelle Revue Française*, July 1953, p. 85.
63. I. I. Novinskii, "Poniatie sviazi v dialekticheskom materializme i voprosy biologii," unpublished dissertation for the degree of doctor of philosophical sciences (Moscow State University, 1963), pp. 324–6. Novik, Biriukov, and Tiukhtin also attempted to apply the vocabulary of cybernetics to the laws of dialectics. I. B. Novik, "Kibernetika i razvitie sovremennogo nauchnogo poznaniia," *Priroda*, LII, No. 10 (1963), pp. 3–11; B. V. Biriukov and V. S. Tiukhin, "Filosofskie voprosy kibernetiki," in Berg, *et al.*, eds., *Kibernetika, myshlenie, zhizn'*, pp. 76–108.
64. A. D. Ursul, "O prirode informatsii," *Voprosy filosofii*, No. 3 (1965), p. 134.
65. Ashby, *An Introduction to Cybernetics*, p. 177.
66. I. Novik, *Kibernetika: filosofskie i sotsiologicheskie problemy* (Moscow, 1963).
67. *Ibid.*, p. 60.
68. *Ibid.*, p. 58.
69. N. I. Zhukov, "Informatsii v svete leninskoi teorii otrazheniia," *Voprosy filosofii*, No. 11 (1963), pp. 156–7.
70. E. A. Sedov, "K voprosu o sootnoshenii entropii informatsionnykh protsessov i fizicheskoi entropii," *Voprosy filosofii*, No. 1 (1965), pp. 135–45.
71. A. D. Ursul, "O prirode informatsii," *Voprosy filosofii*, No. 3 (1965), p. 137.
72. Kol'man, "Chuvstvo mery," p. 62; Pavlov, "Avtomaty, zhizn', soznanie," p. 53.
73. B. S. Griaznov, "Kibernetika v svete filosofii," *Voprosy filosofii*, No. 3 (1965), p. 162.
74. N. Musabaeva, *Kibernetika i kategoriia prichinnosti* (Alma-Ata, 1965).
75. A. D. Ursul, *Priroda informatsii* (Moscow, 1968).
76. He applied, in turn, the Laws of the Unity and Struggle of Opposites, the Transition of Quantity into Quality, and the Negation of the Negation to concepts of information theory. See Ursul, *Priroda informatsii,* pp. 147–56.
77. *Ibid.*, p. 285.
78. *Ibid.*, p. 153.
79. *Ibid.*, p. 35. Ursul credited the work of American scholars such as N. Rashevsky and G. Karreman in providing topological approaches to information.
80. *Ibid.*, p. 150.
81. *Ibid.*, pp. 47–8.
82. *Ibid.*, p. 98. Von Bertalanffy is further discussed at the end of the genetics chapter.
83. A brief discussion of some of the more obvious deficiencies is in Rajko Tomovic, "Limitations of Cybernetics," *Cybernetica,* II, No. 3 (1959), 195–8.
84. G. T. Guilbaud, *What is Cybernetics?* (New York, 1960), p. 3.
85. An interesting discussion of the way in which this confusion was transferred

from certain European and American texts to Soviet writings is in Donald P. Bakker, "The Philosophical Debate in the U.S.S.R. on the Nature of 'Information,' " unpublished M.A. thesis (Columbia University, 1966).

86. Stafford Beer, "The Irrelevance of Automation," *Cybernetica,* I, No. 4 (1958), 288.

87. The viewpoint that I have presented in this chapter is submitted to rather pointed criticism in a non-Soviet Russian journal published in West Germany: V.P., "Nauka i ideologiia v sovetskom obshchestve," *Posev,* No. 4 (1968), pp. 58–9. The criticism is also directed at other authors (George Fischer, Richard De George, Herbert Levine) who wrote chapters in a book in which an earlier version of my description of cybernetics in the Soviet Union appeared. See George Fischer, ed., *Science and Ideology in Soviet Society* (New York, 1967).

88. In my opinion the single biggest difference between the Soviet Union and the United States in terms of the quality of the environment is the lack in the Soviet Union of commercial strips along the highways. The absence of private business in the U.S.S.R., and the fewer number of automobiles, has meant that the outskirts of Soviet cities are not visually atrocious to the degree that the outskirts of American cities are. In the areas of water and air pollution both the Soviet Union and the United States are fighting similar problems.

89. Andrei D. Sakharov, the noted physicist, wrote at length on environmental pollution in his "Thoughts on Progress, Peaceful Coexistence and Intellectual Freedom"; he expressed the fear that the "Soviet Union will poison the United States with its wastes and *vice versa.*" *The New York Times,* July 22, 1968, p. 15.

X: Physiology and Psychology

1. Sechenov submitted the manuscript under its original title to the literary and sociopolitical journal *Sovremennik,* where it was stopped by the censor. The very fact that a journal of this nature would attempt publication of a work on physiology reveals the philosophical and political implications seen in Sechenov's interpretation. His treatise eventually appeared in the much more specialized *Meditsinskii vestnik.*

2. E. A. Budilova in 1960 called the main organ of the Moscow Psychological Society, *Voprosy filosofii i psikhologii,* published after 1890, "an organ of reaction in science," a "tribune of militant idealism" for "all twenty-eight years of its existence." E. A. Budilova, *Bor'ba materializma i idealizma v russkoi psikhologicheskoi nauke: vtoraia polovina XIX—nachalo XX v.* (Moscow, 1960), p. 108.

3. Edwin G. Boring, *A History of Experimental Psychology* (New York, 1950), p. 636.

4. Walter A. Rosenblith, "Postscript," in Ivan M. Sechenov, *Reflexes of the Brain* (Cambridge, Mass., 1965), p. 145.

5. Hilaire Cuny, *Ivan Pavlov: The Man and His Theories* (New York, 1965), p. 76.

6. Quoted in Ezras Asratian, *I. P. Pavlov, His Life and Work* (Moscow, 1953), p. 60.

7. Quoted in B. P. Babkin, *Pavlov: A Biography* (Chicago, 1949), pp. 276–7.

8. See my *The Soviet Academy of Sciences and the Communist Party, 1927–1932* (Princeton, 1967), pp. 108, 109, 113, 116–18.

9. P. K. Anokhin, *Ivan Petrovich Pavlov: zhizn', deiatel'nost' i nauchnaia shkola* (Moscow and Leningrad, 1949), p. 352.

10. See *Akademik I. P. Pavlov: Izbrannye trudy po fiziologii vysshei nervnoi deiatel'nosti* (Moscow, 1950), p. 167.

11. See, for example, P. S. Kupalov, "Uchenie o reflekse i reflektornoi deiatel'nosti i perspektivy ego razvitiia," in *Filosofskie voprosy fiziologii vysshei nervnoi deiatel'nosti i psikhologii* (Moscow, 1963), p. 151.

12. F. V. Bassin, "Zakliuchitel'nye slova," in *Filosofskie voprosy fiziologii vysshei nervnoi deiatel'nosti i psikhologii* (Moscow, 1963), p. 720.

13. A. V. Petrovskii, *Istoriia sovetskoi psikhologii* (Moscow, 1967). Raymond A. Bauer, *The New Man in Soviet Psychology* (Cambridge, Mass., 1952). Petrovskii's work, published at a relatively relaxed time, contains interesting discussions of such men as P. P. Blonskii, K. N. Kornilov, and B. M. Bekhterev. These are men whose schools of thought were later criticized by the Communist Party, but Petrovskii gives a fairly sympathetic account of their efforts to reconstruct pre-Revolutionary psychology. Bauer's work is perceptive and scholarly but, surprisingly, does not go very far into the efforts of individual Soviet writers to link their views with Marxism. The implication is that the efforts were hypocritical. There is considerable truth in his observation that "During the twenties the Marxist ideology appears to have functioned primarily as a screening device whereby certain schools of psychology were rejected as unacceptable" (p. 62), but his approach makes it quite difficult to understand why some psychologists, particularly at later dates, took their Marxism quite seriously.

14. K. Kornilov, "Dialekticheskii metod v psikhologii," *Pod znamenem marksizma,* January 1924, p. 108.

15. V. Struminskii, "Marksizm v sovremennoi psikhologii," *Pod znamenem marksizma,* March 1926, p. 213. Struminskii was a militant materialist whose scholarship was careless. See critical reviews of him, and his replies, in *Pod znamenem marksizma,* November–December 1923, pp. 299–304 and March 1924, pp. 250–4, 255–9.

16. See Sigmund Freud, *Civilization and Its Discontents* (1930), James Strachey, ed. (New York, 1961), pp. 59–61.

17. B. Bykhovskii, "O metodologicheskikh osnovaniiakh psikhoanaliticheskogo ucheniia Freida," *Pod znamenem marksizma,* November–December 1923, pp. 169, 176–7.

18. There is no space here to discuss important developments in applied psychology, such as the psychology of labor or psychotechnics. These developments do play an important role, however, in the general history of Soviet psychology.

19. Edwin G. Boring, "Psychology, History of," *Encyclopedia Britannica,* Vol. XVIII (Chicago, 1959), p. 713.

20. Bauer, *op. cit.,* pp. 93–102.

21. Quoted in *ibid.*, p. 124.
22. Quoted in *ibid.*
23. M. G. Iaroshevskii, *Istoriia psikhologii* (Moscow, 1966), p. 540.
24. *Ibid.*, p. 542.
25. Quoted by O. L. Zangwill, "Psychology: Current Approaches," in *The State of Soviet Science* (Cambridge, Mass., 1965), p. 122.
26. A. R. Luria, *Higher Cortical Functions in Man* (New York, 1966), p. 2.
27. A. V. Brushlinskii, "Kul'turno-istoricheskaia teoriia myshleniia," in E. V. Shorokhova, ed., *Issledovaniia myshleniia v sovetskoi psikhologii* (Moscow, 1966), pp. 123–74.
28. Jerome S. Bruner, Introduction to L. S. Vygotsky, *Thought and Language* (Cambridge, Mass., 1962), pp. vi, x.
29. The translators commented, "Although our more compact rendition could be called an abridged version of the original, we feel that the condensation has increased clarity and readability without any loss of thought content or factual information." Eugenia Hanfmann and Gertrude Vakar, Translators' Preface to L. S. Vygotsky, *Thought and Language* (Cambridge, Mass., 1962), p. xii.
30. L. S. Vygotsky, *Izbrannye psikhologicheskie issledovaniia* (Moscow, 1956), pp. 91–2; also see p. 105.
31. Vygotsky, *Thought and Language,* p. 10.
32. Quoted from Piaget in *ibid.*, p. 12.
33. *Ibid.*, p. 13.
34. *Ibid.*, pp. 18–20.
35. *Ibid.*, pp. 135–6.
36. *Ibid.*, pp. 33, 41.
37. *Ibid.*, p. 43.
38. *Ibid.*, pp. 28, 29, 44.
39. *Ibid.*, p. 49.
40. *Ibid.*, p. 51.
41. Joseph Stalin, *Marxism and Linguistics* (New York, 1951), p. 36.
42. Vygotsky, *Thought and Language,* p. 41.
43. Jean Piaget, *Comments on Vygotsky's Critical Remarks* (Cambridge, Mass., 1962).
44. *Scientific Session on the Physiological Teaching of Academician I. P. Pavlov* (Moscow, 1951). For interpretations, see Gustav Wetter, *Dialectical Materialism* (New York, 1958), pp. 473–87; and Robert C. Tucker, "Stalin and the Uses of Psychology," in *The Soviet Political Mind: Studies in Stalinism and Post-Stalin Change* (New York, 1963), pp. 91–121.
45. *Scientific Session on the Physiological Teachings of Academician I. P. Pavlov,* pp. 12–15.
46. *Ibid.*, p. 19.
47. Josef Brožek, "Soviet Contributions to History," *Contemporary Psychology,* XIV, No. 8 (1969), 433. I am grateful to Professor Brožek for a copy of his review.
48. A. V. Petrovskii, *Istoriia sovetskoi psikhologii* (Moscow, 1967), p. 336.
49. *Ibid.*
50. S. L. Rubinshtein, *Printsipy i puti razvitiia psikhologii* (Moscow, 1959), p.

3. Rubinshtein wrote in the 1934 article that his major goal was to find solutions to the main problems of modern psychology with the help of Marx's writings and, on the basis of these solutions, to begin the construction of a Marxist-Leninist psychology. See S. L. Rubinshtein, "Problemy psikhologii v trudakh Karla Marksa," *Sovetskaia psikhotekhnika,* No. 1 (1934), pp. 4, 3–20. For other important publications of Rubinshtein see: *Osnovy psikhologii* (Moscow, 1935); *Osnovy obshchei psikhologii* (Moscow, 1940, 1946); "Uchenie I. P. Pavlova i nekotorye voprosy perestroiki psikhologii," *Voprosy filosofii,* No. 3 (1952), pp. 197–210; *Bytie i soznanie: o meste psikhicheskogo vo vseobshchei vzaimosviazi iavlenii material'nogo mira* (Moscow, 1957); "Filosofiia i psikhologiia," *Voprosy filosofii,* No. 1 (1957), pp. 114–27; "Voprosy psikhologii myshleniia i printsip determinizma," *Voprosy filosofii,* No. 5 (1957), pp. 101–13; *O myshlenii i putiakh ego issledovaniia* (Moscow, 1958); *Printsipy i puti razvitiia psikhologii* (Moscow, 1959).

51. Raymond A. Bauer, *The New Man in Soviet Psychology* (Cambridge, Mass., 1952), p. 118.

52. S. L. Rubinshtein, "Rech'," *Voprosy filosofii,* No. 1 (1947), pp. 420–7.

53. At the 1962 conference on the philosophical problems of physiology and psychology, to be discussed subsequently, E. V. Shorokhova indicated the continuing influence of Rubinshtein: "In our report we are expressing the position of the Psychology Section of the Institute of Philosophy. Our point of view is to a large degree the collective opinion of that group which was created and led, until his death, by S. L. Rubinshtein. We consider it our responsibility to develop Rubinshtein's principles and to defend his views against the mistaken interpretation which they occasionally encounter." *Filosofskie voprosy fiziologii vysshei nervnoi deiatel'nosti i psikhologii* (Moscow, 1963), pp. 730–1.

54. "Sergei Leonidovich Rubinshtein," *Voprosy filosofii,* No. 2 (1960), pp. 179–80.

55. *Osnovy obshchei psikhologii* (Moscow, 1946), p. 5.

56. *Ibid.*

57. *Ibid.,* pp. 5–6.

58. *Ibid.,* p. 7.

59. *Ibid.,* pp. 7–8.

60. *Ibid.,* p. 8.

61. *Ibid.,* p. 18.

62. *Ibid.*

63. *Ibid.,* p. 19.

64. *Ibid.*

65. Kolbanovskii seems to have misunderstood Rubinshtein when he accused him of equating "psychic facts" with material objects, the error for which Lenin once criticized Joseph Dietzgen. Rubinshtein was more vulnerable to the charge that he had removed psychic phenomena from the area of material events than that he had equated them. See V. Kolbanovskii, "Za marksistskoe osveshchenie voprosov psikhologii," *Bol'shevik,* September 15, 1947, p. 57; and Rubinshtein, *Osnovy obshchei psikhologii* (1946), p. 5.

66. Kolbanovskii much later noted this change. His earlier quite critical tone

was now tempered with praise. See V. N. Kolbanovskii, "Zakliuchitel'noe slovo," in *Dialekticheskii materializm i sovremennoe estestvoznanie* (Moscow, 1964), pp. 399–400.

67. S. L. Rubinshtein, "Uchenie I. P. Pavlova i nekotorye voprosy perestroiki psikhologii," *Voprosy filosofii*, No. 3 (1952), p. 201.
68. Rubinshtein, *Bytie i soznanie*, p. 33.
69. *Ibid.*, pp. 10–11.
70. K. Marx, F. Engels, *Sochineniia*, Vol. I (Moscow, 1929), p. 180.
71. *Voprosy filosofii*, No. 2 (1960), p. 180.
72. *Bytie i soznanie*, pp. 11–12.
73. *Ibid.*, pp. 4–5.
74. *Ibid.*, pp. 5–6.
75. *Ibid.*, pp. 8–9.
76. *Ibid.*, p. 34.
77. *Ibid.*, p. 230.
78. *Ibid.*, p. 232.
79. *Ibid.*, p. 34.
80. *Ibid.*, p. 64.
81. S. L. Rubinshtein, *Printsipy i puti razvitiia psikhologii* (Moscow, 1959), p. 8.
82. *Ibid.*, p. 9.
83. *Ibid.*, p. 11.
84. *Ibid.*, p. 8.
85. P. N. Fedoseev, *et al.*, eds., *Filosofskie voprosy fiziologii vysshei nervnoi deiatel'nosti i psikhologii* (Moscow, 1963).
86. *Ibid.*, p. 113.
87. *Ibid.*, pp. 308–12.
88. *Ibid.*, p. 314.
89. *Ibid.*, p. 322.
90. *Ibid.*, p. 43.
91. *Ibid.*, p. 39.
92. *Ibid.*, pp. 47–8.
93. *Ibid.*, pp. 521–5.
94. John Carew Eccles, *The Neurophysiological Basis of Mind* (Oxford, 1952), pp. 278–9.
95. *Filosofskie voprosy fiziologii vysshei nervnoi deiatel'nosti i psikhologii*, pp. 54–5.
96. *Ibid.*, pp. 302–3.
97. *Ibid.*, p. 44.
98. *Ibid.*, pp. 535
99. See the discussion in the philosophy chapter, p. 49.
100. *Filosofskie voprosy fiziologii vysshei nervnoi deiatel'nosti i psikhologii*, pp. 87–8.
101. *Ibid.*, pp. 554–5.
102. *Ibid.*, p. 584.
103. *Ibid.*, pp. 727–8.
104. *Ibid.*, pp. 509–15.
105. Grashchenkov told the conference that after the 1950 Pavlov session he was

labeled the Number One Anti-Pavlovian and that as a result he lost his position and could not publish his works for a number of years. *Ibid.,* p. 736.

106. See *Obshchie osnovy fiziologii truda* (Moscow, 1935), p. 447.

107. See the discussion of Anokhin on pp. 407–25.

108. See, for example, A. Rosenblueth, N. Wiener, and J. Bigelow, "Behavior, Purpose and Teleology," *Philosophy of Science,* January 1943, pp. 18–24; and W. Ross Ashby, *Design for a Brain: The Origin of Adaptive Behaviour* (London, 1960; 1st ed., 1952).

109. *Filosofskie voprosy fiziologii vysshei nervnoi deiatel'nosti i psikhologii,* pp. 646–7.

110. Karl Marx and Friedrich Engels, *Werke,* Vol. XX (Berlin, 1962), p. 325.

111. *Filosofskie voprosy fiziologii vysshei nervnoi deiatel'nosti i psikhologii,* p. 567.

112. Herbert Feigl, "The 'Mental' and the 'Physical'," in Herbert Feigl, Michael Scriven, and Grover Maxwell, eds., *Minnesota Studies in the Philosophy of Science: Concepts, Theories, and the Mind-Body Problem,* Vol. II (Minneapolis, 1958).

113. D. A. Biriukov's similar attempt to classify Soviet authors was very helpful to me; see *Filosofskie voprosy fiziologii vysshei nervnoi deiatel'nosti i psikhologii,* pp. 378–9.

114. V. M. Arkhipov, "O material'nosti psikhiki i predmete psikhologii," *Sovetskaia pedagogika,* No. 7 (1954), pp. 67–79; and I. G. Eroshkin, *Psikhologiia i fiziologiia vysshei nervnoi deiatel'nosti* (Leningrad, 1958).

115. F. F. Kal'sin, *Osnovnye voprosy teorii poznaniia* (Gor'kii, 1957). Kedrov tried to sketch an over-all classification of moving matter in all nature, from nonliving to living, culminating in human society. See his "O sootnoshenii form dvizheniia materii v prirode," in P. N. Fedoseev *et al.,* eds., *Filosofskie problemy sovremennogo estestvoznaniia* (Moscow, 1959), pp. 137–211.

116. M. P. Lebedev, "Materiia i soznanie," *Voprosy filosofii,* No. 5 (1956), pp. 70–84.

117. V. N. Kolbanovskii, *Filosofskie voprosy fiziologii vysshei nervnoi deiatel'nosti i psikhologii,* p. 606.

118. F. I. Georgiev, "Problema chuvstvennogo i ratsional'nogo v poznanii," *Voprosy filosofii,* No. 1 (1955), pp. 28–41.

119. *Filosofskie voprosy fiziologii vysshei nervnoi deiatel'nosti i psikhologii,* pp. 646–7.

120. Sebastian Peter Grossman, *A Textbook of Physiological Psychology* (New York, 1967), p. 288.

121. The nonspecificity of the reticular formation is, of course, relative, and may be only apparent. As Grossman wrote in 1967, "Much of the nonspecificity of the reticular formation is undoubtedly a reflection of our ignorance. We may expect to progress on the road which has led in a relatively short period of 20 years to the demonstration of a number of specific and anatomically distinct functional mechanisms." *Ibid.* It should be remembered that most of the comments by Soviet dialectical materialists discussed here were made at a time when the reticular formation was still regarded as extraordinarily nonspecific.

122. No Soviet scholars attended the symposium. The record of the reports and

discussion is in *Brain Mechanisms and Consciousness* (Springfield, Ill., 1954).

123. *Ibid.,* p. 210.
124. *Ibid.,* p. 219.
125. E. A. Asratian, "Nekotorye aktual'nye voprosy fiziologii vysshei nervnoi deiatel'nosti i ikh filosofskoe znachenie," in S. A. Petrushevskii, *et al.,* eds., *Dialekticheskii materializm i sovremennoe estestvoznanie* (Moscow, 1964), pp. 231–2.
126. O. Ia. Bokser, D. A. Mishakhin, and S. S. Poltyrev, "Filosofskie voprosy fiziologii retikuliarnoi formatsii," *Voprosy filosofii,* No. 12 (1961), p. 82–6.
127. *Filosofskie voprosy fiziologii vysshei nervnoi deiatel'nosti i psikhologii,* p. 520.
128. A. F. Makarchenko, P. G. Kostiuk, and A. E. Khil'chenko, "Filosofskoe i estestvennonauchnoe znachenie ucheniia I. P. Pavlova o vysshei nervnoi deiatel'nosti v sovremennoi fiziologii," in M. F. Gulyi, *et al.,* eds., *Filosofskie voprosy sovremennoi biologii* (Kiev, 1962), pp. 315–20.
129. Several of his important publications are: P. Anokhin and E. Strezh, "Izuchenie dinamiki vysshei nervnoi deiatel'nosti. Soobshchenie III. Narushenie aktivnogo vybora v rezul'tate zameny bezuslovnogo stimula," *Fiziologicheskii zhurnal SSSR,* No. 2 (1933), pp. 280–98; "Problema tsentra i periferii v sovremennoi fiziologii nervnoi deiatel'nosti," in Anokhin, ed., *Problema tsentra i periferii v fiziologii nervnoi deiatel'nosti* (Gor'kii, 1935), pp. 9–70; *Ot Dekarta do Pavlova: trista let teorii refleksa* (Moscow, 1945); "Uzlovye voprosy v izuchenii vysshei nervnoi deiatel'nosti," in *Problemy vysshei nervnoi deiatel'nosti* (Moscow, 1949), pp. 9–128; "Refleks i funktsional'naia sistema kak faktory fiziologicheskoi integratsii," *Biulleten' Moskovskogo obshchestva ispytatelei prirody,* No. 5 (1949), pp. 130–46; *Ivan Petrovich Pavlov* (Moscow, 1949); "Osobennosti afferentnogo apparata uslovnogo refleksa i ikh znachenie dlia psikhologii," *Voprosy psikhologii,* No. 6 (1955), pp. 16–38; *Obshchie printsipy kompensatsii narushennykh funktsii i ikh fiziologicheskoe obosnovanie* (Moscow, 1956); *Vnutrenee tormozhenie kak problema fiziologii* (Moscow, 1958); *A New Conception of the Physiological Architecture of the Conditioned Reflex* (Moscow, 1959); "Kibernetika i integrativnaia deiatel'nost' mozga," *Voprosy psikhologii,* No. 3 (1966), pp. 10–32; "Metodologicheskoe znachenie kiberneticheskikh zakonomernostei," in M. E. Omel'ianovskii *et al.,* eds., *Materialisticheskaia dialektika i metody estestvennykh nauk* (Moscow, 1968), pp. 547–86; *Biologiia i neirofiziologiia uslovnogo refleksa* (Moscow, 1968).
130. P. K. Anokhin, "Metodologicheskii analiz uzlovykh problem uslovnogo refleksa," in Fedoseev *et al.,* eds., *Filosofskie voprosy fiziologii vysshei nervnoi deiatel'nosti i psikhologii,* p. 158.
131. *Ibid.*
132. *Ivan Petrovich Pavlov,* p. 349.
133. P. Anokhin, ed., *Problema tsentra i periferii v fiziologii nervnoi deiatel'nosti* (Gor'kii, 1935), p. 52.
134. "Uzlovye voprosy v izuchenii vysshei nervnoi deiatel'nosti," p. 12.
135. *Ivan Petrovich Pavlov: zhizn', deiatel'nost' i nauchnaia shkola* (Moscow and Leningrad, 1949), p. 351.

136. *Filosofskie voprosy fiziologii vysshei nervnoi deiatel'nosti i psikhologii,* p. 716.
137. *Ivan Petrovich Pavlov: zhizn', deiatel'nost' i nauchnaia shkola,* p. 355.
138. "Uzlovye voprosy v izuchenii vysshei nervnoi deiatel'nosti," p. 9.
139. *Ibid.*
140. *Ibid.,* p. 18.
141. *Ibid.,* p. 19.
142. *Ibid.,* pp. 98, 106–8, 114.
143. Quoted by Anokhin in "Osobennosti afferentnogo apparata uslovnogo refleksa i ikh znachenie dlia psikhologii," *Voprosy psikhologii,* No. 6 (1955), p. 18.
144. *Ibid.,* p. 19.
145. N. A. Bernshtein, whose approach was in many ways similar to Anokhin's, objected to the term "return afferentiation" on the basis of the fact that all afferentiation is "return afferentiation"; a "nonreturn" afferentiation does not exist. N. A. Bernshtein, "Novye linii razvitiia v fiziologii i ikh sootnoshenie s kibernetikoi," in *Filosofskie voprosy fiziologii vysshei nervnoi deiatel'nosti i psikhologii,* p. 303.
146. P. K. Anokhin, "Osobennosti afferentnogo apparata . . . ," p. 21.
147. *Ibid.,* p. 22.
148. *Ibid.*
149. *Ibid.,* p. 25.
150. *Ibid.*
151. *Ibid.,* p. 26.
152. For a short description in classical Pavlovian terms see B. P. Babkin, *Pavlov: A Biography* (Chicago, 1949), p. 311.
153. P. K. Anokhin, "Osobennosti afferentnogo apparata . . . ," p. 30.
154. For Anokhin's description of this experiment, see *ibid.,* p. 32.
155. *Ibid.,* p. 34.
156. *Ibid.*
157. *Ibid.,* p. 37.
158. See Boring, *A History of Experimental Psychology,* 2nd ed. (1957), pp. 562–3.
159. Charles S. Sherrington wrote: "The brain is . . . always that part of the nervous system which is constructed upon and evolved alongside of the distance-receptors. The importance of this conjunction in this matter is that it means ability on the part of the animal to react to an object when still distant and allows an interval for preparatory reactive steps. . . ." "Brain," *Encyclopedia Britannica,* Vol. IV (Chicago, 1959), p. 1.
160. P. K. Anokhin, "Metodologicheskii analiz uzlovykh problem uslovnogo refleksa," in *Filosofskie voprosy fiziologii vysshei nervnoi deiatel'nosti i psikhologii* (Moscow, 1963), pp. 164–5.
161. *Ibid.,* p. 168.
162. *Ibid.,* p. 169.
163. *Ibid.,* p. 71.
164. Anokhin cited A. Dorn as the originator of the phrase in *Printsip smeny funktsii* (Moscow and Leningrad, 1937).
165. And here he opened himself up to the same sort of criticisms directed

against Oparin for his similar belief that natural selection can occur in the nonliving world. See pp. 292–3.

166. Anokhin, "Metodologicheskii analiz . . . ," pp. 173–4.

167. *Ibid.*, p. 176.

168. Anokhin guarded against the charge of reductionism, the criticism that he ignored the transition of quantity into quality: "When I compare that highly specialized form of anticipatory reflection of the external world that we meet in the conditioned reflex of higher animals with anticipatory reflection in primitive living substances, I in no way ignore the qualitative difference between these phenomena. But any kind of comparison of numerous phenomena inevitably requires a main criterion for their evaluation. Therefore, naturally, we must ask ourselves: What is the most characteristic trait of the conditioned reflex?" *Ibid.*, p. 178.

169. Anokhin, "Metodologicheskoe znachenie kiberneticheskikh zakonomernostei," in M. E. Omel'ianovskii, ed., *Materialisticheskaia dialektika i metody estestvennykh nauk* (Moscow, 1968), p. 547.

170. P. K. Anokhin, "Kibernetika i integrativnaia deiatel'nost' mozga," *Voprosy psikhologii*, No. 3 (1966), pp. 10–32; and "Metodologicheskoe znachenie kiberneticheskikh zakonomernostei," in M. E. Omel'ianovskii *et al.*, eds., *Materialisticheskaia dialektika i metody estestvennykh nauk*, pp. 547–86.

171. See, for example, *Filosofskie voprosy fiziologii vysshei nervnoi deiatel'nosti i psikhologii* (Moscow, 1963), pp. 450–1.

172. *Ibid.*, p. 456.

173. *Ibid.*, p. 685.

174. See D. N. Uznadze, *Psikhologicheskie issledovaniia* (Moscow, 1966).

175. *Filosofskie voprosy fiziologii vysshei nervnoi deiatel'nosti i psikhologii*, p. 471.

176. *Ibid.*, pp. 650–3.

177. *Ibid.*, p. 654.

178. Jeffrey A. Gray, "Attention, Consciousness and Voluntary Control of Behaviour in Soviet Psychology: Philosophical Roots and Research Branches," in Neil O'Connor, ed., *Present-day Russian Psychology* (Oxford, 1966), pp. 2–3.

Appendix I

1. See p. 19.

2. Leonard Schapiro, *The Communist Party of the Soviet Union* (New York, 1960), p. 508.

3. Georg von Rauch, *A History of Soviet Russia* (New York, 1962), p. 403.

4. Donald W. Treadgold, *Twentieth Century Russia* (Chicago, 1964), p. 452.

5. A. A. Zhdanov, "Vystuplenie na diskussii po knige G. F. Aleksandrova 'Istoriia zapadnoevropeiskoi filosofii' 24 iiunia 1947 g" (Moscow, 1952).

6. A. A. Zhdanov, "Doklad o zhurnalakh 'Zvezda' i 'Leningrad' " (Moscow, 1946); "29-ia godovshchina velikoi oktiabrskoi sotsialisticheskoi revoliutsii. Doklad na torzhestvennom zasedanii moskovskogo soveta 6 noiabria 1946 goda" (Moscow, 1946); "O mezhdunarodnom polozhenii. Doklad, sdelannyi

na informatsionnom soveshchanii predstavitelei nekotorykh kompartii v Pol'she, v kontse sentiabria 1947 goda" (Moscow, 1947); "Vystupitel'naia rech' i vystuplenie [na soveshchanii deiatelei sovetskoi muzyki v TsK VKP (b),] v kn.: Soveshchanie deiatelei sovetskoi muzyki v TsK VKP (b)" (Moscow, 1948). Also see *Essays on Literature, Philosophy, and Music* (New York, 1950).

7. *O polozhenii v biologicheskoi nauke. Stenograficheskii otchet sessii vsesoiuznoi akademii sel'sko-khoziaistvennykh nauk imeni V. I. Lenina, 31 iiulia–7 avgusta 1948 g.* (Moscow, 1948). Also available in English: *The Situation in Biological Science* (Moscow, 1949).

8. T. D. Lysenko, "On vdokhnovlial nas na bor'bu za dal'neishii rastsvet nauki," *Izvestiia,* September 1, 1948.

9. Iu. A. Zhdanov, "Poniatie gomologii v khimii i ego filosofskoe znachenie," dissertation for degree of *kandidat* (Institute of Philosophy, Academy of Sciences of the U.S.S.R., Moscow, 1948). On Engels, see my discussion on pp. 53–4.

10. Iu. A. Zhdanov, "Metody sinteza i svoistva uglerod-zameshchennykh uglevodov," dissertation for degree of *doktor* (Moscow State University, Moscow, 1959). In the interim years Iurii Zhdanov wrote another *kandidat* dissertation in chemistry at the University of Rostov on the Don, probably indicating something of a change in his career plans.

11. Iu. A. Zhdanov, "O kritike i samokritike v nauchnoi rabote," *Bol'shevik,* No. 21 (1951), 28–43; "V chem zakliuchaetsia ateisticheskoe znachenie ucheniia I. P. Pavlova?" *Voprosy filosofii,* No. 5 (1954), 234–6; "O perekhode kachestvennykh izmenenii v izmeneniia kolichestvennye," *Voprosy filosofii,* No. 6 (1956), 206–10; "Dialektika tozhdestva i razlichiia v khimii," *Filosofskie nauki,* No. 4 (1958), 168–75; *Lenin i estestvoznanie* (Moscow, 1959); "Kriterii praktiki v khimii," in *XXI s' 'ezd KPSS o teoreticheskikh voprosakh stroitel'stva kommunizma v SSSR* (Kharkov, 1959), pp. 181–215; "Vzaimootnoshenie chasti i tselogo v khimii," *Filosofskie nauki,* No. 1 (1960), 82–9; "Molekula i stroenie veshchestva," *Priroda,* No. 2 (1960), 15–24; "Estestvoznanie i gumanizm," *Priroda,* No. 5 (1962), 63–74; "Khimiia i estetika," *Priroda,* No. 10 (1964), 8–13. More technical articles, which show that Iurii Zhdanov became interested in the cybernetic approach to nature were: "Entropiia informatsii kak mera spetsifichnosti v reaktsiiakh aromaticheskogo zameshcheniia," *Zhurnal fizicheskoi khimii,* No. 3 (1965), 777–9; "Entropiia informatsii v reaktsiiakh aromaticheskogo zameshcheniia," *Zhurnal organicheskoi khimii,* No. 9 (1965), 1521–5.

12. Iu. A. Zhdanov, "O nauchnom tvorchestve molodezhi," *Komsomol'skaia pravda,* May 25, 1948; and his "Tovarishchu I. V. Stalinu," *Pravda,* August 7, 1948.

13. "Tovarishchu I. V. Stalinu," *Pravda,* August 7, 1948.

14. *Voprosy filosofii,* No. 1 (1947).

15. M. A. Markov, "O prirode fizicheskogo znaniia," *Voprosy filosofii,* No. 2 (1947), 140–76.

16. I. I. Schmalhausen, "Predstavleniia o tselom v sovremennoi biologii," *Voprosy filosofii,* No. 2 (1947), 177–83.

17. "Za boevoi filosofskii zhurnal," *Pravda,* September 7, 1949.

18. "Privetstvie akademiku T. D. Lysenko v sviazi s ego 50-letiem," p. 412; "Torzhestvo sovetskoi biologicheskoi nauki," pp. 121–32; I. E. Glush-chenko, "Protiv idealizma i metafiziki v nauke o nasledstvennosti," pp. 133–47; and V. N. Stoletov, "Printsipy ucheniia I. V. Michurina," pp. 148–70; all in *Voprosy filosofii*, No. 2 (1948).

19. "Za bol'shevistskuiu partiinost' v filosofii," *Voprosy filosofii*, No. 3 (1948), p. 11.

20. Iu. Zhdanov, "O kritike i samokritike v nauchnoi rabote," *Bol'shevik*, No. 21 (1951), pp. 28–43, and esp. 38–9.

21. Iu. Zhdanov, "Protiv sub' 'ektivistskikh izvrashchenii v estestvoznanii," *Pravda*, January 16, 1953.

22. The background to this development involves a long discussion of the place of the natural sciences in the base-superstructure scheme, a troublesome issue in the Soviet Union since the days of the proletarian culture movement. See my discussion in "Reorganization of the Academy of Sciences of the USSR," in Peter H. Juviler and Henry W. Morton, eds., *Soviet Policy-Making: Studies of Communism in Transition* (New York, 1967), pp. 151–3. The important point for the consideration of Zhdanov's position is that after several decades of considering science to be associated with the superstructure, Stalin in his *Letters on Linguistics* announced something new: There are intellectual fields that are neither a part of the base nor the superstructure, and science is one of these. He continued this approach in his *Economic Problems of Socialism in the USSR*.

23. Iu. A. Zhdanov, *Lenin i estestvoznanie* (Moscow, 1959), pp. 35–6.

24. Zhores A. Medvedev, *The Rise and Fall of T. D. Lysenko*, trans. I. Michael Lerner (New York, 1969), pp. 112–13.

25. Svetlana Alliluyeva, *Twenty Letters to a Friend* (New York, 1967), p. 198.

26. *Ibid.*, p. 245.

27. Svetlana Alliluyeva, *Only One Year*, trans. Paul Chavchavadze (New York and Evanston, Ill., 1969), p. 380.

28. Augustine, *De Genesi ad Litteram*, I, 18.

Appendix II

1. There is, however, at least one other omission of the same type. See H. J. Muller, "Nauka proshlogo i nastoiashchego i chem ona obiazana Marksu," in *Marksizm i estestvoznanie* (Moscow, 1933), pp. 204–7.

2. See p. 213.

3. Quoted in Tracy M. Sonneborn, "H. J. Muller: Tribute to a Colleague," *The Review* (Alumni Association, Indiana University), Fall 1968, p. 21.

4. Zhores A. Medvedev, *The Rise and Fall of T. D. Lysenko* (New York, 1969), p. 34.

5. I. T. Frolov, *Genetika i dialektika* (Moscow, 1968), pp. 80, 169, 171, 182, 246.

Bibliography

The following bibliography is restricted to sources cited in the footnotes. I hope eventually to publish separately my much larger working bibliography on dialectical materialism and science, in the hope that it will be useful to other scholars.

Abetti, Giorgio, *The History of Astronomy*. London and New York, 1952.

Acton, H. B., *The Illusion of the Epoch: Marxism-Leninism as a Philosophical Creed*. Boston, 1957.

Adams, Frank Dawson, *The Birth and Development of the Geological Sciences*. Dover reprint edition, New York, 1954.

Adams, Mark B., "The Founding of Population Genetics: Contributions of the Chetverikov School 1924–1934." *Journal of the History of Biology*, Spring 1968, pp. 23–29.

Agassi, Joseph, *Towards an Historiography of Science, History and Theory: Studies in the Philosophy of History*. Beiheft 2, The Hague, 1963.

Agranovskii, Anatolii, "Nauka na veru ne prinimaet." *Literaturnaia gazeta*, January 23, 1965, p. 2.

Akademik I. P. Pavlov, Izbrannye trudy po fiziologii vysshei nervnoi deiatel'nosti. Moscow, 1950.

Akhlibinskii, B. V., and N. I. Khralenko, *Chudo nashego vremeni: kibernetika i problemy razvitiia*. Leningrad, 1963.

Aksenov, I. Ia., "O vtorom mezhdunarodnom kongresse po kibernetike," in V. A. Il'in, *et al.*, eds., *Filosofskie voprosy kibernetiki*. Moscow, 1960, pp. 359–73.

———, "Transportnye problemy kibernetiki," in A. I. Berg, ed., *Kibernetika na sluzhbu kommunizma*. Moscow, 1961, pp. 158–96.

Aleksandr Ivanovich Oparin. Moscow and Leningrad, 1949 (AN SSSR. Materialy k biobibliografii uchenykh SSSR. Seriia biokhimii, vyp. 3).

Aleksandrov, A. D., "Dialektika i nauka." *Vestnik akademii nauk SSSR*, No. 6 (1957), pp. 3–17.

———, *Die innere Geometrie der konvexen Flächen*. Berlin, 1955.

———, "Filosofskoe soderzhanie i znachenie teorii otnositel'nosti." *Voprosy filosofii*, No. 1 (1959), pp. 67–84.

———, "Ob idealizme v matematike." *Priroda*, No. 7 (1951), pp. 3–11, and No. 8 (1951), pp. 3–9.

————, "Po povodu nekotorykh vzgliadov na teoriiu otnositel'nosti." *Voprosy filosofii*, No. 5 (1953), pp. 225–45.

————, "Teoriia otnositel'nosti kak teoriia absoliutnogo prostranstvavremeni," in I. V. Kuznetsov and M. E. Omel'ianovskii, eds., *Filosofskie voprosy sovremennoi fiziki*. Moscow, 1959, pp. 269–323.

————, *Vnutrennaia geometriia vypuklykh poverkhnostei*. Moscow and Leningrad, 1948.

————, "Zamechaniia k osnovam teorii otnositel'nosti." *Vestnik Leningradskogo universiteta*, No. 11 (1953), pp. 95–109.

Allilueva, Svetlana, *Only One Year*. New York and Evanston, Ill., 1969.

————, *Twenty Letters to a Friend*. New York, 1967.

Alperovitz, Gar, *Atomic Diplomacy: Hiroshima and Potsdam; the use of the atomic bomb and the American confrontation with Soviet power*. New York, 1965.

Alpher, R. A., H. Bethe, and G. Gamow, "The Origin of the Chemical Elements." *Physical Review* LXXIII (April 1, 1948), 803–4.

Ambartsumian, V. A., "Contemporary Natural Science and Philosophy," in *Papers for XIV International Wien Philosophical Congress, Section 7*. Moscow, 1968, pp. 41–72.

————, "Kosmos, kosmogoniia, kosmologiia." *Nauka i religiia*, No. 12 (1968), pp. 2–37.

————, "Nekotorye metodologicheskie voprosy kosmogonii," in P. N. Fedoseev, et al., eds., *Filosofskie problemy sovremennogo estestvoznaniia*. Moscow, 1959, pp. 268–90.

————, "Nekotorye osobennosti sovremennogo razvitiia astrofiziki," in *Oktiabr' i nauchnyi progress*, Vol. I. Moscow, 1967, pp. 73–85.

————, "Nekotorye voprosy kosmogonicheskoi nauki." *Kommunist*, No. 8 (1959), pp. 86–96.

————, "Nestatsionarnye ob' 'ekty vo Vselennoi i ikh znachenie dlia kosmogonii," in V. A. Ambartsumian, ed., *Problemy sovremennoi kosmogonii*. Moscow, 1969, pp. 5–18.

————, "Problema proiskhozhdeniia zvezd." *Priroda*, No. 9 (1952), pp. 8–18.

————, "Stars of T Tauri and UV Ceti Types and the Phenomenon of Continuous Emission," in George H. Herbig, ed., *Non-Stable Stars*. International Astronomical Union Symposium No. 3, Cambridge, Eng., 1957, pp. 177–85.

————, "Vvodnyi doklad na simpoziuma po evoliutsii zvezd," in V. A. Ambartsumian, *Nauchnye trudy*, Vol. II. Erevan, 1960, pp. 143–63.

————, ed., *Problemy evoliutsii Vselennoi*. Erevan, 1968.

————, and V. V. Kaziutinskii, "Revoliutsiia v sovremennoi astronomii." *Priroda*, No. 4 (1970), pp. 16–26.

Andrade e Silva, J., and G. Lochak, *Quanta*. New York and Toronto, 1969.

Andreev, N. D., and D. A. Kerimov, "Vozmozhnosti ispolzovaniia kiberneticheskoi tekhniki pri reshenii nekotorykh pravovykh problem," in A. I. Berg, ed., *Kibernetika na sluzhbu kommunizma*. Moscow, 1961, pp. 234–41.

Andriushchenko, M. N., "Otvet tovarishcham V. C. Borshchevu, V. V. Il'inu, F. Z. Rokhline." *Filosofskie nauki*, III, No. 4 (1960), 108–10.

Anokhin, P. K., *A New Conception of the Physiological Architecture of the Conditioned Reflex*. Moscow, 1959.

————, *Biologiia i neirofiziologiia uslovnogo refleksa.* Moscow, 1968.

————, *Ivan Petrovich Pavlov: zhizn', deiatel'nost' i nauchnaia shkola.* Moscow and Leningrad, 1949.

————, "Kibernetika i integrativnaia deiatel'nost' mozga." *Voprosy psikhologii,* No. 3 (1966), pp. 10–32.

————, "Metodologicheskii analiz uzlovykh problem uslovnogo refleksa," in P. N. Fedoseev *et al.*, eds., *Filosofskie voprosy fiziologii vysshei nervnoi deiatel'nosti i psikhologii.* Moscow, 1963, pp. 156–214.

————, "Metodologicheskoe znachenie kiberneticheskikh zakonomernostei," in M. E. Omel'ianovskii, ed., *Materialisticheskaia dialektika i metody estestvennykh nauk.* Moscow, 1968, pp. 547–87.

————, *Obshchie printsipy kompensatsii narushennykh funktsii i ikh fiziologicheskoe obosnovanie.* Moscow, 1956.

————, "Osobennosti afferentnogo apparata uslovnogo refleksa i ikh znachenie dlia psikhologii." *Voprosy psikhologii,* No. 6 (1955), pp. 16–38.

————, *Ot Dekarta do Pavlova: trista let teorii refleksa.* Moscow, 1945.

————, "Problema tsentra i periferii v sovremennoi fiziologii nervnoi deiatel'nosti," in P. K. Anokhin, ed., *Problema tsentra i periferii v fiziologii nervnoi deiatel'nosti.* Gor'kii, 1935, pp. 9–70.

————, "Refleks i funktsional'naia sistema kak faktory fiziologicheskoi integratsii." *Biulleten' Moskovskogo obshchestva ispytatelei prirody,* No. 5 (1949), pp. 130–46.

————, "Uzlovye voprosy v izuchenii vysshei nervnoi deiatel'nosti," in *Problemy vysshei nervnoi deiatel'nosti.* Moscow, 1949, pp. 9–128.

————, *Vnutrenee tormozhenie kak problema fiziologii.* Moscow, 1958.

————, ed., *Problema tsentra i periferii v fiziologii nervnoi deiatel'nosti.* Gor'kii, 1935.

Anokhin, P. K., and E. Strezh, "Izuchenie dinamiki vysshei nervnoi deiatel'nosti. Soobshchenie III. Narushenie aktivnogo vybora v rezul'tate zameny bezuslovnogo stimula." *Fiziologicheskii zhurnal SSSR,* No. 2 (1933), pp. 280–98.

Antipenko, L. G., "Razvitie poniatiia material'nogo ob' 'ekta v fizike mikromira." *Voprosy filosofii,* No. 1 (1967), pp. 104–13.

Antonov, N. P., and A. N. Kochergin, "Priroda myshleniia i problema ego modelirovaniia." *Filosofskie nauki,* No. 2 (1963), pp. 37–47.

Arab-ogly, E. A., "Sotsiologiia i kibernetika." *Voprosy filosofii,* No. 5 (1958), pp. 138–51.

Arkhipov, V. M., "O material'nosti psikhiki i predmete psikhologii." *Sovetskaia pedagogika,* No. 7 (1954), pp. 67–79.

Armstrong, E. F., *Chemistry in the Twentieth Century.* London, 1924.

Arsen'ev, A. S., "Nekotorye metodologicheskie voprosy kosmogonii." *Voprosy filosofii,* No. 3 (1955), pp. 32–44.

————, "O gipoteze rasshireniia metagalaktiki i 'krasnom smeshchenii.' " *Voprosy filosofii,* No. 8 (1958), pp. 187–90.

————, "O sub' 'ektivizme v sovremennoi kosmogonii." *Priroda,* No. 6 (1954), pp. 47–56.

Ashby, W. Ross, *An Introduction to Cybernetics.* London, 1956.

————, "Design for a Brain." *Electronic Engineering,* XX, No. 250, 382–3.

————, *Design for a Brain: the Origin of Adaptive Behaviour.* London, 1960.

Asratian, Ezras A., *I. P. Pavlov, His Life and Work*. Moscow, 1963.

————, "Nekotorye aktual'nye voprosy fiziologii vysshei nervnoi deiatel'nosti i ikh filosofskoe znachenie," in S. A. Petrushevskii, V. N. Kolbanovskii, *et al.*, eds., *Dialekticheskii materializm i sovremennoe estestvoznanie*. Moscow, 1964, pp. 195–233.

Babkin, B. P., *Pavlov: A Biography*. Chicago, 1949.

Bakker, Donald P., "The Philosophical Debate in the U.S.S.R. on the Nature of 'Information.'" Unpublished M.A. thesis, Columbia University, 1966.

Barber, Bernard, "Resistance by Scientists to Scientific Discovery." *Science*, CXXXIV (1961), 596–602.

Bastian, H. C., *The Beginning of Life*. New York, 1872.

Batuev, M. I., "K voprosu o sopriazhenii v benzole." *Zhurnal obshchei khimii*, XXIX (November 1958), 3147–54.

Bauer, Raymond A., *The New Man in Soviet Psychology*. Cambridge, Mass., 1952.

Bauer, Raymond A., Alex Inkeles, and Clyde Kluckhohn, *How the Soviet System Works: Cultural, Psychological and Social Themes*. Cambridge, Mass., 1956.

Bazhenov, L. B., and N. N. Nutsubidze, "K diskussii o probleme beskonechnosti Vselennoi," in *Beskonechnost' i Vselennaia*. Moscow, 1969, pp. 129–36.

Beadle, George and Muriel, *The Language of Life: An Introduction to the Science of Genetics*. Garden City, N. Y., 1966.

Beer, Stafford, "The Irrelevance of Automation." *Cybernetica*, I, No. 4 (1958), 280–95.

Belkin, V. D., "Kibernetika i ekonomika," in A. I. Berg, ed., *Kibernetika na sluzhbu kommunizma*. Moscow, 1961.

Berdyaev, N., *Wahrheit und Lüge des Kommunismus*. Lucerne, 1934.

Berg, A. I., "Nauka velichaishikh vozmozhnostei." *Priroda*, No. 7 (1962), pp. 16–22.

Berg, A. I., and Iu. I. Cherniak, *Informatsiia i upravlenie*. Moscow, 1966.

Berg, A. I., ed., *Kibernetika na sluzhbu kommunizma*. Moscow, 1961.

Berg, A. I., and E. Kol'man, eds., *Vozmozhnoe i nevozmozhnoe v kibernetike*. Moscow, 1964.

Berg, A. I., *et al.*, eds., *Kibernetika, myshlenie, zhizn'*. Moscow, 1964.

Berg, R. L., and N. V. Timofeev-Ressovskii, "Paths of Evolution of the Genotype." *Problems of Cybernetics*, No. 5 (1961) (JPRS 10, 292, January 1962—1016).

Bernal, J. D., *The Origin of Life*. London, 1967.

Bernshtein, N. A., "Novye linii razvitiia v fiziologii i ikh sootnoshenie s kibernetikoi," in *Filosofskie voprosy fiziologii vysshei nervnoi deiatel'nosti i psikhologii*. Moscow, 1963, pp. 299–322.

Beskonechnost' i Vselennaia. Moscow, 1969.

Beutner, R., *Life's Beginning on the Earth*. Baltimore, 1938.

Bialik, B., "Tovarishchi, vy eto ser'ezno?" in A. I. Berg and E. Kol'man, eds., *Vozmozhnoe i nevozmozhnoe v kibernetike*. Moscow, 1964, pp. 77–88.

Bianki, V., and V. Stepanov, "Kto napisala oproverzhenie?" *Komsomol'skaia pravda*, March 16, 1965, p. 2.

Biriukov, B. V., D. I. Koshelevskii, and A. E. Furman, eds., *Dialekticheskii materializm i voprosy estestvoznaniia*. Moscow, 1964.

Biriukov, B. V., and V. S. Tiukhin, "Filosofskie voprosy kibernetiki," in A. I. Berg, *et al.*, eds., *Kibernetika, myshlenie, zhizn'*. Moscow, 1964, pp. 76–108.

Blakeley, Thomas J., *Soviet Theory of Knowledge*. Dordrecht-Holland, 1964.

Blokhintsev, D. I., "Kritika idealisticheskogo ponimaniia kvantovoi teorii." *Uspekhi fizicheskikh nauk*, No. 2 (October 1951), pp. 195–228.

————, "Obsuzhdenie stat'i M. A. Markova." *Voprosy filosofii*, No. 1 (1948), pp. 212–14.

————, *Osnovy kvantovoi mekhaniki*. Moscow and Leningrad, 1949.

————, "Otvet akademiku V. A. Foku." *Voprosy filosofii*, No. 6 (1952), pp. 171–5.

————, *Printsipal'nye voprosy kvantovoi mekhaniki*. Moscow, 1966.

————, "Problemy struktury elementarnykh chastits," in I. V. Kuznetsov and M. E. Omel'ianovskii, eds., *Filosofskie problemy fiziki elementarnykh chastits*. Moscow, 1964, pp. 47–59.

————, *The Philosophy of Quantum Mechanics*. Dordrecht-Holland and New York, 1968.

————, *Vvedenie v kvantovuiu mekhaniku*. Moscow and Leningrad, 1944.

————, "Za leninskoe uchenie o dvizhenii." *Voprosy filosofii*, No. 1 (1952), pp. 181–3.

Boas, Marie, *The Scientific Renaissance 1450–1630*. New York, 1962.

Bochenski, J. M., ed., *Bibliographie der sowjetischen Philosophie*. Dordrecht-Holland, 1959–64, Vols. I–V.

Bogdanov, A. A., *Empiriomonizm; stati po filosofii*, 3 vols. Moscow, 1904–7.

Bogdenko, M. L., and I. E. Zelenin, "Osnovnye problemy istorii kollektivizatsii sel'skogo khoziaistva v sovremennoi sovetskoi istoricheskoi literature," in M. P. Kim *et al.*, eds., *Istoriia sovetskogo krest'ianstva i kolkhoznogo stroitel'stva v SSSR*. Moscow, 1963, pp. 192–222.

Bohm, D., *Causality and Chance in Modern Physics*. New York, 1961.

Bohr, N., *Atomtheorie und Naturbeschreibung*. Berlin, 1931.

————, "Can Quantum Mechanical Description of Physical Reality Be Considered Complete?" *Physical Review*, XLVIII, No. 8 (October 15, 1935), 696–702.

————, "Kvantovaia fizika i filosofiia." *Uspekhi fizicheskikh nauk*, No. 1 (1959).

————, "On the Notion of Causality and Complementarity." *Dialectica*, No. 3–4 (1948), pp. 312–319.

Bok, Bart J., *The Astronomer's Universe*. Cambridge, Eng., 1958.

Bokser, O. Ia., D. A. Mishakhin, and S. S. Poltyrev, "Filosofskie voprosy fiziologii retikuliarnoi formatsii." *Voprosy filosofii*, No. 12 (1961), pp. 80–91.

Bondi, H., *Cosmology*. Cambridge, Eng., 1961.

————, "The Steady-State Theory of the Universe," in H. Bondi, W. B. Bonnor, R. A. Lyttleton, and G. J. Whitrow, *Rival Theories of Cosmology*. London, 1960, pp. 12–21.

Bonnor, William, *The Mystery of the Expanding Universe*. New York, 1964.

Boring, Edwin G., *A History of Experimental Psychology*. New York, 1950, 1957.

————, "Psychology, History of," *Encyclopedia Britannica*, Vol. XVIII. Chicago, 1959, p. 713.

Bovin, A., "Na trassakh gosudarstvennykh lesnykh polos." *Pravda*, May 8, 1950, p. 2.

Brain Mechanisms and Consciousness. Springfield, Ill., 1954.

Braines, S. N., "Nevrokibernetika," in A. I. Berg, ed., *Kibernetika na sluzhbu kommunizma.* Moscow, 1961, pp. 140–52.

De Broglie, Louis, "La Cybernétique." *La Nouvelle Nouvelle Revue Française,* July 1953, pp. 60–85.

———, "La Physique quantique restera-t-elle indéterministe?" *Revue d'Histoire des Sciences et de leurs applications,* V (1952), No. 4 (October–December), 289–311.

Brozek, Josef, "Soviet Contributions to History." *Contemporary Psychology* XIV, No. 8 (1969), 432–4.

Bruner, Jerome, Introduction to L. S. Vygotsky, *Thought and Language.* Cambridge, Mass., 1962, pp. v–x.

Brushlinskii, "Kul'turno-istoricheskaia teoriia myshleniia," in E. V. Shorokhova, ed., *Issledovaniia myshleniia v sovetskoi psikhologii.* Moscow, 1966, pp. 123–74.

Buck, R. C., "On the Logic of General Behavior Systems Theory," in H. Feigl and M. Scriven, eds., *Minnesota Studies in the Philosophy of Science,* Vol. I. Minneapolis, 1956, pp. 223–38.

Budilova, E. A., *Bor'ba materializma i idealizma v russkoi psikhologicheskoi nauke: vtoraia polovina XIX–nachalo XX v.* Moscow, 1960.

Budreiko, N. A., *Filosofskie voprosy khimii.* Moscow, 1970.

Bukharin, Nikolai, *Historical Materialism.* Ann Arbor, Mich., 1969.

Bykhovskii, B., "O metodologicheskikh osnovaniiakh psikhoanaliticheskogo ucheniia Freida." *Pod znamenem marksizma,* November–December, 1923, pp. 158–77.

Caldwell, Oliver J., and Loren R. Graham, "Moscow in May 1963: Education and Cybernetics." Washington, D. C., U.S. Office of Education Bulletin, 1964.

Cannon, Walter F., "The Uniformitarian-Catastrophist Debate." *ISIS,* LI, (March 1960), pp. 38–55.

Chargaff, E., "Nucleic Acids as Carriers of Biological Information," in *Proceedings of the First International Symposium on the Origin of Life on the Earth.* New York, 1959, pp. 297–302.

Chelintsev, G. V., *Ocherki po teorii organicheskoi khimii.* Moscow, 1949.

———, "O teorii khimicheskogo stroeniia A. M. Butlerova i ee novykh uspekhakh." *Zhurnal obshchei khimii,* XXII (February 1952), 350–60.

———, "O vtorom izdanii doklada komissii OKhN AN SSSR 'Sostoianie teorii khimicheskogo stroeniia v organicheskoi khimii.' " *Zhurnal obshchei khimii,* XXVII (August 1957), 2308–10.

Cheremisinov, G. A., "Rech' tovarishcha G. A. Cheremisinova." *Pravda,* February 14, 1964, p. 2.

Chernyshevskii, N. G., *Polnoe sobranie sochinenii.* Moscow, 1939.

Chomsky, Noam, "After Pinkville." *The New York Review of Books,* January 1, 1970.

Chudinov, E. M., "Filosofskie problemy sovremennoi fiziki i astronomii." Moscow, 1969.

————, "Logicheskie aspekty problemy beskonechnosti Vselennoi v reliativistskoi kosmologii," in *Beskonechnost' i Vselennaia.* Moscow, 1969, pp. 181–218.

Chuvashov, B. I., "K voprosu o vozniknovenii zhizni na Zemle." *Voprosy filosofii,* No. 8 (1966), pp. 76–83.

Coleman, William, *Georges Cuvier: Zoologist: A Study in the History of Evolutionary Theory.* Cambridge, Mass., 1964.

Colodny, R. G., ed., *Beyond the Edge of Certainty.* Englewood Cliffs, N. J., 1965.

Comey, David Dinsmore, "Soviet Controversies Over Relativity," in *The State of Soviet Science,* Cambridge, Mass., 1965, pp. 186–99.

Comey, David D, and L. R. Kerschner, "Soviet Publications on Cybernetics" and "Western Translations of Soviet Publications on Cybernetics." *Studies in Soviet Thought,* IV, 2 (February 1964), 142–77.

Crombie, A. C., *Robert Grosseteste and the Origins of Experimental Science, 1100–1700.* Oxford, 1953.

Cuny, Hilaire, *Ivan Pavlov: The Man and His Theories.* New York, 1965.

Danilov, S. N., "A. M. Butlerov—osnovatel' teorii khimicheskogo stroeniia." *Zhurnal obshchei khimii,* XXI (October 1951), 1733–48.

————, "A. M. Butlerov (1828–1886)." *Zhurnal obshchei khimii,* XXIII (October 1953), 1601–12.

Danilov, V. P., *Sozdanie material'no-tekhnicheskikh predposylok kollektivizatsii sel'skogo khoziaistva v SSSR.* Moscow, 1957.

Darlington, C. D., "The Retreat from Science in Soviet Russia," in Conway Zirkle, ed., *Death of a Science in Russia.* Philadelphia, 1949, pp. 67–80.

Deborin, A. M., "Lenin and the Crisis of Contemporary Physics," in *Otchet o deiatel'nosti akademii nauk SSSR za 1929 g.* Leningrad, 1930, Vol. I, Appendix.

Deutsch, Karl, *The Nerves of Government: Models of Political Communication and Control.* New York, 1963.

Dijksterhuis, E. J., *The Mechanization of the World Picture.* London, 1961.

"Diskussii: rasshiriat' i uglubliat' tvorcheskuiu diskussiiu po probleme vida i vidoobrazovaniia." *Botanicheskii zhurnal,* XL, No. 2 (1955), pp. 206–13.

Dobb, Maurice, *Soviet Economic Development Since 1917.* New York, 1966.

Dobrotin, R. B., "Khimicheskaia forma dvizheniia materii." Leningrad, 1967.

Dobzhansky, Theodosius, "N. I. Vavilov, A Martyr of Genetics, 1887–1942." *Journal of Heredity,* XXXVIII, No. 8 (August 1947), 229–30.

————, *The Biological Basis of Human Freedom.* New York, 1956.

Dubinin, N. P., "I. V. Michurin i sovremennaia genetika." *Voprosy filosofii,* No. 6 (1966), pp. 59–70.

Dudintsev, V., "Net, istina." *Kosomol'skaia pravda,* October 23, 1964.

Dühring, E., *Cursus der Philosophie als streng wissenschaftlicher Weltanschauung und Lebensgestaltung.* Leipzig, 1875.

Dunn, L. C., *A Short History of Genetics.* New York, 1965.

Dvoriankin, F. A., *et al.,* eds., *Filosofskie voprosy estestvoznaniia: Filosofskoteoreticheskie voprosy michurinskogo ucheniia.* Moscow, 1958.

Dyshlevyi, P. S., *Lenin i filosofskie problemy reliativistskoi fiziki.* Kiev, 1969.

————, "Prostranstvo-vremennye predstavleniia obshchei teorii otnositel'nosti," in I. Z. Shtokalo *et al.*, eds., *Filosofskie voprosy sovremennoi fiziki.* Kiev, 1964, pp. 57–115.

Dyshlevyi, P S., and A. Z. Petrov, eds., *Filosofskie problemy teorii tiagoteniia Einshteina i relativistskoi kosmologii.* Kiev, 1965.

Dyson, Freeman J., "Mathematics in the Physical Sciences." *Scientific American,* September 1964, pp. 129–46.

Eccles, John Carew, *The Neurophysiological Basis of Mind.* Oxford, 1952.

Egorshin, V. P., "Estestvoznanie i klassovaia bor'ba." *Pod znamenem marksizma,* No. 6 (1926), pp. 108–36.

Einstein, Albert, *Sobranie nauchnykh trudov,* Vols. I–IV. Moscow, 1965–7.

Einstein, A., B. Podolsky, and N. Rosen, "Can Quantum Mechanical Description of Physical Reality be Considered Complete?" *Physical Review,* XLVII, No. 10 (May 15, 1935), 777–80.

Einstein, A., and E. G. Straus, "The Influence of the Expansion of Space on the Gravitation Fields Surrounding the Individual Stars." *Reviews of Modern Physics,* XVII, No. 2–3 (1945), 120–4.

Emme, A. M., "Neskol'ko zamechanii po voprosu o protsesse vozniknoveniia zhizni." *Voprosy filosofii,* No. 1 (1956), pp. 155–8.

Engels, Friedrich, *Anti-Dühring.* Berlin, 1962.

————, *Anti-Dühring.* Moscow, 1959.

————, *Dialectics of Nature.* New York, 1940.

————, *Dialectics of Nature.* Moscow, 1954.

————, *Ludwig Feuerbach.* New York, 1935.

Eroshkin, I. G., *Psikhologiia i fiziologiia vysshei nervnoi deiatel'nosti.* Leningrad, 1958.

Fedoseev, P. N., *et al.*, eds., *Filosofskie problemy sovremennogo estestvoznaniia.* Moscow, 1959.

————, *Filosofskie voprosy fiziologii vysshei nervnoi deiatel'nosti i psikhologii.* Moscow, 1963.

Feigl, Herbert, "The 'Mental' and the 'Physical,'" in Herbert Feigl, Michael Scriven, and Grover Maxwell, eds., *Minnesota Studies in the Philosophy of Science: Concepts, Theories, and the Mind-Body Problem,* Vol. II. Minneapolis, 1958, pp. 370–497.

Fesenkov, V. G., "Evoliutsiia i vozniknovenie zvezd v sovremennoi galaktike." *Voprosy filosofii,* No. 4 (1952), pp. 110–24.

————, "O kosmogonicheskoi gipoteze akademika O. Iu. Shmidta i o sovremennom sostoianii kosmogonicheskoi problemy." *Voprosy filosofii,* No. 4 (1951), pp. 134–47.

————, "Problema kosmogonii solnechnoi sistemy." *Priroda,* No. 4 (1940), pp. 7–15.

Feyerabend, Paul K., "Dialectical Materialism and Quantum Theory." *Slavic Review,* XXV (September 1966), 414–17.

————, "Problems of Microphysics," in R. G. Colodny, ed., *Frontiers of Science and Philosophy.* Pittsburgh, 1962, pp. 189–284.

Filipchenko, Iu. A., "Spornye voprosy evgeniki." *Vestnik kommunisticheskoi akademii,* No. 20 (1927), pp. 212–54.

Filosofiia estestvoznaniia. Moscow, 1966.

Filosofskie voprosy estestvoznaniia: Geologo-geograficheskie nauki. Moscow, 1960.

Filosofskie voprosy fiziki i khimii. Sverdlovsk, 1959.

Findlay, Alexander, *A Hundred Years of Chemistry.* New York, 1937.

Fischer, George, ed., *Science and Ideology in Soviet Society.* New York, 1967.

Fock, V. A., "Comments." *Slavic Review,* September 1966, pp. 411–13.

———, "K diskussii po voprosam fiziki." *Pod znamenem marksizma,* No. 1 (1938), pp. 149–59.

———, "K stat'e Nikol'skogo 'Printsipy kvantovoi mekhaniki.'" *Uspekhi fizicheskikh nauk,* XVII, No. 4 (1937), 554.

———, "Kritika vzgliadov Bora na kvantovuiu mekhaniku." *Uspekhi fizicheskikh nauk,* XLV, No. 1 (September 1951), 3.

———, "La physique quantique et les idéalisations classiques." *Dialectica,* No. 3–4 (1965), p. 223–45.

———, "Les principes mécaniques de Galilée et la théorie d'Einstein." *Atti del convegno sulla relativita generale: problemi dell'energia e onde gravitazionali,* Florence, 1965, pp. 1–12.

———, "Nil's Bor v moei zhizni." *Nauka i chelovechestvo 1963,* Vol. II, Moscow, 1963, pp. 518–19.

———, "Ob interpretatsii kvantovoi mekhaniki," in P. N. Fedoseev *et al.,* eds., *Filosofskie problemy sovremennogo estestvoznaniia.* Moscow, 1959, pp. 212–36.

———, "O roli printsipov otnositel'nosti i ekvivalentnosti v teorii tiagotenii Einshteina." *Voprosy filosofii,* No. 12 (1961), pp. 45–52.

———, "Osnovnye zakony fiziki v svete dialekticheskogo materializma." *Vestnik Leningradskogo universiteta,* No. 4 (1949), pp. 34–47.

———, "O tak nazyvaemykh ansambliakh v kvantovoi mekhanike." *Voprosy filosofii,* No. 4 (1952), pp. 170–4.

———, "Poniatiia odnorodnosti, kovariantnosti i otnositel'nosti v teorii prostranstva i vremeni." *Voprosy filosofii,* No. 4 (1955), pp. 131–5.

———, "Protiv nevezhestvennoi kritiki sovremennoi fizicheskikh teorii." *Voprosy filosofii,* No. 1 (1953), pp. 168–74.

———, *The Theory of Space, Time and Gravitation.* New York, 1959.

Fock, V. A., A. Einstein, B. Podolsky, N. Rosen, and N. Bohr, "Mozhno li schitat', chto kvantovomekhanicheskoe opisanie fizicheskoi real'nosti iavliaetsia polnym?" *Uspekhi fizicheskikh nauk,* XVI, No. 4 (1936), 436–57.

Fox, Sidney W., *The Origins of Prebiological Systems and of Their Molecular Matrices.* New York and London, 1965.

Fraenkel-Conrat, H., and B. Singer, "The Infective Nucleic Acid from Tobacco Mosaic Virus," in *Proceedings of the First International Symposium on the Origin of Life on the Earth.* New York, 1959, pp. 303–6.

Frank, Philipp, *Between Physics and Philosophy.* Cambridge, 1941.

———, "Einstein, Mach and Logical Positivism," in Edward H. Madden, ed., *The Structure of Scientific Thought.* Boston, 1960, pp. 84–93.

———, *Foundations of Physics.* Chicago, 1946.

French, Sidney J., *The Drama of Chemistry.* New York, 1937.

Freud, Sigmund, *Civilization and Its Discontents.* New York, 1961.

Fridman (Friedmann), A. A., *Mir kak prostranstvo i vremia*. Moscow, 1965.
Frolov, I. T., *Genetika i dialektika*. Moscow, 1968.
Galileo Galilei, *Dialogue Concerning the Two Chief World Systems—Ptolemaic and Copernican*, trans. Stillman Drake. Berkeley and Los Angeles, 1962.
Gamow, George, *The Creation of the Universe*. New York, 1955.
Gastev, Iu. A., "O metodologicheskikh voprosakh ratsionalizatsii obucheniia," in A. I. Berg *et al.*, eds., *Kibernetika, myshlenie, zhizn'*. Moscow, 1964, pp. 459–72.
Georgiev, F. I., "Problema chuvstvennogo i ratsional'nogo v poznanii." *Voprosy filosofii*, No. 1 (1955), pp. 28–41.
Gerasimovich, B. P., *Vselennaia pri svete teorii otnositel'nosti*. Khar'kov, 1925.
Gevorkian, O. S., "Bor'ba osnovnykh filosofskikh napravlenii v sovremennoi kosmogonii," in P. N. Gapochka, B. M. Kedrov, and M. N. Maslina, *Voprosy dialekticheskogo i istoricheskogo materializma*. Moscow, 1956, pp. 244–82.
Gillispie, Charles Coulston, "Lamarck and Darwin in the History of Science," in Bentley Glass, Owsei Temkin, and William L. Straus, Jr., eds., *Forerunners of Darwin: 1745–1859*. Baltimore, 1959, pp. 265–91.
———, *The Edge of Objectivity*. Princeton, 1960.
Glass, Bentley, "Maupertuis, Pioneer of Genetics and Evolution," in Bentley Glass, Owsei Temkin, William L. Straus, Jr., eds., *Forerunners of Darwin: 1745–1859*. Baltimore, 1959, pp. 51–83.
Glushchenko, I. E., "Protiv idealizma i metafiziki v nauke o nasledstvennosti." *Voprosy filosofii*, No. 2 (1948), pp. 133–47.
Glushkov, V. M., A. A. Dorodnitsyn, and N. Fedorenko, "O nekotorykh problemakh kibernetiki." *Izvestiia*, September 6, 1964, p. 4.
Gödel, K., "An Example of a New Type of Cosmological Solutions of Einstein's Field Equations of Gravitation." *Reviews of Modern Physics*, XXI (July 1949), 447–50.
Gonseth, F., "Remarque sur l'idée de complémentarité." *Dialectica*, II, No. 3–4 (1948), 413–20.
Graham, Loren R., "A Soviet Marxist View of Structural Chemistry." *ISIS*, LV (March 1964), 20–31.
———, "Cybernetics," in George Fischer, ed., *Science and Ideology in Soviet Society*. New York, 1967, pp. 83–106.
———, "Cybernetics in the Soviet Union," in *The State of Soviet Science*. Cambridge, Mass., 1965, pp. 3–18.
———, "Quantum Mechanics and Dialectical Materialism." *Slavic Review*, September 1966, pp. 381–410.
———, "Reorganization of the Academy of Sciences," in Peter H. Juviler and Henry W. Morton, *Soviet Policy-Making*. New York, 1967, pp. 133–61.
———, *The Soviet Academy of Sciences and the Communist Party, 1927–1932*. Princeton, 1967.
Gray, Jeffrey A., "Attention, Consciousness and Voluntary Control of Behaviour in Soviet Psychology: Philosophical Roots and Research Branches," in Neil O'Connor, ed., *Present-day Russian Psychology*. Oxford, 1966, pp. 1–38.
Greniewski, Henry, *Cybernetics Without Mathematics*. Warsaw, 1960.
Griaznov, B. S., "Kibernetika v svete filosofii." *Voprosy filosofii*, No. 3 (1965), pp. 161–5.

Grossman, Sebastian Peter, *A Textbook of Physiological Psychology*. New York, 1967.

Grünbaum, Adolf, "Geometry, Chronometry, and Empiricism," in Herbert Feigl and Grover Maxwell, eds., *Minnesota Studies in the Philosophy of Science*. Vol. III, *Scientific Explanation, Space, and Time*. Minneapolis, 1962, pp. 405–526.

Guilbaud, G. T., *What Is Cybernetics?* New York, 1960.

Gulyi, M. F., *et al.*, eds., *Filosofskie voprosy sovremennoi biologii*. Kiev, 1962.

Gur'ev, N. I., "Sovetskaia kosmologiia na sluzhbe ofitsial'nogo mirovozzreniia." *Vestnik instituta po izucheniiu istorii i kul'tury SSSR*, No. 1 (1951), pp. 30–7.

Gurgenidze, G. S., and A. P. Ogurtsov, "Aktual'nye problemy dialektiki." *Voprosy filosofii*, No. 10 (1965), pp. 130–64.

Haldane, J. B. S., *The Inequality of Man*. London, 1932.

———, *The Marxist Philosophy and the Sciences*. New York, 1939.

———, "The Origin of Life," in J. D. Bernal, *The Origin of Life*. London, 1967, pp. 242–9.

Hanson, Norwood Russell, "Contra-Equivalence: A Defense of the Originality of Copernicus." *ISIS*, LV (September 1964), 308–25.

———, "Five Cautions for the Copenhagen Critics." *Philosophy of Science*, XXVI (1959), 325–37.

Hartley, R. V. L., "Transmission of Information." *The Bell System Technical Journal*, VII (1928), 535–63.

Hegel, G. W. F., *Encyclopedia of Philosophy*, trans. G. A. Mueller. New York, 1959.

———, *Science of Logic*, trans. W. H. Johnston and L. G. Struthers. London, 1951.

Hempel, Carl G., "General System Theory and the Unity of Science." *Human Biology*, XXIII (1951), 313–27.

Hesse, Mary, "Models in Physics." *The British Journal for the Philosophy of Science*, November 1953, pp. 198–214.

———, "Models and Analogy in Science," in *The Encyclopedia of Philosophy*, Vol. V. New York, 1967, pp. 354–9.

Houllevigue, Louis, *L'Évolution des Sciences*. Paris, 1914.

Howard, Walter L., "Luther Burbank: A Victim of Hero Worship." *Chronica Botanica*, IX, No. 5–6 (1945).

Hoyle, Fred, *The Nature of the Universe*. New York, 1960.

Huckel, Walter, *Structural Chemistry of Inorganic Compounds*, trans. L. H. Long. New York, 1950.

Hudson, P. S., and R. H. Richens, *The New Genetics in the Soviet Union*. Cambridge, Eng., 1946.

Hunsberger, I. M.; "Theoretical Chemistry in Russia." *Journal of Chemical Education*, XXXI (October 1954), 504–14.

Huxley, Julian, *Heredity East and West (Lysenko and World Science)*. London, 1949.

Iaroshevskii, M. G., *Istoriia psikhologii*. Moscow, 1966.

———, "Kibernetika—'nauka' mrakobesov." *Literaturnaia gazeta*, April 5, 1952, p. 4.

Ignatov, A. I., "Mezhdunarodnyi simpozium po proiskhozhdeniiu zhizni na zemle." *Voprosy filosofii*, No. 1 (1958), pp. 152–7.

Il'in, M. M., "Filogenez pokrytosemennykh s pozitsii michurinskoi biologii." *Botanicheskii zhurnal*, No. 1 (1953), pp. 97–118.

Il'in, V. A., V. N. Kolbanovskii, and E. Kol'man, eds., *Filosofskie voprosy kibernetiki*. Moscow, 1960.

Istoriia vsesoiuznoi kommunisticheskoi partii (Bol'shevikov): *Kratkii kurs*. Moscow, 1938.

————. Moscow, 1945.

Jacques, J., "Boutlerov, Couper et la Société chimique de Paris (Notes pour servir à l'histoire des théories de la structure chimique)." *Bulletin de la Société chimique de France*, 1953, pp. 528–30.

Jammer, Max, *The Conceptual Development of Quantum Mechanics*. New York, 1966.

Jeffreys, Sir Harold, *Scientific Inference*. Cambridge, Eng., 1957.

Joravsky, David, *Soviet Marxism and Natural Science, 1917–1932*. New York, 1961.

————, "The First Stage of Michurinism," in J. S. Curtis, ed., *Essays in Russian and Soviet History*. New York, 1963, pp. 120–32.

————, *The Lysenko Affair*. Cambridge, Mass., 1970.

————, "The Vavilov Brothers." *Slavic Review*, September 1965, pp. 381–94.

Jordan, P., *Die Herkunft der Sterne*. Stuttgart, 1947

————, *Physics of the Twentieth Century*. New York, 1944.

Jordan, Z. A., *Marxism: An Historical and Critical Study*. New York, 1961.

Juviler, Peter H., and Henry W. Morton, eds., *Soviet Policy-Making*. New York, 1967.

"K diskussii po stat'e M. A. Markova." *Voprosy filosofii*, No. 1 (1948), pp. 225–32.

"K 70-letiiu so dnia rozhdeniia I. V. Stalina." *Zhurnal fizicheskoi khimii*, No. 12 (1949), pp. 1385–6.

Kabachnik, M. I., "Orientatsiia v benzolnom kol'tse." *Uspekhi khimii*, XVII (January 1948), 96–131.

Kaganov, B. M., and G. V. Platonov, eds., *Problema prichinnosti v sovremennoi biologii*. Moscow, 1961.

Kalandiia, A. A., "Otvet na stat'iu G. V. Tsitsishvili po povodu raboty A. A. Kalandiia 'Raschet molekuliarnykh ob' 'emov neorganicheskikh soedinenii tipa AnBmOs.' " *Zhurnal obshchei khimii*, January 1955, pp. 193–6.

Kal'sin, F. F., *Osnovnye voprosy teorii poznaniia*. Gor'kii, 1957.

Kapp, Reginald, *Towards a Unified Cosmology*. London, 1960.

Karasev, M., and V. Nozdrev, "O knige M. E. Omel'ianovskogo 'V. I. Lenin i fizika XX veka.' " *Voprosy filosofii*, No. 1 (1949), pp. 338–42.

Kard, P. G., "O teorii otnositel'nosti." *Voprosy filosofii*, No. 5 (1952), pp. 240–7.

Karpov, M. M., "O filosofskikh vzgliadakh A. Einshteina." *Voprosy filosofii*, No. 1 (1951), pp. 130–41.

Kazanskii, B. A., and G. V. Bykov, "K voprosu o sostoianii teorii khimicheskogo stroeniia v organicheskoi khimii." *Zhurnal obshchei khimii*, January 1953, 168–76.

Kaziutinskii, V. V., "Astronomiia i dialektika (k 100-letiiu so dnia rozhdeniia V. I. Lenina)," in *Astronomicheskii kalendar' 1970*. Moscow, 1969, pp. 138–70.

Kedrov, B. M., "O sootnoshenii form dvizheniia materii v prirode," in P. N. Fedoseev *et al.*, eds., *Filosofskie problemy sovremennogo estestvoznaniia*. Moscow, 1959, pp. 137–211.

———, *Tri aspekta atomistiki:* Vol. I, *Paradoks Gibbsa;* Vol. II, *Uchenie Dal'tona;* Vol. III, *Zakon Mendeleeva*. Moscow, 1969.

———, and D. N. Trifonov, *Zakon periodichnosti i khimicheskie elementy*. Moscow, 1969.

Kekulé, August, "Sur la constitution des substances aromatiques." *Bulletin de la Société Chimique*, III (1865), 98–110.

———, "Untersuchungen über aromatische Verbindungen." *Justus Leibig's Annalen der Chimie*, CXXXVII (1866), 129–96.

Keosian, John, *The Origin of Life*. New York, 1964, 1968.

Khil'mi, G. F., "O vozmozhnosti zakhvata v probleme trekh tel." *Doklady akademii nauk SSSR*, LXII, No. 1 (1948), 39–42.

Khrushchev, N. S., "Rech' tov. N. S. Khrushcheva na soveshchanii rabotnikov sel'skogo khoziaistva Gor'kovskoi, Arzamasskoi, Kirovskoi oblastei, Mariiskoi, Mordovskoi i Chuvashskoi ASSR 8 Aprelia 1957 goda v gorode Gor'kom." *Pravda*, April 10, 1957, p. 2.

Khudiakova, E., "Cybernetics and Medical Diagnosis." *The Soviet Review*, February 1961, pp. 58–61.

Kitov, A. I., "Kibernetika i ekonomika," in A. I. Berg, ed., *Kibernetika na sluzhbu kommunizma*. Moscow, 1961.

Klickstein, Herbert, "A Cumulative Review of Bibliographies of the Published Writings by Albert Einstein." *Journal of the Albert Einstein Medical Center*, July 1962, pp. 141–9.

Knuniants, I. L., B. M. Kedrov, and L. Ia. Bliakher, "A Book That Does Not Deserve High Appraisal." *Pravda*, January 24, 1965, p. 2. From *Current Digest of the Soviet Press*, XVII, No. 4 (1965), 7.

Knuniants, I., and L. Zubkov, "Shkoly v nauke," *Literaturnaia gazeta*, January 11, 1955, p. 1.

Kobushkin, P. K., "Nekotorye filosofskie problemy reliativistskoi kosmologii," in I. Z. Shtokalo *et al.*, eds., *Filosofskie voprosy sovremennoi fiziki*. Kiev, 1964, pp. 116–48.

Kol', A., "Prikladnaia botanika ili leninskoe obnovlenie zemli." *Ekonomicheskaia zhizn'*, January 29, 1931, p. 2.

Kolbanovskii, V. N., "O nekotorykh spornykh voprosakh kibernetiki," in V. A. Il'in *et al.*, eds., *Filosofskie voprosy kibernetiki*. Moscow, 1960, pp. 227–61.

———, "Spornye voprosy genetiki i selektsii (obshchii obzor soveshchaniia)." *Pod znamenem marksizma*, No. 11 (1939).

———, "Za marksistskoe osveshchenie voprosov psikhologii (Ob uchebnike S. L. Rubinshteina 'Osnovy obshchei psikhologii)." *Bol'shevik*, No. 17 (1947), pp. 50–7.

———, *Dialekticheskii materializm i sovremennoe estestvoznanie*. Moscow, 1964.

———, *et al.*, eds., *Nekotorye problemy dialekticheskogo i istoricheskogo materializma*. Moscow, 1958.

Koldanov, V. Ia., "Nekotorye itogi i vyvody po polezashchitnomu lesorazvedeniiu za istekshie piat' let." *Lesnoe khoziaistvo*, No. 3 (1954), pp. 10–18.

Kol'man, E., "Chto takoe kibernetika?" *Voprosy filosofii*, No. 4 (1955), pp. 148–59.

————, "Chuvstvo mery," in A. I. Berg and E. Kol'man, eds., *Vozmozhnoe i nevozmozhnoe v kibernetike.* Moscow, 1964, pp. 51–64.

————, "Kibernetika stavit voprosy." *Nauka i zhizn'*, XXVIII, No. 5 (1961), 43–5.

————, "O filosofskikh i sotsial'nykh problemakh kibernetiki," in V. A. Il'in, V. N. Kolbanovskii, and E. Kol'man, eds., *Filosofskie voprosy kibernetiki.* Moscow, 1960, pp. 86–109.

————, "Sovremennaia fizika v poiskakh dal'neishei fundamental'noi teorii." *Voprosy filosofii*, No. 2 (1965), pp. 111–22.

Kolmogorov, A., "Avtomaty i zhizn'," in A. I. Berg and E. Kol'man, eds., *Vozmozhnoe i nevozmozhnoe v kibernetike.* Moscow, 1964, pp. 10–29.

Konikova, A. S., and M. G. Kritsman, "K voprosu o nachal'noi forme proiavleniia zhizni." *Voprosy filosofii*, No. 1 (1954), pp. 210–16.

————, "Zhivoi belok v svete sovremennykh issledovanii biokhimii." *Voprosy filosofii*, No. 1 (1953), pp. 143–50.

Kopnin, P. V., "Razvitie kategorii dialekticheskogo materializma—vazhneishee uslovie ukrepleniia soiuza filosofii i estestvoznaniia," in P. S. Dyshlevyi and A. Z. Petrov, eds., *Filosofskie problemy teorii tiagoteniia einshteina i reliativistskoi kosmologii.* Kiev, 1965, pp. 5–14.

Körner, S., ed., *Observation and Interpretation in the Philosophy of Physics.* New York, 1957.

Kornilov, K., "Dialekticheskii metod v psikhologii." *Pod znamenem marksizma*, January 1924, pp. 107–13.

Koyré, Alexandre, *Études Galiléennes.* Paris, 1966.

————, *From the Closed World to the Infinite Universe.* Baltimore, 1957.

————, *Metaphysics and Measurement.* London, 1968.

Kozhevnikov, A. V., "O nekotorykh usloviiakh vozniknoveniia zhizni na Zemle." *Voprosy filosofii*, No. 1 (1956), pp. 149–52.

Kratkii slovar' po filosofii. Moscow, 1966.

Krogdahl, W. S., *The Astronomical Universe.* New York, 1962.

Kropotkin, P. A., *Mutual Aid, a Factor of Evolution.* London, 1902.

Krylov, N. S., *Raboty po obosnovaniiu statisticheskoi fiziki.* Moscow and Leningrad, 1950.

Kucher, R. V., "Metodologicheskie problemy razvitii teorii v khimii." *Voprosy filosofii*, No. 6 (1969), pp. 78–85.

Kudriavtseva, N. L., "K voprosu o vozniknovenii zhizni." *Voprosy filosofii*, No. 2 (1954), pp. 218–21.

Kuhn, Thomas, *The Copernican Revolution: Planetary Astronomy in the Development of Western Thought.* Cambridge, Mass., 1957.

————, "The Function of Dogma in Scientific Research," in A. C. Crombie, ed., *Scientific Change.* New York, 1963, pp. 347–69.

————, *The Structure of Scientific Revolutions.* Chicago and London, 1962.

Kukarkin, B. V., and A. G. Masevich, "Sovetskie astronomy na VIII s' 'ezde mezhdunarodnogo astronomicheskogo soiuza v Rime." *Voprosy filosofii*, No. 1 (1953), pp. 222–30.

Kupalov, P. S., "Uchenie o reflekse i reflektornoi deiatel'nosti i perspektivy ego razvitiia," in *Filosofskie voprosy fiziologii vysshei nervnoi deiatel'nosti i psikhologii.* Moscow, 1963, pp. 106–55.

Kuroedov, Aleksandr S., "Rol' sotsialisticheskoi sel'sko-khoziaistvennoi praktiki v razvitii michurinskoi biologii." Unpublished dissertation for the degree of *kandidat*, Moscow State University, 1952.

Kursanov, D. N., *et al.*, "K voprosu o sovremennom sostoianii teorii khimicheskogo stroeniia." *Uspekhi khimii*, XIX, No. 5 (1950), 529–44.

Kursanov, D. N., and V. N. Setkina, "O vzaimodeistvii chetvertichnikh solei ammoniia s prostimi efirami." *Doklady akademii nauk SSSR*, LXV (April 21, 1949), 847–55.

Kursanov, G. A., "Dialekticheskii materializm o prostranstve i vremeni." *Voprosy filosofii*, No. 3 (1950), pp. 173–91.

———, "K kriticheskoi otsenke teorii otnositel'nosti." *Voprosy filosofii*, No. 1 (1952), pp. 169–74.

———, "K otsenke filosofskikh vzgliadov A. Einshteina na prirodu geometricheskikh poniatii," in I. V. Kuznetsov and M. E. Omel'ianovskii, eds., *Filosofskie voprosy sovremennoi fiziki*. Moscow, 1959, pp. 393–410.

———, "O mirovozzrencheskom znachenii dostizhenii sovremennoi astronomii." *Voprosy filosofii*, No. 3 (1960), pp. 61–74.

Kuznetsov, B. G., *Einstein*. Moscow, 1962.

Kuznetsov, I. V., *Liudi russkoi nauki: ocherki o vydaiushchikhsia deiateliakh estestvoznaniia i tekhniki, matematika, mekhanika, astronomiia, fizika, khimiia.* Moscow, 1961.

———, "Sovetskaia fizika i dialekticheskoi materializm," in A. A. Maksimov, ed., *Filosofskie voprosy sovremennoi fiziki*. Moscow, 1952, pp. 31–86.

Kuznetsov, I. V., and M. E. Omel'ianovskii, eds., *Filosofskie problemy fiziki elementarnykh chastits*. Moscow, 1964.

———, *Filosofskie voprosy sovremennoi fiziki*. Moscow, 1958.

———, *Filosofskie voprosy sovremennoi fiziki*. Moscow, 1959.

Kuznetsov, V. I., *Evoliutsiia predstavlenii ob osnovnykh zakonakh khimii*. Moscow, 1967.

———, *Razvitie ucheniia o katalize*. Moscow, 1964.

Landau, L. D., and E. M. Lifshits, *Statisticheskaia fizika, klassicheskaia i kvantovaia*. Moscow, 1951.

Langdon-Davies, John, *Russia Puts the Clock Back*. London, 1949.

Lebedev, M. P., "Materiia i soznanie." *Voprosy filosofii*, No. 5 (1956), pp. 70–84.

Leicester, Henry M., "Alexander Mikhailovich Butlerov." *Journal of Chemical Education*, XVII (May 1940), 208–9.

Lenin, V. I., *Filosofskie tetradi*. Moscow, 1938.

———, *Filosofskie tetradi*. Moscow, 1965.

———, *Materialism and Empirio-Criticism: Critical Comments on a Reactionary Philosophy*. Moscow, 1952.

———, *Polnoe sobranie sochinenii*. Moscow, 1958–66, 55 vols. plus index.

Lepeshinskaia, O. B., *Kletka: ee zhizn' i proiskhozhdenie.* Moscow, 1952.

Levy, Edwin Jr., "Interpretations of Quantum Theory and Soviet Thought." Unpublished Ph.D. dissertation, Indiana University, 1969.

Liapunov, A. A., and A. G. Malenkov, "Logical Analysis of Hereditary Information." *Problems of Cybernetics,* No. 8 (1962) (JPRS 21, 048, November 1963 —19,390).

Lichtheim, George, *Marxism: An Historical and Critical Study.* New York, 1961.

Lovejoy, Arthur O., *The Great Chain of Being: A Study of the History of an Idea.* Cambridge, Mass., 1936.

Luria, A. R., *Higher Cortical Functions in Man.* New York, 1966.

Lyell, Charles, *Principles of Geology,* 12th ed., Vol. I. London, 1875.

Lysenko, T. D., *Agrobiologiia.* Moscow, 1949.

————, *Agrobiologiia.* Moscow, 1952.

————, *Agrobiology.* Moscow, 1954.

————, "Gnezdovaia kul'tura lesa." *Ogonek,* No. 10 (March 1949), pp. 6–7.

————, *Heredity and its Variability,* trans. Th. Dobzhansky. New York, 1946.

————, "Iarovizatsiia—eto milliony pudov dobavochnogo urozhaia." *Izvestiia,* February 15, 1935, p. 4.

————, "Interesnye raboty po zhivotnovodstvu v Gorkakh Leninskikh." *Pravda,* July 17, 1957, pp. 5–6.

————, *Izbrannye sochineniia.* Moscow, 1958.

————, "Novoe v nauke o biologicheskom vide." *Botanicheskii zhurnal,* XXXVIII, No. 1 (1953), 44–56.

————, "Novoe v nauke o biologicheskom vide." *Pravda,* November 3, 1950, p. 2.

————, "Novoe v nauke o biologicheskom vide—o rabotakh deisvitel'nogo chlena meditsinskikh nauk SSSR O. B. Lepeshinskoi." Moscow, 1952.

————, "Obnovlennye semena: beseda s akademikom T. D. Lysenko." *Sotsialisticheskoe zemledelie,* September 16, 1935, p. 1.

————, "On vdokhnovlial nas na bor'bu za dal'neishii rastsvet nauki." *Izvestiia,* September 1, 1948.

————, "Po povodu stat'i akademika N. I. Vavilova." *Iarovizatsiia,* No. 1 (1939).

————, "Posev polezashchitnykh lesnykh polos gnezdovym sposobom." Moscow, 1950.

————, "Rech' tovarishcha T. D. Lysenko." *Pravda,* February 26, 1956, p. 9.

————, "Shire primeniat' v nechernozemnoi polose organomineral'nye smesy." *Izvestiia,* April 27, 1957, p. 2.

————, "Teoreticheskie osnovy napravlennogo izmeneniia nasledstvennosti sel'skokhoziaistvennykh rastenii." *Pravda,* January 29, 1963, pp. 3–4.

————, "Teoreticheskie uspekhi agronomicheskoi biologii." *Izvestiia,* December 8, 1957, p. 5.

————, "Vazhnye rezervy kolkhozov i sovkhozov." *Pravda,* March 14, 1959, pp. 2–3.

————, "Vliianie termicheskogo faktora na prodolzhitel'nost' faz razvitiia rastenii." *Trudy azerbaidzhanskoi tsentral'noi opytno-selektsionnoi stantsii,* No. 3, 1928, pp. 1–169.

——, ed., *I. V. Michurin: Sochineniia v chetyrekh tomakh*, Vols. I–IV. Moscow, 1948.

McCrea, W. H., "Cosmology After Half a Century." *Science*, CLX (June 21, 1968), 1295–9.

McTaggart, J. M. E., *A Commentary on Hegel's Logic*. Cambridge, 1910.

McTighe, Thomas P., "Galileo's 'Platonism': a reconsideration," in Ernan McMullin, ed., *Galileo: Man of Science*. New York, 1968, pp. 365–87.

McVittie, G. C., *Fact and Theory in Cosmology*. New York, 1961.

Mach, Ernst, *The Analysis of Sensations and the Relation of the Physical to the Psychical*. New York, 1959.

Makarchenko, A. F., P. G. Kostiuk, and A. E. Khil'chenko, "Filosofskoe i estestvenno-nauchnoe znachenie ucheniia I. P. Pavlova o vysshei nervnoi deiatel'nosti v fiziologii," in M. F. Gulyi *et al.*, eds., *Filosofskie voprosy sovremennoi biologii*. Kiev, 1962, pp. 294–325.

Maksimov, A. A., "Bor'ba za materializm v sovremennoi fizike." *Voprosy filosofii*, No. 1 (1953), pp. 175–95.

——, "Marksistskii filosofskii materializm i sovremennaia fizika." *Voprosy filosofii*, No. 3 (1948), pp. 105–24.

——, "Ob odnom filosofskom kentavre." *Literaturnaia gazeta*, April 10, 1948, p. 3.

——, "Obsuzhdenie stat'i M. A. Markova (Okonchanie)." *Voprosy filosofii*, No. 3 (1948), pp. 222–8.

——, "O filosofskikh vozzreniiakh akad. V. F. Mitkevich i putiakh razvitiia sovetskoi fiziki." *Pod znamenem marksizma*, No. 7 (1937), pp. 25–55.

——, "Protiv reaktsionnogo einshteinianstva v fizike." *Krasnyi flot*, June 23, 1952.

——, "Teoriia otnositel'nosti i materializm." *Pod znamenem marksizma*, No. 4–5 (1923), pp. 140–56.

Maksimov, A. A., *et al.*, eds., *Filosofskie voprosy sovremennoi fiziki*. Moscow, 1952.

Maksimov, N. A., "Fiziologicheskie faktory, opredeliaiushchie dlinu vegetatsionnogo perioda." *Trudy po prikladnoi botanike, genetike i selektsii*, XX (1929), 169–212.

Maksimov, N. A., and M. A. Krotkina, "Issledovaniia nad posledstviem ponizhennoi temperatury na dlinu vegetatsionnogo perioda." *Trudy po prikladnoi botanike, genetike i selektsii*, XXIII (1929–30), vyp. 2, 427–73.

Mangelsdorf, Paul C., "Hybrid Corn," in L. C. Dunn, ed., *Genetics in the 20th Century*. New York, 1951, pp. 555–71.

Markov, M. A., "O prirode fizicheskogo znaniia." *Voprosy filosofii*, No. 2 (1947), pp. 140–76.

Markov, N. V., "Znachenie geometrii Lobachevskogo dlia fiziki," in A. A. Maksimov *et al.*, eds., *Filosofskie voprosy sovremennoi fiziki*. Moscow, 1952, pp. 186–215.

Marx, Karl, *Capital*. New York, 1967, Vol. I.

——, *Matematicheskie rukopisi*. Moscow, 1968.

Marx, Karl, and Friedrich Engels, *Izbrannye pis'ma*. Moscow, 1947.

——, *Karl Marx, Friedrich Engels, Werke*. Berlin, 1956–67, 40 vols. in 41.

————, K. Marks i F. Engel's: iz rannikh proizvedenii. Moscow, 1956.

————, Selected Works. Moscow, 1958.

"Materialist" (pseud.), "Komu sluzhit kibernetika." *Voprosy filosofii,* No. 5 (1953), pp. 210–19.

Materialy XXII s' 'ezda KPSS. Moscow, 1962.

Medvedev, Zhores A., "Errors in the Reproduction of Nucleic Acid and Proteins and their Biological Significance." *Problems of Cybernetics,* No. 9 (1963) (JPRS 21,448, November 1963—19,750).

————, *The Rise and Fall of T. D. Lysenko.* New York and London, 1969.

Medvedev, Zhores, and V. Kirpichnikov, "Perspektivy sovetskoi genetiki." *Neva,* No. 3 (1963), pp. 165–75.

Mehl, Lucien, "Les sciences juridiques devant l'automation." *Cybernetica,* No. 1, 2 (1960), pp. 22–40, 142–70.

Meliukhin, S. T., "O filosofskom ponimanii beskonechnosti prostranstva i vremeni," in P. S. Dyshlevyi and A. Z. Petrov, eds., *Filosofskie problemy teorii tiagoteniia einshteina i reliativistskoi kosmologii.* Kiev, 1965, pp. 292–7.

————, *Problema konechnogo i beskonechnogo.* Moscow, 1958.

Mikulak, Maxim W., "Relativity Theory and Soviet Communist Philosophy (1922–1960)." Unpublished Ph.D. dissertation, Columbia University, 1965.

————, "Trofim Denisovich Lysenko," in George W. Simmonds, ed., *Soviet Leaders.* New York, 1967, pp. 248–59.

Minkowski, Hermann, *Gesammelte Abhandlungen von Hermann Minkowski.* Leipzig and Berlin, 1911.

Molodtsov, V. S., and A. Ia. Il'in, *Metodologicheskie problemy sovremennoi nauki.* Moscow, 1964.

Mora, Peter T., "The Folly of Probability," in Sidney W. Fox, ed., *The Origins of Prebiological Systems and of Their Molecular Matrices.* New York and London, 1965, pp. 39–64.

Morgenthau, Hans J., "Arguing About the Cold War." *Encounter,* May 1967, pp. 37–41.

Muller, H. J., "Lenin's Doctrines in Relation to Genetics," in *Pamiati V. I. Lenina: sbornik statei k desiatiletiiu so dnia smerti, 1924–1934.* Moscow and Leningrad, 1934, pp. 565–79.

————, "Nauka proshlogo i nastoiashchego i chem ona obiazana Marksu," in *Marksizm i estestvoznanie.* Moscow, 1933, pp. 204–7.

————, *Out of the Night: A Biologist's View of the Future.* New York, 1935.

Muller-Markus, Siegfried, *Einstein und die Sowjetphilosophie,* 2 vols. Dordrecht-Holland, 1960, 1966.

Musabaeva, N., *Kibernetika i kategoriia prichinnosti.* Alma-Ata, 1965.

Naan, G. I., "Gravitatsiia i beskonechnost'," in P. S. Dyshlevyi and A. Z. Petrov, eds., *Filosofskie problemy teorii tiagoteniia Einshteina i relativistskoi kosmologii.* Kiev, 1965, pp. 268–85.

————, "K voprosu o printsipe otnositel'nosti v fizike." *Voprosy filosofii,* No. 2 (1951), pp. 57–77.

————, "O beskonechnosti vselennoi." *Voprosy filosofii,* No. 6 (1961), pp. 93–105.

————, "O sovremennom sostoianii kosmologicheskoi nauki." *Voprosy kosmogonii,* VI (Moscow, 1958), pp. 277–329.

————, "Poniatie beskonechnosti v matematike i kosmologii," in *Beskonechnost' i Vselennaia*. Moscow, 1969, pp.7–77.

————, "Sovremennyi 'fizicheskii' idealizm v SShA i Anglii na sluzhbe popovshchiny i reaktsii." *Voprosy filosofii*, No. 2 (1948), pp. 287–308.

Nagel, Ernest, "The Causal Character of Modern Physical Theory," in S. W. Baron, E. Nagel, and K. S. Pinson, eds., *Freedom and Reason: Studies in Philosophy and Jewish Culture*. Glencoe, Ill., 1951, pp. 244–68.

————, *The Structure of Science: Problems in the Logic of Scientific Explanation*. New York, 1961.

"Na uchenom sovete instituta organicheskoi khimii AN SSSR," *Izvestiia akademii nauk: otdelenie khimicheskikh nauk*, No. 4 (1950), pp. 438–44.

"Nauka i ideologiia v sovetskom obshchestve." *Posev*, No. 4 (1968), pp. 58–9.

"Nauka i zhizn'." *Kommunist*, No. 5 (1954), pp. 3–13.

Nekrasova, I. M., *Leninskii plan elektrifikatsii strany i ego osushchestvlenie v 1921–1931 gg*. Moscow, 1960.

Nesmeianov, A. N., " 'Contact Bonds' and the 'New Structural Theory.' " *Bulletin of the Academy of Sciences of the U.S.S.R.: Division of Chemical Sciences* (Consultants Bureau), No. 1 (1952), pp. 215–21.

Nesmeianov, A. N., R. Kh. Freidlina, and A. E. Borisova, "O kvazikompleksnikh metalloorganicheskikh soedineniiakh." *Izvestiia akademii nauk: otdelenie khimicheskikh nauk*, Jubilee Bulletin (1945), pp. 239–250.

Nesmeianov, A. N., and M. I. Kabachnik, "Dvoistvennaia reaktsionnaia sposobnost' i tautomeriia." *Zhurnal obshchei khimii*, XXV (January 1955), pp. 41–87.

Nett, Roger, and Stanley A. Hetzler, *An Introduction to Electronic Data Processing*. New York, 1963.

Newton, Isaac, *Principia*, trans. Florian Cajori. Berkeley and Los Angeles, 1966.

Nicolle, Jacques, *Louis Pasteur: A Master of Scientific Enquiry*. London, 1961.

Nikol'skii, K. V., *Kvantovye protsessy*. Moscow and Leningrad, 1940.

————, "Otvet V. A. Foku." *Uspekhi fizicheskikh nauk*, XVII, No. 4 (1937), 555.

————, "Printsipy kvantovoi mekhaniki." *Uspekhi fizicheskikh nauk*, XVI, No. 5 (1936), 537–65.

"Norbert Viner v redaktsii nashego zhurnala." *Voprosy filosofii*, No. 9 (1960), pp. 164–8.

Nordenskiöld, Erik, *The History of Biology*. New York, 1935.

Novik, I. B., *Kibernetika: filosofskie i sotsilogicheskie problemy*. Moscow, 1963.

————, "Kibernetika i razvitie sovremennogo nauchnogo poznaniia." *Priroda*, LII, No. 10 (1963), 3–11.

Novikov, I. D., "O povedenii sfericheski-simmetrichnykh raspredelenii mass v obshchei teorii otnositel'nosti." *Vestnik MGU*, Seriia III, No. 6 (1962), pp. 66–72.

Novinskii, I. I., "Poniatie sviazi v dialekticheskom materializme i voprosy biologii." Unpublished dissertation for the degree of doctor of philosophical sciences, Moscow State University, 1963.

Nozhin, N. D., "Nasha nauka i uchenye." *Knizhnyi vestnik*, April 15, 1866, pp. 175–8.

————, "Po povodu statei 'Russkago Slova' o nevol' 'nichestvo." *Iskra,* No. 8 (1865), pp. 115–17.

Nudel'man, Z. N., "O probleme belka." *Voprosy filosofii,* No. 2 (1954), pp. 221–6.

"Ob agrobiologicheskoi nauke i lozhnykh positsiiakh 'Botanicheskogo zhurnala.' " *Pravda,* December 14, 1958, pp. 3–4.

Obshchie osnovy fiziologii truda. Moscow, 1935.

Obshchie voprosy kosmogonii: trudy shestogo soveshchaniia po voprosam kosmogonii. Moscow, 1959.

"O fluktuatsionnoi gipoteze Bol'tsmana." *Voprosy filosofii,* No. 4 (1959), pp. 138–40.

Ol'shanskii, M., "Protiv fal'sifikatsii v biologicheskoi nauke." *Sel'skaia zhizn',* August 18, 1963, pp. 2–3.

Omel'ianovskii, M. E., Borot'ba materiializmu proty idealizmu v suchasnii fizytsi. Kiev, 1947.

————, "Dialekticheskii materializm i tak nazyvaemyi printsip dopolnitel'nosti Bora," in A. A. Maksimov *et al.,* eds., *Filosofskie voprosy sovremennoi fiziki.* Moscow, 1952, pp. 396–431.

————, "Fal'sifikatory nauki: ob idealizme v sovremennoi fizike." *Voprosy filosofii,* No. 3 (1948), pp. 143–62.

————, "Filosofskaia evoliutsiia kopengagenskoi shkoly fizikov." *Vestnik akademii nauk SSSR,* No. 9 (1962), pp. 86–96.

————, "Filosofskie aspekty teorii izmereniia," in *Materialisticheskaia dialektika i metody estestvennykh nauk.* Moscow, 1968.

————, *Filosofskie voprosy kvantovoi mekhaniki.* Moscow, 1956.

————, "Lenin i filosofskie problemy sovremennoi fiziki." Moscow, 1968.

————, "Lenin o prichinnosti i kvantovaia mekhanika." *Vestnik akademii nauk SSSR,* No. 4 (1958), pp. 3–12.

————, "Lenin o prostranstve i vremeni i teoriia otnositel'nosti Einshteina." *Izvestiia akademii nauk SSSR (Seriia istorii i filosofii),* No. 4 (1946), pp. 297–308.

————, "Problema elementarnosti chastits v kvantovo fizike." *Filosofskie problemy fiziki elementarnykh chastits,* Moscow, 1963, pp. 60–73.

————, "The Concept of Dialectical Contradiction in Quantum Physics," in *Philosophy, Science and Man: The Soviet Delegation Reports for the XIII World Congress of Philosophy.* Moscow, 1963.

————, *V. I. Lenin i fizika XX veka.* Moscow, 1947.

————, "Zadachi razrabotki problemy 'Dialekticheskii materializm i sovremennoe estestvoznanie.' " *Vestnik akademii nauk SSSR,* No. 10 (1956), pp. 3–11.

————, ed., *Lenin i sovremennaia nauka.* Moscow, 1970.

————, ed., *Materialisticheskaia dialektika i metody estestvennykh nauk.* Moscow, 1968.

————, *et al.,* eds., *Filosofskie voprosy sovremennoi fiziki.* Kiev, 1956.

Omel'ianovskii, M. E., and I. V. Kuznetsov, eds., *Dialektika v naukakh o nezhivoi prirode.* Moscow, 1964.

"On 'Nonresonance' Between East and West." *Chemical and Engineering News,* XXX (June 16, 1952), 2474.

Oparin, A. I., "K voprosu o vozniknovenii zhizni." *Voprosy filosofii*, No. 1 (1953), pp. 138–42.
————, *Life, its nature, origin and development*. Edinburgh, 1961.
————, *Proiskhozhdenie zhizny*. Moscow, 1924.
————, *The Origin and Initial Development of Life* (NASA TTF–488). Washington, D. C. 1968.
————, *The Origin of Life*, trans. S. Morgulis. New York, 1938, 1953.
————, "The Origin of Life," in J. D. Bernal, *The Origin of Life*. London, 1967, pp. 199–234. Translation by Ann Synge of A. I. Oparin, *Proiskhozhdenie zhizny*. Moscow, 1924.
————, *The Origin of Life on the Earth*, 3rd revised and enlarged ed., trans. Ann Synge. London, 1957.
————, *Zhizn', ee priroda, proiskhozhdenie i razvitie*. Moscow, 1960.
————, "Znachenie trudov tovarishcha I. V. Stalina po voprosam iazikoznaniia dlia razvitiia sovetskoi biologicheskoi nauki." Moscow, 1951.
Oparin, A. I., and V. G. Fesenkov, *Life in the Universe*. New York, 1961.
————, *Zhizn' vo Vselennoi*. Moscow, 1956.
Oparin, A. I., *et al.*, eds., *Proceedings of the 1st International Symposium on The Origin of Life on the Earth*. London, Pergamon Press, 1959.
O polozhenii v biologicheskoi nauke. Stenograficheskii otchet sessii vsesoiuznoi akademii sel'sko-khoziastvennykh nauk imeni V. I. Lenina, 31 iiulia–7 avgusta 1948 g. Moscow, 1948.
Oppenheimer, J. Robert, *The Open Mind*. New York, 1955.
Oppenheimer, J. Robert, and G. M. Volkoff, "On Massive Neutron Cores." *Physical Review*, LV (February 15, 1939), 374–81.
"O rezul'tatakh proverki deiatel'nosti bazy 'gorki leninskie.'" *Vestnik akademii nauk*, No. 11 (1965).
Papaleksi, N. D., *et al.*, *Kurs fiziki*. Moscow and Leningrad, 1948.
Partington, J. R., *A Short History of Chemistry*. London, 1948.
Partridge, R. B., "The Primeval Fireball Today." *American Scientist*, LVII, No. 1 (1969), 37–74.
Pauling, Linus, "Teoriia rezonansa v khimii." *Zhurnal vsesoiuznogo khimicheskogo obshchestva im. D. I. Mendeleeva*, No. 4 (1962), pp. 462–7.
————, *The Nature of the Chemical Bond*. Ithaca, N. Y., 1960.
Pavlov, Todor, "Avtomaty, zhizn', soznanie." *Nauchnye doklady vysshei shkoly: filosofskie nauki*, No. 1 (1963), pp. 49–57.
Perfil'ev, V. V., "O knige M. E. Omel'ianovskogo 'V. I. Lenin i fizika XX veka.'" *Voprosy filosofii*, No. 1 (1948), pp. 311–12.
Petrovskii, A. V., *Istoriia sovetskoi psikhologii*. Moscow, 1967.
Petrushenko, L. A., "Filosofskoe znachenie poniatiia 'obratnaia sviaz' v kibernetike." *Vestnik leningradskogo universiteta: seriia ekonomiki, filosofii, i prava*, XVII (1960), 76–86.
Petrushevskii, S. A., V. N. Kolbanovskii, *et al.*, eds., *Dialekticheskii materializm i sovremennoe estestvoznanie*. Moscow, 1964.
Piaget, Jean, "Comments on Vygotsky's Critical Remarks." Cambridge, Mass., 1962.
Pinter, Ferents, "Aktual'nye voprosy vzaimootnosheniia marksistskoi filosofii i

genetiki." Dissertation for the degree of *kandidat* of the philosophical sciences, Moscow State University, 1965.

Pisarev, D. I., *Selected Philosophical, Social and Political Essays.* Moscow, 1958.

Planck, Max, *Scientific Autobiography and Other Papers.* New York, 1949.

Platonov, G. V., "Dogmy starye i dogmy novye." *Oktiabr',* No. 8 (1965), pp. 149–65.

Plekhanov, G. V., *Izbrannye filosofskie proizvedeniia,* Vol. I. Moscow, 1956.

———, *Protiv filosofskogo revizionizma.* Moscow, 1935.

Plotkin, I. P., "O fluktuatsionnoi gipoteze Bol'tsmana." *Voprosy filosofii,* No. 4 (1959), pp. 138–40.

Pod znamenem marksizma, No. 11 (1939).

Polianskii, D. S., "O dal'neishem razvitii sel'skogo khoziaistva: doklad predsedatelia soveta ministrov RSFSR tovarishcha D. S. Polianskogo." *Pravda,* December 23, 1959, pp. 1–2.

Polianskii, V. I., and Iu. I. Polianskii, *Sovremennye problemy evoliutsionnoi teorii.* Leningrad, 1967.

"Postanovlenie prezidiuma akademii nauk SSSR ot 26 avgusta 1948 g. po voprosu o sostoianii i zadachakh biologicheskoi nauki v institutakh i uchrezhdeniiakh akademii nauk SSSR." *Pravda,* August 27, 1948, p. 1.

Potkov, L. L., "Obsuzhdenie raboty M. A. Markova 'O mikromire.' " *Voprosy filosofii,* No. 2 (1947), pp. 381–2.

———, "Teoriia khimicheskogo stroeniia A. M. Butlerova." *Zhurnal fizicheskoi khimii,* XXXVI, No. 3 (1962), 417–28.

Prilezhaeva, E. N., Ia. K. Syrkin, and M. V. Volkenshtein, "Ramaneffekt galoidoproizvodnikh etilena i elektronni rezonansa." *Zhurnal fizicheskoi khimii,* XIV (1940), 1396–418.

"Privetstvie akademiku T. D. Lysenko v sviazi s ego 50-letiem." *Voprosy filosofii,* No. 2 (1948), p. 412.

Problemy metodologii sistemnogo issledovaniia. Moscow, 1970.

Programma kommunisticheskoi partii Sovetskogo Soiuza. Moscow, 1961.

Purvis, O. N., "The Physiological Analysis of Vernalization," in W. H. Ruhland, ed., *Encyclopedia of Plant Physiology,* Vol. XVI, Berlin, 1961, pp. 76–117.

Putnam, H., "A Philosopher Looks at Quantum Mechanics," in R. G. Colodny, ed., *Beyond the Edge of Certainty.* Englewood Cliffs, N. J., 1965, pp. 75–101.

Quine, William Van Orman, *From a Logical Point of View: Logico-Philosophical Essays.* New York, 1963.

Raychaudhuri, Amalkumar, "Relativistic Cosmology I." *Physical Review,* IIC (May 15, 1955), 1123–6.

Reichenbach, Hans, "The Principle of Anomaly in Quantum Mechanics." *Dialectica,* II, No. 3–4 (1948), 337–50.

———, *The Philosophy of Space and Time.* New York, 1958.

Reutov, O. A., "O nekotorikh voprosakh teorii organicheskoi khimii." *Zhurnal obshchei khimii,* XXI (January 1951), 186–99.

Robb, Alfred A., *The Absolute Relations of Time and Space.* Cambridge, 1921.

Rogers, James Allen, "Charles Darwin and Russian Scientists." *Russian Review,* XIX (1960), 371–83.

————, "Darwinism, Scientism and Nihilism." *Russian Review*, XIX (1960), 10–23.

Ronchi, Vasco, *Histoire de la lumière*, trans. Jean Taton. Paris, 1956.

Rosenblith, Walter A., "Postscript," in Ivan M. Sechenov, *Reflexes of the Brain*. Cambridge, Mass., 1965.

Rosenblueth, A., N. Wiener, and J. Bigelow, "Behavior, purpose and teleology." *Philosophy of Science*, January 1943, pp. 18–24.

Rubashevskii, A. A., *Filosofskoe znachenie teoreticheskogo nasledstva I. V. Michurina*. Moscow, 1949.

Rubinshtein, S. L., *Bytie i soznanie: o meste psikhicheskogo vo vseobshchei vzaimosviazi iavlenii material'nogo mira*. Moscow, 1957.

————, "Filosofiia i psikhologiia." *Voprosy filosofii*, No. 1 (1957), pp. 114–27.

————, *O myshlenii i putiakh ego issledovaniia*. Moscow, 1958.

————, *Osnovy obshchei psikhologii*. Moscow, 1946.

————, *Osnovy psikhologii*. Moscow, 1935.

————, *Printsipy i puti razvitiia psikhologii*. Moscow, 1959.

————, "Problemy psikhologii v trudakh Karla Marksa." *Sovetskaia psikhotekhnika*, No. 1 (1934), pp. 3–20.

————, "Rech'." *Voprosy filosofii*, No. 1 (1947), pp. 420–7.

————, "Uchenie I. P. Pavlova i nekotorye voprosy perestroiki psikhologii." *Voprosy filosofii*, No. 3 (1952), pp. 197–210.

————, "Voprosy psikhologii myshleniia i printsip determinizma." *Voprosy filosofii*, No. 5 (1957), pp. 101–13.

Rukhkian, A. A., "Ob opisannom S. K. Karapetianom sluchae porozhdeniia leshchiny grabom." *Botanicheskii zhurnal*, XXXVIII, No. 6 (1953), 885–91.

Russell, H. N., *The Solar System and Its Origin*. New York, 1935.

Rutkevich, M. N., *Praktika—osnova poznaniia i kriterii istiny*. Moscow, 1952.

Rybnikov, K. A., *et al.*, eds., *Filosofskie voprosy estestvoznaniia: Nekotorye filosofskoteoreticheskie voprosy fiziki, matematiki i khimii*. Moscow, 1959.

Safronov, V. S., "Problema proiskhozhdeniia zemli i planet: soveshchanie po voprosam kosmogonii solnechnoi sistemy." *Vestnik akademii nauk SSSR*, No. 10 (1951), pp. 94–102.

Sakharov, Andrei D., "Thoughts on Progress, Peaceful Coexistence and Intellectual Freedom," *The New York Times*, July 22, 1968.

Sartre, Jean-Paul, *Search for a Method*. New York, 1963.

Schapiro, Leonard, *The Communist Party of the Soviet Union*. New York, 1959.

Schlesinger, Arthur, Jr., "Origins of the Cold War." *Foreign Affairs*, October, 1967, pp. 22–52.

Schmalhausen, I. I., "Evolution in the Light of Cybernetics." *Problems of Cybernetics*, No. 13 (1965) (JPRS 31,567, October 1967—16,376).

————, "Fundamentals of the Evolutionary Process in the Light of Cybernetics." *Problems of Cybernetics*, No. 4 (1960) (JPRS 7402, March 1961—4686).

————, "Hereditary Information and Its Transformations." *Transactions of the Academy of Sciences, USSR, Biological Sciences Section*, 120, 1 (1958), American Institute of Biological Sciences, Washington, D. C.

————, "Natural Selection and Information." *News of the Academy of Science, USSR*, Biology Series, Vol. 25–1 (1960) (JPRS 2815, September 1960—14,090).

————, "Predstavleniia o tselom v sovremennoi biologii." *Voprosy filosofii*, No. 2 (1947), pp. 177–83.

Schmidt, O. Iu., *A Theory of Earth's Origin.* Moscow, 1958.

————, "O vozmozhnosti zakhvata v nebesnoi mekhanike." *Doklady akademii nauk SSSR*, LVII (1947), 213–16.

————, "Problema proiskhozhdeniia zemli i planet." *Voprosy filosofii*, No. 4, (1951), pp. 120–33.

————, *Proiskhozhdenie zemli i planet.* Moscow, 1949.

Schmidt, O. Iu., and G. F. Khil'mi, "Problema zakhvata v zadache o trekh telakh." *Uspekhi matematicheskikh nauk,* No. 26 (1948), pp. 157–9.

Schrödinger, Erwin, "Die gegenwärtige Situation in der Quantenmechanik." *Die Naturwissenschaften*, XXIII, No. 48 (November 29, 1935), 807–12.

————, *What Is Life? Other Scientific Essays.* Garden City, N. Y., 1956.

Scientific Session on the Physiological Teaching of Academician I. P. Pavlov. Moscow, 1951.

Sedov, E. A., "K voprosu o sootnoshenii entropii informatsionnykh protsessov i fizicheskoi entropii." *Voprosy filosofii,* No. 1 (1965), pp. 135–45.

Selsam, Howard, and Harry Martel, eds., *Reader in Marxist Philosophy.* New York, 1963.

Semenov, A. A., "Ob itogakh obsuzhdeniia filosofskikh vozzrenii akademika L. I. Mandel'shtama." *Voprosy filosofii,* No. 3 (1953), pp. 199–206.

Semenov, N. N., "Marksistsko-leninskaia filosofiia i voprosy estestvoznaniia." *Vestnik akademii nauk*, No. 8 (1968), pp. 24–40.

————, "Nauka ne terpit sub' 'ektivizma." *Nauka i zhizn'*, No. 4 (1965), pp. 38–43.

Semkovskii, S. Iu., *Dialekticheskii materializm i printsip otnositel'nosti.* Moscow and Leningrad, 1926.

Serebrovskii, A. S., "Antropogenetika." *Mediko-biologicheskii zhurnal,* No. 5 (1929), p. 18.

"Sergei Leonidovich Rubinshtein." *Voprosy filosofii,* No. 2 (1960), pp. 179–80.

Severnyi, A. B., and V. V. Sobolev, "Viktor Amazaspovich Ambartsumian (K shestidesiatiletiiu so dnia rozhdeniia)." *Uspekhi fizicheskikh nauk,* September 1968, pp. 181–3.

Shakhparanov, M. I., *Dialekticheskii materializm i nekotorye problemy fiziki i khimii.* Moscow, 1958.

Shaliutin, S. M., "O kibernetike i sfere ee primeneniia," in V. A. Il'in, V. N. Kolbanovskii, and E. Kol'man, eds., *Filosofskie voprosy kibernetiki.* Moscow, 1960, pp. 6–85.

Shannon, Claude, and Warren Weaver, *The Mathematical Theory of Communication.* Urbana, Ill., 1949.

Shapere, Dudley, "The Structure of Scientific Revolutions." *The Philosophical Review,* LXXIII (July 1964), 383–94.

Shapley, Harlow, *The Inner Metagalaxy.* New Haven, 1957.

Shatskii, A. L., "K voprosu o summe temperatur, kak sel'skokhoziaistvennoklimaticheskom indekse." *Trudy po sel'sko-khoziaistvennoi meteorologii,* XXI, No. 6 (1930), pp. 261–3.

Shelinskii, G. I., *Khimicheskaia sviaz' i izuchenie ee v srednei shkole.* Moscow, 1969.

Sherrington, Charles S., "Brain," in *Encyclopedia Britannica*, Vol. IV. Chicago, 1959.

Shirokov, M. F., "Filosofskie voprosy teorii otnositel'nosti," in V. N. Kolbanov-skii *et al.*, eds., *Dialekticheskii materializm i sovremennoe estestvoznanie*. Moscow, 1964, pp. 58–80.

———, "O materialisticheskoi sushchnosti teorii otnositel'nosti," in I. V. Kuznetsov and M. E. Omel'ianovskii, eds., *Filosofskie voprosy sovremennoi fiziki*. Moscow, 1959, pp. 324–69.

———, "O preimushchestvennykh sistemakh otscheta v n'iutonovskoi mekhanike i teorii otnositel'nosti." *Voprosy filosofii*, No. 3 (1952), pp. 128–39.

Shklovskii, I. S., and Carl Sagan, *Intelligent Life in the Universe*. San Francisco, 1966.

Shorokhova, E. V., ed., *Issledovaniia myshleniia v sovetskoi psikhologii*. Moscow, 1966.

Shteinman, R. Ia., "O reaktsionnoi roli idealizma v fizike." *Voprosy filosofii*, No. 3 (1948), pp. 163–73.

———, "Za materialisticheskuiu teoriiu bystrykh dvizhenii," in A. A. Maksimov *et al.*, eds., *Filosofskie voprosy sovremennoi fiziki*. Moscow, 1952, pp. 234–98.

Shtern, V., *Erkenntnistheoretische Probleme der Modernen Physik*. Berlin, 1952.

———, "K voprosu o filosofskoi storone teorii otnositel'nosti." *Voprosy filosofii*, No. 1 (1952), pp. 175–81.

Shtokalo, I. Z., *et al.*, eds., *Filosofskie voprosy sovremennoi fiziki*. Kiev, 1964.

Shvyrev, V., "Materialisticheskaia dialektika i problemy issledovaniia nauchnogo poznaniia." *Kommunist*, XVII (November 1968), 40–51.

Sinnott, E. W., L. C. Dunn, and Th. Dobzhansky, *Principles of Genetics*. New York, 1950.

Skabichevskii, A. P., "Problema vozniknoveniia zhizni na Zemle i teoriia akad. A. I. Oparina." *Voprosy filosofii*, No. 2 (1953), pp. 150–5.

Sobolev, S., "Da, eto vpolne ser'ezno," in A. I. Berg and E. Kol'man, eds., *Vozmozhnoe i nevozmozhnoe v kibernetike*. Moscow, 1964, pp. 77–88.

Sobolev, S. L., A. I. Kitov, and A. A. Liapunov, "Osnovnye cherty kibernetiki." *Voprosy filosofii*, No. 4 (1955), pp. 136–48.

Sokolov, B., "Ob organizatsii proizvodstva gibridnykh semian kukuruzy." *Izvestiia*, February 2, 1956, p. 2.

Sonneborn, Tracy M. "H. J. Muller, Crusader for Human Betterment." *Science*, November 15, 1968, p. 772.

———, "H. J. Muller: Tribute to a Colleague." *The Review* (Alumni Association, Indiana University), Fall 1968, pp. 19–23.

Sostoianie teorii khimicheskogo stroeniia: vesesoiuznoe soveshchanie 11–14 iiunia 1951 g.: stenograficheskii otchet. Moscow, 1952.

Sostoianie teorii khimicheskogo stroeniia v organicheskoi khimii. Moscow, 1954.

"Soveshchanie rabotnikov sel'skogo khoziaistva oblastei iugo-vostoka." *Pravda*, March 19, 1955, p. 2.

"Soviets Blast Pauling, Repudiate Resonance Theory." *Chemical and Engineering News*, XXIX (September 10, 1951), 3713.

Stalin, I. V., *Anarkhizm ili sotsializm?* Moscow, 1950.

———, "Ekonomicheskie problemy sotsializma v SSSR." *Bol'shevik*, No. 18 (1952), pp. 1–50.

Stalin, Joseph, *Marxism and Linguistics*. New York, 1951.

Staniukovich, K. P., S. M. Kolesnikov, and V. M. Moskovkin, *Problemy teorii prostranstva, vremeni i materii*. Moscow, 1968.

Stanley, Wendell H., "On the Nature of Viruses, Genes and Life," in *Proceedings of the First International Symposium on the Origin of Life on the Earth*. New York, 1959, pp. 313–21.

Stoletov, V. N., "Printsipy ucheniia I. V. Michurina." *Voprosy filosofii*, No. 2 (1948), pp. 148–70.

Storchak, L. I., "Za materialisticheskoe osveshchenie osnov kvantovoi mekhaniki." *Voprosy filosofii*, No. 3 (1951), pp. 202–5.

———, "Znachenie idei Lobachevskogo v razvitii predstavlenii o prostranstve i vremeni." *Voprosy filosofii*, No. 1 (1951), pp. 142–8.

Struminskii, V., "Marksizm v sovremennoi psikhologii." *Pod znamenem marksizma*, March 1926, pp. 207–33.

Struve, Otto, *The Universe*. Cambridge, Mass., 1962.

———, *Stellar Evolution*. Princeton, 1950.

Struve, Otto, and Velta Zebergs, *Astronomy of the 20th Century*. New York, 1962.

Stukov, A. P., and S. A. Iakushev, "O belke kak nositele zhizni." *Voprosy filosofii*, No. 2 (1953), pp. 139–49.

Sturtevant, A. H., *A History of Genetics*. New York, 1965.

Sukhachev, V. N., "O vnutrividovykh i mezhvidovykh vzaimootnosheniiakh sredi rastenii." *Botanicheskii zhurnal*, XXXVIII, No. 1 (1953).

Sukhachev, V. N., and N. D. Ivanov, "K voprosam vzaimootnoshenii organizmov i teorii estestvennogo otbora." *Zhurnal obshchei biologii*, XV, No. 4 (July–August 1954), 303–19.

Sullivan, Walter, "Moon Soil Indicates Clue to Life Origin," *The New York Times*, January 7, 1970.

———, "The Death and Rebirth of a Science," in Harrison E. Salisbury, ed., *The Soviet Union:The Fifty Years*. New York, 1967, pp. 276–98.

Sviderskii, V. I., *Filosofskoe znachenie prostranstvenno-vremennykh predstavlenii v fizike*. Leningrad, 1956.

———, "O dialektiko-materialisticheskom ponimanii konechnogo i beskonechnogo," in P. S. Dyshlevyi and A. Z. Petrov, eds., *Filosofskie problemy teorii tiagoteniia Einshteina i relativistskoi kosmologii*. Kiev, 1965, pp. 261–8.

Swanson, James M., "The Bolshevization of Scientific Societies in the Soviet Union; An Historical Analysis of the Character, Function and Legal Position of Scientific and Scientific Technical Societies in the USSR 1929–1936." Unpublished dissertation, Indiana University, Bloomington, Ind., 1967.

Syrkin, Ia. K., "Sovremennoe sostoianie problemy valentnosti." *Uspekhi khimii*, XXVIII (August 1959), 903–20.

Syrkin, Ia. K., and M. E. Diatkina, *Khimicheskaia svaiz' i stroenie molekul*. Moscow, 1946.

———, *The Structure of Molecules and the Chemical Bond*. New York, 1950.

Sysoev, A. F., "Samoobnovlenie belka i svoistvo razdrazhimosti—vazhneishie zakonomernosti zhiznennykh iavlenii." *Voprosy filosofii*, No. 1 (1956), pp. 152–5.

T. D. Lysenko. Moscow, 1953.

Tables of Interatomic Distances and Configuration in Molecules and Ions. London, 1958.

Takach, Laslo, "K voprosu o vozniknovenii zhizni." *Voprosy filosofii,* No. 3 (1955), pp. 147–50.

Targul'ian, O. M., ed., *Spornye voprosy genetiki i selektsii: raboty IV sessii akademii 19–27 dekabriia 1936 goda.* Moscow and Leningrad, 1937.

Tatevskii, V. M., and M. I. Shakhparonov, "Ob odnoi makhistskoi teorii v khimii i ee propagandistakh." *Voprosy filosofii,* No. 3 (1949), pp. 176–92.

Terletskii, Ia. P., "Obsuzhdenie stat'i M. A. Markova." *Voprosy filosofii,* No. 3 (1948), pp. 228–31.

The Situation in Biological Science: Proceedings of the Lenin Academy of Agricultural Sciences of the U.S.S.R., July 31–August 7, 1948, Complete Stenographic Report. New York, 1949.

Tiapkin, A. A., "K razvitiiu statisticheskoi interpretatsii kvantovoi mekhaniki na osnove sovmestnogo koordinatno-impul'snogo predstavleniia," in *Filosofskie voprosy kvantovoi fiziki.* Moscow, 1970, pp. 139–80.

Timofeev-Ressovskii, N. V., "Some Problems of Radiation Biogeocenology." *Problems of Cybernetics,* No. 12 (1964) (JPRS 31,214, September 1965—14,503).

Tolman, Richard C., "Static Solutions of Einstein's Field Equations for Spheres of Fluid." *Physical Review,* LV (February 15, 1939), 364–73.

Tomovic, Rajko, "Limitations of Cybernetics." *Cybernetica,* II, No. 3 (1959), 195–8.

"Torzhestvo sovetskoi biologicheskoi nauki." *Voprosy filosofii,* No. 2 (1948), pp. 121–32.

Treadgold, Donald W., *Twentieth Century Russia.* Chicago, 1964.

Trincher, K. S., *Biology and Information—Elements of Biological Thermodynamics.* Moscow, 1964 (JPRS 28,969, April 1965—6475).

Tsitsishvili, G. B., "Ob oshibakh A. A. Kalandiia v stat'e 'Raschet molekuliarnikh ob''emov neorganicheskikh soedinenii tipa AnBmOs' i ego popytakh ukrepit porognuiu kontseptsiiu rezonansa." *Zhurnal obshchei khimii,* XXII (December 1952), 2240–5.

Tucker, Robert C., *The Soviet Political Mind: Studies in Stalinism and Post-Stalin Change.* New York, 1963.

Tugarinov, V. P., "Sootnoshenie kategorii dialekticheskogo materializma." *Voprosy filosofii,* No. 3 (1956), pp. 151–60.

Tugarinov, V. P., and L. E. Maistrov, "Protiv idealizma v matematicheskoi logike." *Voprosy filosofii,* No. 3 (1950), pp. 331–9.

Turing, A. M., "Computing Machinery and Intelligence." *Mind,* LIX (October 1950), 433–60.

Turkevich, John, *Soviet Men of Science.* Princeton, 1963.

Uemov, A. I., *Veshchi, svoistva, otnosheniia.* Moscow, 1963.

"Ukaz prezidiuma verkhovnogo soveta SSSR." *Pravda,* September 29, 1958, p. 1.

Ukraintsev, B. S., "O vozmozhnostiakh kibernetiki v svete svoistva otobrazheniia materii," in V. A. Il'in, V. N. Kolbanovskii, and E. Kol'man, eds., *Filosofskie voprosy kibernetiki.* Moscow, 1960, pp. 110–33.

Ursul, A. D., "O prirode informatsii." *Voprosy filosofii,* No. 3 (1965), pp. 131–40.

————, *Priroda informatsii.* Moscow, 1968.

Uznadze, D. N., *Psikhologicheskie issledovaniia.* Moscow, 1966.

"V akademii nauk SSSR." *Vestnik akademii nauk SSSR,* No. 5 (1938), pp. 72–3.

Vaucouleurs, Gerard de, *Discovery of the Universe.* New York, 1957.

Venikov, V. A., "Primenenie kibernetiki v elektricheskikh sistemakh," in A. I. Berg, ed., *Kibernetika na sluzhbu kommunizma.* Moscow, 1961.

Viatkin, Iu. S., and A. S. Mamzin, "Sootnoshenie strukturno-funktsional'nogo i istoricheskogo podkhodov v izuchenii zhivykh sistem." *Voprosy filosofii,* No. 11 (1969), pp. 46–56.

Vigier, J. P., "The concept of probability in the frame of the probabilistic and the causal interpretation of quantum mechanics," in S. Körner, ed., *Observation and Interpretation in the Philosophy of Physics.* New York, 1957, pp. 71–7.

Voinov, M. S., *Akademik T. D. Lysenko.* Moscow, 1950.

Vorontsov, N., "Zhizn' toropit: nuzhny sovremennye posobiia po biologii." *Komsomol'skaia pravda,* November 11, 1964, p. 3.

Voskresenskii, A. D., and A. I. Prokhorov, "Problemy kibernetiki v meditsine," in A. I. Berg, ed., *Kibernetika na sluzhbu kommunizma.* Moscow, 1961, pp. 126–39.

Vygotsky, L. S., *Izbrannye psikhologicheskie issledovaniia.* Moscow, 1956.

————, *Thought and Language.* Cambridge, Mass., 1962.

Waddington, C. H., "That's Life." *New York Review of Books,* February 29, 1968, p. 19.

Wallace, William, trans., *The Logic of Hegel.* Oxford, 1892.

Watson, James D., *The Double Helix.* New York, 1968.

Weizsäcker, C. F. von, "Über die Entstehung des Planetensystems." *Zeitschrift für Astrophysik,* XXII (1943), 319–55.

Wetter, Gustav A., *Der dialektische Materialismus und der Problem der Entstehung des Lebens. Zur theorie von A. I. Oparin.* Munich, 1958.

————, *Dialectical Materialism: A Historical and Systematic Survey of Philosophy in the Soviet Union,* trans. Peter Heath. New York, 1958.

————, *Soviet Ideology Today.* New York, 1966.

Wheland, G. W., *Resonance in Organic Chemistry.* New York, 1955.

Whitrow, G. J., *The Structure and Evolution of the Universe.* London, 1959.

Whittaker, E. T., *The Beginning and End of the World.* London, 1943.

Wiener, Norbert, *Cybernetics, or Control and Communication in the Animal and the Machine.* Cambridge, Mass., 1962.

————, *The Human Use of Human Beings: Cybernetics and Society.* New York, 1954.

Williams, L. Pearce, *The Origins of Field Theory.* New York, 1966.

Wright, Sewall, "Dogma or Opportunism?" *Bulletin of the Atomic Scientists,* May 1949, pp. 141–2.

"Za boevoi filosofskii zhurnal." *Pravda,* September 7, 1949.

"Za bol'shevistskuiu partiinost' v filosofii." *Voprosy filosofii,* No. 3 (1948), p. 11.

Zaitsev, V. A., *Izbrannye sochineniia.* Moscow, 1934.

Zak, S. E., "Kachestvennye izmeneniia i struktura." *Voprosy filosofii,* No. 1 (1967), pp. 50–8.

"Zakon o dal'neishem sovershenstvovanii organizatsii upravleniia promyshlennost'iu i stroitel'stvom." *Pravda,* May 11, 1957, pp. 1–2.

Zangwill, O. L., "Psychology: Current Approaches," in *The State of Soviet Science.* Cambridge, Mass., 1965, pp. 119–25.

Zel'manov, A. L., "K postanovke kosmologicheskoi problemy." *Trudy vtorogo s' 'ezda vsesoiuznogo astronomo-geogezicheskogo obshchestva 25–31 Ianvaria 1955 g.,* Moscow, 1960, pp. 72–84.

————, "Khronometricheskie invarianty i soputstvyiushchie koordinaty v obshchei teorii otnositel'nosti." *Doklady akademii nauk SSSR,* CVII, No. 6 (1956), 815–18.

————, "K reliativistskoi teorii anizotropnoi neodnorodnoi vselennoi." *Trudy shestogo soveshchaniia po voprosam kosmogonii.* Moscow, 1959, pp. 144–73.

————, "Kosmos, kosmogoniia, kosmologiia." *Nauka i religiia,* No. 12 (1968), pp. 2–37.

————, "Metagalaktika i Vselennaia," in *Nauka i chelevechestvo 1962.* Moscow, 1963, pp. 383–405.

————, "Mnogoobrazie material'nogo mira i problema beskonechnosti Vselennoi," in *Beskonechnost' i Vselennaia.* Moscow, 1969, pp. 274–324.

————, "Nereliativistskii gravitatsionnyi paradoks i obshchaia teoriia otnositel'nosti." *Nauchnye doklady vysshei shkoly: fiziko-matematicheskie nauki,* No. 2 (1958), pp. 124–7.

————, "O beskonechnosti material'nogo mira," in M. E. Omel'ianovskii and I. V. Kuznetsov, *Dialektika v naukakh o nezhivoi prirode.* Moscow, 1964, pp. 227–69.

Zhdanov, A. A., "Doklad o zhurnalakh 'Zvezda' i 'Leningrad.' " Moscow, 1946.

————, *Essays on Literature, Philosophy, and Music.* New York, 1950.

————, "O mezhdunarodnom polozhenii. Doklad, sdelannyi na informatsionnom soveshchanii predstavitelei nekotorykh kompartii v Pol'she, v kontse sentiabria 1947 goda." Moscow, 1947.

————, "Vystupitel'naia rech' i vystuplenie [na soveshchanii deiatelei sovetskoi muzyki v TsK VKP (b),] v kn.: Soveshchanie deiatelei sovetskoi muzyki v TsK VKP (b)." Moscow, 1948.

————, "Vystuplenie na diskussii po knige G. F. Aleksandrova 'Istoriia zapadnoevropeiskoi filosofii,' 24 iiunia 1947 g." Moscow, 1947, 1951, 1952.

————, "29-ia godovshchina velikoi oktiabrskoi sotsialisticheskoi revoliutsii. Doklad na torzhestvennom zasedanii moskovskogo soveta 6 noiabria 1946 goda." Moscow, 1946.

Zhdanov, Iu. A., "Dialektika tozhdestva i razlichiia v khimii." *Filosofskie nauki,* No. 4 (1958), pp. 168–75.

————, "Entropiia informatsii v reaktsiiakh aromaticheskogo zameshcheniia." *Zhurnal organicheskoi khimii,* No. 9 (1965), pp. 1521–5.

————, "Entropiia informatsii kak mera spetsifichnosti v reaktsiiakh aromaticheskogo zameshcheniia." *Zhurnal fizicheskoi khimii,* No. 3 (1965), pp. 777–9.

————, "Estestvoznanie i gumanizm." *Priroda,* No. 5 (1962), pp. 7–12.

————, "Khimiia i estetika." *Priroda,* No. 10 (1964), pp. 8–13.

————, *Lenin i estestvoznanie.* Moscow, 1959.

————, "Metody sinteza i svoistva uglerodzameshchennykh uglevodov." Dissertation for degree of *doktor*, Moscow State University, Moscow, 1959.

————, "Molekula i stroenie veshchestva." *Priroda,* No. 5 (1962), pp. 63–74.

————, "Obrashchenie metoda v organicheskoi khimii." Rostov, 1963.

————, *Ocherki metodologii organicheskoi khimii.* Moscow, 1960.

————, "O kritike i samokritike v nauchnoi rabote." *Bol'shevik,* No. 21 (1951), pp. 28–43.

————, "O nauchnom tvorchestve molodezhi." *Komsomol'skaia pravda,* May 25, 1948.

————, "O perekhode kachestvennykh izmenenii v izmeneniia kolichestvennye." *Voprosy filosofii,* No. 6 (1956), pp. 206–10.

————, "Poniatie gomologii v khimii i ego filosofskoe znachenie." Dissertation for degree of *kandidat*, Institute of Philosophy, Academy of Sciences of the U.S.S.R., Moscow, 1948.

————, "Protiv sub' 'ektivistskikh izvrashchenii v estestvozanii." *Pravda,* January 16, 1953, p. 3.

————, "Tovarishchu I. V. Stalinu." *Pravda,* August 7, 1948.

————, "V chem zakliuchaetsia ateisticheskoe znachenie ucheniia I. P. Pavlova?" *Voprosy filosofii,* No. 5 (1954), pp. 234–6.

————, "Kriterii praktiki v khimii," in *XXI s' 'ezd KPSS o teoreticheskikh voprosakh stroitel'stva kommunizma v SSSR.* Kharkov, 1959, pp. 181–215.

————, "Vzaimootnoshenie chasti i tselogo v khimii." *Filosofskie nauki,* No. 1 (1960), pp. 82–9.

————, "Znachenie leninskikh idei dlia razrabotki metodologicheskikh voprosov khimii." *Filosofskie nauki,* No. 2 (1970), pp. 80–90.

Zhukov, N. I., "Informatsii v svete leninskoi teorii otrazheniia." *Voprosy filosofii,* No. 11 (1963), pp. 153–61.

Zirkle, Conway, ed., *Death of a Science in Russia.* Philadelphia, 1949.

————, *Evolution, Marxian Biology and the Social Scene.* Philadelphia, 1959.

Index

Index

iii

A NOTE ON THE AUTHOR

Loren R. Graham was born in Indiana in 1933 and took his Bachelor of Science degree at Purdue University. Studies in his special field, the history of science with emphasis on Soviet science, were continued at Columbia (M.A., 1960, Ph.D., 1964) and at Moscow University, 1960–61. Professor Graham has been a Fulbright-Hays Fellow and Guggenheim Fellow, and was a member during 1969–70 of the Institute for Advanced Study at Princeton University. He has published one previous book, *The Soviet Academy of Sciences and the Communist Party, 1927–1932*, with the Princeton University Press, 1967, and has been associate professor of history at Columbia University since 1967. Professor Graham lives with his wife and daughter in New York City.

A NOTE ON THE TYPE

The text of this book was set on the Linotype in a face called Times Roman, designed by Stanley Morison for *The Times* (London) and first introduced by that newspaper in 1932.

Among typographers and designers of the twentieth century, Stanley Morison has been a strong forming influence, as typographical adviser to the English Monotype Corporation, as a director of two distinguished English publishing houses, and as a writer of sensibility, erudition, and keen practical sense.

Typography and binding design by Paula Filancia.

This book was composed, printed, and bound by H. Wolff Book Manufacturing Co., Inc., New York.